Metal Oxides in Energy Technologies

The Metal Oxides Book Series Edited by Ghenadii Korotcenkov

Forthcoming Titles

- Metal Oxide Powder Technologies, Yarub Al-Douri, 9780128175057
- Metal Oxide Glass Nanocomposites, Sanjib Bhattacharya, 9780128174586
- Tin Oxide Materials, Marcelo Ornaghi Orlandi, 9780128159248
- Cerium Oxide, Salvatore Scire, Leonardo Palmisano, 9780128156612
- Solution Processed Metal Oxide Thin Films for Electronic Applications, Zheng Cui, 9780128149300
- Metal Oxides for Non-volatile Memory, Panagiotis Dimitrakis, Ilia Valov, 9780128146293
  Colloidal Metal Oxide Nanoparticles, Sabu Thomas, Anu Tresa Sunny, Prajitha V, 9780128133576
- Gallium Oxide, Stephen Pearton, Fan Ren, Michael Mastro, 9780128145210
- Metal Oxide Nanostructures, Daniela Nunes, Lidia Santos, Ana Pimentel, Pedro Barquinha, Luis Pereira, Elvira Fortunato, Rodrigo Martins, 9780128115121
- Gas Sensors Based on Conducting Metal Oxides, Nicolae Barsan, Klaus Schierbaum, 9780128112243
- Metal Oxide Nanostructured Phosphors, H. Nagabhushana, Daruka Prasad, S.C. Sharma, 9780128118528

Recently Published Titles

- Metal Oxide-Based Photocatalysis, Adriana Zaleska-Medynska, 9780128116340
- Metal Oxides in Heterogeneous Catalysis, Jacques C. Vedrine, 9780128116319
- Magnetic, Ferroelectric, and Multiferroic Metal Oxides, Biljana Stojanovic, 9780128111802
- Iron Oxide Nanoparticles for Biomedical Applications, Sophie Laurent, Morteza Mahmoudi, 9780081019252
- The Future of Semiconductor Oxides in Next-Generation Solar Cells, Monica Lira-Cantu, 9780128111659
- Metal Oxide-Based Thin Film Structures, Nini Pryds, Vincenzo Esposito, 9780128111666
- Metal Oxides in Supercapacitors, Deepak Dubal, Pedro Gomez-Romero, 9780128111697
- Transition Metal Oxide Thin Film-Based Chromogenics and Devices, Pandurang Ashrit, 9780081018996

Metal Oxide Series

Metal Oxides in Energy Technologies

Edited by

Yuping Wu

Series Editor

Ghenadii Korotcenkov

ELSEVIER

Elsevier
Radarweg 29, PO Box 211, 1000 AE Amsterdam, Netherlands
The Boulevard, Langford Lane, Kidlington, Oxford OX5 1GB, United Kingdom
50 Hampshire Street, 5th Floor, Cambridge, MA 02139, United States

Library of Congress Cataloging-in-Publication Data
A catalog record for this book is available from the Library of Congress

British Library Cataloguing-in-Publication Data
A catalogue record for this book is available from the British Library

ISBN: 978-0-12-811167-3

For information on all Elsevier publications
visit our website at https://www.elsevier.com/books-and-journals

Publisher: Matthew Deans
Acquisition Editor: Kayla Dos Santos
Editorial Project Manager: Jennifer Pierce
Production Project Manager: Sojan P. Pazhayattil
Cover Designer: Miles Hitchen

Typeset by SPi Global, India

Contents

List of contributors **xi**
About the Series Editor **xv**
About the Author **xvii**
Preface to the series **xix**
Preface **xxiii**

1 Introduction: Energy technologies and their role in our life **1**
Yuping Wu, Teunis van Ree
1.1 Preparation methods of metal oxides and energy technologies **1**
1.2 Energy generation **2**
1.3 Energy storage **5**
1.4 Energy conversion **8**
1.5 Energy and environment **11**
1.6 About this book **12**
Acknowledgment **12**
References **13**

2 Metal oxides in fuel cells **17**
Peng Wang, Jing Wang, Lijun Fu, Yuping Wu, Teunis van Ree
2.1 Introduction **17**
2.2 Metal oxides in SOFCs **19**
2.3 Metal oxides in other fuel cells **39**
2.4 Outlook **43**
Acknowledgment **44**
References **44**
Further reading **47**

3 Metal oxide-based thermoelectric materials **49**
Lian Ji
3.1 Introduction **49**
3.2 Na_xCoO_2 and $Na_xCo_2O_4$ **54**
3.3 $Ca_3Co_4O_9$ **60**
3.4 $SrTiO_3$ **62**
3.5 $CaMnO_3$ **64**
3.6 ZnO **68**
3.7 Conclusion **69**
References **69**

4 **Mixed oxides in nuclear fuels** **73**
Bingwei Chen, Nengfei Yu, Lijun Fu, Yusong Zhu, Teunis van Ree, Yuping Wu
4.1 Introduction **73**
4.2 U-Pu MOX nuclear fuel **74**
4.3 U-Th MOX nuclear fuel **77**
4.4 Other MOXs **81**
4.5 Outlook **86**
Acknowledgment **86**
References **87**
Further reading **89**

5 **Piezoelectric energy harvesting systems with metal oxides** **91**
Kenji Uchino
5.1 Background **91**
5.2 Mechanical-to-mechanical energy transfer **108**
5.3 Mechanical-electrical energy transduction **110**
5.4 Electrical-to-electrical energy transfer **116**
5.5 Summary of the total energy flow **121**
5.6 Hybrid energy harvesting—Magnetoelectric devices **122**
5.7 Summary and outlook **123**
References **125**
Further reading **126**

6 **Metal oxides in batteries** **127**
Faxing Wang, Yi Zhang, Nengfei Yu, Lijun Fu, Yusong Zhu, Yuping Wu, Teunis van Ree
6.1 Metal oxides in Ni//MH batteries **127**
6.2 Metal oxides in LIBs **133**
6.3 Metal oxides in aqueous rechargeable lithium batteries **150**
6.4 Metal oxides in Na-ion batteries **155**
6.5 Metal oxides in Mg-ion batteries **158**
6.6 Metal oxides in Al-ion batteries **161**
6.7 Outlook **161**
Acknowledgments **162**
References **162**

7 **Metal oxides in supercapacitors** **169**
Lei Zhou, Chunyang Li, Xiang Liu, Yusong Zhu, Yuping Wu, Teunis van Ree
7.1 Introduction **169**
7.2 Outline of electrochemical capacitors **170**
7.3 Metal oxides in supercapacitors **176**
7.4 Summary and outlook **188**
Acknowledgments **189**

References 189
Further reading 203

8 Metal oxide semiconductors for solar water splitting **205**
Jing Wang, Teunis van Ree, Yuping Wu, Peng Zhang, Lian Gao
8.1 Introduction 205
8.2 Semiconductors for solar water splitting 210
8.3 Improvements in efficiency for solar water splitting 240
8.4 Outlook 243
References 245

9 Metal oxides for hydrogen storage **251**
Bingwei Chen, Zaichun Liu, Chunyang Li, Yusong Zhu, Lijun Fu, Yuping Wu, Teunis van Ree
9.1 Introduction 251
9.2 Surface modifiers for metal hydrides 254
9.3 Oxides as host for hydrogen storage 262
9.4 Outlook 271
Acknowledgments 271
References 271

10 Requirements for efficient metal oxide photocatalysts for CO_2 reduction **275**
Jennifer Strunk
10.1 Introduction 275
10.2 Overview of photocatalysts for CO_2 reduction 276
10.3 Basics of heterogeneous photocatalysis 278
10.4 Involved (elementary) reactions and energetic requirements of CO_2 reduction 283
10.5 Some remarks concerning photoreactors and reaction conditions 291
10.6 Examples of non-TiO_2 oxide materials for CO_2 reduction 293
10.7 Concluding remarks 297
References 297

11 Metal oxide catalysts for biodiesel production **303**
Umer Rashid, Soroush Soltani, Saud Ibrahim Al-Resayes, Imededdine Arbi Nehdi
11.1 Introduction 303
11.2 Biodiesel 304
11.3 Problems with homogeneous acid and heterogeneous base catalysts 304
11.4 Heterogeneous acid catalysts 305
11.5 Synthesis methods to produce solid acid catalysts 305
11.6 Effect of synthesis conditions on characteristics 310
11.7 Effect of reaction parameters on the catalytic activity of heterogeneous catalysts 312

11.8 Catalyst stability **313**
11.9 Conclusion **316**
Acknowledgment **316**
References **316**

12 Solar-driven fuel production by metal-oxide thermochemical cycles **321**
Xinhai Yuan, Xiongwei Wu, Jing Wang, Yuping Wu, Teunis van Ree
12.1 Introduction **321**
12.2 Reaction principles of the two-step thermochemical metal-oxide cycles **322**
12.3 Considerations in the design of metal-oxide materials **325**
12.4 Metal oxides for two-step thermochemical metal-oxide cycles **328**
12.5 New concepts to further improve the two-step thermochemical metal-oxide cycles **333**
12.6 Outlook **335**
Acknowledgment **335**
References **336**
Further reading **340**

13 Metal oxides in energy-saving smart windows **341**
Yanfang Wang, Bingwei Chen, Yusong Zhu, Lijun Fu, Yuping Wu, Teunis van Ree
13.1 Thermochromic smart windows **341**
13.2 Electrochromic smart windows **350**
13.3 Outlook **358**
Acknowledgment **359**
References **359**

14 Metal oxide-based superconductors in AC power transportation and transformation **361**
Yanfang Wang, Chunyang Li, Weibin Zhou, Yusong Zhu, Lijun Fu, Yuping Wu, Teunis van Ree
14.1 Some parameters for superconductors **361**
14.2 Early stage metal oxide-based superconductors **362**
14.3 Cuprate-based superconductors **365**
14.4 Other materials **380**
14.5 Applications **381**
14.6 Outlook **386**
Acknowledgments **387**
References **388**

15 Metal oxides for emission control **391**
Xinhai Yuan, Xiongwei Wu, Yuping Wu, Teunis van Ree
15.1 Introduction **391**
15.2 Emission control technologies and metal oxide catalysts **392**
15.3 Catalyst preparation and methodology **408**

15.4 Outlook **409**
Acknowledgments **410**
References **410**
Further reading **414**

Index **415**

List of contributors

Saud Ibrahim Al-Resayes Chemistry Department, College of Science, King Saud University, Riyadh, Saudi Arabia

Bingwei Chen State Key Laboratory of Materials-Oriented Chemical Engineering, School of Energy Science and Engineering, Institute for Electrochemical Energy Storage, Nanjing Tech University, Nanjing, People's Republic of China

Lijun Fu State Key Laboratory of Materials-Oriented Chemical Engineering, School of Energy Science and Engineering, Institute for Electrochemical Energy Storage, Nanjing Tech University, Nanjing, People's Republic of China

Lian Gao School of Materials Science and Engineering, Shanghai Jiaotong University, Shanghai, China

Lian Ji School of Energy Science and Engineering, Nanjing Tech University, Nanjing, China

Chunyang Li State Key Laboratory of Materials-Oriented Chemical Engineering, School of Energy Science and Engineering, Institute for Electrochemical Energy Storage, Nanjing Tech University, Nanjing, People's Republic of China

Xiang Liu State Key Laboratory of Materials-Oriented Chemical Engineering, School of Energy Science and Engineering, Institute for Electrochemical Energy Storage, Nanjing Tech University, Nanjing, People's Republic of China

Zaichun Liu State Key Laboratory of Materials-Oriented Chemical Engineering, School of Energy Science and Engineering, Institute for Electrochemical Energy Storage, Nanjing Tech University, Nanjing, People's Republic of China

Imededdine Arbi Nehdi Chemistry Department, College of Science, King Saud University, Riyadh, Saudi Arabia

Umer Rashid Institute of Advanced Technology, Universiti Putra Malaysia, Serdang, Malaysia; Chemistry Department, College of Science, King Saud University, Riyadh, Saudi Arabia

Soroush Soltani Institute of Advanced Technology, Universiti Putra Malaysia, Serdang, Malaysia

Jennifer Strunk Leibniz Institute for Catalysis at the University of Rostock, Rostock, Germany

Kenji Uchino International Center for Actuators and Transducers, The Pennsylvania State University, University Park, PA, United States

Teunis van Ree Department of Chemistry, University of Venda, Thohoyandou, South Africa

Peng Wang State Key Laboratory of Materials-Oriented Chemical Engineering, School of Energy Science and Engineering, Institute for Electrochemical Energy Storage, Nanjing Tech University, Nanjing, People's Republic of China

Jing Wang State Key Laboratory of Materials-Oriented Chemical Engineering, College of Chemistry and Chemical Engineering, Nanjing Tech University, Nanjing, China

Faxing Wang State Key Laboratory of Materials-Oriented Chemical Engineering, School of Energy Science and Engineering, Institute for Electrochemical Energy Storage, Nanjing Tech University, Nanjing, People's Republic of China

Yanfang Wang State Key Laboratory of Materials-Oriented Chemical Engineering, School of Energy Science and Engineering, Institute for Electrochemical Energy Storage, Nanjing Tech University, Nanjing, People's Republic of China

Yuping Wu State Key Laboratory of Materials-Oriented Chemical Engineering, School of Energy Science and Engineering, Institute for Electrochemical Energy Storage, Nanjing Tech University, Nanjing, People's Republic of China; College of Science, Hunan Agricultural University, Changsha, China

Xiongwei Wu College of Science, Hunan Agricultural University, Changsha, China

Nengfei Yu State Key Laboratory of Materials-Oriented Chemical Engineering, School of Energy Science and Engineering, Institute for Electrochemical Energy Storage, Nanjing Tech University, Nanjing, People's Republic of China

Xinhai Yuan State Key Laboratory of Materials-Oriented Chemical Engineering, School of Energy Science and Engineering, Institute for Electrochemical Energy Storage, Nanjing Tech University, Nanjing, People's Republic of China; College of Science, Hunan Agricultural University, Changsha, China

Yi Zhang State Key Laboratory of Materials-Oriented Chemical Engineering, School of Energy Science and Engineering, Institute for Electrochemical Energy Storage, Nanjing Tech University, Nanjing, People's Republic of China

Peng Zhang School of Materials Science and Engineering, Shanghai Jiaotong University, Shanghai, China

Lei Zhou State Key Laboratory of Materials-Oriented Chemical Engineering, School of Energy Science and Engineering, Institute for Electrochemical Energy Storage, Nanjing Tech University, Nanjing, People's Republic of China

Weibin Zhou State Key Laboratory of Materials-Oriented Chemical Engineering, School of Energy Science and Engineering, Institute for Electrochemical Energy Storage, Nanjing Tech University, Nanjing, People's Republic of China

Yusong Zhu State Key Laboratory of Materials-Oriented Chemical Engineering, School of Energy Science and Engineering, Institute for Electrochemical Energy Storage, Nanjing Tech University, Nanjing, People's Republic of China

About the Series Editor

Ghenadii Korotcenkov earned his PhD in material sciences from the Technical University of Moldova, Chisinau, Moldova in 1976 and his doctor of science degree (doctor habilitate) in physics from the Academy of Science of Moldova in 1990 (Highest Qualification Committee of the USSR, Moscow). He has more than 45 years of experience as a scientific researcher. For a long time, he was the leader of the gas sensor group and manager of various national and international scientific and engineering projects carried out in the Laboratory of Micro- and Optoelectronics, Technical University of Moldova. His research had financial support from international foundations and programs such as the CRDF, the MRDA, the ICTP, the INTAS, the INCO-COPERNICUS, the COST, and the NATO. From 2007 to 2008, he was an invited scientist in the Korea Institute of Energy Research, Daejeon, South Korea. After which, until the end of 2017 G. Korotcenkov was a research professor at the School of Materials Science and Engineering at the Gwangju Institute of Science and Technology, Gwangju, South Korea. Currently G. Korotcenkov is the chief scientific researcher at the Department of Physics and Engineering at the Moldova State University, Chisinau, Republic of Moldova.

Specialists from the former Soviet Union know G. Korotcenkov's research results in the field of study of Schottky barriers, MOS structures, native oxides, and photoreceivers on the basis of III–Vs compounds such as InP, GaP, AlGaAs, and InGaAs. His present scientific interests starting from 1995 include material sciences, focusing on the metal oxide film deposition and characterization, surface science, and the design of thin film gas sensors and thermoelectric convertors. These studies were carried out in cooperation with scientific teams from Ioffe Institute (St. Petersburg, Russia), University of Michigan (Ann Arbor, United States), Kiev State University (Kiev, Ukraine), Charles University (Prague, Czech Republic), St. Petersburg State University (St. Petersburg, Russia), Illinois Institute of Technology (Chicago, United States), University of Barcelona (Barcelona, Spain), Moscow State University (Moscow, Russia), University of Brescia (Brescia, Italy), Belarus State University (Minsk, Belarus), and South-Ukrainian University (Odessa, Ukraine).

G. Korotcenkov is the author or editor of 38 books, including the 11-volume *Chemical Sensors* series published by the Momentum Press (United States), 15-volume *Chemical Sensors* series published by Harbin Institute of Technology Press (China),

3-volume *Porous Silicon: From Formation to Application* published by CRC Press (United States), 2-volume *Handbook of Gas Sensor Materials* published by Springer (United States), and 3-volume *Handbook of Humidity Measurement*, published by CRC Press (United States). In addition, at present, G. Korotcenkov is a series editor of *Metal Oxides* series, which will be published by Elsevier.

G. Korotcenkov is author and coauthor of more than 600 scientific publications, including 25 review papers, 38 book chapters, and more than 250 articles published in peer-reviewed scientific journals [h-factor = 39 (Scopus) and h-factor = 47 (Google Scholar citation)]. He is a holder of 17 patents. He has presented more than 200 reports at national and international conferences, including 15 invited talks. G. Korotcenkov was co-organizer of several international conferences. His name and activities have been listed by many biographical publications, including Who's Who. His research activities were honored by an Award of the Supreme Council of Science and Advanced Technology of the Republic of Moldova (2004); Prize of the Presidents of the Ukrainian, Belarus, and Moldovan Academies of Sciences (2003); and National Youth Prize of the Republic of Moldova in the field of science and technology (1980), among others. G. Korotcenkov also received a fellowship from the International Research Exchange Board (IREX, United States, 1998), Brain Korea 21 Program (2008–12), and Brainpool Program (Korea, 2007–08 and 2015–17).

About the author

Dr. Yuping Wu is a Distinguished Professor of Nanjing Tech University. He got Ph.D. degree from Institute of Chemistry, Chinese Academy of Science, Beijing, China, in 1997. Since then, he worked at Tsinghua University, Waseda University, and Chemnitz University of Technology, for 2 years separately. In 2003, he joined Fudan University as a full professor. In 2015, he moved to Nanjing Tech University.

Since 1994, he has published over 260 papers on peer-reviewed journals. As a main author, he contributed six monographs on lithium batteries, whose sale is above 40,000 copies, and eight chapters. His H-index is 64. Since 2015, he has been selected as Highly Cited Researchers over the world by Thomson Reuters and Clarivate Analytics, and won the World's Most Influential Minds (2015) by Thomson Reuters.

He got a lot of prizes such as National Distinguished Youth Scientists by Natural Science Foundation of China in 2014, Innovative Talent by the Thousand Talents Plan of China in 2016, Marquis Who's Who in the World in 2017, and Albert Nelson Marquis Lifetime Achievement Award in 2018.

His main research work is on energy storage and conversion systems. He is the primary inventor of lead lithium battery, second generation of ARLBs, third generation of ARLBs, and pore-free gel separator. He created the first company over the world to mass produce third generation of gel separator for lithium ion batteries with excellent safety.

Preface to the series

The field of the synthesis, study, and application of metal oxides is one of the most rapidly progressing areas of science and technology. Metal oxides are one of the most ubiquitous compound groups on earth, which has a large variety of chemical compositions, atomic structures, and crystalline shapes. In addition, metal oxides are known to possess unique functionalities that are absent or inferior in other solid materials. In particular, metal oxides represent an assorted and appealing class of materials, the properties of which exhibit a full spectrum of electronic properties—from insulating to semiconducting, metallic, and superconducting. Moreover, almost all the known effects including superconductivity, thermoelectric effects, photoelectrical effects, luminescence, and magnetism can be observed in metal oxides. Therefore, metal oxides have emerged as an important class of multifunctional materials with a rich collection of properties, which have great potential for numerous device applications. Specific properties of the metal oxides, such as the wide variety of materials with different electrophysical, optical, and chemical characteristics, their high thermal and temporal stability, and their ability to function in harsh environments, make metal oxides very suitable materials for designing transparent electrodes, high-mobility transistors, gas sensors, actuators, acoustical transducers, photovoltaic and photonic devices, photo- and heterogeneous catalysts, solid-state coolers, high-frequency and micromechanical devices, energy harvesting and storage devices, nonvolatile memories, and many others in the electronics, energy, and health sectors. In these devices, metal oxides can be successfully used as sensing or active layers, substrates, electrodes, promoters, structure modifiers, membranes, and fibers, that is, they can be used as active and passive components.

Among other advantages of metal oxides are the low fabrication cost and robustness in practical applications. Furthermore, metal oxides can be prepared in various forms such as ceramics, thick films, and thin films. For thin film deposition, deposition techniques that are compatible with standard microelectronic technology can be used. The last factor is very important for large-scale production, because the microelectronic approach promotes low cost for mass production, offers the possibility of manufacturing devices on a chip, and guarantees good reproducibility. Various metal oxide nanostructures, including nanowires, nanotubes, nanofibers, core-shell structures, and hollow nanostructures also can be synthesized. As is known, the field of metal-oxide nanostructured morphologies (e.g., nanowires, nanorods, nanotubes, etc.) has become one of the most active research areas within the nanoscience community.

The ability to create a variety of metal-oxide-based composites and the ability to synthesize various multicomponent compounds significantly expand the range of properties that metal-oxide-based materials can have, making metal oxides a truly versatile multifunctional material for widespread use. As is known, small changes in their chemical composition and atomic structure can be accompanied by spectacular variation in properties and behavior of metal oxides. Even now, advances in synthesizing and characterizing techniques are revealing numerous new functions of metal oxides.

Taking into account the importance of metal oxides for progress in microelectronics, optoelectronics, photonics, energy conversion, sensors, and catalysis, a large number of various books devoted to this class of materials have been published. However, one should note that some books from this list are too general, some books are collections of various original works without any generalizations, and other ones were published many years ago. However, during the past decade great progress has been made on the synthesis as well as on the structural, physical, and chemical characterization and application of metal oxides in various devices, and a large number of papers have been published on metal oxides. In addition, until now many important topics related to metal-oxide study and application have not been discussed. To remedy the situation in this area, we decided to generalize and systematize the results of research in this direction and to publish a series of books devoted to metal oxides.

One should note that the proposed book series "Metal Oxides" is the first one, devoted to the consideration of metal oxides only. We believe that combining books on metal oxides in a series could help readers in searching for required information on the subject. In particular, we plan that the books from our series, which have a clear specialization by their content, will provide interdisciplinary discussion for various oxide materials with a wide range of topics, from material synthesis and deposition to characterizations, processing, and then to device fabrications and applications. This book series is prepared by a team of highly qualified experts, which guarantees it a high quality.

I hope that our books will be useful and comfortable in use. I would also like to hope that readers will consider this "Metal Oxides" book series like an encyclopedia of metal oxides that enables to understand the present status of metal oxides, to estimate the role of multifunctional metal oxides in the design of advanced devices, and then based on observed knowledge to formulate new goals for further research.

The intended audience of the present book series is scientists and researchers, working or planning to work in the field of materials related to metal oxides, that is, scientists and researchers whose activities are related to electronics, optoelectronics, energy, catalysis, sensors, electrical engineering, ceramics, biomedical designs, etc. I believe that this "Metal Oxides" book series will also be interesting for practicing engineers or project managers in industries and national laboratories, who would like to design metal-oxide-based devices, but do not know how to do it, and how to select the optimal metal oxide for specific applications. With many references to the vast resource of recently published literature on the subject, this book series will serve as a significant and insightful source of valuable information, providing scientists and engineers with new insights for understanding and improving existing metal-oxide-based devices and for designing new metal-oxide-based materials with new and unexpected properties.

I believe that this "Metal Oxides" book series would be very helpful for university students, postdocs, and professors. The structure of these books offers a basis for courses in the field of material sciences, chemical engineering, electronics, electrical engineering, optoelectronics, energy technologies, environmental control, and many others. Graduate students could also find the book series to be very useful in their research and in understanding features of metal-oxide synthesis, study, and application of this multifunctional material in various devices. We are sure that all of them will find information useful for their activity.

Finally, I thank all contributing authors and book editors who have been involved in the creation of these books. I am thankful that they agreed to participate in this project and for their efforts in the preparation of these books. Without their participation, this project would not have been possible. I also express my gratitude to Elsevier for giving us the opportunity to publish this series. I especially thank the entire team of the editorial office at Elsevier for their patience during the development of this project and for encouraging us during the various stages of preparation.

Ghenadii Korotcenkov

Preface

From a small oak tree to a big one is not a short and easy process. From the primary use of metal oxides to their understanding, from their understanding to their use in energy technologies is the same. During this long history, a lot of people, even some anonymous ones, contributed to our knowledge. Without their pioneering work, it was impossible for us to finish this book. The first salute should be paid to them.

Another salute is to the editor's late mother, Mrs. Chang'e Zhang, on her 86th birthday. It was her diligence and thrift that kept her poor, large family going, even during very tough times, so that Yuping Wu, his brothers, and sister had a chance to study, to graduate from universities, and to grow up.

Finally, the long-term financial support of some agencies should be greatly appreciated, including the National Materials Genome Project (2016YFB0700600), the China National Funds for Distinguished Young Scientists (51425301), NSFC (U1601214, 21374021, 51772147, and 51502137), MOST (2010DFA61770), Jiangsu Distinguished Professorship Program (2016), Natural Science Foundation of Jiangsu Province (KB20171008), and STCSM (12JC1401200 and 14520721800). This provided the editor's laboratory with great opportunities to carry out long-term research work on energy technologies, providing a solid foundation for the completion of this book.

Introduction: Energy technologies and their role in our life

1

Yuping Wu, Teunis van Ree*[†]
*State Key Laboratory of Materials-Oriented Chemical Engineering, School of Energy Science and Engineering, Institute for Electrochemical Energy Storage, Nanjing Tech University, Nanjing, People's Republic of China, [†]Department of Chemistry, University of Venda, Thohoyandou, South Africa

Metal oxides have long been acknowledged and well used in many applications during the development of humankind though they were not well recognized at first. For example, China has long been famous for ceramics production technologies. Later, with the development of modern chemistry originating from the western countries, metal oxides become a highly topical subject. Their components, structures, preparation methods, characterization and so on have been well investigated. Meanwhile, the demand for energy has become a problem due to rapid economic development. New energy technologies have become important in sustainable human development, with metal oxides playing a dominant role.

1.1 Preparation methods of metal oxides and energy technologies

Metal oxides have a wide range of applications, and they can be prepared by a wide variety of methods, such as the following:

- *High-temperature solid-state (ceramic) method.* A simple method in which the solid precursors are mixed intimately and then reacted at a high temperature [1].
- *Hydrothermal method.* A solution of the precursor is heated in an autoclave to produce a crystalline product [2, 3].
- *Templating method.* Organic materials especially polymers such as polystyrene are used to form a template with a distinct morphology and then a metal oxide is deposited into the empty space to form a porous material after the removal of the template [4, 5].
- *Sol-gel method.* A sol is formed by the precursors, usually by means of hydrolysis and condensation reactions. After the sol polymerizes to form a gel, the gel is dried and heated to form the nanomaterial. In some cases, carriers such as organic complex can be added [6, 7].
- *Electrodeposition.* A suspension of metal ions and insoluble particles is deposited electrochemically on an inert cathode surface [8, 9].
- *Chemical vapor deposition.* The low-deposition temperature combined with a high-deposition rate produces a solid membrane with controllable composition and crystallinity [10, 11].

Metal Oxides in Energy Technologies. https://doi.org/10.1016/B978-0-12-811167-3.00001-8

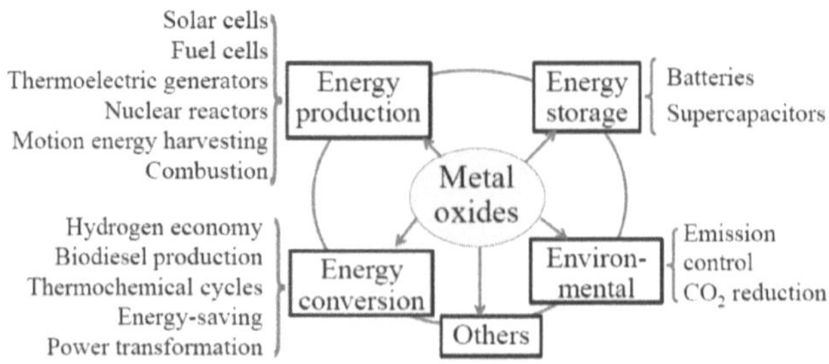

Fig. 1.1 Applications of metal oxides in energy technologies.

- *Chemical precipitation.* An insoluble metal oxide is formed by chemical reaction between different solutions [12, 13].

Of course, morphology or shapes of the prepared metal oxides can also be tailored. They can be used for a wide range of applications [14]. However, one of their most important applications is in energy technologies including energy production, storage, conversion, and emission control (Fig. 1.1).

Metal oxides are widely applied in energy storage technologies, because they generally are able to generate charge carriers when energy is applied. They are especially useful electrode materials, leading to significant performance improvements, because a wide variety of oxidation states is possible for redox charge transfer. Metal oxides are also used as catalysts, for example, for hydrocarbon reforming reactions, in which a substance such as ceria is an outstanding catalyst and is resistant to carbon deposition. In this book, we introduce their applications as electrode materials for energy generation, conversion, and storage in a wide range of applications, such as solar cells, fuel cells, combustion, and nuclear reactors; batteries and supercapacitors; for hydrogen and biodiesel production [15]. Of course, many metal oxides such as catalysts perform very well in emission control, such as spinels (e.g., $ZnAl_2O_4$ and $CuFe_2O_4$), perovskites (such as $SrTiO_3$ and $MgTiO_3$), transition metal oxides (TMOs) (e.g., Mn, Cu, Co, Ti, Zr, and Ce oxides), and mixed metal oxides (MOXs) (such as MnO_x-CeO_2, CoO_x-CeO_2, CuO-CeO_2, and $Ce_{0.5}Pr_{0.5}O_2$) [16].

1.2 Energy generation

Metal oxide catalysts and electrode materials have become essential parts of our society to avoid energy and environmental crises via green energy generation, and degrading environmental organic pollutants [17]. Solar cells, fuel cells, and other developing energy conversion technologies are gradually becoming more useful in terms of efficiency, cost, and long-term stability.

1.2.1 Solar cells

A solar cell converts sunlight directly into electricity. Light shining on a solar panel generates current and voltage to produce electric power. This process requires a material in which the absorption of light excites an electron to a higher energy state, followed by transport of this high-energy electron from the solar cell into an external circuit. The electron drives work in the external circuit and returns to the solar cell. Although many materials and processes can provide photovoltaic energy conversion, most photovoltaic energy conversion is done in practice using semiconductors such as metal oxides and organic perovskites in the form of a *p-n* junction. Solar energy is not only clean, renewable, and plentiful, and an attractive addition or alternative to fossil fuels, but it can also help to relieve concerns about the environment [18] and energy security [19].

Solar water splitting (WS) in photocatalytic and photoelectrochemical systems is based on common semiconductor materials, especially metal oxides such as TiO_2, WO_3, $BiVO_4$, α-Fe_2O_3, and Cu_2O though sulfides are also useful. It is still necessary to improve the efficiency for solar WS, indicating a direction of future research trends.

1.2.2 Fuel cells

Fuel cells have high energy-conversion efficiencies and produce clean emissions, pointing to another promising trend in electric power generation. In fuel cells, chemical energy stored in a fuel, combined with oxygen in the air, is converted directly to electrical power by electrochemical reactions at very high theoretical and practical efficiencies compared to conventional internal combustion engines.

A basic fuel cell consists of three components: an anode, a cathode, and an electrolyte. In the fuel cell, the fuel and oxygen are oxidized and reduced, respectively, with electrons transferred from the anode to the cathode via the external circuit, whereas cations or oxygen anions are transported between the anode and the cathode through the electrolytes. Metal oxides such as cerium oxide have been introduced recently as anode materials, resulting in substantial performance improvement. However, several disadvantages have also emerged, such as low stability, high cost, and complicated processing, necessitating further research to improve performance and stability, and simplify production methods. In the case of solid oxide fuel cells, oxides can act as electrolytes with good stability at high temperature.

1.2.3 Thermoelectric generators

Metal oxides are significant thermoelectric materials for use in a wide range of applications because of their high-temperature stability, tunable electronic and phonon transport properties, and well-established synthesis techniques. Thermoelectric metal oxides can be applied in efficient cooling and refrigeration, energy scavenging, sensing, and thermopower systems. They have also become prominent in conversion processes based on waste heat from engine exhaust systems, furnaces, and industrial processes [20–22]. Thermoelectric (TE) materials are uncomplicated, do not have moving parts and therefore the thermopower system has good advantages such as no vibration, no noise, no pollution, and high stability [23–26].

1.2.4 Nuclear reactors

In the case of nuclear reactors, they are friendly ways to produce green power. Of course, some nuclear wastes should be carefully treated to avoid the possible future side effects on the environment. Currently, China and South Africa are strong international players. The former has excellent technologies, which are superior in safety to the Japanese Technologies such as the Fukushima nuclear reactors, and the latter has rich sources and good technologies to purify the primary materials for use in nuclear reactors.

Fissionable fuels for nuclear reactors should have several desirable characteristics for practical application, including [27] high melting point, high thermal conductivity, and chemical inertness toward oxidizing materials and metal cladding materials. These requirements are best met by ceramic fuels such as UO_2 [28–30], and PuO_2 [31–33].

Because uranium is a limited resource, the plutonium produced during nuclear power generation must be recycled, preferably in fast reactors. Uranium-plutonium MOXs can be used sustainably in fast MOX fuels produced by reprocessing spent fuel, with plutonium replacing fresh uranium oxide. The fission products and minor actinides are separated as high-level waste and vitrified for geological disposal when the used fuel is processed.

The uranium price continues to rise and at the same time the need for weapons-grade plutonium is decreasing, leading to a rising interest in MOX fuels. MOX fuels, such as UO_2-PuO_2 [34, 35], nowadays account for 5% of fuel used in commercial and research reactors, but further progress can be expected. The MOXs ThO_2-UO_2, ThO_2-PuO_2, and CeO_2-UO_2 have also become accessible recently as MOX nuclear fuels.

The fuel cycle can also be closed by recycling all the uranium and plutonium without separating them, and adding some fresh uranium enriched to a higher level than usual. This regenerated mixture (REMIX) fuel [36] is still under development.

1.2.5 Motion energy harvesting

The introduction of miniaturized, implantable, and wearable electronic devices has led to the development of new battery-free and self-powered technologies (so-called "self-sustainable devices") [37] based on motion energy harvesting, for example, using piezoelectric transducers. Everyday activities such as walking can generate significant power [38] using an inertial energy harvester. This needs periodic motion, which is found in most human movements, such as walking, running, and cycling. The output power is optimized when the resonant frequency of the energy harvester is matched to the frequency of motion. Since human motion consists of a varying combination of low-frequency vibrations, the power to be harvested depends on a thorough study of human mobility patterns [39].

Energy harvesting systems typically use a metal oxide piezoelectric, such as lead zirconate titanate (PZT, $PbZr_xTi_{(1-x)}O_3$), to afford efficient harvesting. It is also possible to harvest magnetic field noise, using a composite of piezoelectric and magnetostrictive materials.

1.2.6 Combustion

Combustion of fuels has for ages been the primary means of generating energy for human endeavors, and this can be expected to continue in the foreseeable future, albeit in decreasing measure due to its low utilization efficiency and enhanced environmental pollution. Fuels in use include coal (mostly burned in power plants), gasoline (burned in motor vehicles), heating oil and natural gas (for heating and cooking), which are all nonrenewable, and wood. The rate of use of these natural resources is unsustainable because their supply is finite. The products released when these fuels are burned affect the environment negatively. For example, burning of coal produces soot, CO and CO_2, Hg, SO_x, and NO_x, which all contribute to global warming, acid rain, and poor air and water quality.

To improve the efficiency of coal-fired power stations and decrease harmful emissions, there have been attempts to produce "clean coal" by means of washing to remove sulfur, nitrogen, and certain minerals; gasification to produce water gas, which burns at a temperature lower than coal and does not produce NO_x; wet scrubbing to remove SO_2 using water and limestone; and oxy-fuel combustion, where fossil fuels are burned in a mixture of recirculated flue gas and oxygen, which eliminates nitrogen oxides from the fluent gas, so that CO_2 can be captured more easily. So far, these technologies have not been demonstrated to be successful and cost effective [40].

Biogas provides another alternative fuel for combustion: organic matter is decomposed microbially into methane, with the formation of by-products such as CO_2, small amounts of water, H_2S, and CO. Biogas is usually produced on a relatively small scale and used for cooking, heating, and lighting.

Incineration of garbage is also a feasible source of energy by combustion [41]. Combustion of municipal solid waste must be an integral part of any nonhazardous waste recovery strategy, ranking below source reduction and recycling or reuse. Controlled combustion can reduce the volume of waste going into landfills, and also harvests energy. However, with the identified risks of mercury and dioxin emissions, all facilities need to be fitted with air pollution control systems.

1.3 Energy storage

For energy generation by renewables to flourish and to avoid its associated complications, sustainable large-scale energy storage systems must be developed, which include batteries, capacitors, compressed air, flywheels, pumped-hydro, and other systems.

The nickel metal hydride battery, developed in the 1960s, is still one of the most widely applied aqueous rechargeable batteries, but with the increasing need for higher energy densities, the lithium-ion battery will most likely become the future high-energy storage devices, especially for powering portable electronic devices and electric vehicles (EVs). Post-lithium-ion batteries (such as Na, K, Zn, Mg, and Al-ion batteries) are also explored for energy storage applications because of the natural abundance, low cost, and environmental friendliness of these elements. Metal oxides are widely used in rechargeable batteries as their electrode materials.

Supercapacitors, or electrochemical capacitors, have emerged worldwide as energy storage systems since the 1970s [42, 43]. These exciting new devices can play a bridging role between batteries and conventional capacitors. Supercapacitors have higher power densities than rechargeable batteries, higher energy densities than conventional dielectric capacitors, and long cycle life [44–49]. The promise of higher power and more energy has led to fast development of supercapacitors, especially to be used as hybrid systems for EVs, owing to the steadily rising environmental concerns about the consumption of fossil fuels. More particularly, supercapacitors can provide higher power compensation to meet the acceleration and climbing requirements for EVs and recycle the power or energy during braking. They are already used in certain electronic products for energy storage and in other applications such as buses, and their market share rose gradually with the development of modern society.

Modern carbon materials have large surface areas, ideal for charge storage, but the physical charge storage on these particles in porous electrode layers is generally limited. For advanced supercapacitor electrodes, metal oxides seem to be the most likely material for the next generation of supercapacitors because of their unique physical and chemical properties and capacitance higher than carbon materials [49].

1.3.1 Batteries

In a rechargeable battery, chemical energy is converted directly into electrical energy and vice versa by reversible electrochemical oxidation/reduction reactions. Rechargeable battery systems currently used are mostly lead-acid, nickel-cadmium, nickel-metal hydride, and lithium-ion batteries.

Like all batteries, lithium-ion batteries consist of a positive electrode, a negative electrode, and an electrolyte. The electrodes have different lithium chemical potentials and are in contact with an electrolyte, which must be ionically conducting and electronically insulating. Its working principle is surely ascribed to redox reactions since it belongs to a battery. Beside redox reactions in the electrodes are accompanied by intercalation processes, which is its main difference from other batteries. However, it should be not called "concentration differential battery" since so far there has no technologies to realize a rechargeable one. The physical and electrochemical reactions in lithium-ion batteries consist of lithium-ion intercalation inside the electrode, charge transfer at the electrode-electrolyte interface, and lithium-ion migration through the electrolyte. Lithium-ion batteries can be recharged because of the high reversibility of these processes [50].

Since the early days of lithium-ion batteries, metal oxides have been deemed the most favorable electrode materials for rechargeable batteries, but challenges still remain with electronic conductivity and the huge volume changes taking place during cycling. To solve these problems, heteroatom doping can achieve higher electronic conductivity, and surface modification or coating with suitable conductive materials can also address these challenges.

Carbon materials and conductive polymers have been widely studied as buffering and conductive agents for metal oxide electrodes. Graphene-metal oxide nanocomposites have great potential in energy storage and conversion, including capacitors and lithium-ion batteries. Nanosized metal oxides and highly conductive graphene are advantageous to shorten lithium-ion diffusion pathways and enhance electronic conductivity. By dipping a prefabricated graphene aerogel electrode in a metal ion solution, metal oxide nanoparticles were anchored on the surface of the graphene after sintering, and were fully accessible to the electrolyte [51]. This method can be used to deposit most types of metal oxides onto a prefabricated 3D graphene structure, providing high reversible lithium storage capacities.

The lithium storage mechanisms available to metal oxides play decisive roles in determining their capacity. As a result, the design and fabrication of electrodes with different morphologies and architectures have improved the cyclability and rate capability. Of course, there are still some further needs such as higher capacity, longer cycling stability, and better rate capability for metal oxides in rechargeable batteries.

Some commercial metal oxides for lithium-ion batteries include $LiCoO_2$, $LiMn_2O_4$, $Li[Ni_xCo_yMn_{1-x-y}]O_2$ $(0<x,y<1)$, and $Li_4Ti_5O_{12}$, and their main producers are situated in China. The main reason is that China is the largest market of EVs, which will be 1 million in 2018, and is leading in their technologies. On lithium technologies, BRICS is enhancing collaboration among its members so that they can keep the pioneering position in the next 10 years.

Furthermore, metal oxides can be ionic conductors for lithium-ion batteries. Due to the inherent good safety, they are becoming a target though many challenges are remaining such as low ionic conductivity, high interfacial resistance with electrode materials, and poor mechanical processability.

1.3.2 Supercapacitors

Supercapacitors are expected, with batteries, to play an important part in modern high-power energy storage systems [52]. Although electrochemical capacitors generally have a high power density and good cycling life, their energy density is still quite low [53]. It therefore remains a high priority to develop new, high energy density electrode materials; these currently include novel carbon materials, metal oxides, and conducting polymers [54–59]. Each kind of electrode material has its merits and drawbacks, but all metal oxides suffer from poor conductivity. Therefore, novel electrode materials are being designed that link metal oxides with carbon materials to afford better electrodes for supercapacitors. It is crucial to incorporate metal oxide nanoparticles in nanostructured carbon materials to improve the utilization of active materials and enhance the electrical conductivity and mechanical strength of these composites. Some important oxides are briefly discussed in the following paragraph.

Manganese dioxide (MnO_2) is frequently regarded the best electrode material for electrochemical capacitors because it is inexpensive, nontoxic, abundant, and environmentally acceptable. Cobalt oxide (Co_3O_4) is popular for use as supercapacitor electrode material because of its good reversible redox behavior and cycle stability, large surface area, and great corrosion stability [60, 61]. But the electrochemical behavior of Co_3O_4 depends on factors such as morphology, specific surface area, and electrical conductivity, which need appropriate construction methodology. Nickel oxide (NiO) has a high theoretical specific capacitance (about $3750\,F\,g^{-1}$) and low cost, is environmentally acceptable, and is relatively easy to prepare [62–64], but it has poor electrical conductivity and low specific surface area, which complicate its application in practice. TiO_2 shows good intercalation/deintercalation behavior of Li^+ and Na^+ ions without forming solid electrolyte interface by-products and volume changes causing electrode collapse. TiO_2 is suitable for use as negative electrode material in organic electrolyte for hybrid supercapacitors with high-power capacity and long lifespan [65–70].

Recently, Ni-Co-Mn metal oxides with various morphologies for high-performance supercapacitor electrodes were synthesized on carbon cloth. The structures consist of nanosheets and nanoneedles and had a maximum specific capacitance of $1434\,F\,g^{-1}$ at a current density of $2\,mA\,cm^{-2}$, good rate capability, and excellent cycling stability [71].

1.4 Energy conversion

Hydrogen is a candidate not only as a clean-burning fuel, but also for energy storage. The use of hydrogen could lead to a low carbon industry, sometimes called the Hydrogen Economy [72]. Meanwhile, the production and use of biodiesel is increasing. Biodiesel is made by hydrolysis and esterification of natural, renewable sources, such as vegetable oils and animal fats, and can be used as pure fuel or blended with petroleum. Ethanol is another renewable fuel, but is more expensive than gasoline. There are also indications that it takes more energy to produce a liter of ethanol than that is obtained by burning it [73].

1.4.1 Hydrogen economy

Hydrogen is an industrial raw material used in wide sectors of industry and construction, and potentially, the use of hydrogen could lead to a low carbon industry, also called the Hydrogen Economy [72]. This refers to hydrogen storage technologies as a long-term solution to global energy storage challenges. Hydrogen storage under ambient conditions remains a desirable target, which currently seems unreachable in the near future. Hydrogen has a high gravimetric energy density and a low volumetric energy density, so for use in motor cars the volumetric density becomes a critical factor because the storage system must fit inside the vehicle without adding too much weight. In stationary applications a low volumetric density would be unproblematic because the weight would not influence the system's efficiency much.

Magnesium hydride has been proposed as a hydrogen storage material, but due to its slow reaction kinetics and high dissociation temperature its practical application is limited. On the other hand, metal oxides with good electrical and ionic conductivity, reversible capacity, and mechanical stability are promising hydrogen storage materials, because they can be doped with magnesium hydride to enhance the hydrogen storage capacity. Many attempts have been made to improve metal oxide sorption properties, for example, in the form of nanocrystalline powders, composite metal hydrides, and catalysts. Novel material-based and solid-state storage systems are still being intensively pursued and depend on the availability of new advanced materials. Good hydrogen storage capacities, thermostability and hydro-stability can be achieved using TMO and magnesium hydride-TMO materials. Ways of achieving better performance of TMO-based physical sorption at ambient temperatures are especially conspicuous.

1.4.2 Biodiesel production

Biodiesel is generally produced in the presence of a suitable catalyst to accelerate the conversion of triglycerides to methyl esters. When homogeneous catalysts are used the products must be separated and washed, while the interaction of heterogeneous basic catalysts with free fatty esters results in soap formation. Heterogeneous acid catalysts generally have high acid site strength and minimal leaching of the active sites, and therefore few undesirable side reactions. They are also more easily regenerated and recycled.

Heterogeneous acid catalysts are continuously being improved so that free fatty acids are esterified [74] optimally using waste and cheap feedstocks. Mesostructured catalysts are especially attractive due to their large surface area, flexible pore size, and high thermal stability. This combination simplifies the diffusion of reactants into the catalyst, increasing the accessibility of the reactants to the active sites. Mesoporous catalysts are synthesized mainly by means of self-assembly, sol-gel, incipient wetness impregnation, coprecipitation, and hydrothermal processes [75]. The hydrothermal autoclaving method results in high processing speeds due to the high surface contact and fast mass transfer.

1.4.3 Thermochemical cycles

Synthetic fuels produced by high-temperature thermochemical processes are an important new way to harvest and store solar energy. Hydrogen production by WS and carbon dioxide splitting (CDS), and CO or syngas production by combined CO_2/H_2O splitting are typical examples. Currently, solar-thermochemical cycles, using the sunlight directly to sustain endothermic reactions [76–78] seem to be simpler and more efficient than photoelectrocatalytic processes [76, 77, 79]. In solar-thermochemical cycles, concentrated solar radiation is used to generate H_2 by splitting H_2O and generate CO by splitting CO_2. Metal oxide catalyzed thermochemical cycles exemplify the simplest possible system and are currently the most advanced pathways for solar-driven fuel production. Research on WS and CDS has produced several novel catalysts, such as SnO_2/SnO, CeO_2/Ce_2O_3, GeO_2/GeO, MgO/Mg, and Cr_2O_3/CeO_2 [80, 81].

1.4.4 Energy-saving smart windows

Substantial energy savings are possible if the heat exchange between the interior and exterior of buildings through windowpanes could be more efficiently controlled. Coating thin films on windows to minimize the effects of solar radiation and black-body radiation would be an effective strategy for use of energy in large buildings. Window glass with a high solar transmittance but low thermal emittance on cold days would allow visible light to enter and prevent loss of heat. At the same time, windows should be transparent to visible radiation and reflect infrared (IR) radiation on hot days, to keep the inside of the building cool. Such "smart windows" can be designed for intelligent control of the amount of light and heat (mainly in the near IR region) passing through the window, in response to an external stimulus such as light (photochromic), heat (thermochromic), and electricity (electrochromic).

VO_2 is a prominent metal oxide used in thermochromic smart windows [82, 83]. When the window's temperature increases and reaches a critical value, a semiconductor to metal transition takes place, so that IR radiation is reflected. To lower the transition temperature, W and F dopants are added. Metal oxides such as WO_3, TiO_2, MoO_3, MoO_3-TiO_2, NiO_x, NiO-WO_3, and V_2O_5 [84–90] are used in electrochromic smart windows. In this case, a low DC voltage is applied to change the color and optical properties of the glass, causing a change in the transmittance of visible and near-IR light [91].

With the continuing development of nanotechnology, nanomaterials may offer improved charge capacity, higher coloration efficiency, and more resilience. The development of smart windows has been slower than, for example, information and communication technology, but the evolution of nanotechnology presents an opportunity to reduce considerably the cost of smart windows.

1.4.5 Superconductors in AC power transportation and transformation

Superconductors are materials that lose their electrical resistance below a certain temperature. The magnetic properties of superconductors are also remarkable. In type I superconductors the current flows at the surface of the material, and superconductivity is negated by weak fields, whereas type II superconductors can carry bulk supercurrent at high fields [92]. Metal oxide-based superconductors, such as the early stage metal oxides (e.g., $LiTiO_3$, $SrTiO_3$, and $BaPb_{1-x}Bi_xO_3$), the cuprate-based oxides (e.g., La-Ba-Cu-O, Y-Ba-Cu-O, Bi-Sr-Ca-Cu-O, Tl-Ba-Ca-Cu-O, and Hg-Ba-Ca-Cu-O), and other oxides (e.g., oxy-pnictides, $Na_xCoO_2{\cdot}yH_2O$) have high-power applications, including superconducting wires/tapes, generators and transformers; low-power applications, especially as superconducting computer devices; and diamagnetic applications, like the maglev train and nuclear fusion reactor. Typically, conductors for power applications are wires or tapes in which many superconducting filaments are embedded in a matrix of a conducting metal, such as copper or silver, which provides protection against magnetic flux jumps and thermal quenching.

Superconductors were first used to build computers using cryotron switches. More recently, superconductors have been used to make microwave filters for mobile phone base stations. Other applications are possible where the efficiency, size, and weight advantages of devices based on high-temperature superconductivity compensate the additional costs involved, for example, their application in wind turbines could lead to savings in construction and tower costs, offsetting the higher costs for the generator [93].

Other exciting developments are in high-performance smart grids, electric power transmission, power storage devices, magnetic levitation devices, and enhancing spintronic devices with superconducting materials [94]. However, applications using alternating current (e.g., transformers) will be more difficult to develop than those that use direct current. Superconducting transmission lines would be more efficient and require much less space than conventional lines, leading to lower environmental impact [95].

The search for novel high-temperature materials, and modification of existing materials to find new materials with higher critical temperature and better performance, continues. The rapid development of both materials and devices over the past decades is exciting and new, valuable applications in diverse fields can be expected.

1.5 Energy and environment

Carbon dioxide, produced by combustion of organic molecules, is a common by-product in the energy, transport and industry sectors. The CO_2 concentration is creeping toward the 410 ppm level, a record high [96]. It is widely accepted that CO_2 is one of the main contributors to global warming, as the natural carbon cycle cannot balance the anthropogenic CO_2 emissions. Thus, inventive ways have to be found to transform CO_2 to harmless, and even useful, chemicals. Recycling carbon dioxide to fuel or innovative materials for the chemical industry would be an ideal way to reduce carbon emissions and supply renewable resources [97, 98].

TiO_2 is already widely used as a photocatalyst for the degradation of organic pollutants. Pt catalysts and material dopants are added to increase the overall rate of reaction. Pt enables O_2 adsorption and reduction by providing high-energy electron density which can form an O_2—Pt bond, increasing the photocatalytic rate of TiO_2 [99].

1.5.1 Emission control

Diesel exhaust emissions are among the main sources of air pollutants. Commercial catalysts used to control harmful emissions are expensive platinum group metals, which generally suffer from poisoning. Development of low cost and high-efficiency metal oxide catalysts, other than precious metals, is therefore crucial for sustainable emission control.

Most air pollutants have been shown to have negative effects on public health and the natural environment. Outdoor air pollutants include sulfur dioxide (SO_2), volatile organic compounds (VOCs), nitrogen oxides (NO_x), and particulate matter (PM) consisting of solid carbon (soot) and uncombusted hydrocarbons. While SO_2 and CO emissions can be readily reduced, PM and NO_x emissions are of major concern and threatening to public health and the natural environment [100]. As the cost of after-treatment devices increases, it is necessary to develop inexpensive and efficient metal oxide catalysts other than precious metals with better catalyst activity for emission control [101]. Novel multicomponent catalysts with advanced nanoshapes with modified structural parameters may bring further progress for diesel emission control technologies.

1.5.2 CO_2 reduction

There is a clear need for better oxide-based photocatalyst materials for CO_2 reduction. To reduce carbon emissions, a few chemical processes are already using CO_2 as reactant for decades, for example, for the synthesis of urea, salicylic acid, and cyclic organic carbonates [102]. But more ways have to be found to convert CO_2 without producing additional carbon emissions. Current photocatalysts that can convert CO_2 have low selectivity, with a variety of products such as carbon monoxide, methane, methanol, formic acid, and even longer hydrocarbon chains often obtained in the same process. The most problematic aspect is the highly negative LUMO energy and high thermodynamic stability of the CO_2 molecule. The exact pathways of CO_2 reduction catalyzed by TiO_2 and other metal oxides need to be clarified to provide a basis for the development of materials suitable for large-scale carbon dioxide recycling to useful chemicals.

1.6 About this book

Over the last four decades, metal oxides have captivated researchers with their potential use in a vast array of applications ranging from energy generation, conversion, and storage, to the hydrogen economy, energy-saving smart windows, the environment, and AC power transportation and transformation. Several breakthroughs have been achieved, in spite of the inherent difficulties of fabricating devices that rely on intricate, high-cost technologies. In the following chapters, this book aims to outline the latest developments and opportunities offered by the use of metal oxides in these applications.

Acknowledgment

Financial support from the National Materials Genome Project (2016YFB0700600), China National Fund for Distinguished Youth Scientists (51026004), and NSFC (U1601214, 51502137, and 21603103) is gratefully acknowledged.

References

[1] F. Meng, Z. Fang, Z. Li, W. Xu, M. Wang, Y. Liu, J. Zhang, W. Wang, D. Zhao, X. Guo, J. Mater. Chem. A 1 (2013) 7235–7241.
[2] Y. Jiang, D. Chen, J. Song, Z. Jiao, Q. Ma, H. Zhang, L. Cheng, B. Zhao, Y. Chu, Electrochim. Acta 91 (2013) 173–178.
[3] Y. Wang, H. Chang, C. Shi, L. Duan, J. Li, G. Zhang, L. Guo, Y. You, Catal. Commun. 100 (2017) 210–213.
[4] D. Wang, Y. Yu, H. He, J. Wang, W. Zhao, H. Abruna, ACS Nano 9 (2015) 1775–1781.
[5] X.Y. Liu, K.X. Wang, J.S. Chen, Energy Storage Mat. 3 (2016) 1–17.
[6] M.C. Liu, L.B. Kong, C. Lu, X.M. Li, Y.C. Luo, L. Kang, ACS Appl. Mater. Interfaces 4 (2012) 4631–4636.
[7] S. Thiagarajan, A. Sanmugam, D. Vikraman, Facile methodology of sol-gel synthesis for metal oxide nanostructures, in: U. Chandra (Ed.), Recent Applications in Sol-Gel Synthesis, Intechopen, Rijeka, 2017, https://doi.org/10.5772/intechopen.68708. Available from: https://www.intechopen.com/books/recent-applications-in-sol-gel-synthesis/facile-methodology-of-sol-gel-synthesis-for-metal-oxide-nanostructures. Accessed 6 December 2017.
[8] S.L. Chou, J.Z. Wang, S.Y. Chew, H.K. Liu, S.X. Dou, Electrochem. Commun. 10 (2008) 1724–1727.
[9] Y.C. Liu, Electrodeposition and Characterization of Metals and Metal Oxides for Energy Conversion and Storage (Doctoral Dissertation), Missouri University of Science and Technology, Missouri, 2016. http://scholarsmine.mst.edu/doctoral_dissertations/2541. Accessed 6 December 2017.
[10] M.A. Azam, N.S.A. Manaf, E. Talib, M.S.A. Bistamam, Ionics 19 (2013) 1455–1476.
[11] X. Wang, G. Yushin, Energy Environ. Sci. 8 (2015) 1889–1904.
[12] R.R. Salunkhe, J.J. Lin, V. Malgras, S.X. Dou, J.H. Kim, Y. Yamauchi, Nano Energy 11 (2015) 211–218.
[13] S.N. Pusawale, P.R. Deshmukh, J.L. Gunjakar, C.D. Lokhande, Mater. Chem. Phys. 139 (2013) 416–422.
[14] M.M. Khan, S.F. Adil, A. Al-Mayouf, J. Saudi Chem. Soc. 19 (2015) 462–464.
[15] F. Tronel, L.G. Demourgues, M. Menetrier, L. Croguennec, L. Goubault, P. Bernard, C. Delmas, Chem. Mater. 18 (2006) 5840–5851.
[16] N.D. Wasalathanthri, T.M. Santa Maria, D.A. Kriz, S.L. Dissanayake, C.H. Kuo, S. Biswas, S.L. Suib, Appl. Catal. B Environ. 201 (2017) 543–551.
[17] C.L. Muhich, Metal Oxide Catalysts for Renewable Energy Generation and Green Chemistry Purposes (Ph.D. Dissertation), University of Colorado, Boulder, CO, 2014. 261 pp. Available from: https://scholar.colorado.edu/chbe_gradetds/69.
[18] I. Ermanoski, N.P. Siegel, E.B. Stechel, J. Sol. Energy Eng. 135 (2013) 031002.
[19] J. Lapp, J.H. Davidson, W. Lipiński, Energy 37 (2012) 591–600.
[20] L.E. Bell, Science 321 (5895) (2008) 1457–1461.
[21] X. Zhang, L. Zhao, J. Mater. 1 (2015) 92–105.
[22] G. Chen, J. Phys. Conf. Ser. 660 (2015) 012066.
[23] K. Biswas, J. He, I.D. Blum, C. Wu, T.P. Hogan, D.N. Seidman, V.P. Dravid, M. G. Kanatzidis, Nature 489 (7416) (2012) 414–418.
[24] T.M. Tritt, Annu. Rev. Mater. Res. 41 (2011) 433–438.
[25] J.W. Fergus, J. Eur. Ceram. Soc. 32 (2012) 525–540.

[26] I. Nandhakumar, N.M. White, S. Beeby (Eds.), Thermoelectric Materials and Devices, Royal Society of Chemistry, Cambridge, 2017.
[27] B.R.T. Frost, R. Inst. Chem. Rev. 2 (1969) 163–205.
[28] A. Casella, B. Hanson, W. Miller, J. Nucl. Mater. 476 (2016) 45–55.
[29] H.X. Xiao, Prog. Nucl. Energy 90 (2016) 122–126.
[30] H. Wang, J.H. Liu, Y.D. Sun, X.F. Hong, H.W. Huang, R.C. Yin, Y. Zhang, C.G. Yin, J. Eur. Ceram. Soc. 35 (2015) 1081–1087.
[31] B.Y. Ao, R.Z. Qiu, H.Y. Lu, P.H. Chen, Comput. Mater. Sci. 122 (2016) 263–271.
[32] S.A. Moten, R.A. Fynn, A.K. Ray, M.N. Huda, J. Nucl. Mater. 468 (2016) 37–45.
[33] B.J. Lewis, E.N. Onder, A.A. Prudil, Fundamentals of Nuclear Engineering, John Wiley, Hoboken, NJ, 2017.
[34] I. Shaaban, M. Albarhoum, Ann. Nucl. Energy 94 (2016) 478–486.
[35] R.C. Belin, M. Strach, T. Truphemus, C. Gueneau, J.C. Richaud, J. Rogez, J. Nucl. Mater. 465 (2015) 407–417.
[36] E.A. Bobrov, P.N. Alekseev, P.S. Teplov, J. Phys. Conf. Ser. 781 (2017) 012017.
[37] J. Paradiso, T. Starner, IEEE Pervasive Comput. 4 (2005) 18–27.
[38] T. Starner, IBM Syst. J. 35 (1996) 618–629.
[39] M. Gorlatova, J. Sarik, M. Cong, I. Kymissis, G. Zussman, IEEE J. Sel. Areas Commun. 33 (2015) 1624–1639.
[40] J. Condliffe, Clean Coal's Flagship Project Has Failed, MIT Technology Review, June 2017. https://www.technologyreview.com/s/608191/clean-coals-flagship-project-has-failed/. Accessed 10 December 2017.
[41] G. Tomberlin, B. Moorman, IEEE Power Engineering Society General Meeting, 2004.
[42] B.E. Conway, Electrochemical Supercapacitors: Scientific Fundamentals and Technological Applications, Kluwer Academic/Plenum Publishers, New York, 1999.
[43] Z. Wu, Y. Zhu, X. Ji, C.E. Banks, Transition metal oxides as supercapacitor materials, in: K. Ozoemena, S. Chen (Eds.), Nanomaterials in Advanced Batteries and Supercapacitors, Springer, Cham, 2016, , pp. 317–344.
[44] L.Q. Mai, A.M. Khan, X.C. Tian, K.M. Hercule, Y.L. Zhao, Nat. Commun. 4 (2013), 2923. https://doi.org/10.1038/ncomms3923.
[45] L.B. Liu, Y. Yu, C. Yan, K. Li, Z.J. Zheng, Nat. Commun. 6 (2015), 7260. https://doi.org/10.1038/ncomms8260.
[46] Y. He, W. Chen, X. Li, Z. Zhang, J. Fu, C. Zhao, E. Xie, ACS Nano 7 (2013) 174–182.
[47] X.H. Lu, M.H. Yu, T. Zhai, G.M. Wang, S.L. Xie, T.Y. Liu, Nano Lett. 13 (2013) 2628–2633.
[48] Z. Ren, J. Li, Y. Ren, S. Wang, Y. Qiu, J. Yu, Sci. Rep. 6 (2016) 20021.
[49] Z. Qi, S. Huang, A. Younis, D. Chu, S. Li, Nanostructured metal oxides-based electrode in supercapacitor applications, in: Z. Stevic (Ed.), Supercapacitor Design and Applications, Intech, Rijeka, 2016. https://doi.org/10.5772/65155. Available from: https://cdn.intechopen.com/pdfs-wm/52281.pdf. Accessed 11 December 2017.
[50] J.B. Goodenough, Acc. Chem. Res. 46 (2013) 1053–1061.
[51] J. Ye, Y. An, E. Montalvo, P.G. Campbell, M.A. Worsley, I.C. Tran, Y. Liu, B.C. Wood, J. Biener, H. Jiang, M. Tang, Y.M. Wang, J. Mater. Chem. A 4 (2016) 4032–4043.
[52] M. Salanne, B. Rotenberg, K. Naoi, K. Kaneko, P.-L. Taberna, C.P. Grey, B. Dunn, P. Simon, Nat. Energy 1 (2016) 16070.
[53] Q.T. Qu, L.L. Li, S. Tian, G.X. Wang, R. Holze, J. Power Sources 195 (2010) 2789–2794.
[54] Y. Zhu, S. Murali, M.D. Stoller, K.J. Ganesh, W.W. Cai, P.J. Ferreira, A. Pirkle, K. A. Cychosz, M. Thomes, O. Su, E.A. Stach, R.S. Ruoff, Science 332 (2011) 1537–1541.
[55] Y.P. Zhai, Y.Q. Dou, D.Y. Zhao, P.F. Fulvio, S. Dai, Adv. Mater. 23 (2011) 4828–4850.

[56] X. Lang, A. Hirata, T. Fujita, M.W. Chen, Nat. Nanotechnol. 6 (2011) 232–236.
[57] X. Xia, J. Tu, Y. Zhang, X. Wang, C. Gu, X. Zhao, H.J. Fan, ACS Nano 6 (2012) 5531–5538.
[58] B.H. Zhang, Y. Liu, Z. Chang, Y.Q. Yang, Y.P. Wu, R. Holze, J. Power Sources 253 (2014) 98–103.
[59] G.A. Snook, P. Kao, A.S. Best, J. Power Sources 196 (2011) 1–12.
[60] V. Srinivasan, J.W. Weidner, J. Power Sources 108 (2002) 15–20.
[61] Z. Xun, C. Cai, W. Xing, T. Lu, J. Electroanal. Chem. 545 (2003) 19–27.
[62] M.S. Wu, Y.A. Huang, J.J. Jow, W.D. Yang, C.Y. Hsieh, H.M. Tsai, Int. J. Hydrog. Energy 33 (2008) 2921–2926.
[63] R.S. Jayashree, P.V. Kamath, J. Appl. Electrochem. 31 (2001) 1315–1320.
[64] K.W. Nam, K.H. Kim, E.S. Lee, W.S. Yoon, X.Q. Yang, K.B. Ki, J. Power Sources 182 (2008) 642–652.
[65] D. Deng, M.G. Kim, J.Y. Lee, J. Cho, Energy Environ. Sci. 2 (2009) 818–837.
[66] N. Li, G. Liu, C. Zhen, F. Li, L. Zhang, H. Cheng, Adv. Funct. Mater. 21 (2011) 1717–1722.
[67] J. Xu, C. Jia, B. Cao, W. Zhang, Electrochim. Acta 52 (2007) 8044–8047.
[68] H. Xiong, M.D. Slater, M. Balasubramanian, C.S. Johnson, T. Rajh, J. Phys. Chem. Lett. 2 (2011) 2560–2565.
[69] Y. Xu, E.M. Lotfabad, H.L. Wang, B. Farbod, Z.W. Xu, A. Kohandehghan, D. Mitlin, Chem. Commun. 49 (2013) 8973–8975.
[70] Z.H. Bi, M.P. Paranthaman, P.A. Menchhofer, R.R. Dehoff, C.A. Bridges, M.F. Chi, B. K. Guo, X.G. Sun, S. Dai, J. Power Sources 222 (2013) 461–466.
[71] P. He, Q. Huang, B. Huang, T. Chen, RSC Adv. 7 (2017) 24353–24358.
[72] F. Zhang, P.C. Zhao, M. Niu, J. Maddy, Int. J. Hydrog. Energy 41 (2016) 14535–14552.
[73] D. Pimentel, A. Marklein, M.A. Toth, M. Karpoff, G.S. Paul, R. McCormack, J. Kyriazis, T. Krueger, Energies 1 (2008) 41–78.
[74] S. Soltani, U. Rashid, S.I. Al-Resayes, I.A. Nehdi, Energy Convers. Manag. 141 (2017) 183–205.
[75] H. Wang, Z. Yao, X. Zhan, Y. Wu, M. Li, Fuel 158 (2015) 918–926.
[76] J. Yang, D. Wang, H. Han, C. Li, Acc. Chem. Res. 46 (2013) 1900–1909.
[77] N.P. Siegel, J.E. Miller, I. Ermanoski, R.B. Diver, E.B. Stechel, Ind. Eng. Chem. Res. 52 (2013) 3276–3286.
[78] J.E. Miller, A.H. McDaniel, M.D. Allendorf, Adv. Energy Mater. 4 (2014) 1300469.
[79] R. Abe, J Photochem Photobiol C: Photochem Rev 11 (2010) 179–209.
[80] C. Agrafiotis, M. Roeb, C.A. Sattler, Renew. Sust. Energ. Rev. 42 (2015) 254–285.
[81] S. Mostrou, R. Büchel, S.E. Pratsinis, J.A. van Bokhoven, Appl. Catal. A 537 (2017) 40–49.
[82] Z. Chen, Y. Gao, L. Kang, J. Du, Z. Zhang, H. Luo, H. Miao, G. Tang, Sol. Energy Mater. Sol. Cells 95 (2011) 2677–2684.
[83] M. Jiang, S. Bao, X. Cao, Y. Li, S. Li, H. Zhou, H. Luo, P. Jin, Ceram. Int. 40 (2014) 6331–6334.
[84] J. Neto, R.M. Torresi, S. Torresi, J. Electroanal. Chem. 765 (2016) 111–117.
[85] H. Yu, Y. Li, L. Zhao, G. Li, J. Li, H. Rong, Z. Liu, Mater. Lett. 169 (2016) 65–68.
[86] D. Dong, W. Wang, G. Dong, F. Zhang, Y. He, H. Yu, F. Liu, M. Wang, X. Diao, Appl. Surf. Sci. 383 (2016) 49–56.
[87] G. Cai, P. Darmawan, M. Cui, J. Chen, X. Wang, A.L. Eh, S. Magdassi, P.S. Lee, Nanoscale 8 (2016) 348–357.
[88] J. Chu, Z. Kong, D. Lu, W. Zhang, X. Wang, Y. Yu, S. Li, X. Wang, S. Xiong, J. Ma, Mater. Lett. 166 (2016) 179–182.

[89] Z. Tong, N. Li, H. Lv, Y. Tian, H. Qu, X. Zhang, J. Zhao, Y. Li, Sol. Energy Mater. Sol. Cells 146 (2015) 135–143.
[90] Z. Tong, J. Hao, K. Zhang, J. Zhao, B. Su, Y. Li, J. Mater. Chem. C 2 (2015) 3651–3658.
[91] S.D. Rezaei, S. Shannigrahi, S. Ramakrishna, Sol. Energy Mater. Sol. Cells 159 (2017) 26–51.
[92] D. Larbalestier, A. Gurevich, D.M. Feldmann, A. Polyanskii, Nature 414 (2001) 368–377.
[93] M.R. Islam, Y. Guo, J. Zhu, Renew. Sust. Energ. Rev. 33 (2014) 161–176.
[94] J. Linder, J.W.A. Robinson, Nat. Phys. 11 (2015) 307–315.
[95] H. Thomas, A. Marian, A. Chervyakov, S. Stuckrad, D. Salmieri, C. Rubbia, Renew. Sust. Energ. Rev. 55 (2016) 59–72.
[96] Scripps Institute of Oceanography, The Keeling Curve, 2018. https://scripps.ucsd.edu/programs/keelingcurve/, Accessed 18 June 2018.
[97] S.N. Habisreutinger, L. Schmidt-Mende, J.K. Stolarczyk, Angew. Chem. Int. Ed. 52 (2013) 7372–7408.
[98] L. Guo, Y. Wang, T. He, Chem. Rec. 16 (2016) 1918–1933.
[99] C.L. Muhich, Metal Oxide Catalysts for Renewable Energy Generation and Green Chemistry Purposes (Ph.D. dissertation), University of Colorado (Boulder), 2014. [Chemical & Biological Engineering Graduate Theses & Dissertations, 69, https://scholar.colorado.edu/chbe_gradetds/69. Accessed 30 December 2017.
[100] G.C. Dhal, D. Mohan, R. Prasad, Cat. Sci. Technol. 7 (2017) 1803–1825.
[101] B. Choi, K.S. Lee, Chem. Eng. J. 240 (2014) 476–486.
[102] H.J. Freund, M.W. Roberts, Surf. Sci. Rep. 25 (1996) 225–273.

Metal oxides in fuel cells

2

Peng Wang, Jing Wang†, Lijun Fu*, Yuping Wu*, Teunis van Ree‡*
*State Key Laboratory of Materials-Oriented Chemical Engineering, School of Energy Science and Engineering, Institute for Electrochemical Energy Storage, Nanjing Tech University, Nanjing, People's Republic of China, †State Key Laboratory of Materials-Oriented Chemical Engineering, College of Chemistry and Chemical Engineering, Nanjing Tech University, Nanjing, China, ‡Department of Chemistry, University of Venda, Thohoyandou, South Africa

2.1 Introduction

Along with the consumption of fossil fuels and the corresponding environmental deterioration, worldwide investigations focus on efficient, clean and reliable energy generation, storage, and delivery [1,2]. Fuel cells with their high-energy-conversion efficacy and pollution-free emission are regarded one of the most promising alternative electric power-generating technologies. In fuel cells, the chemical energy stored in various fuels combined with oxygen from the air, is converted directly to electrical power by electrochemical reactions occurring on electrodes, which possess extremely high theoretical and practical efficiencies compared with conventional heat-converting engines.

A basic fuel cell system contains the following three components: anode, cathode, and electrolyte. In the operation of fuel cells, fuels are transported to the anode and and oxidants (oxygen or air) to the cathode, respectively. Fuels are oxidized and and oxidants are reduced under specific catalytic conditions, with electrons being transferred from the anode to the cathode through the external circuit, while cations migrate through the electrolyte from the anode to the cathode, and anions from the cathode to the anode (Fig. 2.1). The overall reaction equations of a fuel cell using hydrogen or hydrocarbon such as methane as fuels are as follows:

$$H_2 + \tfrac{1}{2}O_2 \rightarrow H_2O \tag{2.1}$$

$$CH_4 + 2O_2 \rightarrow CO_2 + 2H_2O \tag{2.2}$$

In these reactions, the fuels are oxidized to H_2O and CO_2, which involves release and translation of electrons, releasing electrical energy instead of heat. The released ions in different kinds of fuel cells are shown in Fig. 2.2.

Typical fuel cells can be classified as follows: alkaline fuel cell (AFC), direct methanol fuel cell (DMFC), proton exchange membrane fuel cell (PEMFC), phosphoric acid fuel cell (PAFC), molten carbonate fuel cell (MCFC), and solid oxide fuel cell (SOFC) (Fig. 2.2). In particular, SOFCs, operating at high temperatures (from 500°C

Metal Oxides in Energy Technologies. https://doi.org/10.1016/B978-0-12-811167-3.00002-X

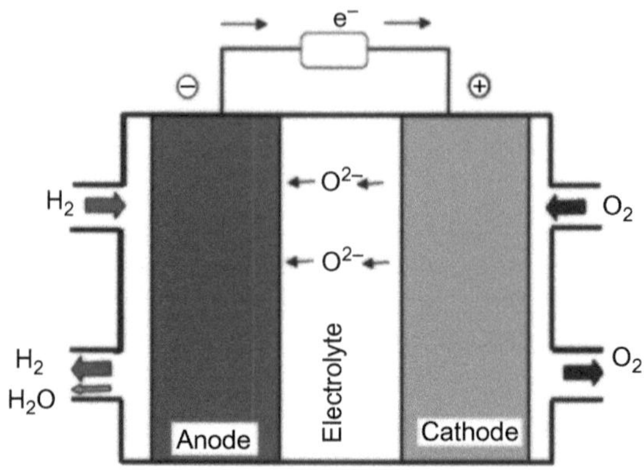

Fig. 2.1 Schematic drawing of a solid oxide fuel cell (SOFC) using hydrogen as fuel. Reproduced by permission from H.A. Taroco, J.A.F. Santos, R.Z. Domingues, T. Matencio, Ceramic materials for solid oxide fuel cells, in: C. Sikalidis (Ed.), Advances in Ceramics—Synthesis and Characterization, Processing and Specific Applications, InTech, 2011. Available from: https://www.intechopen.com/books/advances-in-ceramics-synthesis-and-characterization-processing-and-specific-applications/ceramic-materials-for-solid-oxide-fuel-cells (Last accessed on 28 April 2017).

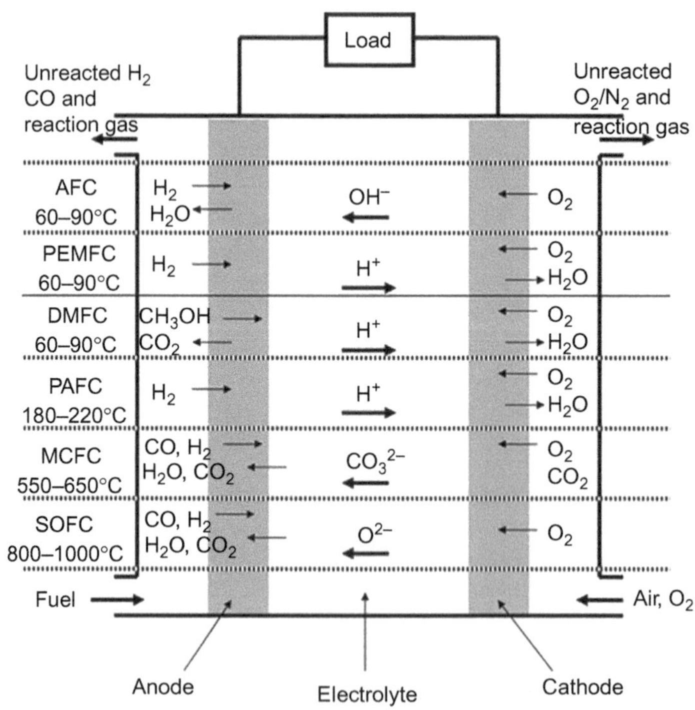

Fig. 2.2 Summary of the reactions and processes that occur in various fuel cell systems. Reproduced by permission from M. Winter, R.J. Brodd, Chem. Rev. 104 (2004) 4245.

to 1000°C), utilize solid oxides, especially metal oxides, as electrodes and electrolytes, which are discussed in this chapter.

2.2 Metal oxides in SOFCs

2.2.1 Anode materials

The anode, also called the fuel electrode, plays a crucial role that provides not only the reaction sites but also the catalyst for the electrochemical oxidation of fuels. Some metals, such as noble metals (platinum, silver) and transition metals (iron, cobalt, and nickel), were used as anode in the early fuel cell development stages, but they suffered from electrochemical performance degradation caused by negligible ion conductivity, thermal expansion mismatch, and low porosity. Consequently, metal oxides such as cerium oxide were introduced as anode materials, leading to significant performance improvements.

2.2.1.1 Ni, Cu, and their modifications

Nickel is a prime candidate as anode material because of its low cost, high activity, and stability in the reducing conditions of fuel gas at extremely high temperatures. To maintain the porous structure of nickel over long periods at high temperatures such as 1000°C and to provide other desired properties for the anode, nickel metal is often dispersed on the surface of an yttria-stabilized zirconia (YSZ) support, which is also the state-of-the-art method to increase the three-phase boundary area [3–5]. The role of the YSZ in the anode is to support the nickel particles, constrain the coarsening of the metallic particles at the operating temperature, and to provide a thermal expansion coefficient acceptably close to those of other cell components.

The SOFC anode cermet is usually made from YSZ and NiO powders. The NiO is reduced in situ to nickel metal when it is exposed to the fuel in the fuel cell. Extensive literature shows that performance and electric properties of Ni/YSZ cermet anodes are critically dependent on the microstructure and distribution of the Ni and YSZ phases [6–8]. Different commercial NiO powders, with varying average grain size and BET surface areas, have been evaluated; a high surface area corresponds to high shrinkage and a low sintering starting temperature (Fig. 2.3) [9]. The effect of the volume ratio of the Ni and YSZ support was also investigated. At concentrations below 30 vol% Ni, the conductivity is mostly ionic. Above 30 vol% Ni, the metallic character dominates, so the conductivity is mostly electronic. The maximum electrical conductivity of the Ni/YSZ anode is reached at approximately 30 vol% Ni [10].

Compared to most other transition metals, Cu is relatively inert as a catalyst for the formation of C—C bonds, possessing good enough stability under high oxygen partial pressure. Cu is another promising candidate for anode material, but it is not as good an electrocatalyst as Ni. Furthermore, a bigger problem is that Cu cermets cannot be produced using the same methods generally used for Ni cermets.

It is possible to prepare Cu-cermet anodes by adding Cu after preparing a porous layer of YSZ on a dense YSZ electrolyte layer. For example, Cu can be added by

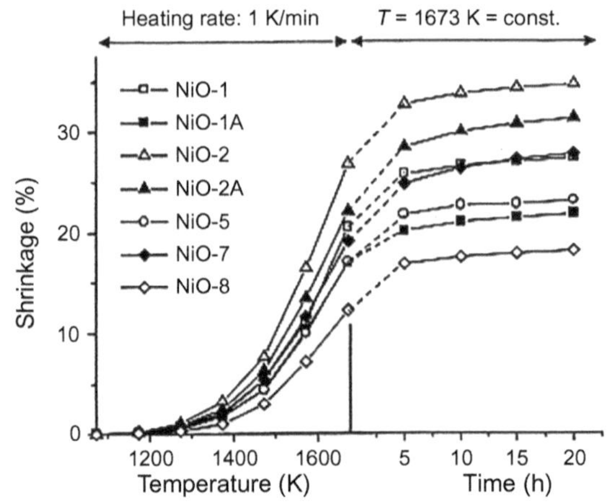

Fig. 2.3 Shrinkage characteristics of different NiO/YSZ substrates at sintering temperatures between 1073 and 1673 K (at 1 K/min) and subsequent dwell time of up to 20 h at 1673 K.
Reproduced by permission from F. Tietz, F.J. Dias, D. Simwonis, D. Stöver, J. Eur. Ceram. Soc. 20 (2000) 1023.

aqueous impregnation with a concentrated solution of $Cu(NO_3)_2$, followed by calcination to decompose the nitrate and form the Cu oxides. Reduction of the oxides by H_2 at 800°C leads to the formation of metallic Cu [11]. Incorporating other components such as Ni [12,13] and ceria [14,15] can increase the activity of copper-based anodes. Because of the relatively low melting point of copper oxides (CuO and Cu_2O), the sintering temperature must be below the melting point, but on the other hand, a low sintering temperature readily results in the separation of the two phases.

2.2.1.2 CeO_2 and its modification

Some metal oxides with mixed ionic and electronic conductivity (MIEC) that have been explored as prospective anode materials for SOFCs, are capable of reforming hydrocarbons. Ceria is an outstanding catalyst for hydrocarbon reforming reactions and is resistant to carbon deposition, which permits a direct supply of dry hydrocarbon fuels to the anode [16–18]. Although ceria has such a remarkable catalytic activity and ion conductivity, its electrical conductivity is relatively low, which decreases the performance of the SOFC. Other components such as Ni, Co, and Pt [19–23] are therefore incorporated with ceria, which can then easily break the C—H bonds in hydrocarbons.

The mixed oxide CuCo-CeO_2 (40% metal load, 1/1 Cu—Co atomic ratio), synthesized by the inverse microemulsion method, has been evaluated as anode material for SOFC at intermediate temperatures (Fig. 2.4) [24]. The substitution of 20% of the copper by cobalt in the anode composition results in higher stability against carbon formation when utilizing methane as fuel. The bimetallic Cu—Co formulation combined with CeO_2 as a SOFC anode shows higher catalytic activity for methane oxidation, favoring charge transfer on the triple phase boundary, and showing potential sulfur tolerance in a feed of methane and two different fuel mixtures simulating biogas (Fig. 2.5) [25].

Yttria and gadolinium can also be doped in ceria to fabricate an SOFC anode for use with carbon-based fuels. These materials have high ionic conductivity in an oxidizing environment and are known to have reasonable MIEC in a reducing

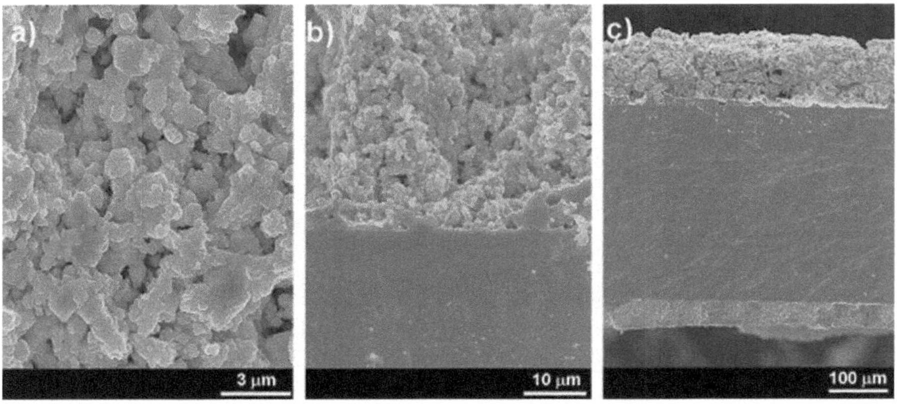

Fig. 2.4 SEM micrographs of a cross section of the tested single cell: (A) anode, (B) anode/electrolyte, and (C) anode/electrolyte/cathode.
Reproduced by permission from A. Fuerte, R.X. Valenzuela, M.J. Escudero, L. Daza, J. Power Sources 196 (2011) 4324.

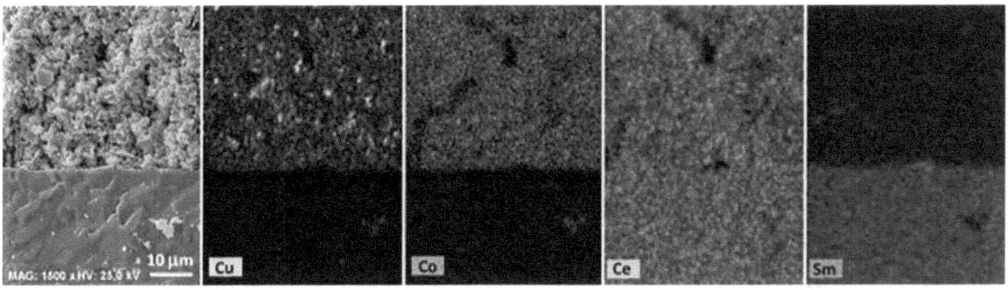

Fig. 2.5 Cross-sectional view and mapping of the Cu-Co-Ce/SDC layer after 740h testing at 1023K using different fuels (H_2, CH_4, H_2S, and simulated biogas).
Reproduced by permission from A. Fuerte, R.X. Valenzuela, M.J. Escudero, L. Daza, Int. J. Hydrogen Energy 39 (2014) 4060.

environment. Both these anode materials can be made by spray coating with coprecipitated nanopowders, and demonstrate satisfactory structural stability in the DCFC test environment using coconut char as fuel [26].

2.2.1.3 Perovskites and their composites

The Ni-based SOFC anodes are limited to hydrogen fuels to some extent as carbon deposition is a main issue for SOFC operating in dry hydrocarbon fuels, whereas perovskite oxides can effectively inhibit such processes with better stability and catalytic activity. Oxides with general perovskite-type formula ABO_3 have also been used as cathode materials. Fig. 2.6 shows their structure, with A and B cations having a total charge of +6. The A-site cations (such as those of La, Sr, Ca, and Pb) have lower

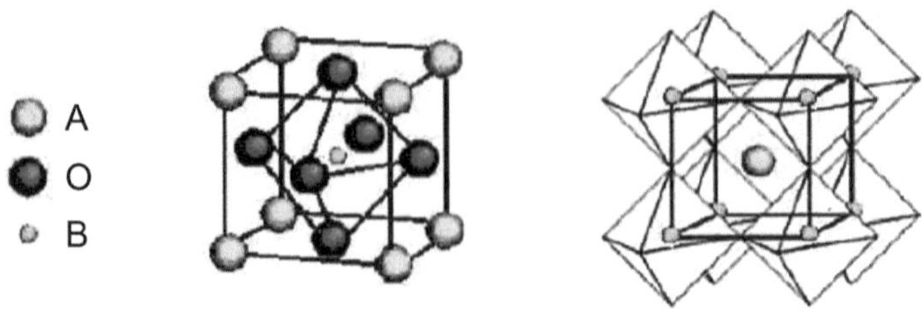

Fig. 2.6 Typical perovskite crystal structure.
Reproduced by permission from H.A. Taroco, J.A.F. Santos, R.Z. Domingues, T. Matencio, Ceramic materials for solid oxide fuel cells, in: C. Sikalidis (Ed.), Advances in Ceramics—Synthesis and Characterization, Processing and Specific Applications, InTech, 2011. Available from: https://www.intechopen.com/books/advances-in-ceramics-synthesis-and-characterization-processing-and-specific-applications/ceramic-materials-for-solid-oxide-fuel-cells (Last accessed on 28 April 2017).

valences and are therefore relatively large and coordinated to 12 oxide anions. The B-site cations (such as those of Ti, Cr, Ni, Fe, Co, and Zr) occupy much less space and are coordinated to six oxide anions.

Cation layered double perovskite oxides, particularly $PrBaCo_2O_5$, display excellent mixed-ionic and electronic conductivity and fast oxygen kinetics and show excellent characteristics when used as SOFC oxygen electrodes. Similarly, the layered perovskite anode $PrBaMn_2O_{5+\delta}$ (PBMO, Fig. 2.7), fabricated by in situ annealing of $Pr_{0.5}Ba_{0.5}MnO_{3-\delta}$ under fuel conditions, showed good redox stability with tolerance to coking and sulfur contamination from hydrocarbon fuels [27]. The layered oxygen-deficient double perovskite $Sr_2Fe_{1.5}Mo_{0.5}O_{6-\delta}$-$Gd_{0.1}Ce_{0.9}O_{2-\delta}$ (SFM-GDC) anode-supported SOFCs can be fabricated by low-temperature co-sintering and wet impregnation techniques; they show excellent sulfur tolerance toward H_2 containing up to 100 ppm H_2S [11,28].

A perovskite-related material, $La_{0.75}Sr_{0.25}Cr_{0.5}Mn_{0.5}O_3$ (LSCM), has been made for use as anode for an SOFC operating in low moisture (3% H_2O) H_2 and CH_4 fuels, showing no significant coking with CH_4 [29]. This material also exhibits low reforming activity for both H_2O and CO_2 but is active for methane oxidation (Fig. 2.8). A Sr-doped perovskite material, $La_{0.75}Sr_{0.25}Cr_{0.5}Fe_{0.5}O_{3-\delta}$ (LSCrF), has also been found to be a good catalyst for methane reforming and oxidation, catalyzing complete stoichiometric methane oxidation and partial methane oxidation with increased oxygen vacancy content (Fig. 2.9) [30].

Double perovskites such as $Sr_2Mg_{1-x}Mn_xMoO_{6-\delta}$ can also meet the requirements for long-term stability with sulfur tolerance, and perform well in a hydrogen-methane single cell (Fig. 2.10) [31]. Furthermore, A-site and B-site modified perovskites have been investigated extensively to improve the performance of anode materials. A-site deficiencies can promote the valence change of B-site elements and the formation of additional oxygen vacancies. For example, A-site-deficient $SrTiO_3$-based perovskite with donor doping on an Sr or Ti site could remarkably increase the electronic conductivity or ionic conductivity [32].

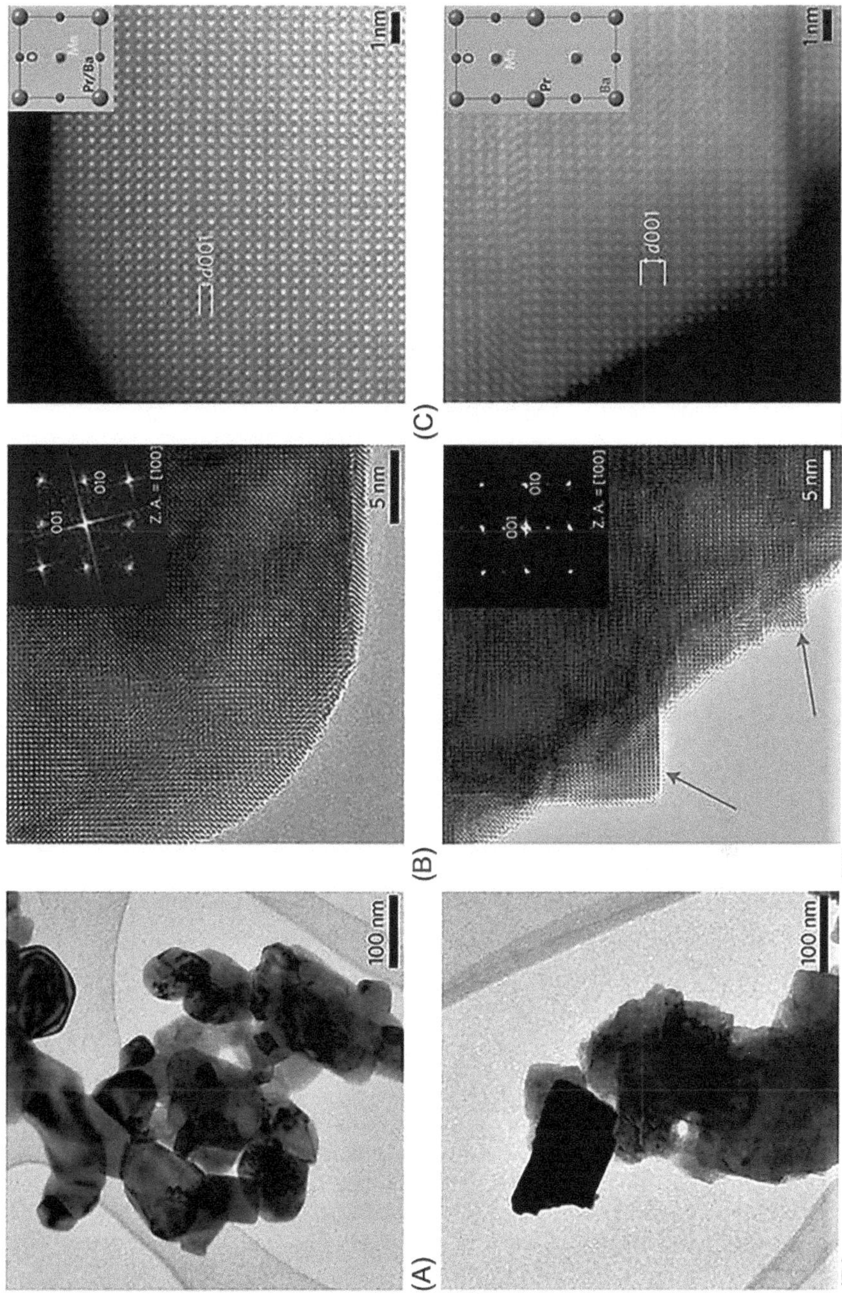

Fig. 2.7 See figure legend on next page.

(Continued)

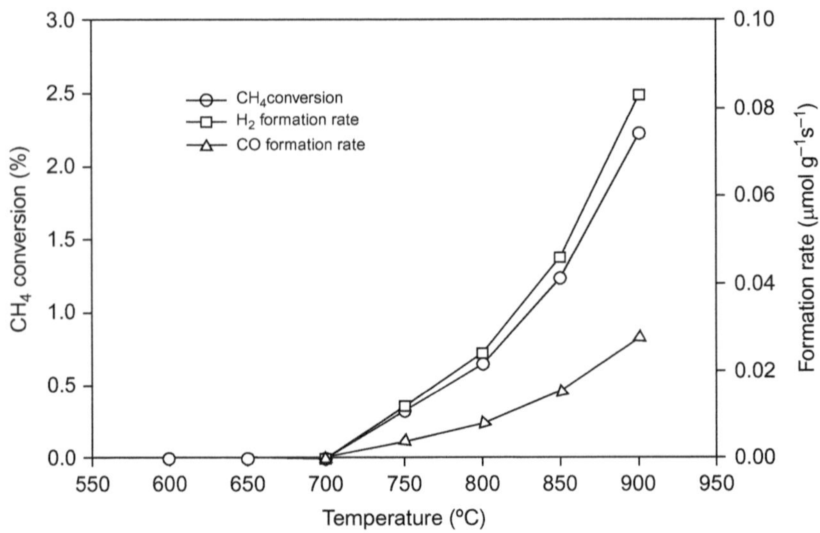

Fig. 2.8 Methane conversion and gas formation rate over LSCM as a function of temperature with $CH_4/H_2O = 1{:}1$.
Reproduced by permission from S.W. Tao, J. Irvine, S.M. Plint, J. Phys. Chem. B 110 (2006) 21771.

2.2.2 *Electrolyte materials*

An electrolyte is a fundamental requirement for an SOFC unit, since it enables the movement of oxide ions from the cathode to the anode where the oxygen reacts with fuel. Oxide ions are conducted by a thermally activated oxygen vacancy hopping process. To accomplish high ionic conductivity, the solid electrolyte's crystal structure should have large available interionic spaces to attain a high level of point defect disorder and low migration enthalpy (normally <1 eV). In addition, electrolytes should be dense enough to avoid gas permeation and should be chemically stable. Some metal oxides were found to satisfy the high-temperature requirements, for example, ZrO_2, CeO_2, fluorite structured Bi_2O_3-based oxides, $LaGaO_3$-based perovskites, etc. [33].

Fig. 2.7, cont'd (A) Bright-field TEM. (B) High-resolution TEM image and corresponding fast-Fourier transformed pattern with zone axis (Z.A.) = [100]. (C) High-angle annular dark-field scanning TEM image of $Pr_{0.5}Ba_{0.5}MnO_3$ and its atomic arrangement viewed in the [100] direction with *d*-spacing 001. (D) Bright-field TEM image. (E) High-resolution TEM image and corresponding fast-Fourier transformed pattern with Z.A. = [100]. (F) High-angle annular dark-field scanning TEM image of A-site layered $PrBaMn_2O_{5+\delta}$ and its atomic arrangement aligned along the [100] direction with *d*-spacing 001.
Reproduced by permission from S. Sengodan, S. Choi, A. Jun, T.H. Shin, Y. Ju, H.Y. Jeong, J. Shin, J.T.S. Irvine, G. Kim, Nat. Mat. 14 (2015) 205.

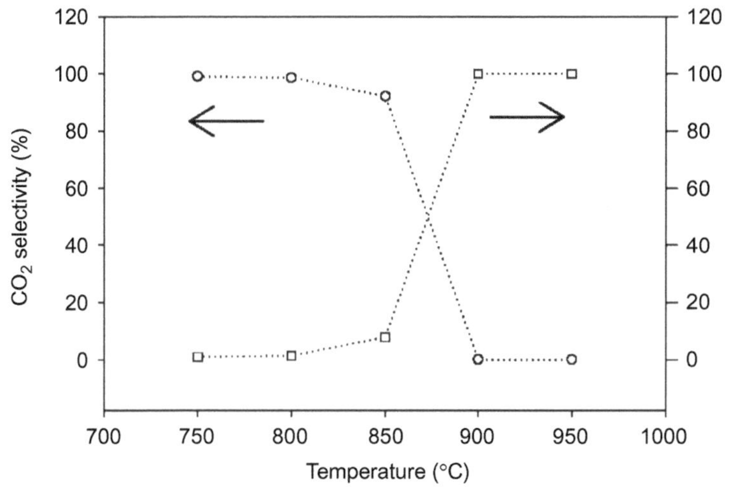

Fig. 2.9 Methane conversion as a function of the pO_2/pCH_4 ratio at 900°C. Reproduced by permission from S.W. Tao, J. Irvine, Chem. Mater. 16 (2004) 4116.

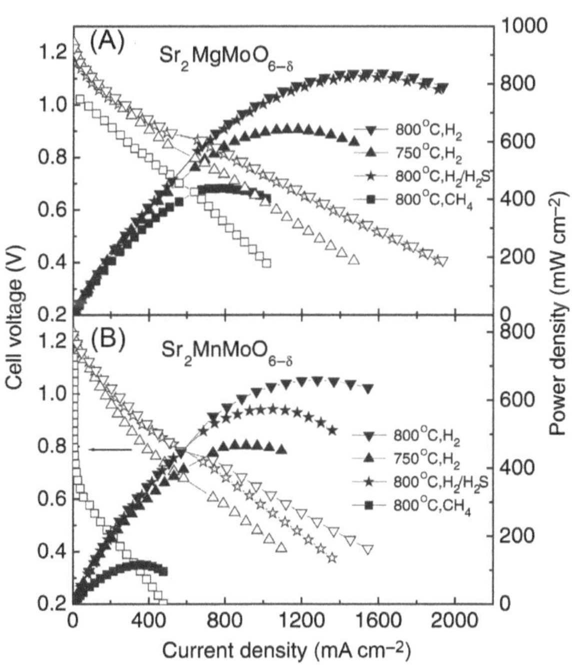

Fig. 2.10 Cell voltage and power density as a function of current density for the single fuel cells with anodes of (A) $Sr_2MgMoO_{6-\delta}$ and (B) $Sr_2MnMoO_{6-\delta}$ with H_2, H_2/H_2S, and CH_4 fuels. Reproduced by permission from Y.H. Huang, R.I. Dass, Z.L. Xing, J.B. Goodenough, Science 312 (2006) 254.

2.2.2.1 ZrO_2 and its composites

Zirconia-based ion-conducting electrolytes are historically widely used for high-temperature SOFCs. Pure ZrO_2 is a stable monoclinic crystal at room temperature, transforming to tetragonal above 1170°C, and ultimately to a cubic fluorite structure at 2370°C, which is stable up to its melting point at 2680°C [34]. These martensitic phase changes are reversed during cooling associated with a large volume change, which results in the formation of cracks on cooling and low thermal shock resistance. The high-temperature tetragonal or cubic phase of ZrO_2 can be stabilized at room temperature by replacing a small amount of Zr^{4+} ions with acceptor dopant cations.

The addition of different ratios of Y_2O_3 can stabilize the ZrO_2 cubic phase at temperatures up to 2500°C. Since the valence of the Y^{3+} dopant is smaller than that of Zr^{4+}, some oxygen vacancies are generated, which stabilizes the cubic ZrO_2 crystal lattice and boosts its oxide ion conductivity [1]. A variety of Y_2O_3-ZrO_2 solid solutions have been prepared, but 8 mol% Y_2O_3-ZrO_2 (8YSZ) seems to have the highest ionic conductivity of 0.1 $S\,cm^{-1}$ at 1000°C, which is the minimum ionic conductivity required for an electrolyte at these temperatures [35]. Most of the anodes used with YSZ electrolyte in SOFC are Ni-based materials because of their thermal compatibility. Continuing research for alternative anode and electrolyte materials is producing new anode materials for use with YSZ electrolyte, such as $La_{0.75}Sr_{0.25}Cr_{0.5}Mn_{0.5}O_{3-\delta}$ [36]. Despite its relatively low electric conductivity when working at intermediate to low temperatures, YSZ is currently the most popular commercial electrolyte material due to its stability in reducing environments, low cost, and simple processing, such as spin coating.

For SOFC applications, the YSZ electrolyte (Fig. 2.11) must be dense (>95% of theoretical density) and impervious to gases, leading to the requirement of high sintering temperatures (>1400°C) [37]. Lowering the sintering temperature of the YSZ electrolyte can avoid the formation of highly resistive materials such as $La_2Zr_2O_7$ and $SrZrO_3$, which are known to be formed due to chemical reaction between the YSZ electrolyte and cathode materials such as $La(Sr)MnO_3$ (LSM) and $La(Sr)CoFO_3$ (LSCF), at processing temperature of around 1200°C [38]. One of the most simple and scalable techniques to reduce the sintering temperature is to add sintering aids, such as strontium gallate, copper, and manganese [39–43].

Among several other doped ZrO_2 systems, Sc_2O_3-doped ZrO_2 (ScSZ) materials have attracted considerable attention for use as electrolyte in SOFCs. The ionic conductivity of 11 mol% Sc_2O_3-doped ZrO_2 at 850°C and 1000°C is about 1.5 and 3 times higher, respectively, than that of 8YSZ [44], as illustrated in Table 2.1. Since ScSZ materials possess high ionic transference (and insignificant electronic conductivity) over a wide range of temperatures and partial oxygen pressures, and outstanding thermochemical properties, they provide a chance to lower the working temperature of SOFCs. However, some disadvantages, such as a complex phase assembly, conductivity aging behavior over time at high temperatures, and poor accessibility exclude their practical application.

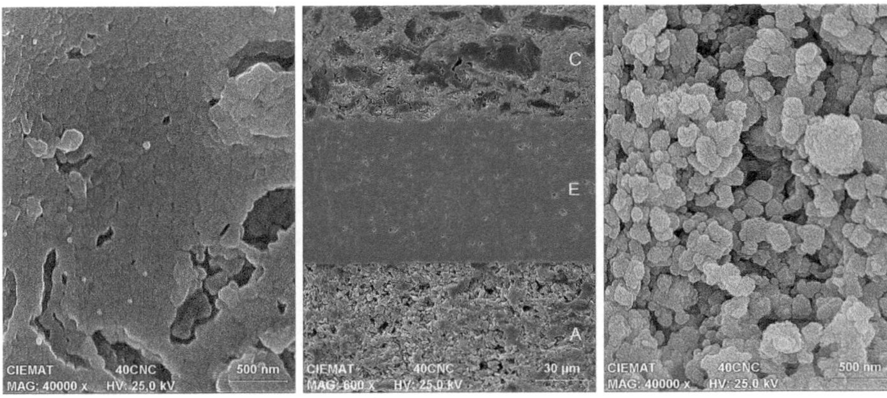

Fig. 2.11 SEM images of a single cell (Cu-Ni-CeO_2/YSZ/LSF cell) with a dense YSZ electrolyte (left: cathode; middle: C = cathode, E = electrolyte, A = anode; right: anode). Reproduced by permission from A. Hornés, M.J. Escudero, L. Daza, A. Martínez-Arias, J. Power Sources 249 (2014) 520.

Table 2.1 **Properties of the ZrO_2-based electrolytes [44]**

Composition	Expansion coefficient ($10^{-6}\ K^{-1}$)	Bend strength at 300 K (MPa)	Conductivity at 1000°C ($S\ cm^{-1}$)
ZrO_2-3 mol% Y_2O_3	10.8	1200	0.05
ZrO_2-8 mol% Y_2O_3	10.5	230	0.09
ZrO_2-11 mol % Sc_2O_3	10.0	255	0.30

2.2.2.2 CeO_2 and its composites

Ceria (CeO_2) has a stable cubic fluorite structure from room temperature up to its melting point (2400°C) at 1 atm pressure. The ceria structure has large unoccupied octahedral sites that provide a route for rapid diffusion of oxide ions, but the oxygen vacancies in the pure CeO_2 are not enough to show significant ionic conductivity. Therefore, to attain high ionic conductivity, oxygen vacancies are brought into the structure by the substitution of the host Ce^{4+} ions with acceptor cations, such as Gd^{3+}, Sm^{3+}, or Y^{3+} [45–49].

Doped CeO_2 compounds such as Gd-doped CeO_2, La-doped CeO_2, and Sm-doped CeO_2 are under investigation for low-temperature (500–650°C) operation of SOFCs

[50–52]. The ionic conductivity of nanocrystalline Gd-doped CeO_2 thin films increases with decreasing activation energy as the grain size decreases, which results in increased electrochemical performance when used as thin film electrolyte [53]. Apart from the effects of grain size and heat treatment, using different compositions in the range 9%–11% $GdO_{1.5}$ improves the performance at 500°C [54]. Both Sm-doped CeO_2 and La-doped CeO_2 electrolytes with large micro-domains over 10 nm in the lattice possess increased ion conductivity and consequently yield high-quality electrolytes with superior performance. On the other hand, highly effective Sm or La and alkaline earth co-doped CeO_2-based electrolytes have controlled small micro-domains of around 1–3 nm in the microstructure [55]. As for co-doped CeO_2, $Ce_{0.8}Gd_{0.18}Sm_{0.02}O_{2-\delta}$ has an ion conductivity of $2.28 \times 10^{-2}\,S\,cm^{-1}$ at 700°C, while $Ce_{0.8}Gd_{0.1}Bi_{0.1}O_{2-\delta}$ (Fig. 2.12) displays a high ion conductivity of $4.46 \times 10^{-2}\,S\,cm^{-1}$ [56]. When the doped species are transition metals such as Mo, $Ce_{0.9}Mo_{0.1}O_{2+\delta}$ exhibits the highest electrical conductivity of the Mo-doped ceria materials, with the values of 2.8×10^{-4} and $5.08 \times 10^{-2}\,S\,cm^{-1}$ at 550°C in air and in wet H_2, respectively [57].

Although doped CeO_2 shows excellent ion conductivity and phase stability while operating over a broad temperature range, Ce^{4+} ions tend to be reduced to Ce^{3+} in the reducing environment or low partial pressure of oxygen. Although this leads to high electronic conductivity (which can result in a short circuit in the SOFC), the end-result is structure expansion (due to the increased size of the Ce^{3+} ions), and therefore poorer mechanical properties.

In attempts to synthesize high-performance SOFC cathodes, strontium-doped $SmCoO_3$ (SSC, a widely used $LnCoO_3$-based oxide) and samarium-doped CeO_2 (SDC) nano-composite powders were prepared by spray pyrolysis [58]. An anode-supported configuration with a thin YSZ electrolyte and SSC-SDC nano-composite cathode deposited on the porous NiO-YSZ anode substrate was fabricated with a thin, dense electrolyte layer that prevented gas cross-leakage (Fig. 2.13A and B). The SSC-SDC cathodes had a fine microstructure, although the grain size of SSC and GDC slightly increased with increasing sintering temperature (Fig. 2.13C–E). The

Fig. 2.12 SEM image of the surface of $Ce_{0.8}Gd_{0.18}Bi_{0.02}O_{2-\delta}$ densified at 1350°C.
Reproduced by permission from S. Dikmen, H. Aslanbay, E. Dikmen, O. Şahin, J. Power Sources 195 (2010) 2488.

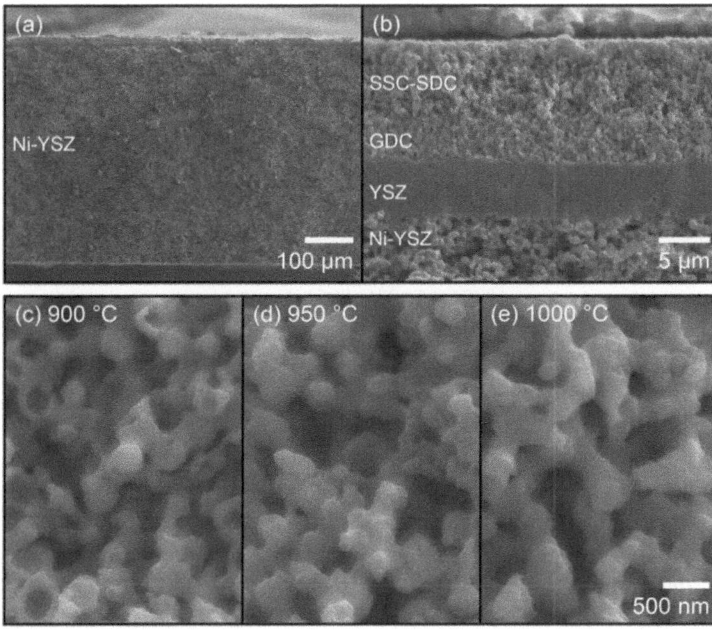

Fig. 2.13 FE-SEM images of cross section of anode-supported cells: (A) overview of a cell, (B) cathode/interlayer/electrolyte interface; SSC-SDC cathodes sintered at (C) 900°C, (D) 950°C, and (E) 1000°C.
Reproduced by permission from H. Shimada, T. Yamaguchi, T. Suzuki, H. Sumi, K. Hamamoto, Y. Fujishiro, J. Power Sources 302 (2016) 308.

individual particles consisted of SSC and SDC nano-sized crystal fragments and abundant nano-sized pores (porosity about 40 vol%) caused by the dehydration and decomposition taking place during spray pyrolysis. The cathodes sintered at different temperatures have all achieved high-power densities; in particular, the cathode sintered at 800°C exhibited the highest value, 3.65 W cm^{-2}, with the others achieving 2.44, 1.43, and 0.76 W cm^{-2} for electrodes sintered at 750°C, 700°C, and 650°C, respectively, using humidified H_2 as fuel and air as oxidant.

2.2.2.3 *Bi_2O_3 and its composites*

The phase structure of pure bismuth oxide (Bi_2O_3) includes α, β, γ, and δ phases [59], of which the face-centered cubic δ-Bi_2O_3 possibly exhibits the best oxide ion conductivity. The high-temperature fluorite structure of δ-Bi_2O_3 polymorph has a high concentration of intrinsic oxygen vacancies (25% of the oxygen sublattice sites), which offer high ionic conductivity. When cooled down to 973 K, the phase of pure Bi_2O_3 is transformed from the cubic (δ) to the monoclinic (α) phase, which results in a simultaneous decrease in ionic conductivity, volume change, and rupture of the structure.

To stabilize the cubic phase at low temperature, some metal oxide dopants, such as CaO, Y_2O_3, and WO_3 can be added in Bi_2O_3 [60–65].

Labeled isotope experiments allowed direct quantification of oxygen diffusivity in the polymorphs of bismuth oxide, showing that δ-Bi_2O_3 is the fastest known pure oxide ion conducting ceramic to date. The low temperature α-Bi_2O_3 polymorph showed a remarkable drop in diffusivity of over 7 orders of magnitude with a temperature drop of just 150°C [61]. A cell with an YDB/GDC composite electrolyte film (Fig. 2.14) showed increased open-circuit voltage of 68–96 mV and achieved maximum power density 0.6–1.5 times higher than that of a single GDC electrolyte film over the operating temperature range of 700–500°C [66]. Furthermore, the phase structure of Bi_2O_3 can be stabilized by introducing an Y_2O_3 dopant ranging from 25 to 43 mol% [67,68]. Samples containing <25 mol% Y_2O_3 show an abrupt increase in the conductivity, due to a phase transition. However, the ionic conductivity of Bi_2O_3 doped with >25 mol% Y_2O_3 does not go through the same transition as observed in pure Bi_2O_3, and thus does not show a jump in conductivity with increasing temperature. The γ-Bi_2O_3 crystal system is obtained by doping $0.01 < \text{mol\%}\ Nb_2O_5 < 0.04$ at 750°C for 48–96 h, showing good thermal behavior and thermal stability. In Nb_2O_5-doped γ-Bi_2O_3 systems with varying dopant content and operating temperatures (Fig. 2.15), the highest conductivity determined was $\sigma_T = 0.016\,\Omega^{-1}\,cm^{-1}$ at 650 °C with 4 mol% Nb_2O_5 added [69]. When doped with Sm_2O_3 and Tb_4O_7, the Bi_2O_3-based ceramic system showed an oxide ion-type electrical conductivity which increases with increasing amounts of Sm_2O_3, with a highest conductivity value of $3.48 \times 10^{-1}\,S\,cm^{-1}$ for the $(Bi_2O_3)_{0.85}(Tb_4O_7)_{0.1}(Sm_2O_3)_{0.05}$ ternary system at 850°C [70].

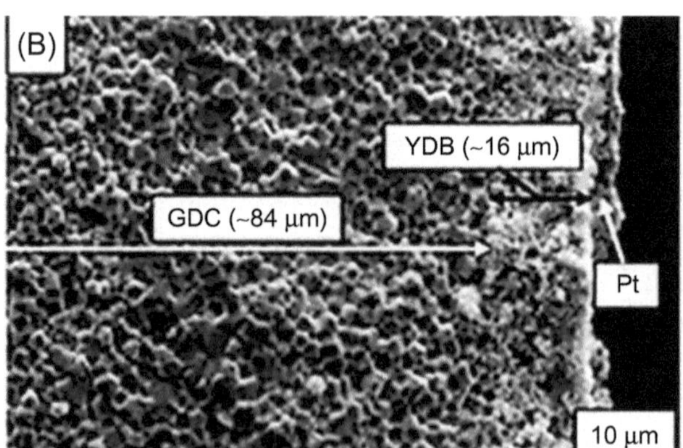

Fig. 2.14 SEM images of cross-sectional view of anode-supported SOFCs with YDB/GDC composite electrolyte (left: GDC/YDB composite electrolyte, right: Pt cathode; Ni-GDC anode not shown in the figure).
Reproduced by permission from Y.J. Leng, S.H. Chan, Electrochem. Solid State Lett. 9 (2006) A56.

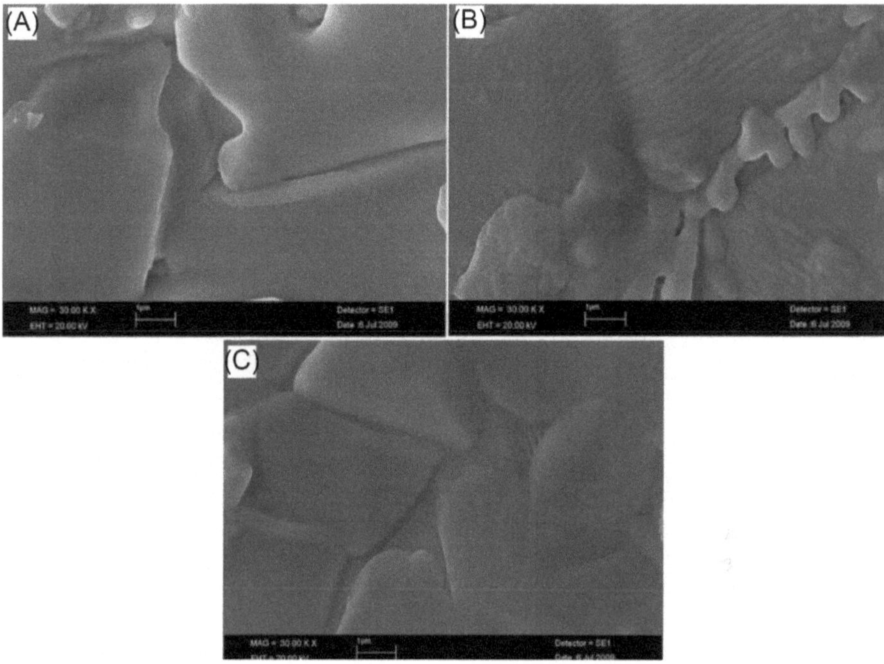

Fig. 2.15 SEM images of γ-Bi_2O_3 at (A) 2mol%, (B) 3mol%, and (C) 4mol%.
Reproduced by permission from H. Ozlu, S. Cakar, C. Bilir, E. Ersoy, O. Turkoglu, Bull. Mater. Sci. 37 (2014) 843.

2.2.2.4 $LaGaO_3$ and its composites

Oxides with the general perovskite-type formula ABO_3 have been explored as electrolyte materials, as discussed in Section 2.2.1.3 [71,72]. At high temperatures in reducing atmospheres, substantial gallium deficits were found on the surface of doped $LaGaO_3$ electrolyte, which was ascribed to the high partial pressure of Ga oxides such as Ga_2O. Adding strontium to the lanthanum sites enhanced the Ga depletion, whereas the addition of Mg to the Ga sites did not. With magnesium doping, the vapor pressure of the metal dopant M (mostly Cr or Ga) does not change on the M-deficient side, whereas with Ca and Sr doping, the vapor pressure of M changes significantly when the A-site dopant (Ca or Sr)-rich phase appears as the second phase. On the other hand, the apparent diffusion coefficient correlates well with the oxygen non-stoichiometry, as illustrated in Table 2.2, signifying that B-site cation diffusion is strongly correlated with the availability of oxide ion vacancies [73].

Oxygen conductors employing lanthanum gallate ($LaGaO_3$) have also been studied as SOFC electrolytes. In these ceramics, the lanthanum is replaced partly by Sr, Ca, Ba, Sm, or Nd, while Ga may also be replaced by Mg, In, Al, or Zn, as in the structure of $La_{1-x}Sr_xGa_{1-y}Mg_yO_{3-\delta}$ (LSGM) [74–81]. Structures containing Sr and Mg exchanges, respectively, for La (between 10% and 20%) and Ga (between 10%

Table 2.2 **Compositional dependence of *D* and *α***

Composition	D ($cm^2 s^{-1}$)	α ($cm s^{-1}$)
$LaGaO_3$	1.9×10^{-16}	1.5×10^{-10}
$LaGa_{0.9}Mg_{0.1}O_{2.95}$	2.6×10^{-16}	1.3×10^{-10}
$La_{0.9}Sr_{0.1}Ga_{0.8}Mg_{0.2}O_{2.85}$	1.3 ± 10^{-15}	6.2×10^{-10}

The samples were treated at 1173 K for 10 h in a H_2-1.2% H_2O gas stream [73].

and 20%) show high ionic conductivities in both oxidizing and reducing conditions. Among these, the $La_{0.8}Sr_{0.2}Ga_{0.83}Mg_{0.17}O_{3-\delta}$ composite shows the highest ionic conductivity of about 0.17 $S cm^{-1}$ at 800°C [82,83]. However, these ceramics are unstable in reducing conditions and Ga ions are lost, resulting in the formation of new phases [33,84]. These observations are unfavorable for the use of doped lanthanum gallate as SOFC electrolyte.

The LSGM powders with four different compositions (Fig. 2.16), synthesized using the glycine nitrate process (GNP), have been used as the electrolyte material

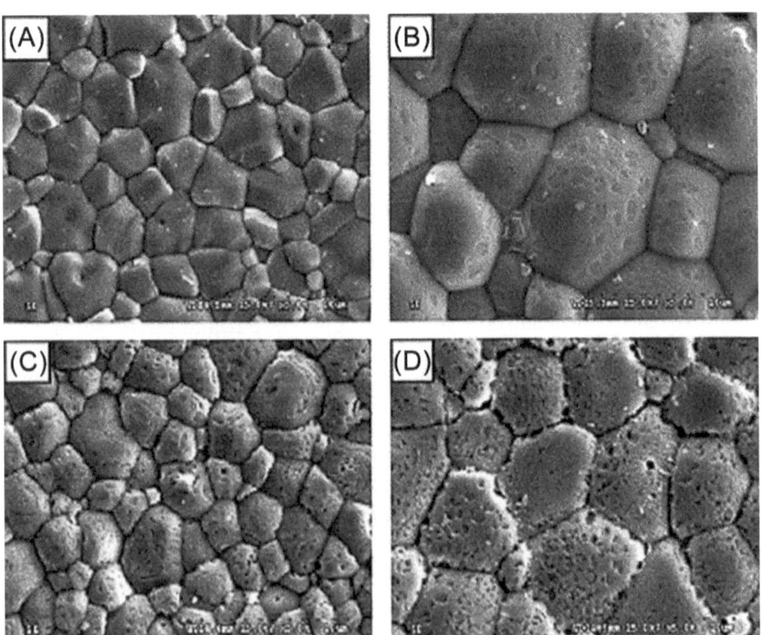

Fig. 2.16 SEM images for (A) LSGM9191, (B) LSGM9182, (C) LSGM8291, and (D) LSGM8282 compositions sintered at 1400°C for 4 h.
Reproduced by permission from D. Lee, J.H. Han, Y. Chun, R.H. Song, D.R. Shin, J. Power Sources 166 (2007) 35.

for an IT-SOFC [79,85]. The electrical conductivity of LSGM8282 (0.105 $S\,cm^{-1}$ at 800°C) turned out to be the highest among the four LSGM compositions and more than twice as high as that of YSZ (~0.045 $S\,cm^{-1}$). A single cell adding an electrolyte-anode interlayer of Sm-CeO_2 showed great improvement in the exchange current density and peak power density, which demonstrates that the realization of an SOFC based on an LSGM thick-membrane electrolyte operating in the range from 600°C to 800°C appears to be a realistic goal [86].

Doping small amounts of Co into Ga sites is effective for increasing the oxide ion conductivity. However, the partial electronic conduction of LSGMC monotonically increases with increasing Co amount, and the P_{O2} at the p-n transition was shifted to lower values. A cell made with an Ni-$Sm_{0.2}Ce_{0.8}O_2$ porous anode and cathode, La-doped CeO_2 interlayer film, and dense LSGMC electrolyte film has shown a fairly large maximum power density (0.21 $W\,cm^{-2}$) at 873 K and even at 773 K, and the maximum power density of the cell as high as 0.12 W/cm^2 was exhibited on the SOFC using a 40-μm thick LSGMC electrolyte, as shown in Fig. 2.17 [87].

Depositing LSGM thin film on NiO anode substrate is quite difficult due to the high reactivities of LSGM and NiO [88]. A thick buffer layer such as lanthanum-doped ceria (LDC) and samarium-doped ceria (SDC) can be applied between the electrolyte and the anode film to avoid the reaction and to provide adequate ion conductivity. A single fuel cell making use of an $LaGaO_3$/SDC bilayer film demonstrated an open-circuit potential of 1.1 V, which was close to the theoretical value; a high power density of 0.6 $W\,cm^{-2}$ at 773 K was obtained with a 5-μm thick $LaGaO_3$ film. The

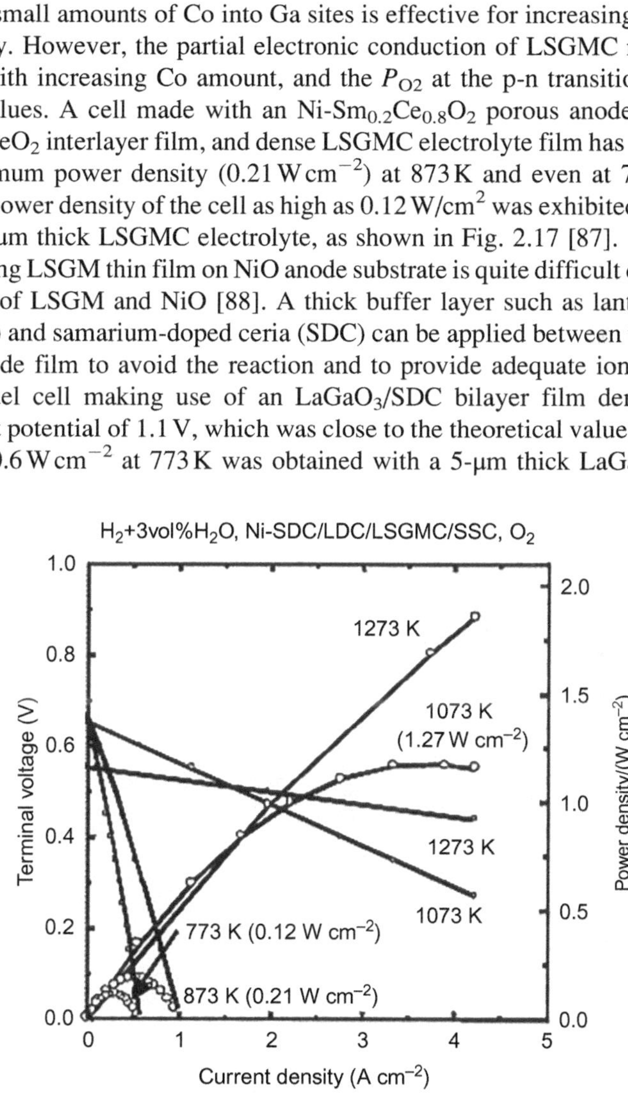

Fig. 2.17 Power generating properties of the cell using a prepared LSGMC film for electrolyte. Reproduced by permission from T. Ishihara, J. Tabuchi, S. Ishikawa, J. Yan, M. Enoki, H. Matsumoto, Solid State Ionics 177 (2006) 1949.

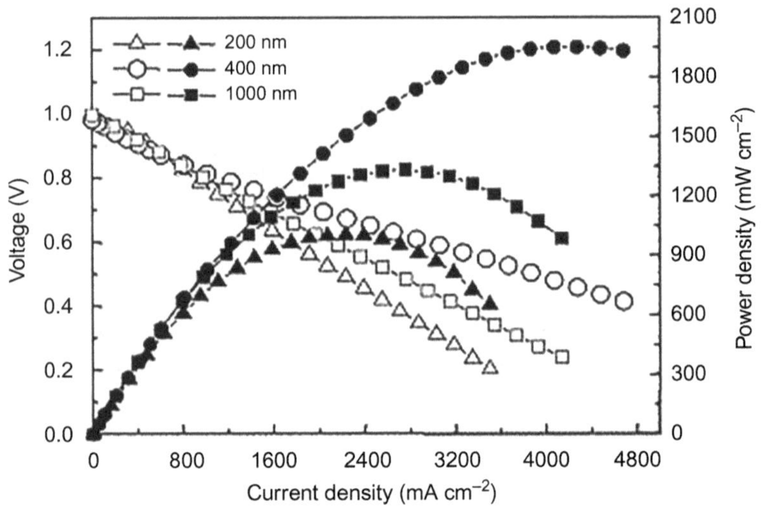

Fig. 2.18 Effect of SDC interlayer thickness on power density of SOFC using LSGM film of 5 μm thickness at 873 K.
Reproduced by permission from J. Yan, H. Matsumoto, T. Akbay, T. Yamada, T. Ishihara, J. Power Sources 157 (2006) 714.

thickness of SDC and $LaGaO_3$ is a key parameter to the reaction between NiO and the $LaGaO_3$ film, as has been shown experimentally by the reaction occurring with an extremely thin film (Fig. 2.18) [89].

2.2.2.5 *Oxyapatite and its composites*

A family of oxide conducting solid electrolytes, the oxyapatites with general formula $A_{10-x}(MO_4)_6O_{2\pm\delta}$, where A = rare earth or alkaline earth and M = Si or Ge, is another interesting candidate for SOFCs due to their conductivity at moderate temperature as compared to YSZ. The apatite lattice's capability to tolerate A-site deficiencies and extensive doping offers a chance to optimize further the ion transport characteristics of such solid electrolytes. Additional oxygen sites that are surrounded by A-site ions form channels running along the *c* axis through the structure. Interstitial migration along these channels contributes strongly to the total ionic transport level. One example of the oxyapatite crystal structure is shown in Fig. 2.19 [90]. Some properties of various apatite ceramics are summarized in Table 2.3 [91].

The A-site deficiency influences relaxation of the SiO_4 tetrahedra and is thought to result in the displacement of the anions from the channels to new interstitial sites, hence creating vacancies with stable oxygen content [92]. For instance, the ionic conductivity is enhanced in the $La_{9.33+x/3}Si_{6-x}Al_xO_{26}$ series, in which aluminum doping is balanced by A-site vacancies without a change in the oxygen content, with maximum conductivity found when $x = 1.5$. Dense ceramics of apatite-type $La_{10-x}Si_{6-y}Al_yO_{27-3x/2-y/2}$ ($x = 0$–0.33; $y = 0.5$–1.5) show oxygen ion transference numbers higher than 0.995 in air and increasing on reducing P_{O2} due to decreasing

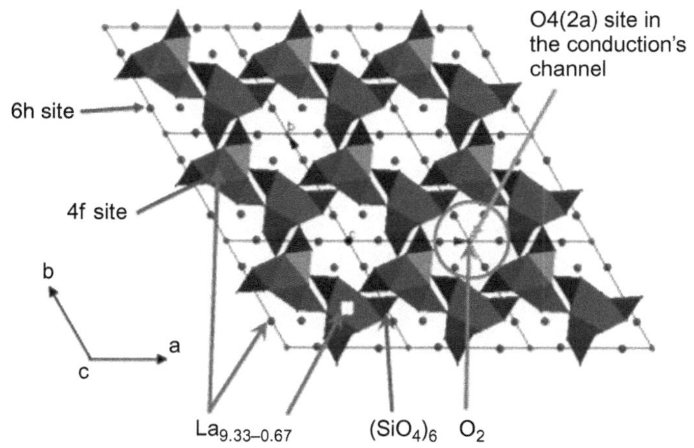

Fig. 2.19 *c*-Axis view of the crystal structure at room temperature of $La_{9.33-0.67}(SiO_4)_6O_2$ showing the SiO_4 groups as tetrahedra and La(4f), La(6h), and O(2a) atoms as balls. Reproduced by permission from E. Béchade, I. Julien, T. Iwata, O. Masson, P. Thomas, E. Champion, K. Fukuda, J. Eur. Ceram. Soc. 28 (2008) 2717.

p-type electronic conduction, in which the main factor determining the level of ionic transport is the total oxygen content governed by cation composition.

2.2.3 Cathode materials

The cathode functions as the site for the electrochemical reduction of oxygen by accepting the incoming electrons from the current collector. To do this effectively, cathode materials are needed with characteristics such as high electronic conductivity, adequate porosity, stability in an oxidizing atmosphere, high catalytic activity, and the thermal expansion coefficient of the electrolyte matched with the interconnect materials [93–95]. Consequently, some oxides with perovskite structure have been the most prominent candidates since the 1960s when doped lanthanum cobaltites came into use, followed in 1973 by lanthanum manganites.

2.2.3.1 $LaMnO_3$ and its modifications

Lanthanum manganites ($LaMnO_3$) are the most commonly used in high-temperature SOFCs due to their compatibility with YSZ electrolytes [71,96]. Doping them at the A sites with cations such as 10–20 mol% Sr^{2+} or 10–30 mol% Ca^{2+} tends to improve their performance as cathodes [71, 72, 84]. Since the Sr^{2+} size matches that of the lanthanum cation, strontium is the most commonly used dopant for lanthanum manganite.

The perovskite lanthanum strontium manganite (LSM) has high electrical conductivity, high electrochemical activity for the O_2 reduction reaction, high thermal stability, and compatibility with YSZ, GDC, and LSGM under the standard operating

Table 2.3 Properties of apatite ceramics [91]

Composition	Space group	Unit cell parameters		Relative density (%)	Average TECs	
		aa (Å)	cc (Å)		*T* (K)	$\overline{\alpha} \times 10^6$ (K^{-1})
$La_{9.67}Si_{4.5}Al_{1.5}O_{25.75}$	$P\overline{3}$ (No 147)	9.733 (3)	7.235 (6)	96.7	373–1173	9.99 ± 0.01
$La_{9.67}Si_5AlO_{26}$	$P6_3$ (No 173)	9.732 (1)	7.223 (6)	97.4	373–1173	9.92 ± 0.01
$La_{9.67}Si_{5.5}Al_{0.5}O_{26.25}$	$P6_3/m$ (No 176)	9.712 (2)	7.207 (6)	93.8	373–1273	9.39 ± 0.02
$La_{9.83}Si_{4.5}Al_{1.5}O_{26}$	$P\overline{3}$ (No 147)	9.734 (3)	7.236 (6)	96.7	373–1173	8.86 ± 0.02
$La_{9.83}Si_{5.5}Al_{0.5}O_{26.5}$	$P6_3/m$ (No 176)	9.709 (4)	7.207 (9)	91.3	373–1273	10.80 ± 0.02
$La_{10}Si_5AlO_{26.5}$	$P6_3/m$ (No 176)	9.729 (5)	7.235 (9)	91.6	473–1173	9.08 ± 0.02
$La_{10}Si_{5.5}Al_{0.5}O_{26.75}$	$P6_3/m$ (No 176)	9.716 (5)	7.212 (7)	90.5	1173–1323	8.68 ± 0.01

Composition	Ion transference numbers in air				Activation energies (kJ/mol)	
	1223 K	1123 K	1073 K	973 K	Ion transport	Hole transport
$La_{9.67}Si_{4.5}Al_{1.5}O_{25.75}$	–	–	–	–	60 ± 1	–
$La_{9.67}Si_5AlO_{26}$	0.9959	0.9959	0.9964	0.9965	67 ± 1	72 ± 3
$La_{9.67}Si_{5.5}Al_{0.5}O_{26.25}$	0.9956	0.9962	0.9966	0.9979	63 ± 2	89 ± 8
$La_{9.83}Si_{4.5}Al_{1.5}O_{26}$	0.9949	0.9951	0.9951	0.9948	60 ± 2	57 ± 2
$La_{9.83}Si_{5.5}Al_{0.5}O_{26.5}$	0.9967	0.9975	0.9982	0.9989	57 ± 3	100 ± 6
$La_{10}Si_5AlO_{26.5}$	0.9959	0.9968	0.9975	0.9985	56 ± 3	96 ± 6
$La_{10}Si_{5.5}Al_{0.5}O_{26.75}$	0.9995	0.9996	0.9996	0.9997	56 ± 3	68 ± 3

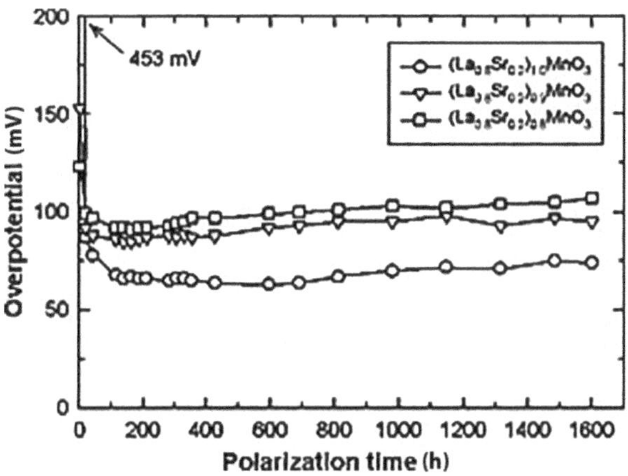

Fig. 2.20 Polarization stability of LSM cathodes for the O_2 reduction reaction under a constant current load of 500 mA cm^{-2} at 1000°C for 1600 h.
Adapted by permission from S.P. Jiang, W. Wang, Solid State Ionics 176 (2005) 1185.

conditions. In addition, LSM is the most important and common cathode material, with excellent microstructural stability and long-term performance stability. As shown in Fig. 2.20, LSM electrodes show no degradation over the period studied and the composition range studied [97].

For the $(La_{0.8}Sr_{0.2})_{0.9}MnO_3$ electrode, the particle size of the LSM grains is 0.88 ± 0.30 μm after sintering at 1000°C under a current load of 500 mA cm^{-2} for 1600 h (Fig. 2.21C), significantly smaller than 1.17 ± 0.25 μm observed for the LSM coating sintered at open circuit under the same conditions (Fig. 2.21D). Similar trends are apparent for the LSM electrodes with different compositions (Fig. 2.21). The grain growth and sinterability of the LSM electrode materials are reduced under cathodic polarization conditions, in contrast to those sintered under open-circuit conditions.

2.2.3.2 $LaCoO_3$ and its modification

Cobalt-containing perovskites have high mixed electronic and ionic conductivities and remarkable electrochemical activity in the oxygen reduction reaction. Unfortunately, they mostly have high thermal expansion coefficients $>20 \times 10^{-6}$ K^{-1} and react with YSZ at relatively low temperatures (700°C) to form insulating phases. An intermediate barrier layer, usually consisting of doped ceria, and modification of the composition or the use of a composite material is required to diminish the effects of thermal expansion.

$La_{1-x}Sr_xCoO_{3-\delta}$ (LSC) perovskites possessing considerable oxide ion conductivity and electronic conductivity perform exceptionally well with ceria-based electrolytes [98]. $La_{1-x}Sr_xCo_{1-y}Fe_yO_3$ (LSCF) has been investigated intensely, from the effect of the La:Sr ratio to the operating temperature range; the peak conductivity occurred at 550°C with a composition of $La_{0.6}Sr_{0.4}Co_{0.8}Fe_{0.2}O_3$ as candidate cathode material [99].

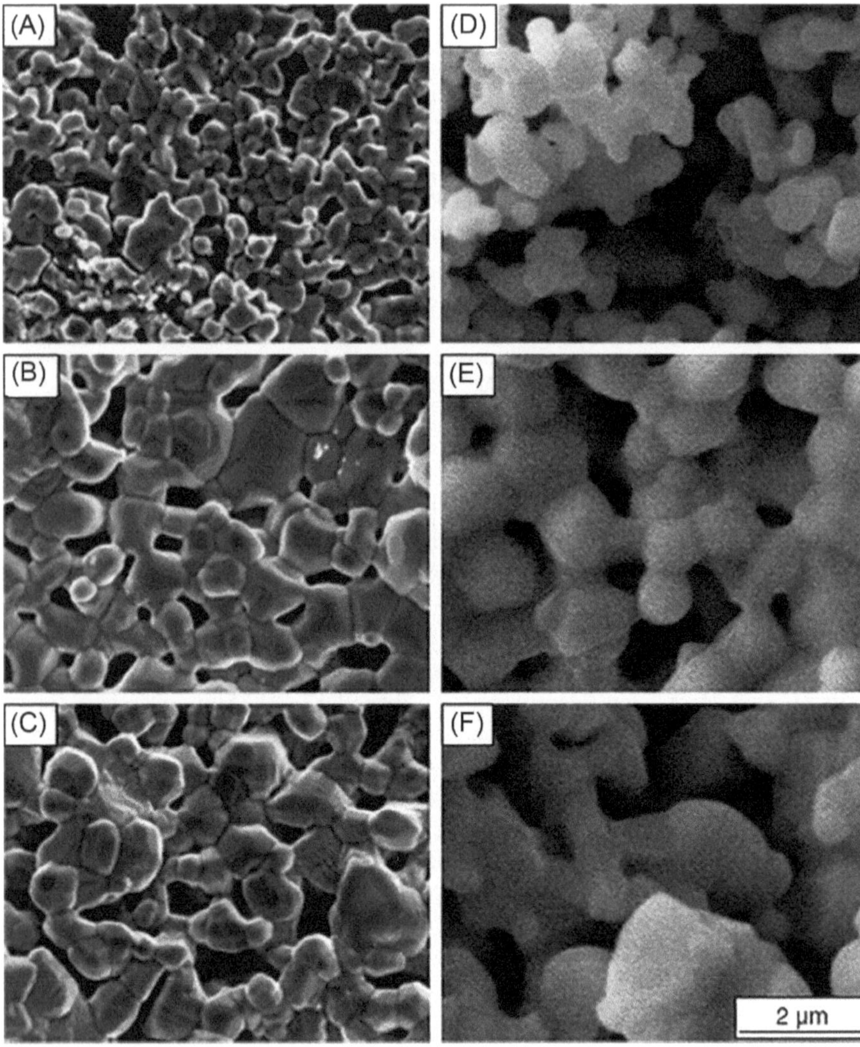

Fig. 2.21 SEM micrographs of the electrode surface of $(La_{0.8}Sr_{0.2})_xMnO_3$ with (A) $x=1.0$, (C) $x=0.9$, and (E) $x=0.8$ sintered under a current load of $500\,mA\,cm^{-2}$ and with (B) $x=1.0$, (D) $x=0.9$, and (F) $x=0.8$ sintered under no current load at 1000°C in air for 1600h. Reproduced by permission from S.P. Jiang, W. Wang, Solid State Ionics 176 (2005) 1185.

2.2.3.3 A_2BO_4 and its modification

Given a reducing operating environment of 600–800°C, it is reasonable to replace conventional lanthanum chromite ceramics with cost-effective high-temperature oxidation-resistant alloys for constructing interconnects in SOFC stacks. In this case, chromium poisoning is an inevitable problem that needs to be solved. To protect SOFCs against chromium poisoning and improve metallic interconnect stability,

Table 2.4 **EDS analysis of Cr content (wt%) in the LSM cathode operated with the bare and $Co_{0.40}Fe_{0.60}$-plated Crofer 22 APU alloy interconnects, respectively [100]**

	Cr content (wt%)			
Interconnect	**Region #1**	**Region #2**	**Region #3**	**Region #4**
Bare alloy	0.19	1.62	1.42	0.51
Plated alloy	0.15	0.34	0.37	0.42

spinel-structured oxide protection layers are useful, such as $Mn_{1.5}Co_{1.5}O_4$ [100–102], $CoFe_2O_4$ [103], $LiNi_{0.5}Mn_{1.5}O_4$ [104], and $Cu_{1.25}Mn_{1.75}O_4$.

Thermally grown layers of $Mn_{1.5}Co_{1.5}O_4$ not only significantly decrease the contact resistance between a cathode and stainless steel interconnect, but also act as a mass barrier to inhibit scale growth on the stainless steel and to prevent Cr outward migration through the coating [105]. The $CoFe_2O_4$ spinel has an electrical conductivity of $0.85\,S\,cm^{-1}$ at 800°C in air and an average coefficient of thermal expansion of $11.80 \times 10^{-6}\,K^{-1}$ from room temperature to 800°C. The presence of the dense, Cr-free $CoFe_2O_4$ spinel layer is effective in blocking the Cr migration/transport and thus contributes to the improvement in cell performance stability, as illustrated in Table 2.4 [100].

2.3 Metal oxides in other fuel cells

In other fuel cells, such as PEMFCs and DMFCs, metal oxides are normally induced in electrode and electrolyte materials as additive and catalyst support material to improve the stability of the fuel cells. For instance, introducing TiO_2 into a proton exchange membrane can increase the wettability and improve the mechanical properties of the membrane.

2.3.1 TiO_2 and its composites

The operation of PEMFCs at elevated temperatures can not only significantly reduce CO poisoning of the electrode catalysts that harms the combined performance of PEMFCs, but also enhances the electrode reaction kinetics and, consequently, boost the fuel cell efficiency. However, it presents a challenge for membrane development, as the membrane material should be capable of maintaining proper hydration when the water evaporation rate is high. One possible method is to add TiO_2 to the Nafion membrane to get particularly self-humidifying performance. In particular, Nafion-Pt-TiO_2 membranes show significantly higher water uptake and lower fuel crossover than unmodified Nafion membranes [105]. The carbon stability and cathode performance in PEMFCs improve as Pt nanoparticles are supported on TiO_2, presumably due to reduced direct contact between Pt and C (Fig. 2.22) [106]. A PEMFC with a

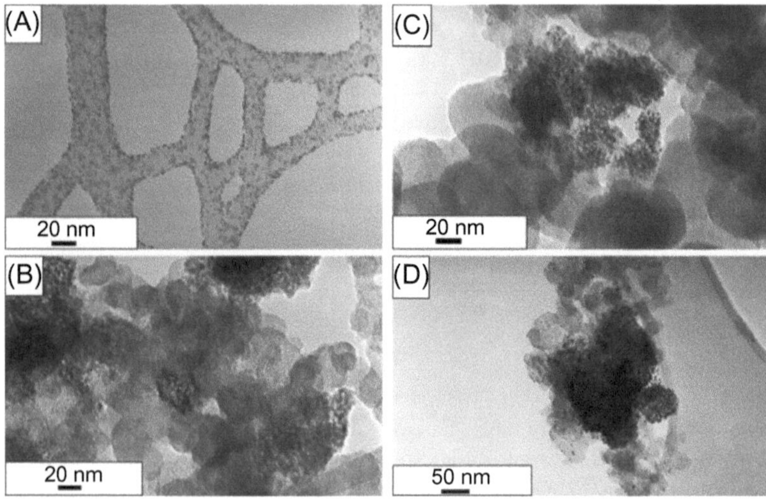

Fig. 2.22 TEM images of the Pt-TiO_2/C cathode material.
Reproduced by permission from S. von Kraemer, J. Wikander, G. Lindbergh, A. Lundblad, A. Palmqvist, J. Power Sources 180 (2008) 185.

Nafion-TiO_2 composite membrane prepared by carrying out an in situ sol-gel reaction of $Ti(OC_4H_9)_4$ with Nafion perfluorosulfonic acid film shows enhanced performance even when running with dry gases [107]. The performance of a PEMFC with a cathode gas diffusion layer produced by atomic-layer deposition shows distinct improvement in low humidity operation [108].

TiO_2 also shows good applicability as anodic photo-electrocatalyst in DMFCs. Coupled with Ni nanoparticles (average diameter of 200 nm), TiO_2 nanotube electrodes (Fig. 2.23) display good electrocatalysis, obtaining a maximum power of 48 μW using different concentrations of MeOH dissolved in 0.3 M KOH solution as fuel [109]. A catalyst consisting of Pt_8Ti and rutile TiO_2 nanoparticles dispersed in nitrogen-doped carbon hierarchical nanostructures exhibited a 50-mV positive onset potential and 10 times higher specific activity than a commercial Pt/C catalyst [110]. A rutile TiO_2 support that is chemically resistant to the acidic electrolyte medium contributes to the excellent stability.

2.3.2 SiO_2 and its composites

To overcome the CO poisoning problem on the noble metal catalyst in PEMFCs, a raised operating temperature can be an effective solution. However, the high evaporation rate of water causes a new problem in that the membrane used in the fuel cells is hard to wet. A fully hydrated membrane is vital for efficient proton conduction in the membrane; therefore, incorporation of hydrophilic metal oxide particles such as SiO_2 in the membrane was introduced. The composite membrane prepared by impregnation of Nafion 115 via sol-gel processing of tetraethoxysilane and by preparing a recast

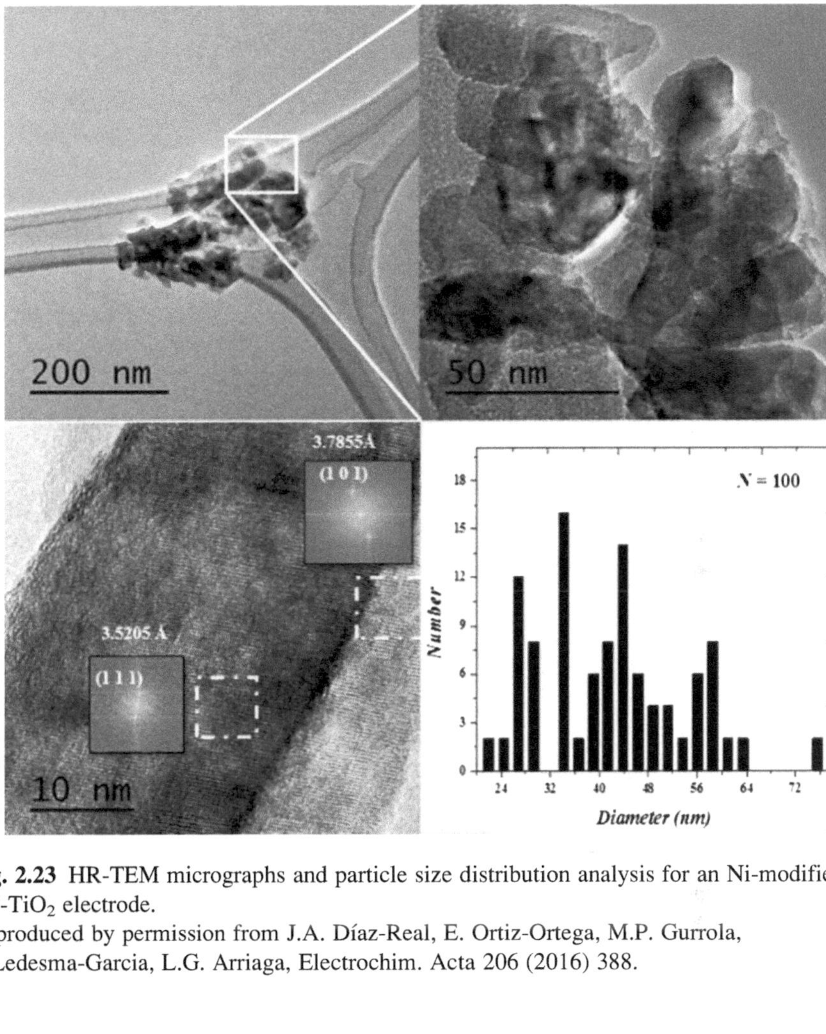

Fig. 2.23 HR-TEM micrographs and particle size distribution analysis for an Ni-modified NT-TiO_2 electrode.
Reproduced by permission from J.A. Díaz-Real, E. Ortiz-Ortega, M.P. Gurrola, J. Ledesma-Garcia, L.G. Arriaga, Electrochim. Acta 206 (2016) 388.

film, using solubilized Nafion 115 and a silicon oxide polymer/gel shows excellent wettability at 80–140°C [111]. Tetraethoxysilane reacts with water in an acidic medium, undergoing polymerization to form a mixture of SiO_2 and siloxane polymer containing hydroxide and ethoxide groups.

Hydrophilic SiO_2 with respect to both proton conductivity and methanol barrier property can be used to fabricate composite membranes with desirable properties for DMFC. When combined with sulfonated poly(arylene ether sulfone), SiO_2 particles with high surface area and small particle size show the best results: high proton conductivity, long membrane life time under oxidative conditions, good dimensional stability, outstanding single cell performance, and reduced methanol crossover [112]. The composite membranes consisting of sulfonated polyether ether ketone polymer and inorganic proton conducting fillers, developed from tungstosilicic

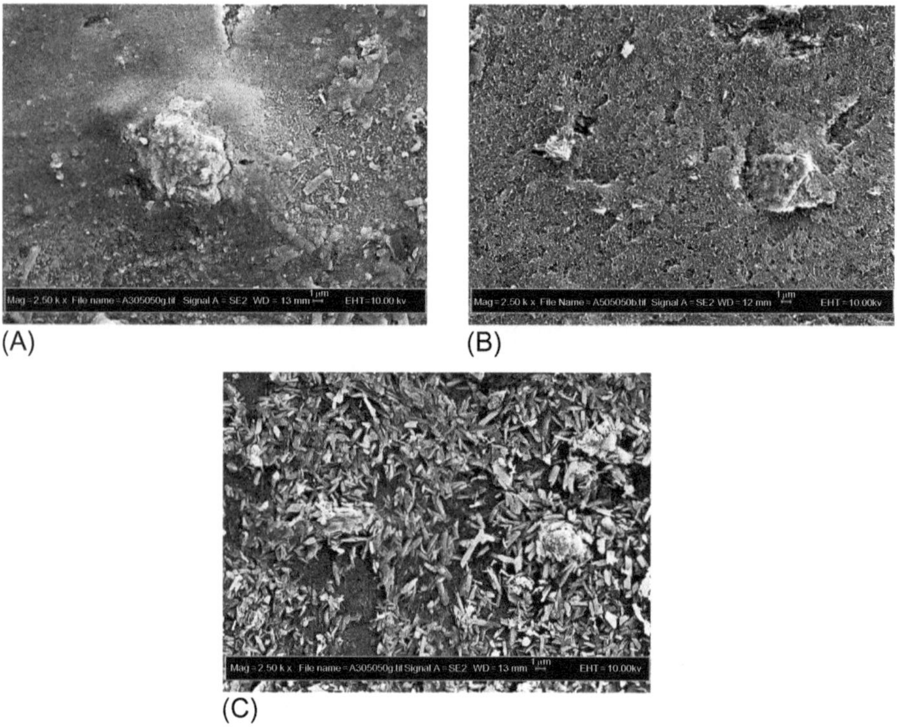

Fig. 2.24 SEM images of composite membrane at (A) 30%; (B) 50%; (C) 70% SiWA content. Reproduced by permission from A.F. Ismail, N.H. Othman, A. Mustafa, J. Memb. Sci. 329 (2009) 18.

acids (SiWA) loaded on silica–aluminum oxide (SiO_2-Al_2O_3) composite (Fig. 2.24), exhibit both high water uptake and proton conductivity (maximum value $6.1 \times 10^{-2}\,S\,cm^{-1}$) [113].

2.3.3 WO_3 and its composites

Just as TiO_2, tungsten oxide (WO_3)) has also been extensively investigated for a wide range of applications, mainly due to its unique electrochromic properties and its electrocatalytic activities [114,115]. The electrochromism allows WO_3 to intercalate/deintercalate ions such as H^+ into its structure in the formation of tungsten bronzes such as H_xWO_3, where $0 < x < 1$. Thin films of WO_x and Pt on WO_x were evaporated onto the microporous layer of a gas diffusion layer and served as model electrodes in PEMFCs; the WO_x increased the Pt utilization for the hydrogen oxidation reaction and increased the CO tolerance of the anode catalyst [116]. Cathodic electrodes based on electrodeposited Pt-WO_3 material enhanced the deposition of Pt in a single cell PEMFC configuration due to the presence of WO_3. These Pt-WO_3 electrodes show improved stability and good response in single cells for up to 1500h [117].

Fig. 2.25 SEM images of (A and B) HMTTS and (C and D) Pt/HMTTS.
Reproduced by permission from Y. Zhou, X. Hu, Y. Xiao, Q. Shu, Electrochim. Acta 111 (2013) 588.

Hybrid materials with Pt nanoparticles supported on WO_3/C can be used for electrocatalytic electrooxidation of methanol. It turns out that uniformly distributed WO_3 particles with smaller particle size display much better methanol oxidation activity when applied in DMFCs [118,119]. Hollow mesoporous WO_3 microspheres with an average size of 5 nm synthesized by spray drying method combined with Pt (Fig. 2.25) show electrocatalytic activity and good stability during methanol oxidation [120].

2.4 Outlook

The applications of metal oxides cover almost all characteristics of SOFCs, and involve several specific fields in other types of fuel cells such as PEMFCs. Although extensive investigations indicate that some metal oxides possess impressive and outstanding performance in some application respects, various disadvantages emerge simultaneously, such as poor stability, high cost, and complicated processing. Further research may focus on exploring many more metal oxide types, enhancing performance and stability, and simplifying processing methods.

Acknowledgment

Financial support from the National Materials Genome Project (2016YFB0700600), China National Fund for Distinguished Youth Scientists (51026004), and NSFC (U1601214, 51502137, and 21603103) is gratefully acknowledged.

References

[1] N.Q. Minh, J. Am. Ceram. Soc. 76 (1993) 563.
[2] D. Ding, X. Li, S.Y. Lai, K. Gerdes, M. Liu, Energy Environ. Sci. 7 (2014) 552.
[3] D. Simwonis, F. Tietz, D. Stover, Solid State Ionics 132 (2000) 241.
[4] W.Z. Zhu, S.C. Deevi, Mater. Sci. Eng. A 362 (2003) 228.
[5] S.P. Jiang, S.H. Chan, J. Mater. Sci. 39 (2004) 4405.
[6] B. Iwanschitz, L. Holzer, A. Mai, M. Schutze, Solid State Ionics 211 (2012) 69.
[7] J. Van Herle, D. Perednis, K. Nakamura, S. Diethelm, M. Zahid, A. Asianides, T. Somekawa, Y. Baba, K. Horiuchi, Y. Matsuzaki, M. Yoshimoto, O. Bucheli, J. Power Sources 182 (2008) 389.
[8] A. Utz, A. Leonide, A. Weber, E. Ivers-Tiffee, J. Power Sources 196 (2011) 7217.
[9] F. Tietz, F.J. Dias, D. Simwonis, D. Stöver, J. Eur. Ceram. Soc. 20 (2000) 1023.
[10] R.S. Amado, L.F. Brum Malta, F.M.S. Garrido, M.E. Medeiros, Quim Nova 30 (2007) 189.
[11] R.J. Gorte, S. Park, J.M. Vohs, C. Wang, Adv. Mater. 12 (2000) 1465.
[12] A. Rismanchian, J. Mirzababaei, S. Chuang, Catal. Today 245 (2015) 79.
[13] M.Q. Liu, S.L. Wang, T. Chen, C. Yuan, Y.C. Zhou, S.R. Wang, J. Huang, J. Power Sources 274 (2015) 730.
[14] R.J. Gorte, J.M. Vohs, J. Catal. 216 (2003) 477.
[15] J.S. Qiao, K.N. Sun, N.Q. Zhang, D.R. Zhou, Rare Metal Mater. Eng. 363 (2007) 113.
[16] M.D. Gross, J.M. Vohs, R.J. Gorte, J. Electrochem. Soc. 154 (2007) B694.
[17] V.V. Kharton, A.A. Yaremchenko, A.A. Valente, E.V. Frolova, M.I. Ivanovskaya, J. R. Frade, F. Marques, J. Rocha, Solid State Ionics 177 (2006) 2179.
[18] E. Ramírez-Cabrera, A. Atkinson, D. Chadwick, Solid State Ionics 136 (2000) 825.
[19] D.P. Wilkinson, J. Zhang, R. Hui, J. Fergus, X. Li, Proton Exchange Membrane Fuel Cells: Materials Properties and Performance, CRC Press, Boca Raton, 2009.
[20] T. Horita, K. Yamaji, N. Sakai, M. Ishikawa, H. Yokokawa, M. Dokiya, Ionics 3 (1997) 67.
[21] H. Lv, D.J. Yang, X.M. Pan, J.S. Zheng, C.M. Zhang, W. Zhou, J.X. Ma, K.A. Hu, Mater. Res. Bull. 44 (2009) 1244.
[22] J.S. Qiao, K.N. Sun, N.Q. Zhang, B. Sun, J.R. Kong, D.R. Zhou, J. Power Sources 169 (2007) 253.
[23] M.J. Escudero, A. Fuerte, Fuel Cells 16 (2016) 340.
[24] A. Fuerte, R.X. Valenzuela, M.J. Escudero, L. Daza, J. Power Sources 196 (2011) 4324.
[25] A. Fuerte, R.X. Valenzuela, M.J. Escudero, L. Daza, Int. J. Hydrog. Energy 39 (2014) 4060.
[26] A. Kulkarni, S. Giddey, S. Badwal, J. Solid State Electrochem. 19 (2015) 325.
[27] S. Sengodan, S. Choi, A. Jun, T.H. Shin, Y. Ju, H.Y. Jeong, J. Shin, J.T.S. Irvine, G. Kim, Nat. Mater. 14 (2015) 205.
[28] Z. Han, Y. Wang, Y. Yang, L. Li, Z. Yang, M. Han, J. Alloys Compd. 703 (2017) 258.
[29] S.W. Tao, J. Irvine, S.M. Plint, J. Phys. Chem. B 110 (2006) 21771.

[30] S.W. Tao, J. Irvine, Chem. Mater. 16 (2004) 4116.
[31] Y.H. Huang, R.I. Dass, Z.L. Xing, J.B. Goodenough, Science 312 (2006) 254.
[32] O.A. Marina, N.L. Canfield, J.W. Stevenson, Solid State Ionics 149 (2002) 21.
[33] V.V. Kharton, F. Marques, A. Atkinson, Solid State Ionics 174 (2004) 135.
[34] M.A. Taylor, M. Kilo, G. Borchardt, S. Weber, H. Scherrer, J. Eur. Ceram. Soc. 25 (2005) 1591.
[35] T.H. Etsell, S.N. Flengas, Chem. Rev. 3 (1970) 339.
[36] Y.Q. Li, Z.H. Wang, J.W. Li, X.B. Zhu, Y.H. Zhang, X.Q. Huang, Y.J. Zhou, L. Zhu, Z. Lu, J. Alloys Compd. 698 (2017) 794.
[37] A. Hornés, M.J. Escudero, L. Daza, A. Martínez-Arias, J. Power Sources 249 (2014) 520.
[38] H.J. Ko, J. Myung, S. Hyun, J.S. Chung, J. Appl. Electrochem. 42 (2012) 209.
[39] J. Lee, J. Electroceram. 17 (2006) 709.
[40] G.S. Lewis, A. Atkinson, B. Steele, J. Mater. Sci. Lett. 20 (2001) 1155.
[41] M.C. Tucker, G.Y. Lau, C.P. Jacobson, S.J. Visco, L.C. De Jonghe, J. Power Sources 195 (2010) 3119.
[42] M. Feng, J.B. Goodenough, J. Am. Ceram. Soc. 77 (1994) 1954.
[43] T.S. Zhang, S.H. Chan, W. Wang, K. Hbaieb, L.B. Kong, J. Ma, Solid State Ionics 180 (2009) 82.
[44] S. Badwal, F.T. Ciacchi, D. Milosevic, Solid State Ionics 136–137 (2000) 91.
[45] C. Sanchez-Bautista, A.J. Dos Santos-Garcia, J. Pena-Martinez, J. Canales-Vazquez, Bol. Soc. Esp. Ceram. Vidrio 49 (2010) 7.
[46] R.O. Fuentes, R.T. Baker, Int. J. Hydrog. Energy 33 (2008) 3480.
[47] J.A. Ding, J.A. Liu, G.Q. Yin, J. Membr. Sci. 371 (2011) 219.
[48] M. Morales, S. Pinol, M. Segarra, J. Power Sources 194 (2009) 961.
[49] J. Patakangas, Y. Ma, Y.F. Jing, P. Lund, J. Power Sources 263 (2014) 315.
[50] R. Farra, M. Garcia-Melchor, M. Eichelbaum, M. Hashagen, W. Frandsen, J. Allan, F. Girgsdies, L. Szentmiklosi, N. Lopez, D. Teschner, ACS Catal. 3 (2013) 2256.
[51] S. Cho, D.E. Fowler, E.C. Miller, J.S. Cronin, K.R. Poeppelmeier, S.A. Barnett, Energy Environ. Sci. 6 (2013) 1850.
[52] K.C. Anjaneya, G.P. Nayaka, J. Manjanna, G. Govindaraj, K.N. Ganesha, J. Alloys Compd. 578 (2013) 53.
[53] T. Tsuchiya, K. Terabe, M. Aono, Appl. Phys. Lett. 103 (2013) 73110.
[54] B.C.H. Steele, Solid State Ionics 129 (2000) 95.
[55] T. Mori, J. Drennan, J. Lee, J. Li, T. Ikegami, Solid State Ionics 154–155 (2002) 461.
[56] S. Dikmen, H. Aslanbay, E. Dikmen, O. Şahin, J. Power Sources 195 (2010) 2488.
[57] Q. Li, V. Thangadurai, Fuel Cells 9 (2009) 684.
[58] H. Shimada, T. Yamaguchi, T. Suzuki, H. Sumi, K. Hamamoto, Y. Fujishiro, J. Power Sources 302 (2016) 308.
[59] E.M. Levin, R.S. Roth, J. Res. Natl. Bur. Stand. Sect. A 68 (1964) 197.
[60] D.S. Aidhy, S.B. Sinnott, E.D. Wachsman, S.R. Phillpot, J.C. Nino, J. Solid State Chem. 182 (2009) 1222.
[61] R.D. Bayliss, S.N. Cook, S. Kotsantonis, R.J. Chater, J.A. Kilner, Adv. Energy Mater. 4 (10) (2014). https://doi.org/10.1002/aenm.201301575.
[62] S. Durmus, V. Corumlu, T. Cifci, I. Ermis, M. Ari, Ceram. Int. 39 (2013) 5241.
[63] W.M. Guo, J.A. Liu, J. Power Sources 195 (2010) 8185.
[64] M. Hirano, T. Oda, K. Ukai, Y. Mizutani, J. Am. Ceram. Soc. 85 (2002) 1336.
[65] R.K. Datta, J.P. Meehan, Z. Anorg. Allg. Chem. 383 (1971) 328.
[66] Y.J. Leng, S.H. Chan, Electrochem. Solid-State Lett. 9 (2006) A56.
[67] N.M. Sammes, G.A. Tompsett, H. Näfe, F. Aldinger, J. Eur. Ceram. Soc. 19 (1999) 1801.

[68] T. Takahashi, H. Iwahara, Mater. Res. Bull. 13 (1978) 1447.
[69] H. Ozlu, S. Cakar, C. Bilir, E. Ersoy, O. Turkoglu, Bull. Mater. Sci. 37 (2014) 843.
[70] O. Ermiş, M. Sertkol, V. Çorumlu, Y. Dagdemir, M. Arı, Synth. React. Inorg., Met.-Org., Nano-Met. Chem. 48 (2018) 91–95.
[71] Y. Sun, C. Wang, Y. Wen, L. Chu, H. Pan, M. Nie, M. Tang, J. Am. Ceram. Soc. 93 (2010) 2178.
[72] S.C. Singhal, K. Kendall, High-Temperature Solid Oxide Fuel Cells: Fundamentals, Design and Applications, Elsevier, Oxford, 2003.
[73] K. Yamaji, H. Negishi, T. Horita, N. Sakai, H. Yokokawa, Solid State Ionics 135 (2000) 389.
[74] X.G. Zhang, S. Ohara, R. Maric, K. Mukai, T. Fukui, H. Yoshida, M. Nishimura, T. Inagaki, K. Miura, J. Power Sources 83 (1999) 170.
[75] X.G. Zhang, S. Ohara, R. Maric, H. Okawa, T. Fukui, H. Yoshida, T. Inagaki, K. Miura, Solid State Ionics 133 (2000) 153.
[76] X.G. Zhang, S. Ohara, H. Okawa, R. Maric, T. Fukui, Solid State Ionics 139 (2001) 145.
[77] B. Rambabu, S. Ghosh, W.C. Zhao, H. Jena, J. Power Sources 159 (2006) 21.
[78] J. Pena-Martinez, D. Marrero-Lopez, D. Perez-Coll, J.C. Ruiz-Morales, P. Nunez, Electrochim. Acta 52 (2007) 2950.
[79] D. Lee, J.H. Han, Y. Chun, R.H. Song, D.R. Shin, J. Power Sources 166 (2007) 35.
[80] K. Sasaki, M. Muranaka, A. Suzuki, T. Terai, Solid State Ionics 179 (2008) 1268.
[81] R.K. Raghvendra, P. Singh, Solid State Ionics 262 (2014) 428.
[82] S. Badwal, S.P. Jiang, J. Love, J. Nowotny, M. Rekas, E.R. Vance, Ceram. Int. 27 (2001) 419.
[83] S. Badwal, S.P. Jiang, J. Love, J. Nowotny, M. Rekas, E.R. Vance, Ceram. Int. 27 (2001) 431.
[84] K.C. Wincewicz, J.S. Cooper, J. Power Sources 140 (2005) 280.
[85] D. Lee, J. Han, Electrochem. Solid-State Lett. 14 (2011) B85.
[86] K.Q. Huang, R. Tichy, J.B. Goodenough, J. Am. Ceram. Soc. 81 (1998) 2581.
[87] T. Ishihara, J. Tabuchi, S. Ishikawa, J. Yan, M. Enoki, H. Matsumoto, Solid State Ionics 177 (2006) 1949.
[88] Z.H. Bi, B.L. Yi, Z.W. Wang, Y.L. Dong, H.J. Wu, Y.C. She, M.J. Cheng, Electrochem. Solid-State Lett. 7 (2004) A105.
[89] J. Yan, H. Matsumoto, T. Akbay, T. Yamada, T. Ishihara, J. Power Sources 157 (2006) 714.
[90] E. Béchade, I. Julien, T. Iwata, O. Masson, P. Thomas, E. Champion, K. Fukuda, J. Eur. Ceram. Soc. 28 (2008) 2717.
[91] A.L. Shaula, V.V. Kharton, F.M.B. Marques, J. Solid State Chem. 178 (2005) 2050.
[92] E.J. Abram, D.C. Sinclair, A.R. West, J. Mater. Chem. 11 (2001) 1978.
[93] A.C. van Veen, M. Rebeilleau, D. Farrusseng, C. Mirodatos, Chem. Commun. (2003) 32.
[94] S. Li, Z. Lü, B. Wei, X. Huang, J. Miao, G. Cao, R. Zhu, W. Su, J. Alloys Compd. 426 (2006) 408.
[95] W. Kao, M. Lee, Y. Chang, T. Lin, C. Wang, J. Chang, J. Power Sources 195 (2010) 6468.
[96] R. Belardi, J. Deseure, M.C. Brant, T. Matencio, R.Z. Domingues, Ionics 15 (2009) 227.
[97] S.P. Jiang, W. Wang, Solid State Ionics 176 (2005) 1185.
[98] M. Godickemeier, K. Sasaki, L.J. Gauckler, I. Riess, Solid State Ionics 86–88 (1996) 691.
[99] L.W. Tai, M.M. Nasrallah, H.U. Anderson, D.M. Sparlin, S.R. Sehlin, Solid State Ionics 76 (1995) 273.
[100] Z. Yang, G. Xia, X. Li, J.W. Stevenson, Int. J. Hydrog. Energy 32 (2007) 3648.

[101] Z. Yang, G. Xia, J.W. Stevenson, Electrochem. Solid-State Lett. 8 (2005) A168.
[102] Z. Yang, G. Xia, S.P. Simner, J.W. Stevenson, J. Electrochem. Soc. 152 (2005) A1896.
[103] Z.H. Bi, J.H. Zhu, J.L. Batey, J. Power Sources 195 (2010) 3605.
[104] Y.K. Sun, Y.S. Lee, M. Yoshio, K. Amine, Electrochem. Solid-State Lett. 5 (2002) A99.
[105] E. Chalkova, M.B. Pague, M.V. Fedkin, D.J. Wesolowski, S.N. Lvov, J. Electrochem. Soc. 152 (2005). A1035.
[106] S. von Kraemer, J. Wikander, G. Lindbergh, A. Lundblad, A. Palmqvist, J. Power Sources 180 (2008) 185.
[107] H. Yamada, T. Hirai, I. Moriguchi, T. Kudo, J. Power Sources 164 (2007) 538.
[108] M. Choun, S. Chung, H. Jeon, S. Uhm, J. Lee, Electrochem. Commun. 24 (2012) 108.
[109] J.A. Díaz-Real, E. Ortiz-Ortega, M.P. Gurrola, J. Ledesma-Garcia, L.G. Arriaga, Electrochim. Acta 206 (2016) 388.
[110] J. Sanetuntikul, K. Ketpang, S. Shanmugam, ACS Catal. 5 (2015) 7321.
[111] K.T. Adjemian, S.J. Lee, S. Srinivasan, J. Benziger, A.B. Bocarsly, J. Electrochem. Soc. 149 (2002) A256.
[112] C.H. Lee, K.A. Min, H.B. Park, Y.T. Hong, B.O. Jung, Y.M. Lee, J. Membr. Sci. 303 (2007) 258.
[113] A.F. Ismail, N.H. Othman, A. Mustafa, J. Membr. Sci. 329 (2009) 18.
[114] E. Antolini, E.R. Gonzalez, Appl. Catal. B Environ. 96 (2010) 245.
[115] J.B. Christian, S. Smith, M.S. Whittingham, H.D. Abruna, Electrochem. Commun. 9 (2007) 2128.
[116] B. Wickman, M. Wesselmark, C. Lagergren, G. Lindbergh, Electrochim. Acta 56 (2011) 9496.
[117] A.J. Martin, A.M. Chaparro, L. Daza, J. Power Sources 196 (2011) 4187.
[118] Z. Cui, L. Feng, C. Liu, W. Xing, J. Power Sources 196 (2011) 2621.
[119] L. Feng, L. Yan, Z. Cui, C. Liu, W. Xing, J. Power Sources 196 (2011) 2469.
[120] Y. Zhou, X. Hu, Y. Xiao, Q. Shu, Electrochim. Acta 111 (2013) 588.

Further reading

[121] H.A. Taroco, J.A.F. Santos, R.Z. Domingues, T. Matencio, Ceramic materials for solid oxide fuel cells, in: C. Sikalidis (Ed.), Advances in Ceramics-Synthesis and Characterization, Processing and Specific Applications, InTech, 2011. Available from: https://www.intechopen.com/books/advances-in-ceramics-synthesis-and-characterization-processing-and-specific-applications/ceramic-materials-for-solid-oxide-fuel-cells. (Accessed 28 April 2017).
[122] M. Winter, R.J. Brodd, Chem. Rev. 104 (2004) 4245.

Metal oxide-based thermoelectric materials

3

Lian Ji
School of Energy Science and Engineering, Nanjing Tech University, Nanjing, China

3.1 Introduction

3.1.1 Background

With the rapid development of industrialization, the shortage and exhaustion of energy sources such as fossil fuels, has become a serious problem which seriously restricts the long-term stable development of the society [1–3]. The increasing demand for energy provides impetus to the research and evolution of renewable and sustainable energy sources that are crucial in addressing the power needs of the future. Solar cells, fuel cells, and other emerging energy conversion technologies are still inadequate in terms of efficiency, cost, and long-term stability, while thermoelectric devices can convert waste heat from engine exhaust systems, furnaces, and industrial processes into useful electricity by taking advantage of the Seebeck effect [2, 4–8]. Therefore, thermoelectric (TE) materials are of great significance in the search for sustainable energy-efficient technologies with the advantages of simple configuration, no moving parts, no vibration, no noise, no pollution, and high stability [9–13]. In addition, TE materials can reduce the negative effects of global warming and the resulting climate change to some extent [14, 15], and offer opportunities for the development of smart sensors, energy harvesting, the new concept of thermopower wave sources, automobiles, woodstoves, and diesel power plants [13, 16–24].

3.1.2 Thermoelectric effect

In 1821, Seebeck observed that when two dissimilar materials are joined together and the junctions are held at different temperatures, a voltage difference developed that was proportional to the temperature difference (the Seebeck effect), as illustrated in Fig. 3.1. The ratio of the voltage developed in response to the temperature gradient is related to an intrinsic property of the materials called the Seebeck coefficient (S) or the thermopower. Later (in 1834), the Peltier effect was discovered, which states that when an electrical current is passed through a junction of two dissimilar materials, heat is either absorbed or given off at the junction, depending on the direction of the current, as shown in Fig. 3.2. This effect is due to the difference in Fermi energies of the two materials [7]. In the 1850s, William Thomson established the relationship between the Seebeck effect and the Peltier effect and predicted a third thermoelectric effect—the Thomson effect. The Seebeck effect provides the basic requirement for thermoelectric power

Metal Oxides in Energy Technologies. https://doi.org/10.1016/B978-0-12-811167-3.00003-1

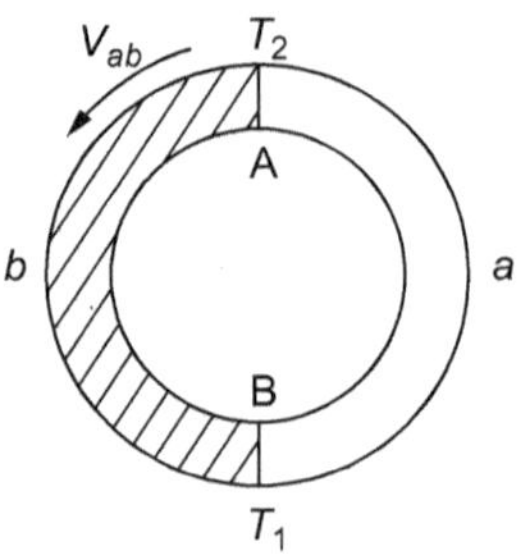

Fig. 3.1 A schematic illustrating the Seebeck effect at a junction of two dissimilar materials (*a* and *b*), where a voltage difference (V_{ab}) is generated that is proportional to the temperature gradient between the ends of the junction ($T_2 - T_1$).

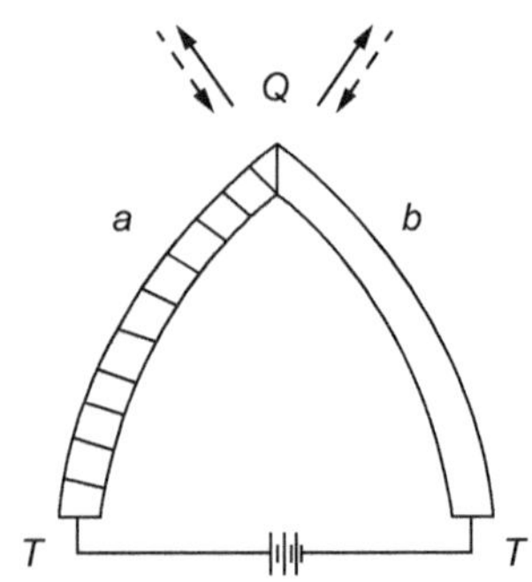

Fig. 3.2 A schematic illustrating the Peltier effect between two dissimilar materials (*a* and *b*). As the current flows through the junction of the two materials, heat is either absorbed or produced.

devices and the Peltier effect provides the basis for many modern-day thermoelectric refrigeration devices. In 1911 the term "thermoelectric figure of merit" was first introduced by Altenkirch and he suggested that, for best performance, TE materials should have low thermal conductivity, high electrical conductivity, and a large Seebeck coefficient [25]. In 1949, Ioffe proposed the thermoelectric figure of merit (*ZT*) of a material as a quantity expressing its thermoelectric performance [26].

3.1.2.1 Power conversion efficiency

Thermoelectric power conversion efficiency has undergone a resurgence of attention, being based on a class of materials known as TE materials that can recover waste heat and convert it into electrical power. The power conversion efficiency η of a thermoelectric device is typically defined as

$$\eta = \left(\frac{T_2 - T_1}{T_2}\right)\left(\frac{\sqrt{1+ZT_m} - 1}{\sqrt{1+ZT_m} + \frac{T_1}{T_2}}\right) \tag{3.1}$$

where T_1 and T_2 are the cold- and hot-side temperatures, respectively, T_m the mean temperature, the first factor is the Carnot efficiency, and ZT_m is the figure of merit for the device [14, 27, 28].

3.1.2.2 Thermoelectric figure of merit

The charge carriers in thermoelectric generators are conveyed by p-n junctions comprising p- and n-type materials, with electrons performing as the working "fluid" (Fig. 3.3). The gradient of charge carriers propagated from the heat source to the heat sink is generated by the applied temperature. Meanwhile, the applied temperature produces an electrostatic potential (ΔV), which is directly proportional to the temperature difference ΔT ($\Delta V = S \cdot \Delta T$, where S is the Seebeck coefficient). The parameters that determine the thermoelectric transport are the diffusivity and mobility (flow) of the charge carriers [29–31]. The figure of merit ZT used to evaluate the properties of thermoelectric transport in materials [29, 31] can be defined as follows:

$$ZT = \left(\frac{\sigma S^2}{k}\right) T \tag{3.2}$$

where σ is the electrical conductivity, S the Seebeck coefficient, k the thermal conductivity, and T is temperature [6, 32–34].

From Eq. (3.2) it follows that the thermoelectric figure of merit (ZT) of a material can be improved by increasing the power factor or by reducing the thermal conductivity; therefore, a high Seebeck coefficient, high electrical conductivity, and low thermal conductivity are necessary to create a large potential difference across the junction. Then,

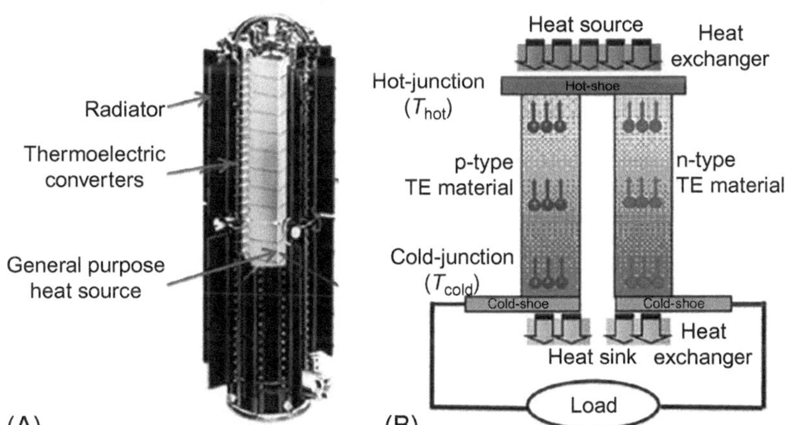

Fig. 3.3 (A) The long-life thermoelectric power source used on deep space science exploration mission and (B) a thermoelectric couple operating across a temperature differential and generating electrical power.
Reproduced with permission from J.P. Fleurial, JOM. 61 (2009) 79–85.

the Joule heating effect can be minimized so that a large temperature gradient can be maintained [29, 35].

Eq. (3.1) implies that thermoelectric energy conversion depends on the material's figure of merit and the operating temperature gradient ($T_2 - T_1$). Therefore, if we assume that T_1 is room temperature, then the thermoelectric conversion efficiency and the *ZT* value of the material increases with increasing hot-side temperature (T_2). We can then deduce that, for example, when $T_1 = 300\,K$, $ZT = 1$ implies 5% conversion efficiency at $T_2 = 400\,K$, and nearly 15% conversion efficiency at $T_2 = 900\,K$. Most TE materials have maximum *ZT* values between 1 and 2 over their respective operating temperature ranges, due to the interdependence between σ, S, and k, although for ideal TE materials, *ZT* should be at least 1 to achieve a conversion efficiency above 10%. This clearly shows that the stability of TE materials at high operating temperatures is just as important as high *ZT* values to attain better heat to electrical energy conversion efficiencies [8].

3.1.2.3 Thermopower or Seebeck coefficient

The thermopower or Seebeck coefficient S is defined as the difference of electrochemical potential per unit temperature difference developing across an electrically isolated sample exposed to a temperature gradient [14]. Accordingly, an ideal TE material would have a Seebeck coefficient $>200\,\mu V\,K^{-1}$ for best voltage generating ability. The Mott relation for the thermopower or Seebeck coefficient (S) is given as

$$S = \frac{\pi^2}{3}\frac{k_B}{q}k_B T\left(\frac{1}{n}\frac{dn(E)}{dE} + \frac{1}{\mu}\frac{d\mu(E)}{dE}\right)_{E=E_F} \tag{3.3}$$

where k_B is the Boltzmann constant, T is the temperature, $n(E)$ is the carrier density at energy E, $\mu(E)$ is the mobility at energy E, E_F is the Fermi energy, and q is the electronic charge [36]. The Seebeck coefficient is determined by the carrier concentration, the crystal structure of the conductor, and the temperature [14].

3.1.2.4 Electrical conductivity of metal oxide

The electrical conductivity (σ) is related to the density of charge carriers (n) and their mobility (μ), typically given by

$$\sigma = e(\mu_e \cdot n + \mu_h \cdot p) \tag{3.4}$$

where μ_e, μ_h, n, and p represent the electron mobility, hole mobility, number density of electrons, and the number density of holes, respectively.

The mobility of electrons and holes depends on lattice and impurity scattering. With increasing temperature, the lattice vibrations increase and the mobility decreases as a consequence. However, impurity scattering decreases with an increase in temperature and results in an increase in mobility. The electrical conductivity of TE materials is affected not only by the type of material, doping, impurities, and temperature, but also by structural characteristics such as grain size, strain, and lattice constant.

3.1.2.5 Thermal conductivity

Thermal conductivity in TE materials is the result of contributions from two sources: one from the transportation of charge carriers including electrons and holes, and the other from lattice vibrations. When a single type of charge carrier is predominant in a material, the total thermal conductivity k is the sum of the lattice thermal conductivity k_{lattice} and the charge-carrier thermal conductivity k_C, that is, $k = k_C + k_{\text{lattice}}$ [8, 29, 30]. Intensive research is still in progress in efforts to reduce the lattice thermal conductivity, so that the total thermal conductivity will be reduced by maximizing the k_C/k_{lattice} ratio [8, 29, 30, 36]. The charge-carrier thermal conductivity k_C can be obtained from the Wiedemann-Franz law $k_C = L\sigma T$, where L is the Lorenz constant ($\sim 2.45 \times 10^{-8}\ \text{V}^2\ \text{K}^{-2}$) [7].

The lattice thermal conductivity is related to the transfer of vibrational energy from one atom to the next, which can be considered as the propagation of a wave. A phonon is defined as a quantized vibrational wave, which is regarded as the energy carrier responsible for heat propagation in a lattice. The vibrational spectrum of a three-dimensional lattice consists of three acoustic modes (two transverse and one longitudinal) and $(3n - 3)$ optical modes (where n is the number of atoms per primitive unit). Among the three acoustic modes, the acoustic phonons are the main contributors to the lattice thermal conductivity, while the optical phonons oscillate back and forth from their standing positions instead of moving through the lattice. The lattice thermal conductivity corresponds to the transportation of phonons in three dimensions through the crystal lattice and is defined by

$$k_{\text{lattice}} = \frac{1}{3} C_V v_s l_{ph} = \frac{1}{3} C_V v_s^2 \tau \tag{3.5}$$

where C_V is the specific heat per unit volume, t is the phonon relaxation time, l_{ph} is the phonon mean free path, and v_s is the velocity of sound, where $v_s = l_{ph}/\tau$.

The lattice thermal conductivity is determined by the combination of phonon-phonon scattering and point defect scattering, or (in nanostructured materials) by boundary scattering, and is expressed as [37]

$$\tau^{-1} = \sum_i \tau_i^{-1} \tag{3.6}$$

or as

$$\frac{1}{l_{ph}} = \frac{1}{l_{\text{defect}}} + \frac{1}{l_{\text{boundary}}} + \frac{1}{l_{\text{phonon}}} \tag{3.7}$$

A reduction in k_{lattice} is desirable to improve the performance of TE materials applied in cooling and energy scavenging. When the phonon mean free path is not more than the interatomic spacing of the constituent atoms, the minimum $k_{lattice}$ can be achieved in theory [7]. Therefore, creating low-dimensional materials to limit l_{ph} by restraining phonons can assist in reaching this objective.

3.1.3 Thermoelectric materials

Since 1990, various TE materials have been developed, and can be classified into two categories. One consists of intermetal compounds, such as skutterudite (CoSb, etc.), clathrate (SrGaGe, etc.), half Heusler (FeVSb, etc.), and silicide (FeSi, MgSi, etc.). The other category is metal oxide-based TE materials, including many different types of compounds. Well-known oxides are $NaCo_2O_4$ [38], Y_2O_3 [39], $Ca(Mn,In)O_3$ [40], (Zn,Al)O [41], and Fe_2O_3 [42]. Oxide research originated in Japan and now a large variety of materials is investigated by many researchers. There are several reasons for selecting a metal oxide as TE material: (a) stability at high temperatures, (b) short mean free path of phonons, and (c) low cost [15].

3.2 Na_xCoO_2 and $Na_xCo_2O_4$

Na_xCoO_2 has a hexagonal layered structure with alternating stacks of CdI_2-type CoO_2 layers and charge-balancing Na^+ ions along the *c* axis, a structure that was first discovered by Jansen and Hoppe in 1974 [43]. In Na_xCoO_2, the CoO_2 layers are separated by a layer of sodium ions, as illustrated in Fig. 3.4. Each CoO_2 layer is rhombohedrally distorted, with edge-sharing CoO_6 octahedra stacked in alternating fashion. Two different types of trigonal prismatic sites are generated and are partially occupied by randomly distributed Na^+ ions. Electrons and phonons inside the layered structure follow

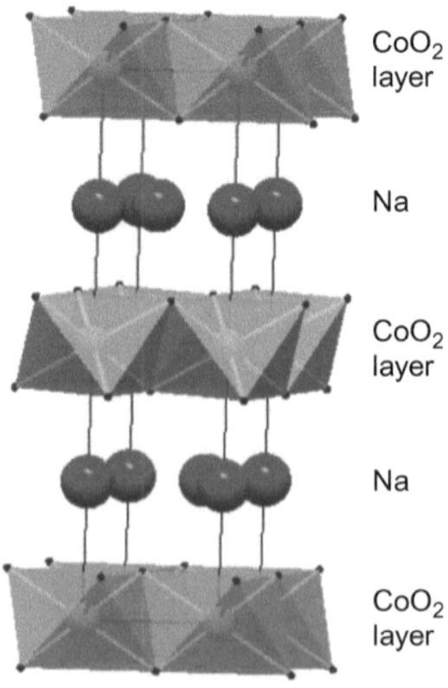

Fig. 3.4 Crystal structure of Na_xCoO_2.
Reproduced with permission from A. Nag, V. Shubha, J. Electron. Mater. 43 (2014) 962–977.

different paths; hence, the lattice thermal conductivity can be controlled by selecting an appropriate number of insulating block layers. The electron crystal-phonon glass model is a manifestation of material design, called "nanoblock integration" [8]. In other words, the layered Na_xCoO_2 crystal structure can be envisioned as a prototype phonon glass electronic crystal, where the ordered CoO_2 layers are electronic transport layers and the disordered Na^+ blocks are phonon-scattering regions causing low thermal conductivity. Moreover, the CdI_2-type CoO_2 blocks in layered cobalt oxides prefer the low-spin state of Co^{3+} and Co^{4+}, leading to large thermopower at high temperatures [14, 44, 45].

The fabrication cost of a $Na_xCo_2O_4$ single crystal is rather high; therefore, several attempts have been made to improve the thermoelectric properties of synthesized $Na_xCo_2O_4$ polycrystals. Polycrystalline samples of $(Na_{1-y}M_y)_xCo_2O_4$ (where M=K, Sr, Y, Nd, Sm, and Yb; y=0.01–0.35) were prepared by a solid-state reaction (SSR) method [46]. Na_2CO_3, Co_3O_4, K_2CO_3, $SrCO_3$, $Y_2(CO_3)_3{\cdot}3H_2O$, $Nd_2(CO_3)_3{\cdot}2.3H_2O$, $Sm_2(CO_3)_3{\cdot}3H_2O$, and $Yb_2(CO_3)_3{\cdot}4H_2O$ powders were used as starting materials. Each mixture in the desired mole ratio was pressed into a pellet under a pressure of 5.6×10^6 MPa. The pellets were calcined at 1153 K for 20 h. The calcined pellets were ground into powder, and then the powders were again pressed into pellets and sintered at 1193 K for 20 h. In the sintering process, the sample was sintered in the powder bed with the same composition as the sample. Also, 20% excess Na_2CO_3 was added during each calcination and sintering process since Na tends to evaporate during the heating process.

Polycrystalline $Na_xCo_2O_4$ samples can also be synthesized using a sol-gel method. Thermoelectric $Na_xCo_2O_4$ samples were produced using a powder precursor made by a polymerized complex (PC) process [47]. After dissolving citric acid in ethylene glycol at 473 K in the proportion of 4:150 mol, respectively, for each mole of metal cation, $Co(NO_3)_2{\cdot}6H_2O$ (99.5% purity) and $NaNO_3$ (99.9% purity) were added in an appropriate mole ratio corresponding to $Na_{1.4}Co_2O_4$. The resulting mixture was stirred for 6 h at 553 K to promote the formation of a polymer between ethylene glycol and metal citrate complexes. With concentration of the colloidal suspension it became highly viscous, and this viscous product was heated at 623 K for 1 h in air in a heating mantle to form a dark precursor mass. The precursor was ground and calcined for 5 h at 1073 K to yield a γ-$Na_xCo_2O_4$ phase. The calcined powder precursor was compacted under a pressure of 560 MPa and then subjected to unpressurized sintering in air for 20 h at 1193 or 1153 K.

Fig. 3.5A and B displays the X-ray diffraction (XRD) patterns of the SSR and the PC sample, respectively, sintered at 1193 K. The PC sample contained a small amount of CoO and Co_3O_4 phases, while the SSR sample consisted of $Na_xCo_2O_4$ without any impurity phase. The formation of cobalt oxide impurities was thought to be due to Na evaporation during sintering. Based on this concern, the PC powder was then sintered at 1153 K. Fig. 3.5C shows the XRD pattern of the resulting product. The XRD pattern still showed several small peaks of the Co_3O_4 phase, indicating that the Na content of the starting mixture should probably be higher than $Na_{1.4}Co_2O_4$ for proper single-phase $Na_xCo_2O_4$ formation. Consequently, PC powders with nominal compositions of $Na_{1.6}Co_2O_4$ and $Na_{1.7}Co_2O_4$ were synthesized and then sintered for 20 h in air at 1153 K.

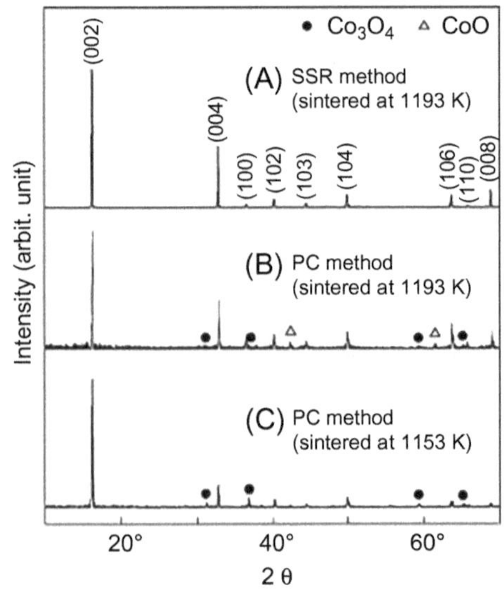

Fig. 3.5 XRD patterns of (A) the SSR sample sintered at 1193 K, (B) the PC sample sintered at 1193 K, and (C) the PC sample sintered at 1153 K.
Reproduced with permission from M. Ito, T. Nagira, D. Furumoto, S. Katsuyama, H. Nagai, Scr. Mater. 48 (2003) 403–408.

The Na_xCoO_2 single crystal displays a high thermopower of 100 μV K^{-1}, low electrical resistivity of 200 μΩ cm, a power factor of 50 μW K^{-2} cm^{-1}, and thermal conductivity between 4 and 5 Wm^{-1} K^{-1} at 300 K [38, 48]. The *ZT* of Na_xCoO_2 single crystal is about 1 at 800 K, while polycrystalline Na_xCoO_2 has a high resistivity resulting in a low *ZT* value of 0.8 [47, 49]. Na_xCoO_2 is unstable in air above 800°C due to its hygroscopicity and volatility, therefore, its application as a high-temperature thermoelectric oxide is limited. The properties of some Na_xCoO_2 samples are summarized in Fig. 3.6 [12].

Suitable doping elements can be used to improve the *ZT* value of Na_xCoO_2. The addition of silver as a dopant or a second phase results in a rise in electrical conductivity and thermopower, with a concomitant lowering of thermal conductivity; although the figure of merit is enhanced it does not change much. Seetawan et al. [50] synthesized polycrystalline samples of $Na_{1.5}Co_{2-x}Ag_xO_4$ ($x=0$, 0.1, and 0.2) from powder precursors prepared by a PC method. The power factor was calculated from the measured thermoelectric power and electrical resistivity. The values of the electrical resistivity, thermoelectric power, and thermoelectric power factor of the Ag-doped samples (where $x=0.2$) are 66 μΩ, 101 μV/K, and 157 μWm^{-1} K^{-2} at 300 K, respectively. The thermoelectric properties of $Na_xCo_{2-y}Ag_yO_4$ ($x\sim1.5$, $y=0$, 0.1, 0.2, 0.3, 0.4, 0.5) were also investigated [51]. The Ag-doped samples ($x=0.2$) were found to be composed of significantly fine grains, and Ag_2O particles around 10 μm were precipitated on the grain boundary. The electrical resistivity value

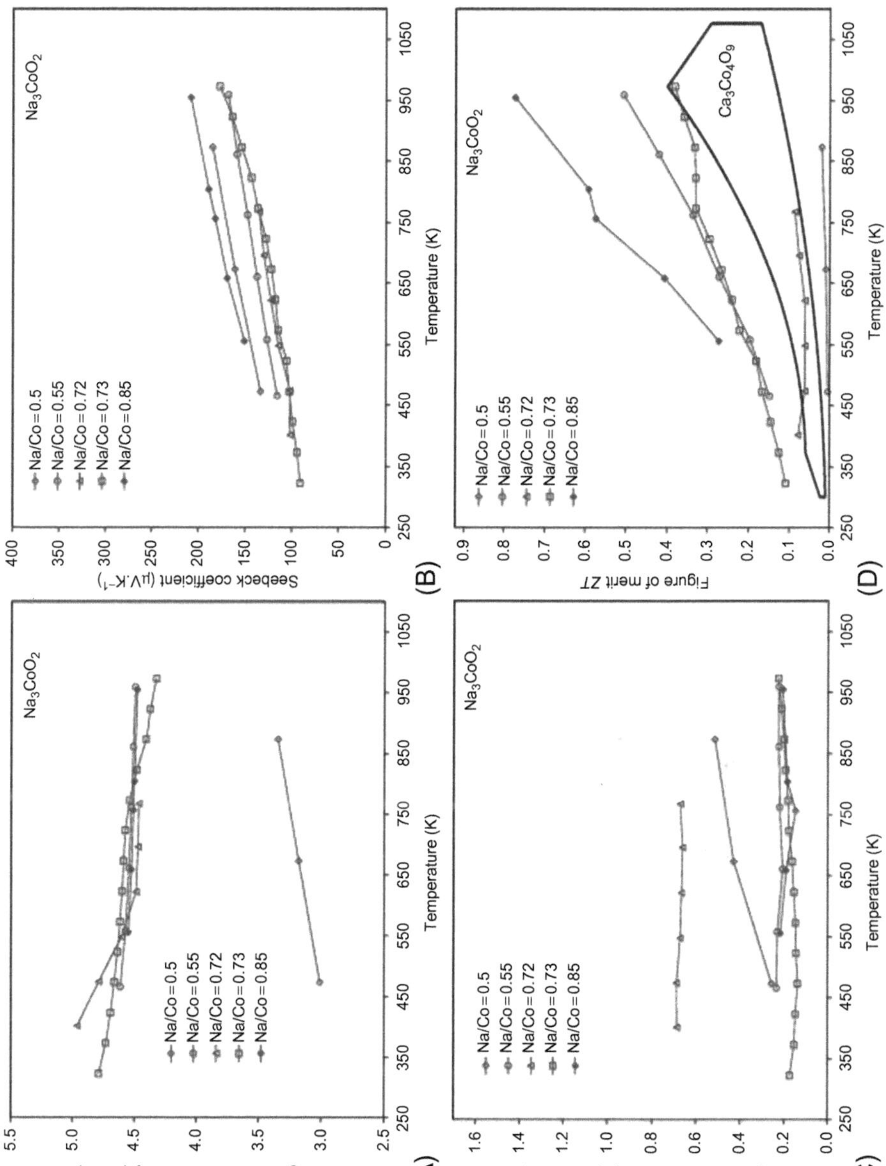

Fig. 3.6 (A) Electrical conductivity, (B) Seebeck coefficient, (C) thermal conductivity, and (D) figure of merit (ZT) for Na_xCoO_2. Reproduced with permission from J.W. Fergus, J. Eur. Ceram. Soc. 32 (2012) 525–540.

of the Ag-doped sample ($x=0.2$) was less than that for the non-doped sample, and the figure of merit (ZT) reached a maximum value of 0.124 at 973 K. The Seebeck coefficient of an Ag-doped sample ($x=0.1$) was greater than that for the non-doped sample, while the lowest thermal conductivity value of the Ag-doped sample ($x=0.1$) was 2.2 $Wm^{-1} K^{-1}$ at 300 K. The composite $Na_xCo_2O_4$/Ag samples were synthesized by mechanical milling and subsequent sintering by Ito et al. [52]. Agglomeration and coarsening of Ag particles occurred during the preparation process, and coarse Ag particles around 10 μm were dispersed in the $Na_xCo_2O_4$ matrix. The Ag addition was significantly effective in increasing the Seebeck coefficient, which is assumed to be influenced by the compositional homogeneity as compared to the sample without dopants, and decreasing the electrical resistivity. Moreover, it has been shown that addition of silver increases the thermal conductivity, but the figure of merit of $Na_xCo_2O_4$ could not be improved.

Other metallic dopants for Na_xCoO_2, such as Cu [53, 54] and Zn [55, 56], increase the electrical conductivity and thermopower, while Ni increases the thermopower but lowers the electrical conductivity [53, 57, 58]. However, the conductivity decreases when Ni and Cu are added together [53]. Copper addition was investigated by Park et al. [54], focusing on $Na(Co_{1-y}Cu_y)_2O_4$ ($0 \leq y \leq 0.2$) polycrystalline samples prepared using a SSR method. The Cu dopants led to a decrease in the porosity and an increase in the grain size. Well-defined plate-like grains were observed in the sintered bodies and the bronze-type layered structure of $Na(Co_{1-y}Cu_y)_2O_4$ was the dominant phase. The electrical conductivity of $Na(Co_{1-y}Cu_y)_2O_4$ was shown to decrease with increasing temperature, while the Seebeck coefficient increased with increasing temperature. For all samples, the sign of the Seebeck coefficient was always positive, which demonstrates that the holes were the dominant conductivity carriers. Both the electrical conductivity and the Seebeck coefficient of Cu-doped $Na(Co_{1-y}Cu_y)_2O_4$ were much higher than those of undoped $Na_xCo_2O_4$, indicating that the Cu dopants lead to an enhanced increase in the power factor. The power factor increased with an increase in temperature and the maximum value of 3.08×10^{-3} $Wm^{-1} K^{-2}$ was attained at 1073 K.

Nickel additions were also investigated by Park et al. [53], concentrating on $Na(Co_{1-y-z}Cu_yNi_z)_2O_4$ ($0 \leq y \leq 0.15$ and $0 \leq z \leq 0.15$) samples using a SSR process. The electrical conductivity increased due to the Cu additions, which was caused by the substitution of divalent Cu^{2+} for trivalent Co^{3+} with the increasing hole concentration of the system and the increasing density and grain size. However, the electrical conductivity decreased by the addition of Ni in $Na(Co_{1-y}Cu_y)_2O_4$ ($0 \leq y \leq 0.15$), mainly due to a decrease in density and grain size. Furthermore, the Cu and/or Ni substitutions led to a significant increase in the Seebeck coefficient. The power factor was improved by the addition of Cu and Ni in $NaCo_2O_4$. The maximum power factor of $Na(Co_{0.7}Cu_{0.15}Ni_{0.15})_2O_4$ of 3.27×10^{-3} $Wm^{-1} K^{-2}$ occurred at 1073 K. Wang et al. [57] investigated the nickel-doped polycrystal synthesized by the citric acid complex (CAC) process and subsequent unpressurized sintering. The electrical resistivity increased and thermal conductivity decreased due to a decrease in grain size. However, the Ni dopants caused an increase in the Seebeck coefficient. The maximum figure of merit of 0.176 was obtained at 673 K for $NaCo_{0.9}Ni_{0.1}O_2$.

Addition of ZnO to $Na(Co_{1-x}Zn_x)_2O_4$ ($0 \leq x \leq 0.1$) samples fabricated by the tape casting method was also studied by Park et al. [55]. The solid solution of the constituent oxides with a bronze-type layered structure was the dominant phase occurring in the as-sintered $Na(Co_{1-x}Zn_x)_2O_4$ ($0 \leq x \leq 0.1$) bodies. The electrical conductivity increased with an increase in ZnO content. For all samples, the Seebeck coefficient was positive over the whole temperature range (723–1073 K). The power factor of $Na(Co_{0.95}Zn_{0.05})_2O_4$ was $1.7 \times 10^{-3} Wm^{-1} K^{-2}$ at 1073 K, which was about four times that of undoped $NaCo_2O_4$. This phenomenon may be caused by the large Seebeck coefficient (415 μV/K) and high conductivity ($127\ \Omega^{-1}\ cm^{-1}$) at 1073 K. Tsai et al. [56] also investigated the thermoelectric properties of $Na_{0.8}Zn_xCo_{1-x}O_2/(ZnO)_y$ ($x \leq 0.01$, $0 \leq y \leq 0.14$). It was found that doping divalent Zn ions leads to a simultaneous reduction in resistivity and enhancement of thermopower. It is inferred that the reduction in resistivity may be caused by improved mobility of carriers, while the enhancement of thermopower may be attributed to the geometric relaxation of distorted CoO_6 octahedra caused by Zn dopants. The improvement in power factor results from a narrower bandwidth.

Fig. 3.7 depicts the influence of other dopants on the Na_xCoO_2 figure of merit. It can be concluded that the doped materials have figures of merit close to those of the matching undoped material. The most significant change was found when doped with nickel, which was caused by differences in the grain sizes of the materials [57]. Strontium additions increase the electrical conductivity and the figure of merit of Na_xCoO_2, while potassium addition has little effect on it [46]. It has been reported that with

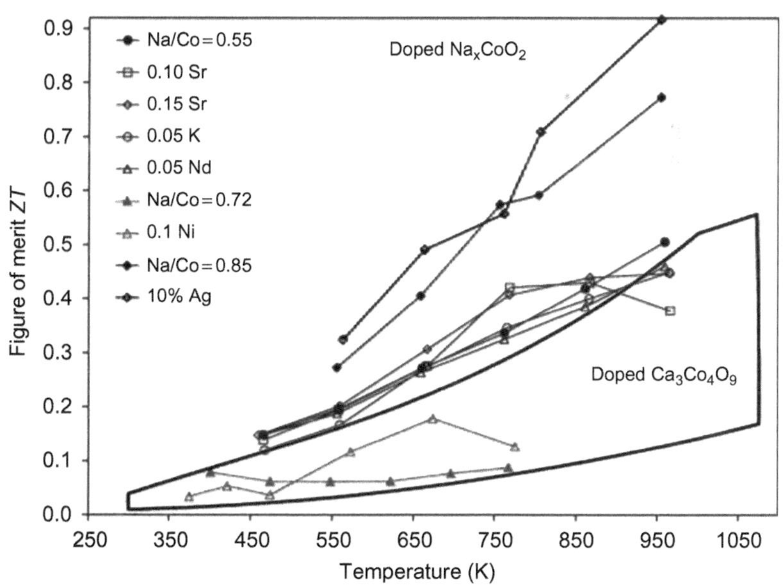

Fig. 3.7 Figure of merit (*ZT*) for doped Na_xCoO_2.
Reproduced with permission from J.W. Fergus, J. Eur. Ceram. Soc. 32 (2012) 525–540.

lanthanide dopants the ytterbium Seebeck coefficient increases while the dysprosium electrical conductivity increases [59]; moreover, addition of lanthanides such as neodymium, samarium, or ytterbium, and also yttrium, decreases the thermal as well as electrical conductivity with only slight influence on the value of *ZT* [46].

3.3 $Ca_3Co_4O_9$

Among the misfit-layered cobalt oxides, which are a class of cobaltite, $Ca_3Co_4O_9$ has interesting thermoelectric properties. The compound $Ca_3Co_4O_9$ shown in Fig. 3.8 can be represented as ($[Ca_2CoO_3]^{RS}[CoO_2]_{1.62}$), where RS denotes the block of rock salt-type layers. In contrast with the Na_xCoO_2 lattice, each sublattice of RS and CoO_2 layers is incommensurate with respect to the other monoclinic b lattice parameter. The coexistence of two monoclinic subsystems consisting of a CoO_2 layer and an RS block was showed by the electron diffraction patterns, which have the same a, c, and β parameters but different b unit cell parameters, b_1 and b_2, with incommensurability ratio b_1/b_2. The CoO_2 layer is responsible for electric conduction while the rock salt block stabilizes the crystal structure by ionic interaction similar to Na_xCoO_2 [14].

Vapor phase methods that include molecular beam epitaxy (MBE), pulsed laser deposition (PLD), and electron beam evaporation (EBE) have conventionally been employed for the growth of epitaxial $Ca_3Co_4O_9$ films [60]. To summarize, the thermoelectric properties of undoped $Ca_3Co_4O_9$ are fairly consistent with a conductivity of around 10^4 S m^{-1}, a Seebeck coefficient of around 150 μV K^{-1}, a thermal conductivity of around 2 W $K^{-1}m^{-1}$, and a figure of merit of around 0.3 at up to 1000K (Fig. 3.9). Studies by Satake et al. [61] showed that the lattice thermal conductivity is determined by the rock salt layer. With increasing rock salt layers, the in-plane

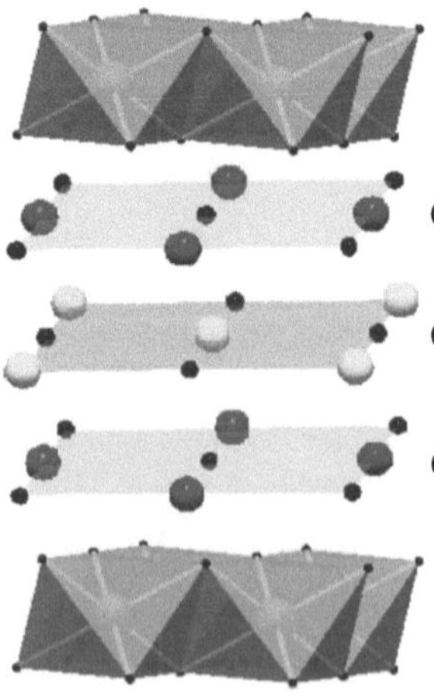

Fig. 3.8 Crystal structure of $Ca_3Co_4O_9$.
Reproduced with permission A. Nag, V. Shubha, J. Electron. Mater. 43 (2014) 962–977.

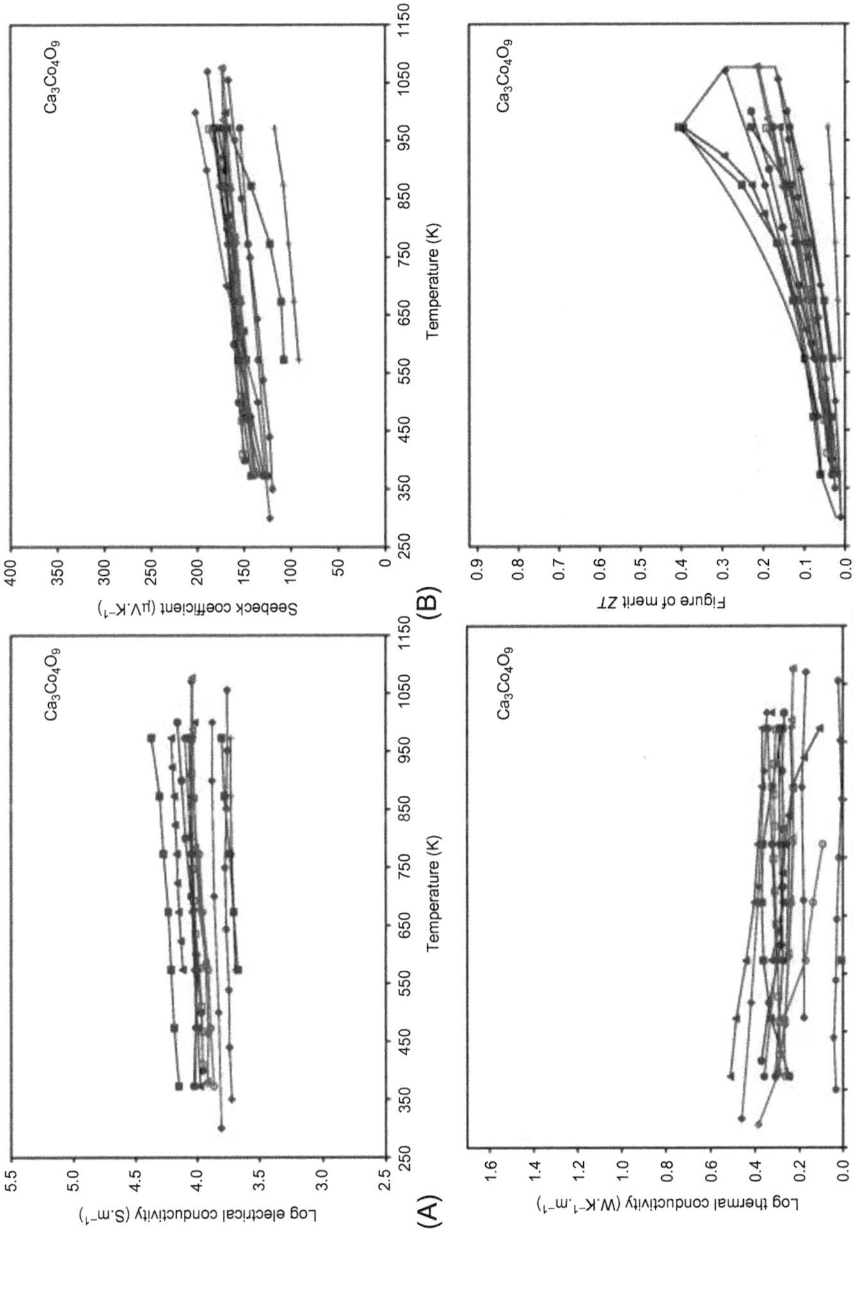

Fig. 3.9 (A) Electrical conductivity, (B) Seebeck coefficient, (C) thermal conductivity, and (D) figure of merit (*ZT*) for $Ca_3Co_4O_9$. Reproduced with permission from J.W. Fergus, J. Eur. Ceram. Soc. 32 (2012) 525–540.

thermal conductivity decreases while the electronic properties of the CoO_2 blocks remain undisturbed. Therefore, it can be inferred that when more RS layers are added, the thermal conductivity will decrease and the thermoelectric properties will improve.

Addition of silver as a dopant or as a second phase appears to have the most significant effect on the figure of merit. Generally, Ag doping increases the electrical conductivity, carrier concentration, and mobility. The large atomic mass of Ag results in a decrease in the thermal conductivity. However, the addition of silver has been shown to either increase or decrease the Seebeck coefficient [12, 62–64].

Bismuth doping increases the electrical conductivity as well as the Seebeck coefficient, while decreasing the thermal conductivity. The increase in electrical conductivity results from an increase in carrier mobility instead of carrier concentration, which is valuable as the increase in carrier concentration tends to decrease the Seebeck coefficient. The enhancements in the electrical conductivity have also been attributed to differences in the microstructure, while the decrease in thermal conductivity has been a consequence of the larger size and mass of bismuth as compared to calcium [65–71].

Copper is consistently reported to increase the electrical conductivity while decreasing the Seebeck coefficient [70, 72–74]. Copper occupies sites in the Ca_2CoO_3 rock salt layer instead of the CoO_2 layer of $Ca_3Co_4O_9$ [72]. However, Cu is not stable at the calcium sites as a second phase, which has been observed by XRD. However, it seems clear that the thermoelectric properties may be affected by copper.

$Ca_3Co_4O_9$ has also been doped with transition metals such as titanium, nickel, manganese, and iron [72, 74, 75], and in most cases, the beneficial effect of transition metal dopants has been shown to be an increase in thermopower, rather than increase in electrical conductivity. Rhodium doping increases electrical conductivity but decreases thermopower [76]. However, the heavy transition metal tantalum increases thermopower but decreases the electrical conductivity [77]. Gallium or lead addition has been shown to increase both the electrical conductivity and the thermopower [77]. Gallium also increases the figure of merit and decreases the thermal conductivity [78, 79]. Doping with lanthanum and lanthanide elements has also been investigated, however, TE enhancement was not impressive. Fig. 3.10 depicts the influence of other dopants on the $Ca_3Co_4O_9$ figure of merit [12].

3.4 $SrTiO_3$

$SrTiO_3$ has the cubic perovskite ABO_3 structure as illustrated in Fig. 3.11. It can be seen that the large Sr^{2+} ions occupy the cube center, while the smaller Ti^{4+} ions are at the cube corners, being surrounding by six O^{2-} to form the TiO_6 octahedral building units. Stoichiometric $SrTiO_3$ is an insulator with a band gap of 3.2 eV. However, it becomes a decent electronic conductor after doping with higher valence ions [80].

Many different methods, such as traditional solid state, sol-gel, and combustion synthesis, have been used to prepare La-doped $SrTiO_3$ ceramics. The lanthanum dopant increases the electrical conductivity and Seebeck coefficient, but relatively small changes arise in the thermal conductivity, so that the figure of merit is improved [75,

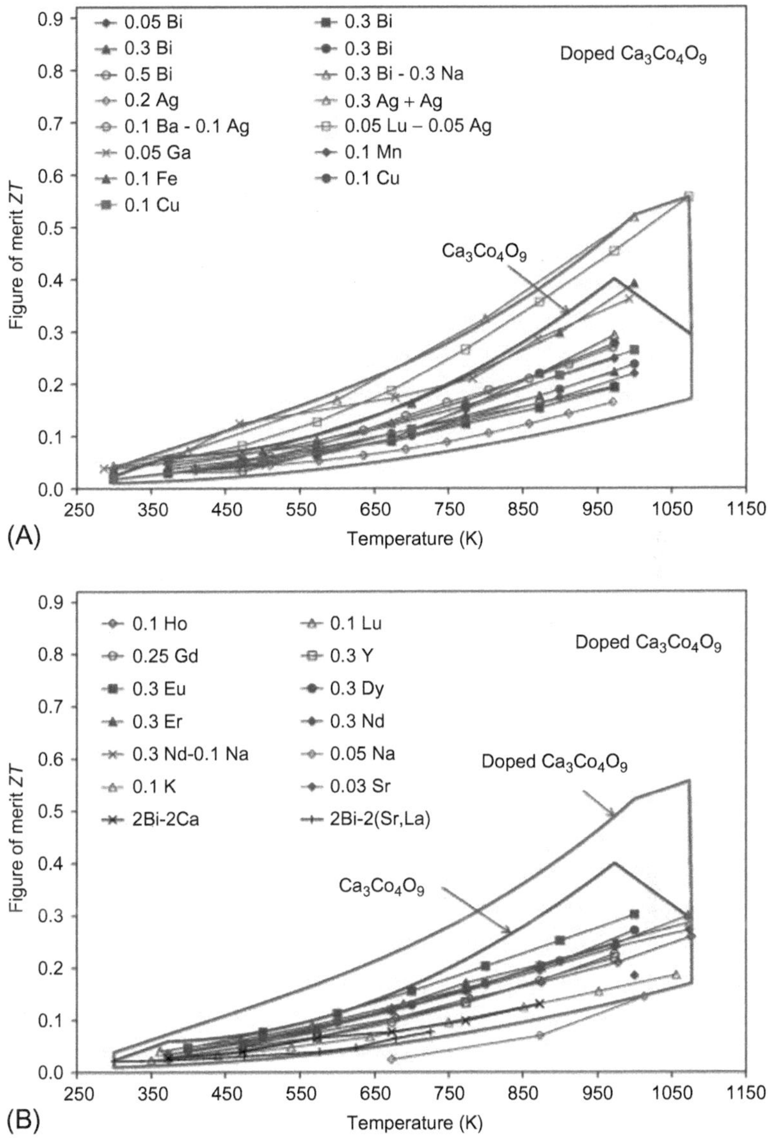

Fig. 3.10 Figure of merit (*ZT*) for $Ca_3Co_4O_9$ (A) doped with transition metals and (B) doped with other dopants.
Reproduced with permission from J.W. Fergus, J. Eur. Ceram. Soc. 32 (2012) 525–540.

81–84]. When strontium titanate is doped, its perovskite crystal structure may change into a superlattice Ruddlesden-Popper structure. With this structure a mechanism may arise to reduce the thermal conductivity, as seen in layered cobaltites. Furthermore, the formation of the Ruddlesden-Popper phase causes a decrease in the electrical conductivity [85]. The highest ZT for La-doped $SrTiO_3$ has been reported to be 0.1 at room

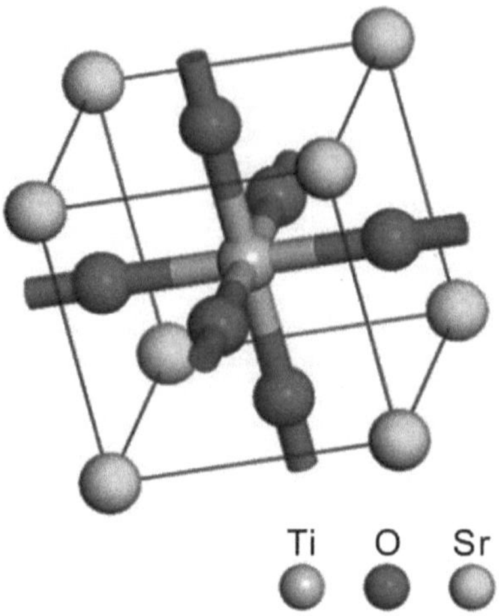

Fig. 3.11 Crystal structure of $SrTiO_3$.
Reproduced with permission from H. Wang, W. Su, J. Liu, C. Wang, J. Mater. 2 (2016) 225–236.

temperature and 0.27 at 1000 K, as shown in Fig. 3.12 [9, 82, 86, 87]. A series of studies on the thermoelectric properties of La-doped $SrTiO_3$ thin films has been conducted. The figures of merit are limited because of its high thermal conductivity, about 11.0 $Wm^{-1}K^{-1}$ at 300 K and 3.2 $Wm^{-1}K^{-1}$ at 1000 K [88].

Other lanthanide dopants are also used with strontium titanate. When compared with several lanthanide dopants, dysprosium showed the highest figure of merit [89]. When strontium is replaced by dysprosium, the electrical conductivity and Seebeck coefficient increase, while the thermal conductivity decreases. When lanthanum is replaced with dysprosium, both the conductivity and the Seebeck coefficient decrease [90, 91]. The beneficial influence of yttrium is due to an increase in the Seebeck coefficient [83].

When niobium is added, the electrical conductivity increases, while the Seebeck coefficient remains similar or is lowered [82, 84, 92, 93]. The effect of niobium on the thermal conductivity is still unclear, as an increase [84], a decrease [9], and no significant changes [82, 92] have been reported.

Addition of potassium titanate (KTO) increases the electrical conductivity, while the thermal conductivity is lowered, which enhances the figure of merit [94]. A solid solution of lanthanum-doped barium titanate-strontium titanate has been evaluated for its thermoelectric properties, and it was found that a composition containing 30% barium (i.e., $Ba_{0.3}Sr_{0.6}La_{0.1}TiO_3$) had the best performance [95].

3.5 $CaMnO_3$

$CaMnO_3$ also has a perovskite structure as shown in Fig. 3.13. The orthorhombic crystal structure of $CaMnO_3$ belongs to the Pnma space group. The Seebeck coefficient of $CaMnO_3$ ceramics is high, but the figure of merit remains low at high temperatures

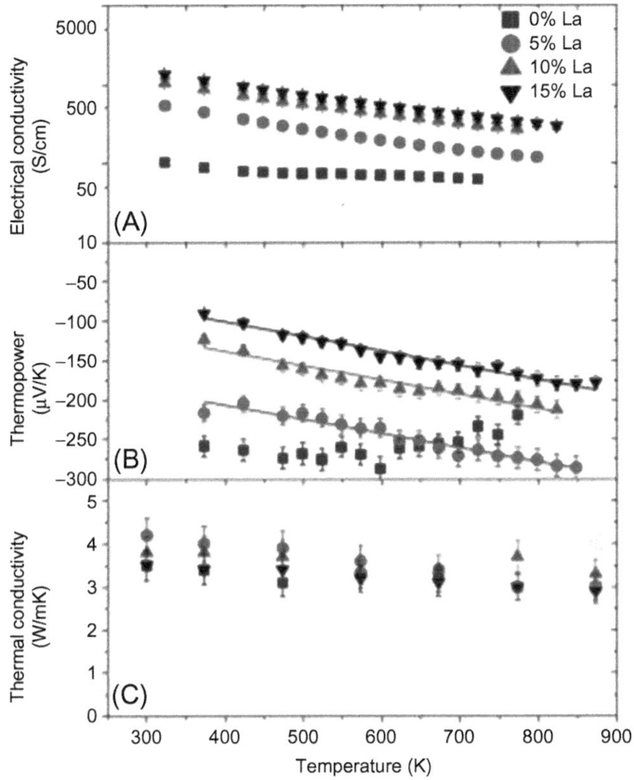

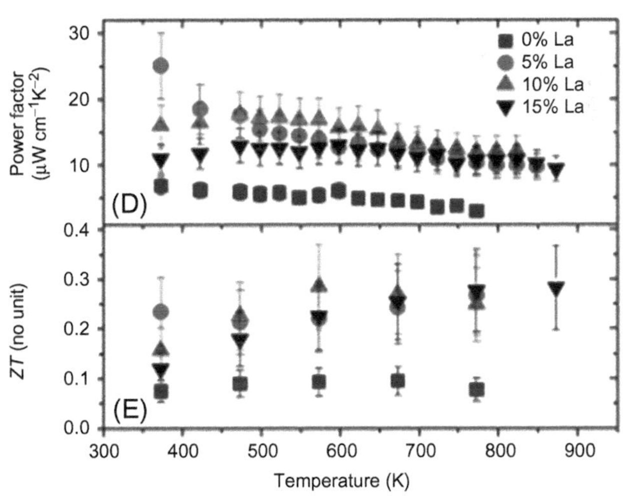

Fig. 3.12 (A) Electrical conductivity, (B) Seebeck coefficient, (C) thermal conductivity, (D) TPF, and (E) *ZT* versus temperature for different concentrations of La dopant in $SrTiO_3$ thin films.

Reproduced with permission from S. Walia, S. Balendhran, H. Nili, S. Zhuiykov, G. Rosengarten, Q.H. Wang, M. Bhaskaran, S. Sriram, M.S. Strano, K. Kalantar-zadeh, Prog. Mater. Sci. 58 (2013) 1443–1489.

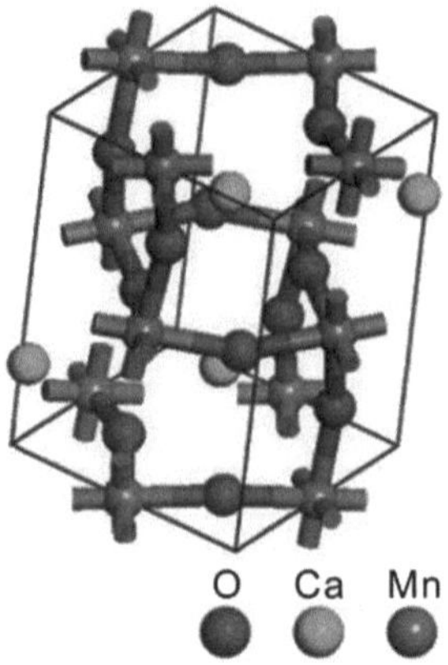

Fig. 3.13 Crystal structure of $CaMnO_3$.
Reproduced with permission from H. Wang, W. Su, J. Liu, C. Wang, J. Mater. 2 (2016) 225–236.

because of their large resistivities [80]. However, the addition of some tri- or tetravalent elements decreases the resistivity of $CaMnO_3$ and hence moderate Seebeck coefficients are obtained. Generally, the thermoelectric properties of $CaMnO_3$ materials are improved by adding these dopants [80].

The thermoelectric properties of $CaMnO_3$ with ytterbium and niobium addition are summarized in Fig. 3.14. Ytterbium-doped $CaMnO_3$ materials have been shown to have a higher electrical conductivity and Seebeck coefficient, leading to a higher figure of merit. Compared with $SrTiO_3$, $CaMnO_3$ has a lower thermal conductivity, lower electrical conductivity, and a lower Seebeck coefficient. It can be concluded that addition of ytterbium effectively increases the figure of merit. Since the phonon contributions to the thermal conduction decrease, the thermal conductivity of ytterbium-doped $CaMnO_3$ material decreases along with it. Niobium doping increases the electrical conductivity and decreases the thermal conductivity, whereas it decreases the Seebeck coefficient. Similar phenomena were seen with tantalum dopants, but niobium dopants caused higher electrical conductivities [96–99].

Other lanthanide oxide dopants are also used for $CaMnO_3$. Lanthanum [100, 101] and dysprosium [98, 99, 102] dopants increase the electrical conductivity and thermopower of $CaMnO_3$, whereas praseodymium dopants result in a lower thermal conductivity [103, 104]. This phenomenon can be utilized to design more effective thermoelectric materials. The addition of strontium increases the electrical conductivity and figure of merit of ytterbium-doped $CaMnO_3$ [105]. Single dopants such as dysprosium and niobium are also effective in some cases. Silver addition leads to an increase in the Seebeck coefficient [106] while bismuth, either alone [107] or co-doped with vanadium [108] or niobium [109], increases the electrical conductivity and the figure of merit.

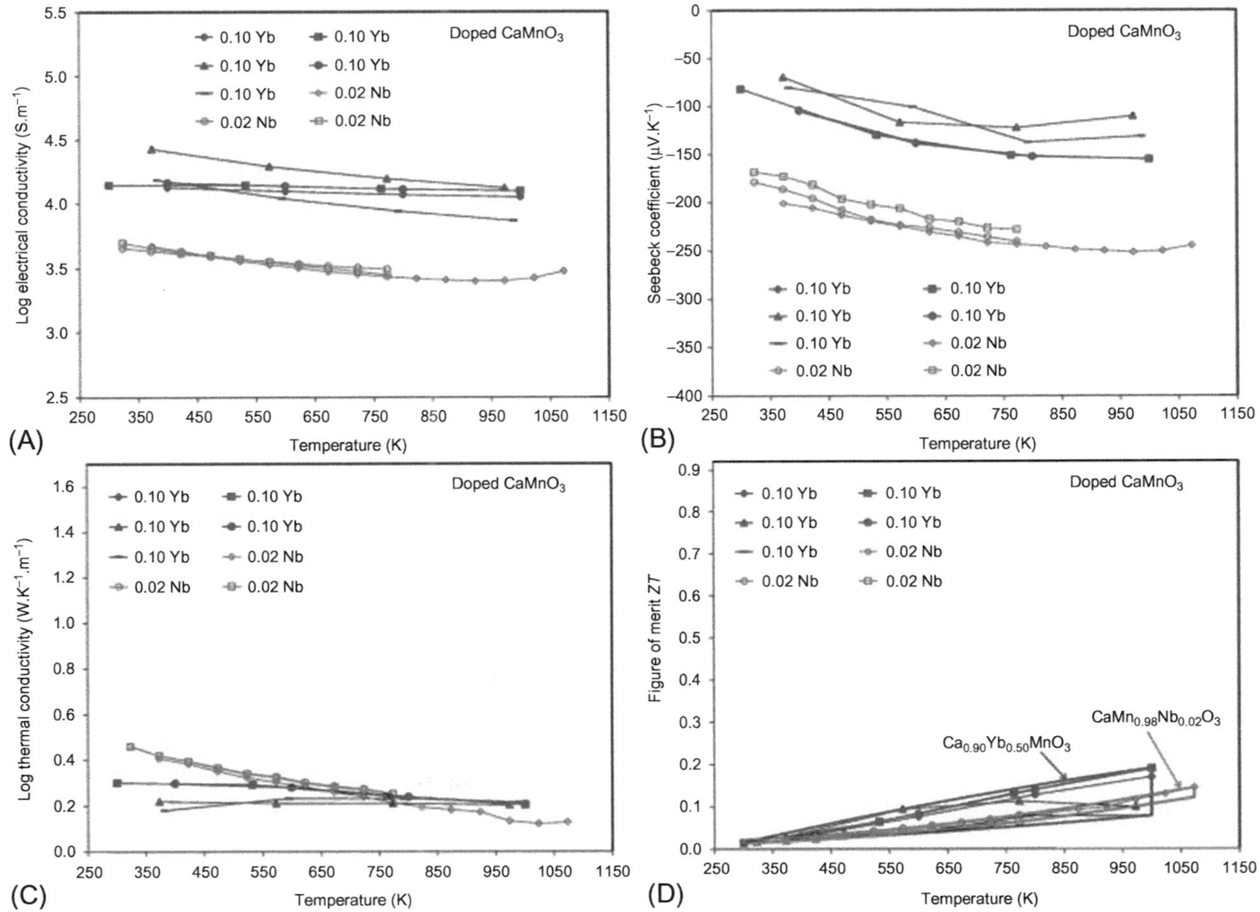

Fig. 3.14 (A) Electrical conductivity, (B) Seebeck coefficient, (C) thermal conductivity, and (D) figure of merit (*ZT*) for doped $CaMnO_3$. Reproduced with permission from J.W. Fergus, J. Eur. Ceram. Soc. 32. (2012) 525–540.

3.6 ZnO

ZnO occurs in a hexagonal wurtzite, cubic zinc-blende, or rarely rock salt structure. In the wurtzite structure, the Zn atoms are each surrounded by four oxygen atoms, while in the zinc blende structure, each oxygen atom is similarly surrounded by four Zn atoms. Wurtzite is a stable polymorph at ambient temperature and pressure, but when ZnO is grown on cubic substrates, it attains the stable zinc blende structure [60].

ZnO has potential as a high-temperature TE material and can be synthesized using a number of liquid and vapor phase fabrication techniques. It can also be synthesized in various morphologies, such as epitaxial thin films, single and polycrystalline ceramics, nanobelts, nanorings, nanorods, and hierarchical nanostructures. Liquid-phase fabrication techniques generally employed for preparing ZnO are hydrolysis, hydrothermal synthesis, sol-gel synthesis, spray pyrolysis, and electrochemical techniques such as anodization and electrodeposition. Furthermore, vapor deposition techniques such as atomic layer deposition (ALD), RF/DC sputtering, MBE, chemical vapor deposition (CVD), and metal-organic chemical vapor deposition (MOCVD) are also used to prepare doped ZnO thin films, while spark plasma sintering (SPS) and PLD are used to prepare densified ZnO [60].

Pure ZnO has a high thermal conductivity of 40 kW/K at ambient temperatures but only about 5 kW/K at 1000 K. It has been argued that it is the big and largely anisotropic thermal expansion of ZnO that may lead to an increase in phonon scattering. For nanograined ZnO ceramics, it is possible to reach low thermal conductivities of 2–3 kW/K at room temperature, but these structures have low electrical conductivities σ and figures of merit ZT. It is inferred that the defects at grain boundaries mainly contribute to the decrease in electrical conductivity. In general, samples with larger grain sizes have a higher electrical conductivity [110].

The value of ZT for doped and co-doped ZnO is 0.65 but the conductivity of doped ZnO can be as high as that of strontium titanate or even higher with increasing temperature, which becomes a drawback for the use of ZnO. When small amount of aluminum is added (e.g., 0.25%) ZnO has a high electrical conductivity, low thermal conductivity, and low thermopower due to the increase in carrier mobility and decrease in the phonon mean free path in the crystal lattice; however, this changes when larger amounts (e.g., 2%) are added. The figure of merit has been shown to be as high as 0.30 at 1273 K [111, 112]. The addition of aluminum contributes to a decrease in the c/a ratio of the crystal [113]. However, transmission electron microscopy has shown that second phases are present at the grain boundaries, even in samples that seem to be single phase [114]. Grain boundaries also have an effect on the grain size and thus on thermoelectric properties. The thermal and electrical conductivities of ZnO change obviously with grain size, and the figure of merit for large grains is more satisfactory [110].

When nickel is added to ZnO, the electrical conductivity and thermopower both increase, and the performance of ZnO can be further improved with aluminum as a co-dopant [115, 116]. Adding titanium to ZnO causes an increase in electrical conductivity, but the thermopower decreases [117]. Moreover, titanium also improves the performance of ZnO as a co-dopant with aluminum, just as nickel. The addition of tin and antimony separately or simultaneously can increase the conductivity and Seebeck coefficient of ZnO [118–120].

3.7 Conclusion

In summary, a relatively comprehensive analysis of the most important metal oxide-based thermoelectric materials has been presented. The synthesis and thermoelectric performance of several useful oxides was described briefly. Modification strategies based on doping were discussed to obtain promising materials with low thermal conductivity, high electrical conductivity, and large Seebeck coefficient. Thermoelectric materials can be used in various highly efficient cooling and refrigeration, energy scavenging, sensing, and thermopower systems, even at the micro- and nanoscale. However, many issues still remain to be solved and a thorough understanding of the fundamental thermoelectric parameters needs to be developed.

References

[1] F.J. DiSalvo, Science 285 (5428) (1999) 703–706.
[2] L.E. Bell, Science 321 (5895) (2008) 1457–1461.
[3] G.J. Snyder, E.S. Toberer, Nat. Mater. 7 (2008) 105–114.
[4] J. He, M.G. Kanatzidis, V.P. Dravid, Mater. Today 16 (2013) 166–176.
[5] X. Zhang, L. Zhao, J. Mater. 1 (2015) 92–105.
[6] J. Li, W. Liu, L. Zhao, M. Zhou, NPG Asia Mater. 2 (2010) 152–158.
[7] T. Tritt, Encyclopedia of Materials Science & Technology, Elsevier, Amsterdam, 2002, pp. 1–11.
[8] D.M. Rowe, Thermoelectrics Handbook: Macro to Nano, CRC Press, Boca Raton, FL, 2005.
[9] H. Ohta, K. Sugiura, K. Koumoto, Inorg. Chem. 47 (2008) 8429–8436.
[10] K. Biswas, J. He, I.D. Blum, C. Wu, T.P. Hogan, D.N. Seidman, V.P. Dravid, M. G. Kanatzidis, Nature 489 (7416) (2012) 414–418.
[11] T.M. Tritt, Annu. Rev. Mater. Res. 41 (2011) 433–438.
[12] J.W. Fergus, J. Eur. Ceram. Soc. 32 (2012) 525–540.
[13] S. Kim, B. Won, S. Rhi, S. Kim, J. Yoo, J. Jang, J. Electron. Mater. 40 (2011) 778–783.
[14] A. Nag, V. Shubha, J. Electron. Mater. 43 (2014) 962–977.
[15] S. Sugihara, K. Nishiyama, Y. Igarashi, K. Morikawa, Adv. Quantum Chem. 54 (2008) 227–243.
[16] S. Walia, R. Weber, S. Balendhran, D. Yao, J.T. Abrahamson, S. Zhuiykov, M. Bhaskaran, S. Sriram, M.S. Strano, K. Kalantar-zadeh, Chem. Commun. 48 (2012) 7462–7464.
[17] S. Walia, R. Weber, K. Latham, P. Petersen, J.T. Abrahamson, M.S. Strano, K. Kalantar-zadeh, Adv. Funct. Mater. 21 (2011) 2072–2079.
[18] S.B. Riffat, X. Ma, Appl. Therm. Eng. 23 (2003) 913–935.
[19] S. Walia, S. Balendhran, P. Yi, D. Yao, S. Zhuiykov, M. Pannirselvam, R. Weber, M. S. Strano, M. Bhaskaran, S. Sriram, K. Kalantar-zadeh, J. Phys. Chem. C 117 (2013) 9137–9142.

[20] A. Patyk, J. Electron. Mater. 39 (2010) 2023–2028.
[21] L.I. Anatychuk, Y.Y. Rozver, D.D. Velichuk, J. Electron. Mater. 40 (2011) 1206–1208.
[22] J. Yang, F.R. Stabler, J. Electron. Mater. 38 (2009) 1245–1251.
[23] M.A. Korzhuev, I.V. Katin, J. Electron. Mater. 39 (2010) 1390–1394.
[24] J.P. Doumerc, M. Blangero, M. Pollet, D. Carlier, J. Darriet, R. Berthelot, C. Delmas, R. Decourt, J. Electron. Mater. 38 (2009) 1078–1082.
[25] G.S. Nolas, J. Sharp, H.J. Goldsmid, Thermoelectrics-Basic Principles and New Materials Developments, Springer, Berlin, 2001.
[26] A.F. Ioffe, L.S. Stil'Bans, E.K. Iordanishvili, T.S. Stavitskaya, A. Gelbtuch, G. Vineyard, Phys. Today 12 (1959) 42.
[27] C. Gayner, K.K. Kar, Prog. Mater. Sci. 83 (2016) 330–382.
[28] G.J. Snyder, Electrochem. Soc. Interface (2008) 54–56.
[29] R.P. Chasmar, R. Stratton, J. Electron. Control 7 (1959) 52–72.
[30] F.D. Rosi, Solid State Electron. 11 (1968) 833–868.
[31] J.P. Fleurial, JOM 61 (2009) 79–85.
[32] A. Sussardi, T. Tanaka, A.U. Khan, L. Schlapbach, T. Mori, J. Mater. 1 (2015) 196–204.
[33] D. Nemir, J. Beck, J. Electron. Mater. 39 (2010) 1897–1901.
[34] G. Sebald, D. Guyomar, A. Agbossou, Smart Mater. Struct. 18 (2009) 125006.
[35] D.M. Rowe, G. Min, J. Power Sources 73 (1998) 193–198.
[36] J.P. Heremans, V. Jovovic, E.S. Toberer, A. Saramat, K. Kurosaki, A. Charoenphakdee, S. Yamanaka, G.J. Snyder, Science 321 (5888) (2008) 554–557.
[37] E.S. Toberer, A. Zevalkink, G.J. Snyder, J. Mater. Chem. 21 (2011) 15843.
[38] I. Terasaki, Y. Sasago, K. Uchinokura, Phys. Rev. B 56 (1997) R12685–R12687.
[39] K. Koumoto, W.S. Seo, S. Ozawa, Appl. Phys. Lett. 71 (1997) 1475–1476.
[40] M. Ohtaki, H. Koga, T. Tokunaga, K. Eguchi, H. Arai, J. Solid State Chem. 120 (1995) 105–111.
[41] H. Ohta, W. Seo, K. Koumoto, J. Am. Ceram. Soc. 79 (1996) 2193–2196.
[42] H. Muta, K. Kurosaki, M. Uno, S. Yamanaka, J. Alloys Compd. 335 (2002) 200–202.
[43] M. Jansen, R. Hoppe, Z. Anorg. Allg. Chem. 408 (1974) 104–106.
[44] Y. Wang, N.S. Rogado, R.J. Cava, N.P. Ong, Nature 423 (2003) 425.
[45] W. Koshibae, K. Tsutsui, S. Maekawa, Phys. Rev. B 62 (2000) 6869.
[46] T. Nagira, M. Ito, S. Katsuyama, K. Majima, H. Nagai, J. Alloys Compd. 348 (2003) 263–269.
[47] M. Ito, T. Nagira, D. Furumoto, S. Katsuyama, H. Nagai, Scr. Mater. 48 (2003) 403–408.
[48] T. Tanaka, S. Nakamura, S. Iida, Jpn. J. Appl. Phys. 33 (1994) L581.
[49] K. Fujita, T. Mochida, K. Nakamura, Jpn. J. Appl. Phys. 40 (Part 1, 7) (2001) 4644–4647.
[50] T. Seetawan, V. Amornkitbamrung, T. Burinprakhon, S. Maensiri, K. Kurosaki, H. Muta, M. Uno, S. Yamanaka, J. Alloys Compd. 407 (2006) 314–317.
[51] T. Seetawan, V. Amornkitbamrung, T. Burinprakhon, S. Maensiri, K. Kurosaki, H. Muta, M. Uno, S. Yamanaka, J. Alloys Compd. 414 (2006) 293–297.
[52] M. Ito, D. Furumoto, J. Alloys Compd. 450 (2008) 494–498.
[53] K. Park, K.Y. Ko, J.G. Kim, W.S. Cho, Mater. Sci. Eng. B 129 (2006) 200–206.
[54] K. Park, K.U. Jang, H.C. Kwon, J.G. Kim, W.S. Cho, J. Alloys Compd. 419 (2006) 213–219.
[55] K. Park, J.H. Lee, Mater. Lett. 62 (2008) 2366–2368.
[56] P.H. Tsai, T.S. Zhang, R. Donelson, T.T. Tan, S. Li, J. Alloys Compd. 509 (2011) 5183–5186.
[57] L. Wang, M. Wang, D. Zhao, J. Alloys Compd. 471 (2009) 519–523.
[58] K. Park, K.U. Jang, Mater. Lett. 60 (2006) 1106–1110.

[59] M. Ito, T. Nagira, S. Hara, J. Alloys Compd. 408-412 (2006) 1217–1221.
[60] S. Walia, S. Balendhran, H. Nili, S. Zhuiykov, G. Rosengarten, Q.H. Wang, M. Bhaskaran, S. Sriram, M.S. Strano, K. Kalantar-zadeh, Prog. Mater. Sci. 58 (2013) 1443–1489.
[61] A. Satake, H. Tanaka, T. Ohkawa, T. Fujii, I. Terasaki, J. Appl. Phys. 96 (2004) 931–933.
[62] Y. Wang, Y. Sui, J. Cheng, X. Wang, W. Su, J. Alloys Compd. 477 (2009) 817–821.
[63] Y. Wang, Y. Sui, J. Cheng, X. Wang, W. Su, J. Phys. D. Appl. Phys. 41 (2008) 45406.
[64] F.P. Zhang, Q.M. Lu, J.X. Zhang, X. Zhang, J. Alloys Compd. 477 (2009) 543–546.
[65] Y. Liu, Y. Lin, L. Jiang, C. Nan, Z. Shen, J. Electroceram. 21 (2008) 748–751.
[66] H. Su, Y. Jiang, X. Lan, X. Liu, H. Zhong, D. Yu, Phys. Status Solidi A 208 (2011) 147–155.
[67] T. Sun, H.H. Hng, Q.Y. Yan, J. Ma, J. Appl. Phys. 108 (2010) 83709.
[68] S. Li, R. Funahashi, I. Matsubara, H. Yamada, K. Ueno, S. Sodeoka, Ceram. Int. 27 (2001) 321–324.
[69] Y. Song, Q. Sun, L. Zhao, F. Wang, Z. Jiang, Mater. Chem. Phys. 113 (2009) 645–649.
[70] H. Hao, L. Zhao, X. Hu, J. Mater. Sci. Technol. 25 (2009) 105–108.
[71] G. Xu, R. Funahashi, M. Shikano, I. Matsubara, Y. Zhou, Appl. Phys. Lett. 80 (2002) 3760–3762.
[72] Y. Wang, Y. Sui, X. Wang, W. Su, X. Liu, J. Appl. Phys. 107 (2010) 33708.
[73] Q. Yao, D.L. Wang, L.D. Chen, X. Shi, M. Zhou, J. Appl. Phys. 97 (2005) 103905.
[74] M. Mikami, R. Funahashi, J. Solid State Chem. 178 (2005) 1670–1674.
[75] L. Xu, F. Li, Y. Wang, J. Alloys Compd. 501 (2010) 115–119.
[76] T. Takami, H. Ikuta, J. Appl. Phys. 103 (2008) 7B–701B.
[77] Y. Oide, Y. Miyazaki, Y. Ono, X.Y. Huang, T. Kajitani, Proceedings of 25th International Conference on Thermoelectrics, 2006, pp. 402–405.
[78] N. Van Nong, C.J. Liu, M. Ohtaki, J. Alloys Compd. 491 (2010) 53–56.
[79] N. Van Nong, S. Yanagiya, M. Sonne, N. Pryds, M. Ohtaki, J. Electron. Mater. 40 (2011) 716–722.
[80] H. Wang, W. Su, J. Liu, C. Wang, J. Mater. 2 (2016) 225–236.
[81] J. Liu, C.L. Wang, W.B. Su, H.C. Wang, P. Zheng, J.C. Li, J.L. Zhang, L.M. Mei, Appl. Phys. Lett. 95 (2009) 162110.
[82] S. Ohta, T. Nomura, H. Ohta, K. Koumoto, J. Appl. Phys. 97 (2005) 34106.
[83] N. Okinaka, L. Zhang, T. Akiyama, Trans. Iron Steel Inst. Jpn. 50 (2010) 1300–1304.
[84] A. Kinaci, C. Sevik, T. Çağın, Phys. Rev. B 82 (2010) 115114.
[85] Y.F. Wang, K.H. Lee, H. Ohta, K. Koumoto, Ceram. Int. 34 (2008) 849–852.
[86] T. Okuda, K. Nakanishi, S. Miyasaka, Y. Tokura, Phys. Rev. B 63 (2001) 113104.
[87] J. Ravichandran, W. Siemons, D.W. Oh, J.T. Kardel, A. Chari, H. Heijmerikx, M. L. Scullin, A. Majumdar, R. Ramesh, Phys. Rev. B 82 (2010) 165126.
[88] K. Koumoto, Y. Wang, R. Zhang, A. Kosuga, R. Funahashi, Annu. Rev. Mater. Res. 40 (2010) 363–394.
[89] H. Muta, K. Kurosaki, S. Yamanaka, J. Alloys Compd. 350 (2003) 292–295.
[90] H.C. Wang, C.L. Wang, W.B. Su, J. Liu, Y. Sun, H. Peng, L.M. Mei, J. Am. Ceram. Soc. 94 (2011) 838–842.
[91] H.C. Wang, C.L. Wang, W.B. Su, J. Liu, Y. Zhao, H. Peng, J.L. Zhang, M.L. Zhao, J. C. Li, N. Yin, L.M. Mei, Mater. Res. Bull. 45 (2010) 809–812.
[92] N. Wang, L. Han, H. He, Y. Ba, K. Koumoto, J. Alloys Compd. 497 (2010) 308–311.
[93] K.H. Lee, A. Ishizaki, S.W. Kim, H. Ohta, K. Koumoto, J. Appl. Phys. 102 (2007) 33702.
[94] N. Wang, H. He, Y. Ba, C. Wan, K. Koumoto, J. Ceram. Soc. Jpn. 118 (1383) (2010) 1098–1101.

[95] H. Muta, K. Kurosaki, S. Yamanaka, J. Alloys Compd. 368 (2004) 22–24.
[96] Y. Wang, Y. Sui, H. Fan, X. Wang, Y. Su, W. Su, X. Liu, Chem. Mater. 21 (2009) 4653–4660.
[97] R. Funahashi, A. Kosuga, N. Miyasou, E. Takeuchi, S. Urata, K. Lee, H. Ohta, K. Koumoto, Proceedings of the 26th International Conference on Thermoelectrics, South Korea: IEEE, 2007, p. 124.
[98] Y. Wang, Y. Sui, X. Wang, W. Su, Appl. Phys. Lett. 97 (2010) 52109.
[99] Y. Wang, Y. Sui, W. Su, J. Appl. Phys. 104 (2008) 93703.
[100] J. Lan, Y. Lin, A. Mei, C. Nan, Y. Liu, B. Zhang, J. Li, J. Mater. Sci. Technol. 25 (2009) 535–538.
[101] H. Taguchi, T. Kugi, M. Kato, K. Hirota, J. Am. Ceram. Soc. 93 (2010) 3009–3011.
[102] C. Liu, A. Bhaskar, J.J. Yuan, Appl. Phys. Lett. 98 (2011) 214101–214103.
[103] S. Choi, C. Lim, W. Seo, J. Electron. Mater. 40 (2011) 551–556.
[104] B.T. Cong, T. Tsuji, P.X. Thao, P.Q. Thanh, Y. Yamamura, Phys. B Condens. Matter 352 (2004) 18–23.
[105] A. Kosuga, Y. Isse, Y. Wang, K. Koumoto, R. Funahashi, J. Appl. Phys. 105 (2009) 93717.
[106] A. Kosuga, S. Urata, K. Kurosaki, S. Yamanaka, R. Funahashi, Jpn. J. Appl. Phys. 47 (2008) 6399–6403.
[107] F. Kawashima, X.Y. Huang, K. Hayashi, Y. Miyazaki, T. Kajitani, J. Electron. Mater. 38 (2009) 1159–1162.
[108] X.Y. Huang, Y. Miyazaki, T. Kajitani, Solid State Commun. 145 (2008) 132–136.
[109] J.W. Park, D.H. Kwak, S.H. Yoon, S.C. Choi, J. Alloys Compd. 487 (2009) 550–555.
[110] Y. Kinemuchi, M. Mikami, K. Kobayashi, K. Watari, Y. Hotta, J. Electron. Mater. 39 (2010) 2059–2063.
[111] E. Guilmeau, A. Maignan, C. Martin, J. Electron. Mater. 38 (2009) 1104–1108.
[112] M. Ohtaki, T. Tsubota, K. Eguchi, H. Arai, J. Appl. Phys. 79 (1996) 1816–1818.
[113] J.P. Wiff, Y. Kinemuchi, H. Kaga, C. Ito, K. Watari, J. Eur. Ceram. Soc. 29 (2009) 1413–1418.
[114] N. Ma, J.F. Li, B.P. Zhang, Y.H. Lin, L.R. Ren, G.F. Chen, J. Phys. Chem. Solids 71 (2010) 1344–1349.
[115] K. Park, J.K. Seong, G.H. Kim, J. Alloys Compd. 473 (2009) 423–427.
[116] H. Yamaguchi, Y. Chonan, M. Oda, T. Komiyama, T. Aoyama, S. Sugiyama, J. Electron. Mater. 40 (2011) 723–727.
[117] X. Dong, Y. Gan, Y. Wang, S. Peng, L. Dong, J. Alloys Compd. 581 (2013) 52–55.
[118] K. Park, K.Y. Ko, W.S. Seo, W.S. Cho, J.G. Kim, J.Y. Kim, J. Eur. Ceram. Soc. 27 (2007) 813–817.
[119] K. Park, J.K. Seong, J. Alloys Compd. 464 (2008) 1–5.
[120] K. Park, J.K. Seong, S. Nahm, J. Alloys Compd. 455 (2008) 331–335.

Mixed oxides in nuclear fuels

4

Bingwei Chen, Nengfei Yu*, Lijun Fu*, Yusong Zhu*, Teunis van Ree†, Yuping Wu*,‡*

*State Key Laboratory of Materials-Oriented Chemical Engineering, School of Energy Science and Engineering, Institute for Electrochemical Energy Storage, Nanjing Tech University, Nanjing, People's Republic of China, †Department of Chemistry, University of Venda, Thohoyandou, South Africa, ‡College of Science, Hunan Agricultural University, Changsha, China

4.1 Introduction

Nuclear reactors provide a source of electrical power that may be less expensive over the long term than coal or oil burning power plants [1]. At the core of the reactor a unique, self-sustaining fission process produces copious amounts of heat [2]. A fissionable atom, for example, ^{235}U, which is captured in a pellet consisting of sintered UO_2 crystals, will undergo fission when a neutron having a suitable energy is absorbed in the nucleus. The resulting nucleus is highly unstable and immediately splits into two nuclei of similar size, for example,

$$^{235}U + n^1 \rightarrow {}^{95}X + {}^{139}Y + 2\,n^1 \tag{4.1}$$

The mass difference expressed by Eq. (4.1) corresponds to an energy release of about 200 MeV. The greatest part of this energy is transferred to the fission products, which travel very rapidly from the site of fission in a straight trajectory of about 10 pm before they come to rest. In the process they transfer their energy to the metal oxide lattice, mostly as thermal vibrations, which cause significant damage to the lattice. This energy, added to small amounts produced by other sources, is converted to heat in the nuclear fuel, which must in turn be converted to a more useful form [3].

Until recently this energy conversion was effected using the generated heat to produce high-temperature steam, which drives a turbine connected to an alternator. When used this way, nuclear energy directly substitutes electrical energy generated from coal or oil, and the maximum Carnot efficiency is similarly attained by working at high temperatures [4]. The nuclear plant usually is more expensive to build than a fossil fuel plant, so the electricity costs can only be competitive if relatively inexpensive nuclear fuel is used.

For long-term economic viability, the nuclear fuel must remain dimensionally and chemically stable in a very unusual environment for a long time [5]. The lifetime of the fuel is limited because the fission products created occupy a larger volume than the original fuel material, so the fuel material is damaged during operation. Because 10% of the fission product consists of the noble gases Xe and Kr, these could assemble to form bubbles, which cause an even greater volume change, resulting in damage. Apart

Metal Oxides in Energy Technologies. https://doi.org/10.1016/B978-0-12-811167-3.00004-3

from this complication, the fuel must be used at surface temperatures around 600°C for the highest thermodynamic efficiency, and it should also not react chemically with a gaseous coolant such as CO_2 or a liquid one such as sodium. Considering these and other issues, a nuclear fuel should [6]:

(i) have a high melting point
(ii) have high thermal conductivity
(iii) have a high uranium density
(iv) not contain any highly absorbing species such as boron or gadolinium
(v) be chemically inert toward oxidizing materials and to metal cladding materials

Experience has shown that these requirements are best met by ceramic fuels such as UO_2 [7–11], PuO_2 [12–14], UO_2-PuO_2 [15–17], and (U, Pu)C [18–21]. For several decades now, MOX fuels have been in use in commercial and research reactors, and their development continues.

The MOX fuels are mainly made from reactor-grade plutonium obtained by reprocessing spent fuel. The steeply rising uranium price and the availability of weapons-grade plutonium (a positive outcome of the agreement on the reduction of strategic arms) have contributed to the rising interest in MOX fuels. Conceivably, plutonium from this source will in future be utilized as a MOX fuel in commercial thermal and fast reactors. Recently, ThO_2-UO_2 MOX, ThO_2-PuO_2 MOX, and CeO_2-UO_2 MOX have become available for application as MOX nuclear fuels.

4.2 U-Pu MOX nuclear fuel

4.2.1 Preparation method

4.2.1.1 MIcronized MASter blend method

Currently, the MIcronized MASter blend (MIMAS) process [22] is the most widely used industrial process for the production of MOX fuel. Two distinct steps are executed. First, a mixture consisting of plutonium dioxide, uranium dioxide, and scrap with a total plutonium content below 30% is ground to a fine powder. This 'master blend' is then sifted and diluted with uranium dioxide to provide a secondary mixture with the desired plutonium content. This is then pressed into pellets, which are then sintered and machined to the specified diameter by centerless grinding.

4.2.1.2 Optimized co-milling process (OCOM)

This process was originally intended for the manufacture of fast reactor fuel and produced an extremely fine-grained powder mixture [23]. Sintering of the respective pellets resulted in the formation of a continuous series of solid (U, Pu)O_2 solutions, with the Pu fraction typically corresponding to a content of 25%–30% PuO_2. Blending this solution with UO_2 powder for use in a light-water reactor (LWR) resulted in a fuel with final composition of about UO_2–5%PuO_2, depending on the fissile Pu fraction, which in turn depends on the scrap fraction obtained by fabrication of MOX for fast

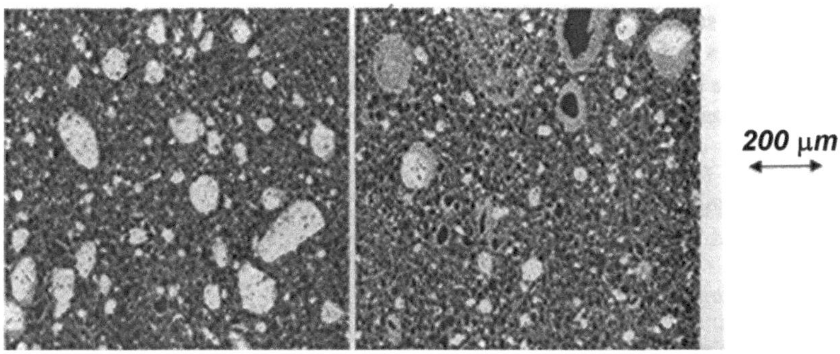

Fig. 4.1 Electron microprobe images of MOX fabricated using two different methods. Reproduced with permission from G. Oudinet, I. Munoz-Viallard, L. Aufore, M.J. Gotta, J.M. Becker, G. Chiarelli, R. Castelli, Characterization of plutonium distribution in MIMAS MOX by image analysis, J. Nucl. Mater. 375 (2008) 86–94; D. Olander, Nuclear fuels–presentand future, J. Nucl. Mater. 389 (2009) 1–22.

reactors. Electron microprobe pictures (Fig. 4.1) of two types of MOX microstructure clearly show the plutonium-rich particles, where most of the nuclear fission takes place, as large white spots.

4.2.2 Performance

UO_2 and PuO_2 both have face centered cubic fluorite (CaF_2) crystal structures and display full solid solubility when the composition ratio of oxygen to metal (O/M) is close to 2. However, for an O/M ratio <2, the composite exists as a two-phase mixture below 723 K and as a single phase at higher temperatures. When PuO_2 is added to UO_2, the thermal conductivity of the resulting MOX fuel decreases slightly with increasing PuO_2 content [24,25], typically with a 4% decrease for 10% Pu/U+Pu. The thermal resistivity is described by an expression of the form

$$1/K = A + BT \tag{4.2}$$

where A is the phonon scattering (which increases with addition of plutonium) and BT is the intrinsic thermal resistivity arising from phonon-phonon interactions (which increases with temperature, T). Laser flash methods are used for thermal conductivity measurements. The conductivity of MOX fuels degrades further with burn up, because of irradiation damage and formation of fission products, a phenomenon also observed with UO_2 [26]. The thermal expansion coefficient of a MOX in an LWR is about 1% higher [27] and the melting point 20°C lower than that of pure UO_2 [28]. The thermal creep, for UO_2 as well as MOX, can be described as the sum of diffusional creep (a linear term proportional to the applied stress) and dislocation climb (which is a term proportional to stress to the power 4.5). The irradiation creep is described as the combination of irradiation-enhanced creep and irradiation-induced creep [29–31].

4.2.3 Recycling

4.2.3.1 Pyrochemical reprocessing

Several processes have been patented. In the American process [6] the oxide fuel is reduced to metal using magnesium as a reductant at ~750°C; the metal is then dissolved in a molten metal such as zinc. The uranium is separated (as U-Zn alloy) from the fission products by fractional crystallization, and the zinc is finally removed by vaporization. In the case of spent MOX, the molten U-Pu alloy may be reacted with molten salts to remove the fission products, and eventually the U-Pu is converted back to oxide or transformed into carbide. A simplified scheme for this process is shown in Fig. 4.2.

4.2.3.2 Electrochemical processing

High-temperature electrochemical fuel processing, or pyroprocessing, is a process whereby the spent nuclear fuel is reprocessed into metal fuel for a fast reactor. In this process the spent oxide fuel from an LWR is treated to produce a metal fuel that subsequently is irradiated in a fast reactor. Recycling the spent fuel will increase the efficiency of uranium use and decrease the radiotoxicity as well as the amount of radioactive waste generated [32–40]. Pyroprocessing consists of head-end processes (consisting of decladding, voloxidation, and oxide feed preparation), electrochemical processes (electrochemical reduction, electrorefining, and electrowinning), and waste treatment processes (Fig. 4.3) [41]. The objective of head-end processing is to convert the spent fuel into a suitable feed material for use in the following electrochemical

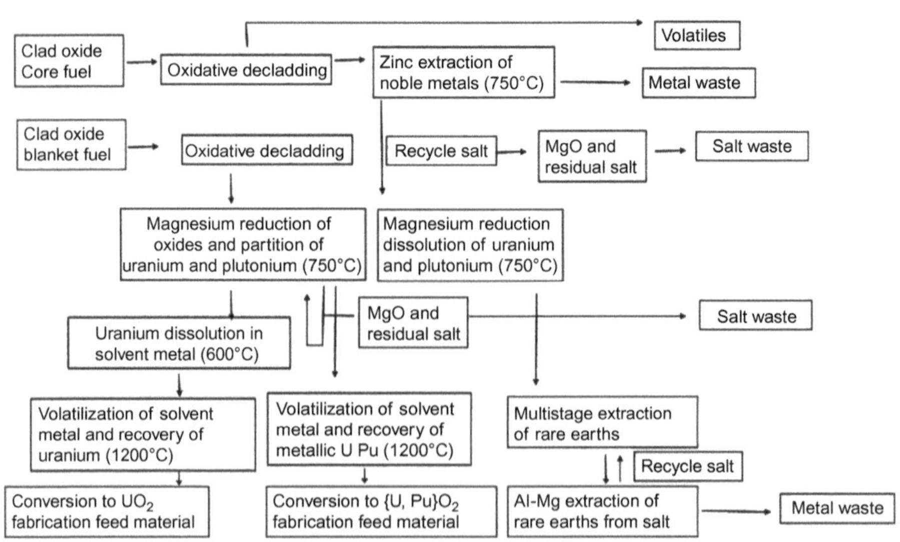

Fig. 4.2 Scheme for pyrochemical reprocessing of oxide and carbide fuels.
Reproduced with permission from B.R.T. Frost, Nuclear fuels, R. Inst. Chem. Rev. 2 (1969) 163–205.

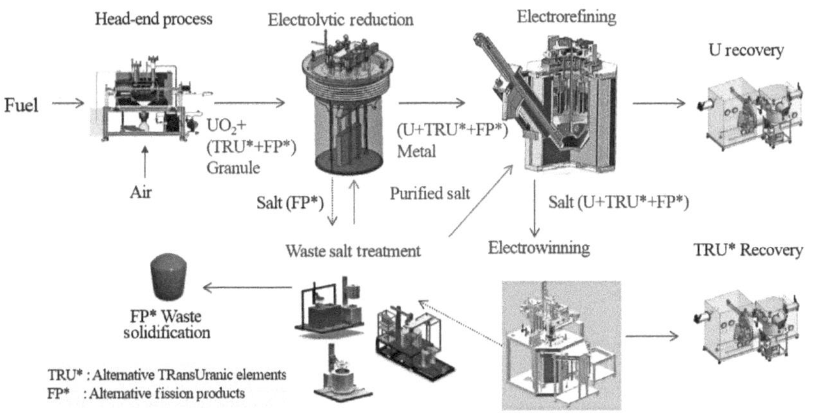

Fig. 4.3 Flow diagram of pyroprocessing.
Reproduced with permission from E.-Y. Choi, S.M. Jeong, Electrochemical processing of spent nuclear fuels: an overview of oxide reduction in pyroprocessing technology, Prog. Nat. Sci.: Mater. Int. 25 (2015) 572–582.

reduction process. After decladding, the voloxidation process oxidizes and pulverizes the oxide pellet obtained from the cladding hull in air at a temperature above 500°C to a powdered form of U_3O_8. The voloxidized U_3O_8 powder is used to produce an oxide feed possessing the right porosity and form required for electrochemical processing. The electrochemical reduction process (also called oxide reduction process), executed in a molten Li_2O-LiCl mixture, reduces the spent oxide fuel to provide a pure metal feed for the electrorefining process. Furthermore, high-heat-load fission products are also dissolved in the molten Li_2O-LiCl mixture, thus decreasing the heat load of the spent fuel.

4.3 U-Th MOX nuclear fuel

4.3.1 Preparation method

4.3.1.1 Impregnation technique

This technique [42] is an attractive alternative to produce MOX fuel pellets based on thoria containing the highly radiotoxic ^{233}U [43]. The procedure for the manufacture of ThO_2–4% UO_2 pellets followed by Bhabha Atomic Research Centre (BARC) involves the following steps (Fig. 4.4) [44]:

1. Fabrication of low density (66% T.D.) ThO_2 pellets in an unshielded facility using the powder route.
2. Impregnation of the pellets with uranyl nitrate solution in vacuum in a shielded facility.
3. Drying and final sintering at 1700°C in a reducing atmosphere.

   The U loading of the ThO_2 pellets can be varied by controlling parameters such as the following:

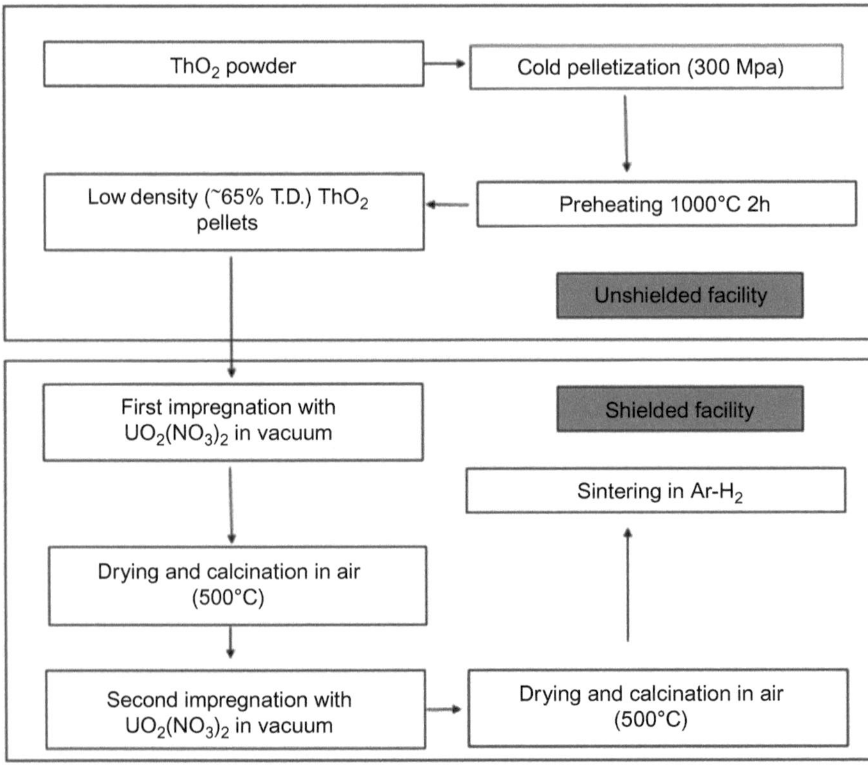

Fig. 4.4 Fabrication of ThO_2-UO_2 pellets by the impregnation process. Reproduced with permission from T.R.G. Kutty, M.R. Nair, P. Sengupta, U. Basak, A. Kumar, H.S. Kamath, Characterization of $(Th,U)O_2$ fuel pellets made by impregnation technique, J. Nucl. Mater. 374 (2008) 9–19.

- **(a)** Density of the presintered ThO_2 pellets.
- **(b)** Concentration of the uranyl nitrate solution.
- **(c)** Period of impregnation.

Fig. 4.5 shows the microstructures of the ThO_2, ThO_2–2% UO_2, and ThO_2–4% UO_2 pellets. The grain sizes of the ThO_2 and ThO_2–4% UO_2 pellets shown in Fig. 4.5A and C are not largely dispersed and fairly uniform. The grains of the ThO_2–2% UO_2 pellets (Fig. 4.5B) are also quite uniform but not as well developed as those of the ThO_2–4% UO_2 pellets. The average grain sizes of the ThO_2 and ThO_2–4% UO_2 pellets are almost the same (12 μm), whereas the ThO_2–2% UO_2 pellets are slightly smaller (10 μm).

4.3.1.2 Sol-gel microsphere pelletization (SGMP) technique

The SGMP technique offers an elegant combination of the advantages of the sol–gel route, which eliminates powder handling, with an increased flexibility to automation by the proven technology of the powder pellet route [45]. The preparation of (Th, U)

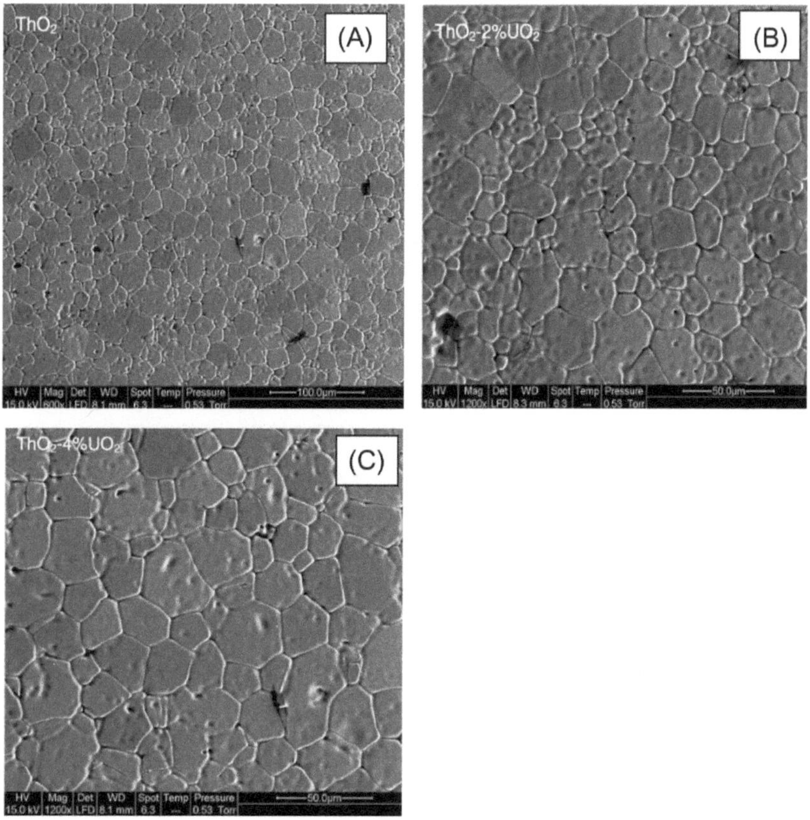

Fig. 4.5 Microstructures of (A) ThO_2, (B) ThO_2–2% UO_2, and (C) ThO_2–4% UO_2 pellets. These pellets were sintered in air and etched thermally.
Reproduced with permission from T.R.G. Kutty, M.R. Nair, P. Sengupta, U. Basak, A. Kumar, H.S. Kamath, Characterization of (Th,U)O_2 fuel pellets made by impregnation technique, J. Nucl. Mater. 374 (2008) 9–19.

O_2 SGMP pellets involves preparation of feed solution, gel microsphere formation, washing and heat treatment of these microspheres, pelletization, and finally sintering to obtain high-density pellets (Fig. 4.6).

The typical microstructure of these pellets has a grain size between 3 and 5 μm, with uniform pores of size <1 μm homogeneously distributed throughout the pellet (Fig. 4.7).

4.3.2 Performance

The trends of the bulk thermal conductivities of UO_2, $(U_{0.95}\,Th_{0.05})O_2$, $(U_{0.90}\,Th_{0.10})O_2$, $(U_{0.75}\,Th_{0.25})O_2$, and $(U_{0.5}\,Th_{0.5})O_2$ versus temperature are shown in Fig. 4.8A. The nonuniform cation sublattice of these oxides results in scattering leading to a reduction

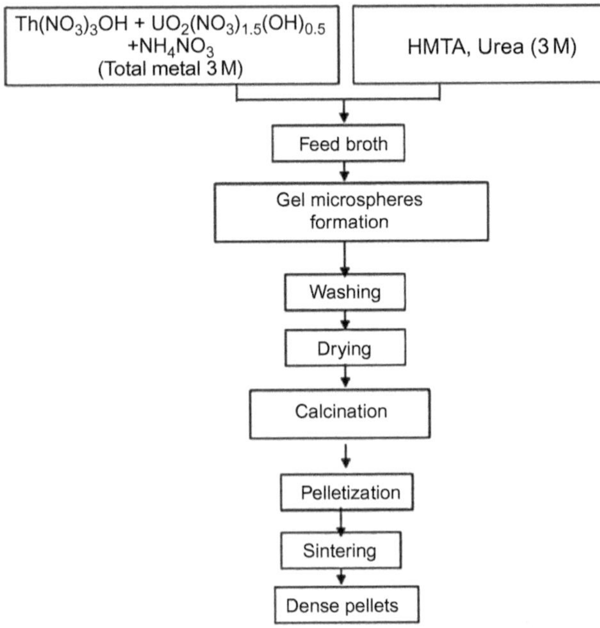

Fig. 4.6 Flow sheet for the preparation of $(Th,U)O_2$ pellets by internal gelation. Reproduced with permission from R.V. Pai, S.K. Mukerjee, V.N. Vaidya, Fabrication of (Th,U) O_2 pellets containing 3 mol% of uranium by gel pelletisation technique, J. Nucl. Mater. 325 (2004) 159–168.

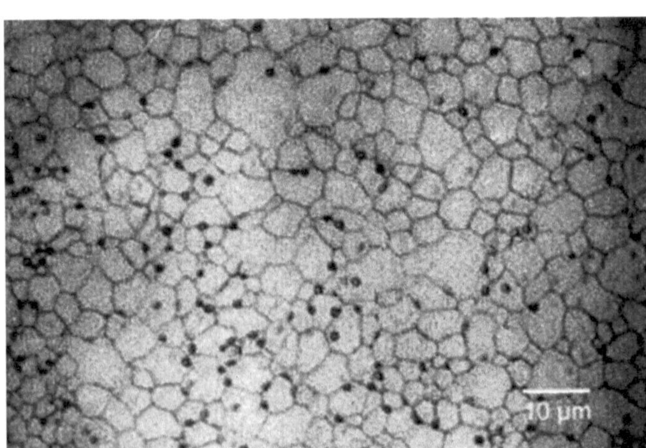

Fig. 4.7 Micrograph of $(Th,U)O_2$ pellet polished and etched in vacuum at 1350°C for 5 h. Reproduced with permission from R.V. Pai, S.K. Mukerjee, V.N. Vaidya, Fabrication of (Th,U) O_2 pellets containing 3 mol% of uranium by gel pelletisation technique, J. Nucl. Mater. 325 (2004) 159–168.

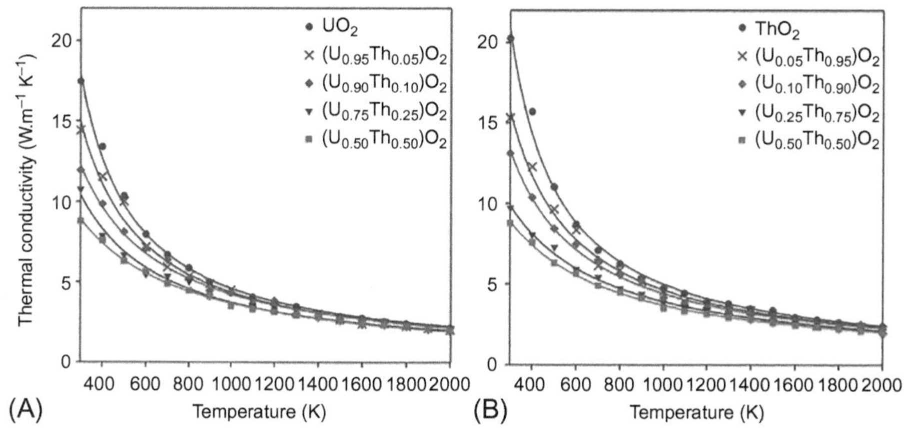

Fig. 4.8 Effects of the addition of (A) Th to UO_2 and (B) U to ThO_2 for a range of $(U_xTh_{1-x})O_2$ compositions, as predicted using the nonequilibrium method. Eq. (4.2) was fitted to the data points from molecular dynamics calculations and the result is shown using *solid lines*. Reproduced with permission from M.W.D. Cooper, S.C. Middleburgh, R.W. Grimes, Modelling the thermal conductivity of $(U_xTh_{1-x})O_2$ and $(U_xPu_{1-x})O_2$, J. Nucl. Mater. 466 (2015) 29–35.

in the phonon mean free path; this causes a drastic decrease in the thermal conductivity, especially at low temperatures [46]. Similarly, although ThO_2 has a higher thermal conductivity than UO_2, there is a large lowering of the thermal conductivity of the MOXs at low temperatures, due to uranium addition by substitution (Fig. 4.8B). But at higher temperatures the proportionate effect of the nonuniform cation sublattice is less pronounced; here, phonon-phonon interactions become the limiting factor in determining the phonon mean free path. Therefore, it is important to consider the designed operating temperature of the fuel. For example, high-temperature reactors, such as the advanced gas-cooled reactor (AGR) or molten salt type, operate at temperatures up to 900–1000 K on the surface of the pellet [47]; here the $(U_{0.5}\ Th_{0.5})O_2$ thermal conductivity is 21% lower than that of UO_2, while the degradation at pressurized water reactor (PWR) operating temperatures is as much as 31%. Likewise, at the peak centerline temperature (which is typically around 1650 K) the conductivity degradation of $(U_{0.5}\ Th_{0.5})O_2$ is about 12% relative to that of pure UO_2.

4.4 Other MOXs

4.4.1 CeO_2-UO_2 MOX nuclear fuel

CeO_2 has been used instead of plutonium oxide to produce a MOX fuel, to avoid several pitfalls encountered when PuO_2 is used directly. The thermal processes involved, and also the material properties of this material, are similar to those of the (U, Pu)O_2 solid solution [48–50].

4.4.1.1 Preparation method

CeO_2 was mixed with UO_2 using a Turbula mixer. The powder mixture was milled with a continuous-type attrition mill with 10–15 passes through the mill, followed by the addition of ZS (zinc stearate, $Zn(C_{17}H_{35}O_2)_2$) and DA (dicarbon amide, $C_2H_4N_2O_2$) or AZB (azodicarbonamide, $C_2H_4N_4O_2$) to the milled powder. A granulated powder was manufactured by several consecutive precompacting, crushing, and sieving steps. The granulated powder was mixed with additional ZS and then pressed into green pellets. These green pellets were sintered using two different sintering processes: process S1 takes place in an atmosphere of $8H_2+92N_2$ at 1873 K for 10 h, whereas process S2 is executed in an oxidizing atmosphere of CO_2 gas at 1723 K for 4 h, followed by 2 h at 1473 K in $8H_2+92N_2$.

4.4.1.2 Performance

Kim et al. [51] fabricated Ce-MOX and Pu-MOX pellets using the same process conditions. Both the Ce-MOX and Pu-MOX pellets made by an oxidative sintering process had the same average grain size of 11 μm, and a density of 95% T.D. When the heat capacity of the Ce-MOX (calculated using Neumann-Kopp's equation) is compared with that of Pu-MOX the heat capacity curve of the Ce-MOX differs only slightly from those of UO_2 and Pu-MOX (Fig. 4.9). The trends in the thermal diffusivities of the Ce-MOX and Pu–MOX samples (calculated from the measured data) differ somewhat (Fig. 4.10). The diffusivity of the Pu-MOX is higher than that of the Ce-MOX at lower temperatures but the trend is reversed above 950 K. The thermal conductivities of the pure

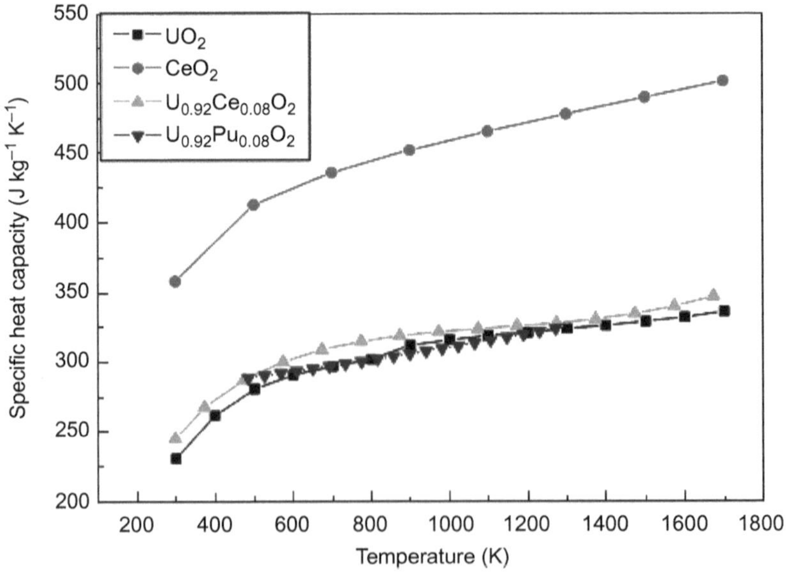

Fig. 4.9 Heat capacity of $U_{0.92}Ce_{0.08}O_2$ and $U_{0.92}Pu_{0.08}O_2$.
Reproduced with permission from H.S. Kim, C.Y. Joung, B.H. Lee, J.Y. Oh, Y.H. Koo, P. Heimgartner, Applicability of CeO_2 as a surrogate for PuO_2 in a MOX fuel development, J. Nucl. Mater. 378. (2008) 98–104.

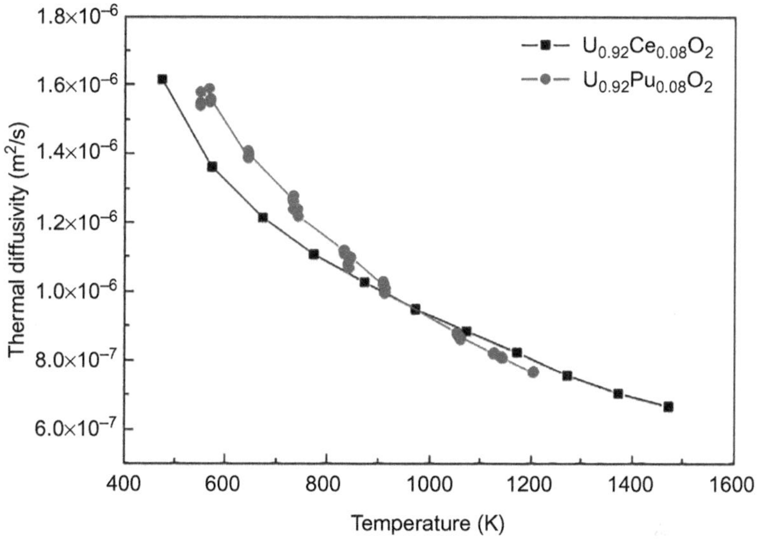

Fig. 4.10 Thermal diffusivities of $U_{0.92}Ce_{0.08}O_2$ and $U_{0.92}Pu_{0.08}O_2$.
Reproduced with permission from H.S. Kim, C.Y. Joung, B.H. Lee, J.Y. Oh, Y.H. Koo, P. Heimgartner, Applicability of CeO_2 as a surrogate for PuO_2 in a MOX fuel development, J. Nucl. Mater. 378. (2008) 98–104.

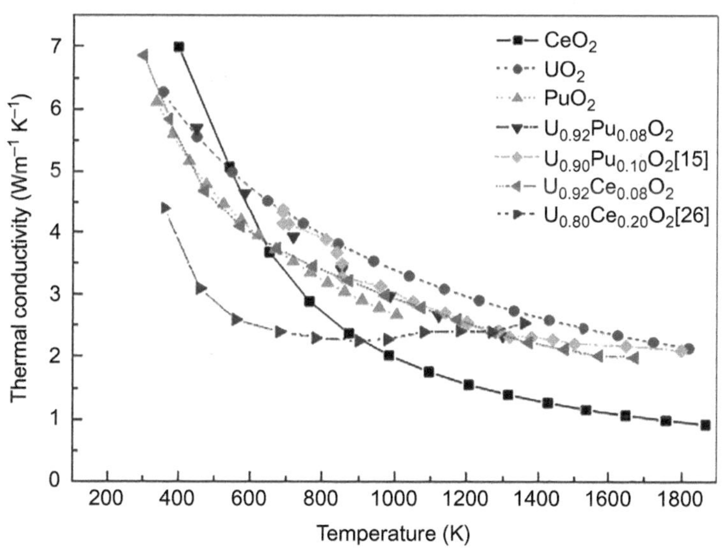

Fig. 4.11 Comparison of the thermal conductivities of pure oxide, Pu-MOX and Ce-MOX. Reproduced with permission from H.S. Kim, C.Y. Joung, B.H. Lee, J.Y. Oh, Y.H. Koo, P. Heimgartner, Applicability of CeO_2 as a surrogate for PuO_2 in a MOX fuel development, J. Nucl. Mater. 378. (2008) 98–104.

oxide and MOX samples also show different trends (Fig. 4.11). Above 600 K the conductivities of both $U_{0.92}Ce_{0.08}O_2$ and $U_{0.90}Pu_{0.10}O_2$ tend to lie between those of pure UO_2 and CeO_2/PuO_2. The conductivity of $U_{0.92}Pu_{0.08}O_2$ is a little higher than that of $U_{0.92}Ce_{0.08}O_2$ at lower temperatures, but the two curves tend to overlap above 900 K. Thus, the thermal properties of Pu-MOX and Ce-MOX vary with the composition and the temperature, resulting in different trends.

4.4.2 ThO_2-PuO_2 MOX nuclear fuel

Thorium mixed oxide (Th-MOX) was introduced as a prospective fuel blended with plutonium oxide for fast and thermal reactors in the 1960s. The PWRs can use (Th, Pu) O_2 fuel without changing the reactor design. Using thorium instead of uranium in future reactor designs, including LWRs, contributes to the trend against nuclear proliferation [52]. The thorium(IV) in thorium dioxide (ThO_2) has its highest known oxidation state and therefore it produces more stable and less high-level waste than UO_2. Furthermore, the abundance of thorium in the Earth's crust is higher than uranium [53,54] and it has several favorable properties boosting its use as a nuclear fuel; for example, ThO_2 has a thermal conductivity higher than that of uranium dioxide [55], and it has the highest melting point of all oxides. It also operates at lower temperatures and more of the fission products, notably gaseous ones, are retained inside the fuel material. With thorium-based fuels a burn up higher than that of uranium-based fuels is attainable. These considerations have inspired new concept designs for advanced thorium/plutonium-fuel cycles; these advanced reactor designs include LWRs, high-temperature gas-cooled reactors (GCRs), and advanced heavy water reactors.

4.4.2.1 Performance

Cozzo et al. [56] determined the thermal diffusivity of ThO_2, PuO_2, and $(Th,Pu)O_2$ with 3, 8, and 30 wt% PuO_2 added, respectively, to calculate their thermal conductivity. At low temperatures, the thermal conductivity of Th-MOX with a PuO_2 content from 0 to 30 wt% decreases with increasing PuO_2 content, similar to the decrease in the thermal diffusivity (Fig. 4.12). At temperatures above 900 K the thermal conductivity of Th-MOX containing 0–8 wt% PuO_2 is nearly independent of the plutonium content; this cannot be attributed to the temperature dependence [57] of the porosity correction factor because its variation was negligible [56]. The high-temperature conductivity of Th-MOX containing 30 wt% PuO_2 is the lowest, while the conductivity of pure PuO_2 is higher than that of the other materials investigated. The thermal diffusivities of samples with 3 and 8 wt% PuO_2 added, respectively, are nearly the same and are slightly lower than that of pure ThO_2 (Fig. 4.13), while the thermal diffusivity of the sample containing 30 wt% Pu (OMICO30) is much lower than that of ThO_2.

4.4.2.2 Recycling

In terms of plutonium and actinide management the Th-MOX looks very promising. Th-232 is a productive radionuclide which is transmuted in a neutron flux to the thermally fissile U-233 through capture reactions, similar to the formation of Pu-239 from

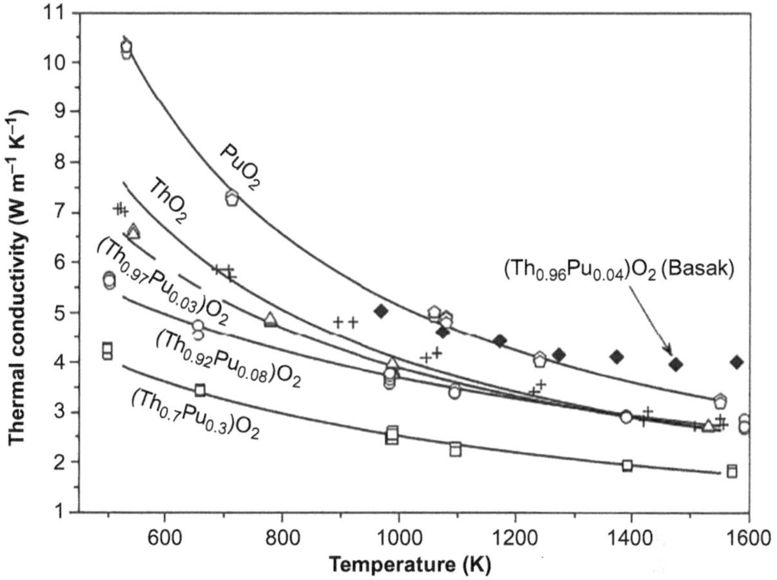

Fig. 4.12 Thermal conductivity of Th-MOX: measured values for 95% of the theoretical density.
Reproduced with permission from H.S. Kim, C.Y. Joung, B.H. Lee, J.Y. Oh, Y.H. Koo, P. Heimgartner, Applicability of CeO_2 as a surrogate for PuO_2 in a MOX fuel development, J. Nucl. Mater. 378. (2008) 98–104.

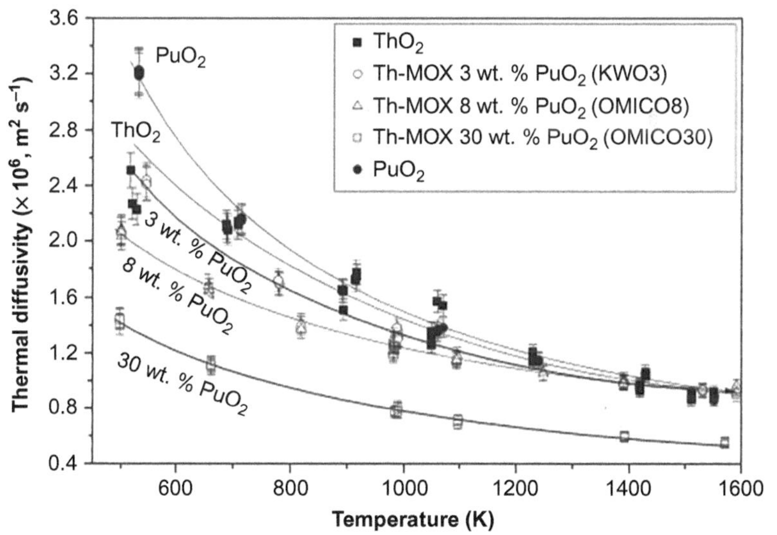

Fig. 4.13 Th-MOX fuels: thermal diffusivity of the 0, 3, 8, and 30 wt% PuO_2 samples for 95% of the theoretical density.
Reproduced with permission from C. Cozzo, D. Staicu, J. Somers, A. Fernandez, R.J.M. Konings, Thermal diffusivity and conductivity of thorium–plutonium mixed oxides, J. Nucl. Mater. 416 (2011) 135–141.

U-238 in a standard matrix. Loading of Th-Pu MOX fuel in a PWR, therefore, will not differ significantly in terms of in-core behavior, from a full U-Pu MOX loading, with the notable exception that much less plutonium and minor actinides is produced. A PWR loaded with Pu MOX fuel could achieve plutonium consumption rates of up to 115 kg/TWh.

4.5 Outlook

With the advent of commercially generated nuclear energy in the 1960s the consensus seems to have been that uranium was a limited resource. Therefore, the need arose to recover, reprocess, and recycle the plutonium produced in the course of nuclear power generation, preferably in fast reactors. From the initial work on the production of plutonium-bearing fuels it soon became clear that uranium-plutonium MOXs could be used sustainably in fast reactors. The term "mixed oxide" generally comprises both thermal and fast reactor fuels. It has now been shown that ThO_2-UO_2 MOX, ThO_2-PuO_2 MOX, and CeO_2-UO_2 MOX can also be used as MOX nuclear fuels. The fourth generation (Gen IV) nuclear plant will probably be the very high-temperature reactor (VHTR), which will be fueled by tiny tristructural-isotropic (TRISO) fuel particles embedded in a graphite block matrix and cooled by helium. Examples of this kind of reactor are the pebble-bed reactor (PBR) and the prismatic-block GCR. This kind of reactor is expected to deliver outlet coolant at temperatures approaching 1000°C, which would make thermochemical hydrogen production or electrolysis of water economical. The UREX-1a reprocessing suite of four distinct solvent-extraction steps is being developed to fulfill the nonproliferation requirements by separating key radionuclides from spent fuel, and to avoid approaching limiting temperatures and site-boundary dose rates in high-level waste repositories. However, the efficacy of UREX-1a must still be demonstrated. Lastly, the sodium-cooled fast reactor (SFR) is envisaged to operate together with one or more LWRs in a fuel cycle that accepts only fresh uranium and delivers only fission products and electricity. The SFR will recycle the minor actinides and plutonium until they are converted fully to fission products. The separation process, called pyroprocessing, makes use of an electro-refiner, which can recover all the constituents of spent fuel, whether from the SFR, the LWR, or other reactors [58]. The separation factors, however, still leave much to be desired.

Acknowledgment

Financial support from the National Materials Genome Project (2016YFB0700600), China National Fund for Distinguished Youth Scientists (51026004), and NSFC (U1601214, 51502137 and 21603103) is gratefully acknowledged.

References

[1] J.P. Zhang, Z.R. Liu, L. Wang, Uranium demand and economic analysis of different nuclear fuel cycles in China, Energ. Strat. Rev. 9 (2016) 50–61.

[2] P.F. Venneri, Y. Kim, A feasibility study on low enriched uranium fuel for nuclear thermal rockets – II: rocket and reactor performance, Prog. Nucl. Energy 87 (2016) 156–167.

[3] Y.S. Kim, H.Y. Kim, J.Y. Bak, J.J. Jeong, J.Y. Jeon, B.J. Yun, Evaluation of thermal hydraulic safety of a nuclear fuel assembly in a mast assembly of nuclear power plant, Ann. Nucl. Energy 92 (2016) 136–149.

[4] J.D. Hales, M.R. Tonks, K. Chockalingam, D.M. Perez, S.R. Novascone, B.W. Spencer, R. L. Williamson, Asymptotic expansion homogenization for multiscale nuclear fuel analysis, Comput. Mater. Sci. 99 (2015) 290–297.

[5] J.M. Gatt, J. Sercombe, I. Aubrun, J.C. Ménard, Experimental and numerical study of fracture mechanisms in UO_2 nuclear fuel, Eng. Fail. Anal. 47 ((2015) 299–311.

[6] B.R.T. Frost, Nuclear fuels, R. Inst. Chem. Rev. 2 (1969) 163–205.

[7] A. Casella, B. Hanson, W. Miller, The effect of fuel chemistry on UO_2 dissolution, J. Nucl. Mater. 476 (2016) 45–55.

[8] Z.C. Chen, G. Subhash, J.S. Tulenko, Raman spectroscopic investigation of graphitization of diamond during spark plasma sintering of UO_2-diamond composite nuclear fuel, J. Nucl. Mater. 475 (2016) 1–5.

[9] H.X. Xiao, Modeling of fission gas swelling in LWR UO_2 fuel under irradiation, Prog. Nucl. Energy 90 (2016) 122–126.

[10] H.X. Xiao, C.S. Long, H.S. Chen, Model for evolution of grain size in the rim region of high burn up UO_2 fuel, J. Nucl. Mater. 471 (2016) 74–79.

[11] H. Wang, J.H. Liu, Y.D. Sun, X.F. Hong, H.W. Huang, R.C. Yin, Y. Zhang, C.G. Yin, Research on preparation technology, sintering performance, microstructure of UO_{2-x} fuel pellets, J. Eur. Ceram. Soc. 35 (2015) 1081–1087.

[12] B.T. Wang, J.J. Zheng, X.T. Qu, W.D. Li, P. Zhang, Thermal conductivity of UO_2 and PuO_2 from first-principles, J. Alloys Compd. 628 (2015) 267–271.

[13] B.Y. Ao, R.Z. Qiu, H.Y. Lu, P.H. Chen, First-principles DFT + U calculations on the energetics of Ga in Pu, Pu_2O_3 and PuO_2, Comput. Mater. Sci. 122 (2016) 263–271.

[14] S.A. Moten, R.A. Fynn, A.K. Ray, M.N. Huda, Size effects on the electronic and magnetic properties of PuO_2 (111) surface, J. Nucl. Mater. 468 (2016) 37–45.

[15] I. Shaaban, M. Albarhoum, Neutronic calculations of the MTR-22 MW research reactor loaded with MOX (PuO_2&UO_2) caramel fuel using the MCNP5 code, Ann. Nucl. Energy 94 (2016) 478–486.

[16] R.C. Belin, M. Strach, T. Truphemus, C. Gueneau, J.C. Richaud, J. Rogez, In situ high temperature X-ray diffraction study of the phase equilibria in the UO_2-PuO_2-Pu_2O_3 system, J. Nucl. Mater. 465 (2015) 407–417.

[17] V.P. Sinha, P.V. Hegde, G.J. Prasad, S. Pal, G.P. Mishra, Development of UO_2/PuO_2 dispersed in uranium matrix CERMET fuel system for fast reactors, J. Nucl. Mater. 427 (2012) 12–17.

[18] F. Mazaudier, C. Tamani, A. Galerie, Y. Marc, On the oxidation of (U, Pu)C fuel: experimental and kinetic aspects, practical issues, J. Nucl. Mater. 406 (2010) 277–284.

[19] F. Le Guyadec, C. Rado, S. Joffre, S. Coullomb, C. Chatillon, E. Blanquet, Thermodynamic and experimental study of UC powders ignition, J. Nucl. Mater. 393 (2009) 333–342.

[20] H.P. Nawada, P. Srirama Murti, G. Seenivasan, S. Anthonysamy, C.K. Mathews, Thermogravimetric study of the oxidation behaviour of uranium dicarbide, J. Therm. Anal. 35 (1989) 1145–1155.

[21] S.K. Mukerjee, G.A. Rama Rao, J.V. Dehadraya, V.N. Vaidya, V. Venugopal, The oxidation of uranium monocarbide microspheres, J. Nucl. Mater. 210 (1994) 97–106.

[22] G. Oudinet, I.M. Viallard, L. Aufore, M.J. Gotta, J.M. Becker, G. Chiarelli, R. Castelli, Characterization of plutonium distribution in MIMAS MOX by image analysis, J. Nucl. Mater. 375 (2008) 86–94.

[23] H. Kleykamp, Post-irradiation studies on LWR-MOX fuel fabricated by the optimized co-milling process, J. Nucl. Mater. 324 (2004) 198–202.

[24] M. Lung, A Present Review of the Thorium Nuclear Fuel Cycles, Nuclear Science and Technology, European Commission, 1997. EUR 17771.

[25] K. Anantharaman, V. Shivakumar, D. Saha, Utilisation of thorium in reactors, J. Nucl. Mater. 383 (2008) 119–121.

[26] S. Bremier, D. Haas, J. Somers, C.T. Walker, Large area quantitative X-ray mapping of (U, Pu)O_2 nuclear fuel pellets using wavelength dispersive electron probe microanalysis, Spectrochim. Acta B 58 (2003) 651–658.

[27] D. Haas, MOX fuel fabrication experience at Belgonucleaire, Use of Pu in LWRs, ENS TOPNUX93, The Hague, 1993.

[28] H.S. Kamath, D.S.C. Purushotham, Mixed oxide (MOX) fuels, in: Reference Module in Materials Science and Materials Engineering – Encyclopedia of Materials: Science and Technology, second ed., Elsevier, New York, 2001, pp. 5687–5689.

[29] M. Strach, D. Manara, R.C. Belin, J. Rogez, Melting behavior of mixed U–Pu oxides under oxidizing conditions, Nucl. Instrum. Meth. B 374 (2016) 125–128.

[30] M. Kato, Y. Ikusawa, T. Sunaoshi, A.T. Nelson, K.J. McClellan, Thermal expansion measurement of (U, Pu)O_{2-x} in oxygen partial pressure-controlled atmosphere, J. Nucl. Mater. 469 (2016) 223–227.

[31] A. Inoue, Invalidity of constructing U–Pu–O ternary phase diagram on the basis of the analysis of air-oxidation products of (U, Pu)O_2, J. Alloys Compd. 468 (2009) L25–L26.

[32] J.J. Laidler, J.E. Battles, W.E. Miller, J.P. Ackerman, E.L. Carls, Development of pyroprocessing technology, Prog. Nucl. Energy 31 (1997) 131–140.

[33] J.P. Ackerman, T.R. Johnson, L.S.H. Chow, E.L. Carls, W.H. Hannum, J.J. Laidler, Treatment of wastes in the IFR fuel cycle, Prog. Nucl. Energy 31 (1997) 141–154.

[34] R.W. Benedict, H.F. McFarlane, EBR-II spent fuel treatment demonstration project status, Radioact. Waste Manage. 5 (1998) 23–30.

[35] T. Inoue, L. Koch, Development of pyroprocessing and its future direction, Nucl. Eng. Technol. 40 (2008) 183–190.

[36] T. Inoue, T. Koyama, Y. Arai, State of the art of pyroprocessing technology in Japan, Energy Procedia 7 (2011) 405–413.

[37] K.M. Goff, J.C. Wass, K.C. Marsden, G.M. Teske, Electrochemical processing of used nuclear fuel, Nucl. Eng. Technol. 43 (2011) 335–342.

[38] K. Nagarajan, B.P. Reddy, S. Ghosh, G. Ravisankar, K.S. Mohandas, U.K. Mudali, K.V. G. Kutty, K.V.K. Viswanathan, C.A. Babu, P. Kalyanasundaram, P.R.V. Rao, B. Raj, Development of pyrochemical reprocessing for spent metal fuels, Energy Procedia 7 (2011) 431–436.

[39] T. Koyama, Y. Sakamura, M. Iizuka, T. Kato, T. Murakami, J.P. Glatz, Development of pyro-processing fuel cycle technology for closing actinide cycle, Procedia Chem. 7 (2012) 772–778.

[40] H. Lee, G.I. Park, J.W. Lee, K.H. Kang, J.M. Hur, J.G. Kim, S. Paek, I.T. Kim, I.J. Cho, Current status of pyroprocessing development at KAERI, Sci. Technol. Nucl. Install. 2013 (2013) 343492–343503.

[41] E.-Y. Choi, S.M. Jeong, Electrochemical processing of spent nuclear fuels: an overview of oxide reduction in pyroprocessing technology, Prog. Nat. Sci.: Mater. Int. 25 (2015) 572–582.
[42] T.R.G. Kutty, M.R. Nair, P. Sengupta, U. Basak, A. Kumar, H.S. Kamath, Characterization of $(Th,U)O_2$ fuel pellets made by impregnation technique, J. Nucl. Mater. 374 (2008) 9–19.
[43] Y. Croixmarie, E. Abonneau, A. Fernandez, R.J.M. Konings, F. Desmouliere, L. Donnet, Fabrication of transmutation fuels and targets: the ECRIX and CAMIX-COCHIX experience, J. Nucl. Mater. 320 (2003) 11–17.
[44] A. Feraday Melville, Preparation of Mixed Oxide Nuclear Fuel, United States Patent 40201311975, 1975.
[45] R.V. Pai, S.K. Mukerjee, V.N. Vaidya, Fabrication of $(Th,U)O_2$ pellets containing 3 mol% of uranium by gel pelletisation technique, J. Nucl. Mater. 325 (2004) 159–168.
[46] M.W.D. Cooper, S.C. Middleburgh, R.W. Grimes, Modelling the thermal conductivity of $(U_xTh_{1-x})O_2$ and $(U_xPu_{1-x})O_2$, J. Nucl. Mater. 466 (2015) 29–35.
[47] IEE, Nuclear Reactor Types, The Institution of Electrical Engineers, London, 2005. Technical report.
[48] J.J. Carabajo, G.L. Yoder, S.G. Popov, V.K. Ivanov, A review of the thermophysical properties of MOX and UO_2 fuels, J. Nucl. Mater. 299 (2001) 181–198.
[49] R.J. White, S.B. Fisher, P.M.A. Cook, R. Stratton, C.T. Walker, I.D. Palmer, Measurement and analysis of fission gas release from BNFL's SBR MOX fuel, J. Nucl. Mater. 288 (2001) 43–56.
[50] T.L. Markin, R.S. Street, E.C. Crouch, The uranium-cerium-oxygen ternary phase diagram, J. Inorg. Nucl. Chem. 32 (1970) 59–75.
[51] H.S. Kim, C.Y. Joung, B.H. Lee, J.Y. Oh, Y.H. Koo, P. Heimgartner, Applicability of CeO_2 as a surrogate for PuO_2 in a MOX fuel development, J. Nucl. Mater. 378 (2008) 98–104.
[52] K.I. Björk, V. Fhager, C. Demaziere, Comparison of thorium-based fuels with different fissile components in existing boiling water reactors, Prog. Nucl. Energy 53 (2010) 618–625.
[53] K. Anantharamanand, V. Shivakumar, D. Saha, Utilisation of thorium in reactors, J. Nucl. Mater. 383 (2008) 119–121.
[54] D.R. Lide, Handbook of Chemistry and Physics, 86th ed., CRC Press Inc., Boca Raton, FL, 2005.
[55] J. Belle, R.M. Berman, Thorium Dioxide: Properties and Nuclear Applications, Technical report DOE/NE-0060, Naval Reactors Office, United States, Department of Energy, 1984.
[56] C. Cozzo, D. Staicu, J. Somers, A. Fernandez, R.J.M. Konings, Thermal diffusivity and conductivity of thorium–plutonium mixed oxides, J. Nucl. Mater. 416 (2011) 135–141.
[57] R. Brandt, G. Neuer, Utilisation of thorium in reactors, J. Non-Equilib. Thermodyn. 1 (1976) 3–23.
[58] D. Olander, Nuclear fuels–present and future, J. Nucl. Mater. 389 (2009) 1–22.

Further reading

[59] G. Oudinet, I. Munoz-Viallard, L. Aufore, M.J. Gotta, J.M. Becker, G. Chiarelli, R. Castelli, Characterization of plutonium distribution in MIMAS MOX by image analysis, J. Nucl. Mater. 375 (2008) 86–94.

Piezoelectric energy harvesting systems with metal oxides

5

Kenji Uchino
International Center for Actuators and Transducers, The Pennsylvania State University, University Park, PA, United States

5.1 Background

5.1.1 Necessity of piezoelectric energy harvesting

21st century is called "The Century of Environmental Management." Though most of the major countries rely on fossil fuel and nuclear power plants at present, the renewable energy development becomes important for complementing the energy deficiency. Energy recovery from wasted or unused power has been the topic of discussion for a long period. Unused power exists in various electromagnetic and mechanical forms such as ambient electromagnetic noise around high power cables, noise vibrations, water flow, wind, human motion, and shock waves. In recent years, industrial and academic research units have focused their attention on harvesting energy from vibrations using piezoelectric transducers. These efforts have provided the initial research guidelines and have brought light to the problems and limitations of implementing the piezoelectric transducer [1].

5.1.2 From passive damping to energy harvesting

Historically, the author's group started the research on passive vibration damping using piezoelectric materials in the 1980s. Fig. 5.1A and B shows the results for damping vibration generated in a bimorph transducer [2]. Resistive shunt was used with the bimorph, and the vibration damping was measured by changing the external resistance. As shown in Fig. 5.1B, the quickest damping was observed with a 6.6-kΩ resistor, which is almost the same value as the electrical impedance of the bimorph (i.e., $R = 1/\omega C$). In these results, mechanical vibration energy was converted to electric energy, which was dissipated through the resistor as Joule heat, leading to the quickest damping effectively. In addition to the resistive shunt, capacitive, inductive, and switch shunts have been successively studied, aiming at adaptive vibration control after our studies [1].

However, after getting into the late 1990s, we decided to save and store this generated electrical energy into a rechargeable battery, instead of dissipating it as the Joule heat. This research target change is schematically illustrated in Fig. 5.2, which fits to the renewable energy boom in good timing. Note that the maximum energy harvesting condition corresponds to the largest vibration damping; "If you run after

Metal Oxides in Energy Technologies. https://doi.org/10.1016/B978-0-12-811167-3.00005-5

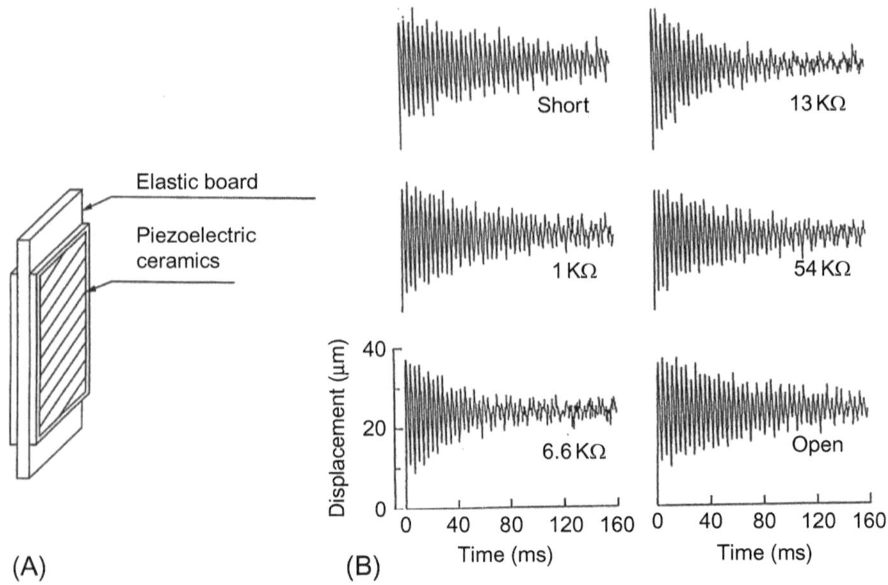

Fig. 5.1 Vibration damping change associated with external resistance change. (A) Bimorph transducer for this measurement. (B) Damped vibration with external resistor [1].

Fig. 5.2 Difference between piezoelectric passive vibration damping and piezoelectric energy harvesting.

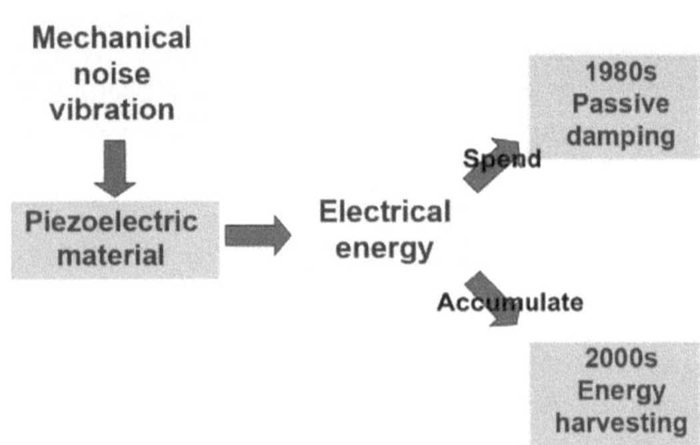

two hares (vibration suppression and energy harvesting), you will catch BOTH," rather than "you will catch NEITHER (the original proverb)," which seems to be the best development strategy! The detailed development process is described in Section 5.1.4.

5.1.3 Recent research trends

The worldwide annual revenue of piezoelectric devices reached \$22 billion in 2012, and will exceed \$37 billion by the end of 2017 [2]. With respect to the marketing category, a category of "actuators & piezo-generators" increases from \$7.2 B (2012) to \$13.7 B (2017), and in particular, the "ecology and energy harvesting" share increased dramatically from almost zero to 7% of the total revenue during these years.

We summarize in this section the research trends primarily after the 2000s, by referencing a book Ref. [3]. First, as one of the pioneers in the piezoelectric energy harvesting, I feel a sort of frustration in 90% of the recent research papers from the following points:

(1) Though the electromechanical coupling factor k is the smallest (i.e., the energy conversion rate from the input mechanical to electric energy is the lowest) among various device configurations, the majority of researchers primarily use the "unimorph" design. Why?
(2) Though the typical noise vibration is in a much lower frequency range, the researchers measure the amplified resonance response (even at a frequency higher than 1 kHz) and report this unrealistically harvested electric energy. Why?
(3) Though the harvested energy is <1 mW, which is less than the required electric energy to operate a typical energy harvesting electric circuit with a DC/DC converter (typically around 2–3 mW), the researchers report the result as an energy "harvesting" system. Does this situation mean actually energy "losing"? Why?
(4) Few papers have reported complete energy flow or exact efficiency from the input mechanical noise energy to the final electric energy in a rechargeable battery via the piezoelectric transducer step by step. Why?

Interestingly, the unanious answer from these researchers to my questions "why" is "because the previous researchers did so."

5.1.3.1 Mechanical engineers' approach

Wallaschek et al. [4] and Smithmaitrie [5] reported comprehensive studies on the bimorph transducer design optimization, using a mechanical equivalent system with mass, spring, and dashpot in the former, and using a membrane/beam theory in the latter, including various references studied previously. The former indicated that the effective electromechanical coupling factor k can be increased by reducing the thickness ratio (PZT thickness/metal shim thickness); while in the latter paper, the voltage generation in the piezoelectric membrane shows the maximum around the resonance frequency (50 Hz). One of the significant problems of the mechanical engineers is the neglect of the piezoelectric ceramics PZT's loss and performance limitation. As we will discuss in Section 5.3, the maximum handling energy level in the current PZT's is only 10–30 W/cm^3. Thus, reducing the PZT thickness in order to increase the electromechanical coupling factor k dramatically reduces the handling energy level (note that the energy conversion amount is given by the product of input energy and k^2). We should understand that though we try to increase the handling electric energy level with the mechanical vibration input, additional input energy will convert to just heat generation due to the hysteresis; that is, the PZT become a ceramic heater. Knowing the material's limitation is primary essential prior to extending the theoretical mechanical analysis.

Machinery vibration (resonance usage)

Beeby and Zhu discussed commercially available energy harvesting products in their article [6]. Products by Midé Technology (US), Adaptivenergy (US), Arveni (France), Advanced Cerametrics (US), and Prepetuum (UK) were compared. These products are based on relatively large piezo-bimorph/unimorph designs (40–200 cm^3) in order to set the resonance frequency around 50 Hz with a narrow bandwidth of $<$10 Hz. Power of 1–20 mW is obtained under the acceleration 0.5–2 G. Most of them can be used to operate sensors of various machinery and health-monitoring systems (the machine should be vibrating constantly at 50 Hz) and to transmit the wireless signals of the monitored data. This energy level is not sufficient for general energy storage applications.

Arms et al. [7] developed a smart system powered by piezoelectrics in a helicopter. The integrated structural health monitoring and reporting (SHMR) includes strain gauges, accelerometers, and load/torque cells. Data were stored in a central location. Wireless sensors were capable of logging 50,000 sample/s consuming 9 mA and 3 V DC (27 mW). The energy was supplied by piezoelectric micro-generators capable of harvesting vibration energy from the helicopter gearbox. Micro-generators (dimensions similar to multiple American quarter coins) working under a constant blade rotation condition can generate 37 mW, more than enough to feed the sensors.

Human motion

NEC-Tokin Corporation, Japan has already commercialized two piezoelectric energy harvesting products in the early 2000s by utilizing an impact-based mechanism [8]. First, finger snapping action on a piezo-bimorph generates the electric energy to illuminate LED lamps equipped at the front of a key holder, so that the user can easily find the key hole on the house or car doors. Second, "piezoelectric windmills" were produced and aligned in the northern part of Japan along the major highway roadsides in tunnels, so that the wind generated by running vehicles or humans rotates the windmills, then the rotation of mills generates the steel ball motion as shown in Fig. 5.3A, which hits the piezo-bimorphs, leading to the electric energy generation for operating an LED traffic light array system which can navigate the vehicle drivers smoothly along the road curb (Fig. 5.3B).

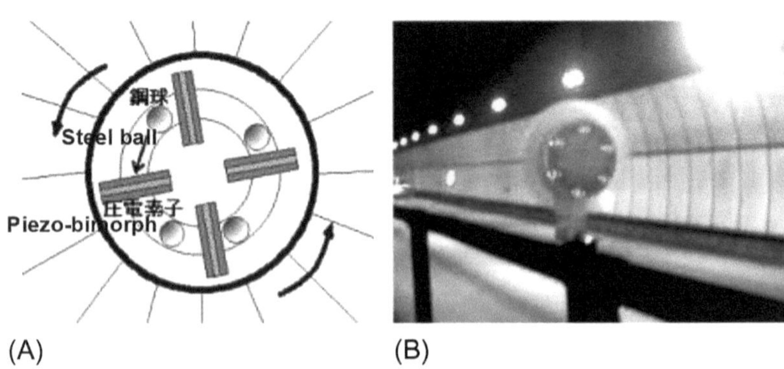

Fig. 5.3 Piezoelectric windmill structure (A) for driving an LED traffic light array system (B). Courtesy by NEC-Tokin.

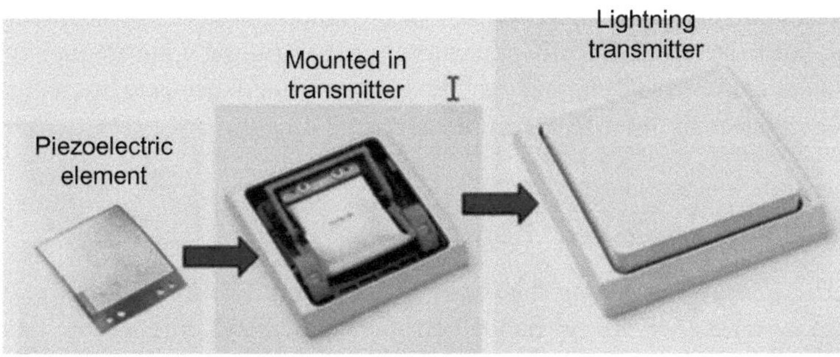

Fig. 5.4 Lightning switch with piezoelectric thunder actuator. Courtesy by Face Electronics.

Another successful product (million sellers) is "Lightning Switch" commercialized by Face Electronics (PulseSwitch Systems), VA, which is a remote switch for room lights, using a unimorph ("Thunder") piezoelectric component. In addition to the living convenience, Lightning Switch (Fig. 5.4) can reduce the house construction cost dramatically, due to a significant reduction in the copper electric wire and the wire aligning labor [9]. Harvesting energy from shoes was also reported using the Thunder component by Face Electronics [10].

Renaud et al. [11] reported a similar impact-based energy harvester, which consists of two piezo-bimorphs and a movable elastic rod. The total volume of the device 25 cm^3 could generate an output power of 60 μW under a motion of 10 Hz with 10 cm linear motion amplitude (which seems to be too quick for a human action).

Since the human action is very slow and rather random, the resonance-type devices are not suitable. Thus, there seems to be two choices: (1) usage of impact-based snap action, and (2) off-resonance, but carefully matching the acoustic/mechanical impedance with soft human tissue and soft piezoelectric devices.

5.1.3.2 Electrical engineers' approach

Because the electric energy obtained by the piezoelectric energy harvesting system exhibits high voltage/low current (high impedance), we need to convert the electric impedance to sufficiently low level ($\sim 50\,\Omega$) in order to charge up a rechargeable battery smoothly or drive an ohmic typical portable device.

Guyomar et al. [12] reviewed studies on "switch" shunt circuits combined with the cyclic piezoelectric energy harvesting systems. They compared initially two cases: the switching device is connected in parallel with the piezoelectric element (parallel synchronized switch harvesting on inductor, parallel SSHI) or in series between the active material and harvesting stage (series SSHI). The discussion was held only on a monochromatic excitation, which narrows the application of this technique.

As mentioned in Section 5.4.1, because of the randomness of the mechanical noise frequency, our strategy is basically to accumulate the original wide-frequency electric energy in a capacitor as DC charge, then the electric impedance is to be changed using a DC/DC converter in order to facilitate to charge a battery.

5.1.3.3 MEMS engineers' approach

Piezoelectric energy harvesting has been ignited significantly in the microelectromechanical systems (MEMS) or nanoelectromechanical systems (NEMS) area, with a new terminology "nano-harvesting." Lopes and Kholkin's paper [13] is a good review to be referred. Designs for the piezoelectric micro-harvesting are currently limited only to the unimorph types: (1) one-end clamp cantilever, and (2) both-end clamp membrane. Because of the combination of this low electromechanical coupling k design and very thin (low volume) piezoelectric film, it is difficult to expect high power or efficiency theoretically. Harvesting energy of 1–10 μW is reported at a resonance frequency higher than 1 kHz, which is 3 orders of magnitude smaller than the practically usable level (10 mW).

A scientifically interesting topic is the usage of nano piezoelectric fiber (i.e., single crystal). Chang et al. [14] reported direct-write, piezoelectric polymeric nanogenerators based on organic nonfibers with higher energy conversion efficiency. These nanofibers are made of poly-vinylidene di-fluoride (PVDF) with high flexibility, minimizing resistance to external mechanical motion at low frequency. The PVDF exhibits reasonable piezoelectric and soft mechanical properties, suitable to the medical applications for realizing mechanical impedance matching with soft human tissues. Chang et al. utilized electro-spinning process with strong electric field ($>10^7$ V/m) and stretching forces from the naturally aligned dipoles in the nanofiber crystal such that the nonpolar R phase is transformed into polar β phase, determining the polarity of the electrospun nanofiber. They reported the energy conversion rate of 12% as average, which corresponds to the electromechanical factor $k=35\%$, which is 3 times higher than the commercial PVDF films.

Wang et al. [15] found another nanowire application in ZnO. High energy can be produced by making an array of ZnO flexible nanowires where each individual wire can produce electricity. Further improvement was achieved by using a zigzag configuration on the top electrode surface. Though they reported the energy harvesting density of 2.7 mW/cm^3, density 5 times higher than the PZT cantilever types, the most serious practical problem is how to sum up the harvesting electric output from each nanofiber in phase without canceling each other.

5.1.3.4 Military application—Programmable air-burst munition

Until the 1960s, the development of weapons of mass destruction (WMD) was the primary focus, including nuclear bombs and chemical weapons. However, based on the global trend for "Jus in Bello (Justice in War)," environment-friendly "green" weapons became the mainstream in the 21st century; that is, minimal destructive weapons with a pinpoint target such as laser guns and rail guns. In this direction, programmable air-burst munition (PABM) was developed from 2003.

After the attack of World Trade Center by Al Qaeda on September 11, 2001, the US military started the "revenge" war against Afghanistan. The US troops initially destroyed all the buildings by bombs, which dramatically increased the cost of restructuring new buildings. Thus, the US Army changed the war strategy; without collapsing the building, just kill the Al Qaeda soldiers inside by using a sort of micro-missile. A micro-missile passes the window by making a small hole on the window glass and explodes in the air 3–5 m (programmable!) inside the window, so that the building structure damages become minimized. For this purpose, each bullet needs to be installed with a microprocessor chip, which navigates the bullet to a certain programmed point. ATK Integrated Weapon Systems, AZ started to produce button-battery operated PABM first [16]. Although they worked beautifully for a couple of months, due to severe weather condition (incredibly hot in daytime in Afghanistan), most of the batteries wasted out in 3 months. No soldier was willing to open a dangerous bullet for exchanging a battery!

Under this circumstance, the Army Research Office contacted Micromechatronics Inc., State College, PA (ICAT/PSU spin-off company) for the development of a compact electric energy source to be embedded in the PABM bullet. The initial shooting impact can be converted to electricity via a piezoelectric device, which should fulfill energy to be spent in the micro-processor for 3–5 min during maneuvering the bullet. Micromechatronics Inc. developed an energy source for a 25-mm caliber "Programmable Ammunition." Instead of a battery, a multilayer PZT piezo-actuator is used for generating electric energy under shot mechanical impact to activate the operational amplifiers which ignite the burst according to the command program (Fig. 5.5). This may be one of the highest revenue-generating products with piezoelectric energy harvesting systems in 13 years. The key development is the multilayer

Fig. 5.5 Programmable air-bust munition (PABM) developed by micromechatonics (25 mm caliber).

usage for realizing high electromechanical coupling factor k and high capacitance, low electrical impedance.

Since many of the current research trends introduced above are not satisfactory enough with my development philosophy, Uchino decided to introduce my personal development principles in this chapter.

5.1.4 Piezoelectric energy harvesting principle

5.1.4.1 Piezoelectric constitutive equations

We assume the linear *constitutive piezoelectric equations* in the relationship between the extensive strain x, electric displacement D (or almost the same as polarization P in ferroelectrics), and the intensive stress X and electric field E:

$$x = s^{E} X + d E, \tag{5.1}$$

$$D = d X + \varepsilon_0 \varepsilon^{X} E, \tag{5.2}$$

where s^E and ε^X are elastic compliance and relative permittivity (ε_0: vacuum permittivity), and d is the piezoelectric constant. Note that the direct piezoelectric effect (Eq. 5.2) and the converse piezoelectric effect (Eq. 5.1) keeps the same d constant.

5.1.4.2 Piezoelectric figures of merit

There are five important figures of merit in piezoelectrics: the *piezoelectric strain constant d*, the *piezoelectric voltage constant g*, the *electromechanical coupling factor k*, the *mechanical quality factor* Q_m, and the *acoustic impedance Z*. These figures of merit are introduced below.

Piezoelectric strain constant d

The magnitude of the induced strain x by an external electric field E is represented by this figure of merit (an important figure of merit for actuator applications):

$$x = d E. \tag{5.3}$$

Piezoelectric voltage constant g

The induced electric field E is related to an external stress X through the piezoelectric voltage constant g (an important figure of merit for sensor applications):

$$E = g X. \tag{5.4}$$

Taking into account the relation, D (or P) $= d\ X$, we obtain an important relation between g and d:

$$g = d/\varepsilon_0 \varepsilon^{X}. \left(\varepsilon^{X} : \text{relative permittivity}\right) \tag{5.5}$$

Electromechanical coupling factor k

The terms *electromechanical coupling factor*, *energy transmission coefficient*, and *efficiency* are sometimes confused. All are related to the conversion rate between electrical energy and mechanical energy, but their definitions are different [17, 18].

The electromechanical coupling factor *k*

$$k^2 = (\text{Stored mechanical energy}/\text{input electrical energy}) \tag{5.6}$$

or

$$k^2 = (\text{Stored electrical energy}/\text{input mechanical energy}) \tag{5.7}$$

Let us calculate Eq. (5.7), when an external stress X is applied to a piezoelectric material. See Fig. 5.6A first, when the top and bottom electrodes are short-circuited (i.e., E constant condition). Since the input mechanical energy is (1/2) $s^E X^2$ per unit volume and the stored electrical energy per unit volume under zero external electrical impedance (i.e., short-circuit) is given by (1/2) $D^2/\varepsilon_0\varepsilon^X = (1/2)\ (d\ X)^2/\varepsilon_0\varepsilon^X$ [D (or P) is obtained by integrating the current i], k^2 can be calculated as

$$\begin{aligned} k^2 &= \left[(1/2)(dX)^2/\varepsilon_0\varepsilon^X\right] / \left[(1/2)\, s^E X^2\right] \\ &= d^2/\varepsilon_0\varepsilon^X \cdot s^E. \end{aligned} \tag{5.8}$$

When the top and bottom electrodes are open-circuited in Fig. 5.6B, depolarization field E is induced because of the piezoelectrically induced charge in order to satisfy $D=0$ in Eq. (5.2):

$$E = -\left(d/\varepsilon_0\varepsilon^X\right) X. \tag{5.9}$$

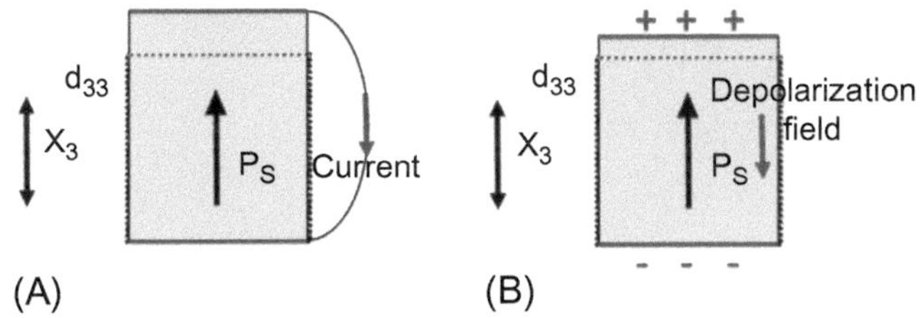

Fig. 5.6 Calculation models of electromechanical coupling factor k for (A) short-circuit and (B) open-circuit conditions.

Note that this depolarization field is valid only for the short time period, by neglecting the charge drift. Thus, from Eq. (5.1), the strain induced under the open-circuit condition should be smaller than that under the short-circuit, as calculated:

$$x = s^E X + dE = s^E X - (d/\varepsilon_0\varepsilon^X) X = s^E [1 - (d/\varepsilon_0\varepsilon^X s^E)] X.$$

If we denote $x = s^D X$ under the open-circuit, we can obtain the following important equation:

$$s^D = s^E (1 - k^2).(k : \text{electromechanical coupling factor}) \tag{5.10}$$

It is worth to note that depending on the electrical load (i.e., impedance), input mechanical energy differs significantly even if we keep the same stress X level, because the elastic compliance of the piezoelectric material changes with the electrical constraint condition. Accordingly, the importance of the energy transmission coefficient transpired in the piezo-energy harvesting system.

The energy transmission coefficient λ_{max} Not all the mechanically stored energy can be actually used under the electrical drive, and the actual work done in the piezoelectric actuator depends on the mechanical load. With zero mechanical load or a complete clamp (no strain), no output work is done. Vice versa, not all the electrically converted energy can be actually spent in the energy harvesting case. With zero impedance (short-circuit) or infinite impedance (open-circuit), no electrical work can be obtained. Thus, the energy transmission coefficient is defined by

$$\lambda_{max} = (\text{Output mechanical energy/input electrical energy})_{max} \tag{5.11}$$

or equivalently,

$$\lambda_{max} = (\text{Output electrical energy/input mechanical energy})_{max} \tag{5.12}$$

The difference of the above from Eqs. (5.6), (5.7) is "stored" or "output/spent."

The energy transmission coefficient can be obtained as

$$\begin{aligned} \lambda_{max} &= \left[(1/k) - \sqrt{(1/k^2) - 1}\right]^2 \\ &= \left[(1/k) + \sqrt{(1/k^2) - 1}\right]^{-2}. \end{aligned} \tag{5.13}$$

Since the detailed derivation process under the electric field drive for actuator application is introduced in a textbook in Ref. [17], only the key points are described here for the energy harvesting under the mechanical stress drive, using Fig. 5.7, which shows an electric energy harvesting model under the external electrical impedance Z on a piezoelectric actuator. Fig. 5.8 summarizes the calculation processes of the input mechanical and output electric energy under various impedance Z's: (a) the

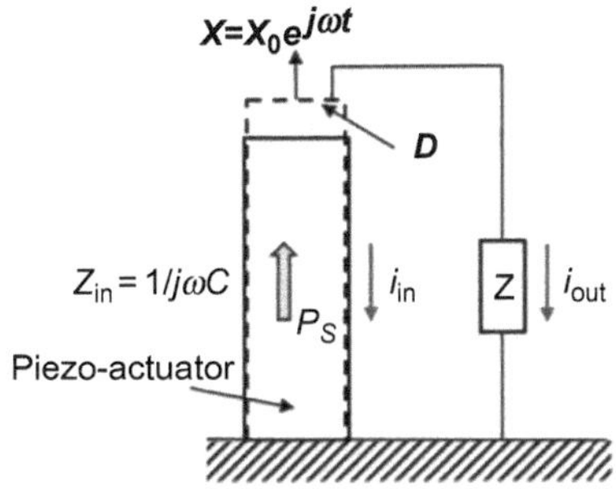

Fig. 5.7 Electric energy harvesting model under the external electrical impedance Z on a piezoelectric actuator.

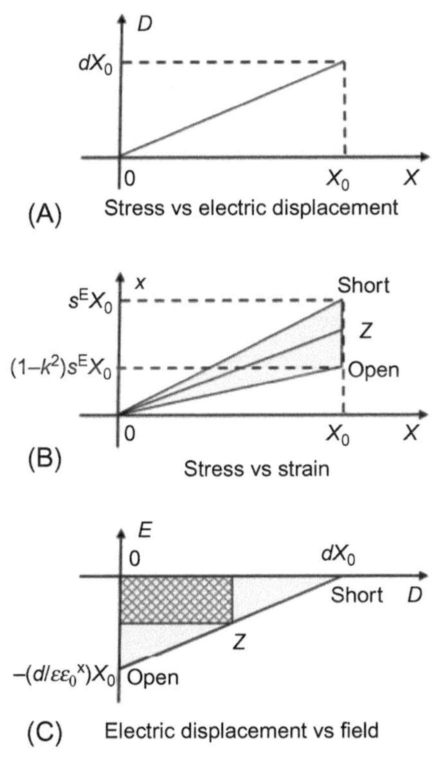

Fig. 5.8 Calculation models of the input mechanical and output electric energy. (A) The stress versus electric displacement relation; (B) the stress versus strain relation to calculate the input mechanical energy; and (C) the electric displacement versus electric field to calculate the output electric energy.

stress versus electric displacement relation; (b) the stress versus strain relation to calculate the input mechanical energy; and (c) the electric displacement versus electric field to calculate the output electric energy.

When we assume sinusoidal input stress $X = X_0 e^{j\omega t}$ and output electric displacement $D = dX_0 e^{j\omega t}$, we can derive the following current and voltage relationships from Fig. 5.7:

$$i = \frac{\partial D}{\partial t} = i_{in} + i_{out} = j\omega dX_0;$$

$$Z_{in} i_{in} = Z i_{out};$$

$$i_{out}(1 + j\omega CZ) = j\omega dX_0.$$

Thus, we can obtain the output electric energy as

$$|P| = \frac{1}{2} Z i_{out}^2 = \frac{1}{2} Z \frac{(\omega dX_0)^2}{\left(1 + (\omega CZ)^2\right)}. \tag{5.14}$$

Fig. 5.9 shows the electric load dependence of the output electric energy, which concludes that the maximum electric energy $|P| = \frac{1}{4}\frac{\omega d^2 X_0^2}{C}$ can be obtained at $Z = 1/\omega C$. In other words, the "stored" electric energy can be spent maximum when the external load impedance matches exactly to the internal impedance.

However, we need to be aware that since the input mechanical energy is changed (even if we keep the stress/force constant) due the elastic compliance change with the external electrical impedance, the condition for realizing the maximum transmission coefficient is slightly off from the electrical impedance matching point. We can also note from Eq. (5.13) that

$$k^2/4 < \lambda_{max} < k^2/2, \tag{5.15}$$

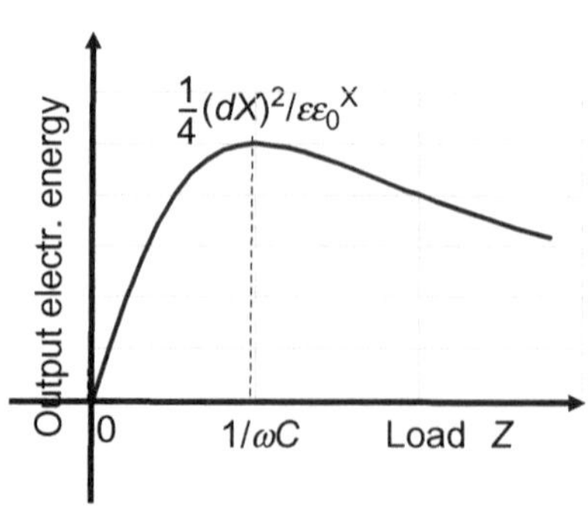

Fig. 5.9 Output electric energy dependence on the external electrical load Z in a piezoelectric energy harvesting system.

depending on the k value. For a small k (<0.3), $\lambda_{max} \approx k^2/4$, and for a large k (~0.7), $\lambda_{max} \approx k^2/3 \sim k^2/2$. When k is unrealistically high (~0.95), λ_{max} approaches to 1.

The efficiency η

$$\eta = (\text{Output mechanical energy})/(\text{Consumed electrical energy}) \tag{5.16}$$

or

$$\eta = (\text{Output electrical energy})/(\text{Consumed mechanical energy}). \tag{5.17}$$

The difference of the efficiency definition from Eqs. (5.11), (5.12) is "input" energy and "consumed" energy in the denominators. In a work cycle (e.g., a mechanical stress cycle), the input mechanical energy is transformed partially into electrical energy and the remaining is stored as mechanical energy (mechanical energy in a piezoelectric spring) in an actuator. In this way, the ineffective mechanical energy can be returned to the vibration source, leading to near 100% efficiency, if the loss is small. Typical values of dielectric loss in PZT are about 1%–3%.

Fig. 5.10 illustrates the difference between electromechanical coupling factor k, energy transmission coefficient λ_{max}, and efficient η for a piezoelectric material with $k_{33} = 70\%$. Note that the efficiency is not related with the electromechanical coupling factor k, as exemplified by a quartz crystal device ($k_t = 9\%$) with 99.99% efficiency.

Mechanical quality factor Q_M

The mechanical quality factor, Q_M, is a parameter that characterizes the sharpness of the electromechanical resonance spectrum. When the motional admittance Y_m is plotted around the resonance frequency ω_0, the mechanical quality factor Q_M is defined with respect to the full width $[2\Delta\omega]$ at $Y_m/\sqrt{2}$ as

$$Q_M = \omega_0/2\Delta\omega. \tag{5.18}$$

Also note that Q_M^{-1} is equal to the intensive mechanical loss ($\tan \phi'$). When we define a complex elastic compliance, $s^E = s^{E\prime} - j\, s^{E\prime\prime}$, the mechanical loss tangent is provided

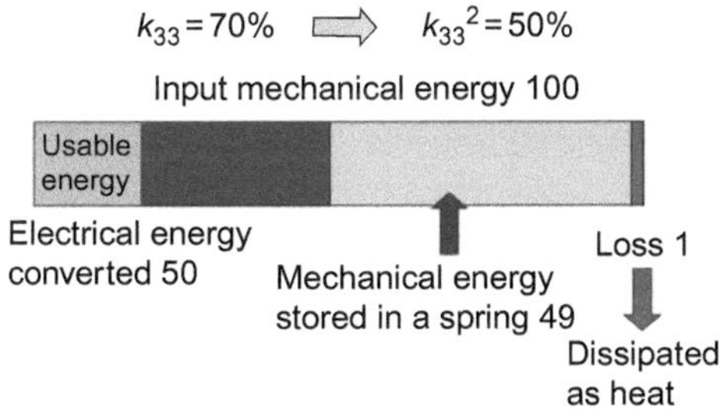

Fig. 5.10 Schematic illustration on how much electric energy can be harvested from the electromechanical coupling factor's viewpoint.

by tan $\phi' = s^{E''}/s^{E'}$. The Q_M value is very important in evaluating the magnitude of the resonant displacement and strain. The vibration amplitude at an off-resonance frequency ($dE \cdot L$, L: length of the sample) is amplified by a factor proportional to Q_M at the resonance frequency. For example, a longitudinally vibrating rectangular plate through the transverse piezoelectric effect d_{31} generates the maximum displacement given by $(8/\pi^2)\ Q_M\, d_{31} E\, L$. Refer to a textbook in Ref. [19] for details on the mechanical quality factors.

Mechanical/acoustic impedance Z

The mechanical impedance Z is a parameter used for evaluating the acoustic/sound wave energy transfer between two materials. It is defined, in general, by

$$Z^2 = (\text{pressure}/\text{volume velocity}). \tag{5.19}$$

In a solid material,

$$Z = \sqrt{\rho c}, \tag{5.20}$$

where ρ is the density and c is the elastic stiffness of the material.

The mechanical work done by one material on the other is evaluated by the product of the applied force F and the displacement ΔL: $W = F \times \Delta L$. This corresponds to "Pushing a curtain," exemplified by the case when the acoustic wave is generated in water directly by a hard PZT transducer. Most of the acoustic energy generated in the PZT is reflected at the interface, and only a small portion of acoustic energy transfers into water. On the other hand, if the material is very hard, the displacement will be very small, again leading to very small W. This corresponds to "Pushing a wall." Polymer piezoelectric PVDF (polyvinylidene di-fluoride) cannot drive a hard steel part efficiently. Therefore, the *mechanical/acoustic impedance* must be adjusted to maximize the output mechanical power:

$$\sqrt{\rho_1 c_1} = \sqrt{\rho_2 c_2}, \tag{5.21}$$

where ρ is the density and c is the elastic stiffness, and the subscripts 1 and 2 denote the two materials. In practice, acoustic impedance matching layers (elastically intermediate materials between PZT and water, such as a polymer. More precisely the acoustic impedance Z should be $\sqrt{Z_1 \cdot Z_2}$.) are fabricated on the PZT transducer to optimize the transfer of mechanical energy to water.

In more advanced discussions, there are three kinds of impedances: specific acoustic impedance (pressure/particle speed), acoustic impedance (pressure/volume speed), and radiation impedance (force/speed). See Ref. [20] for the details.

5.1.4.3 Piezoelectric passive damper

The principle of the piezoelectric vibration damper is explained based on a piezoelectric ceramic single plate in Fig. 5.11A [21], as it was a prior product to the energy harvesting devices. When an external impulse force is applied to the piezo-plate, an electric charge

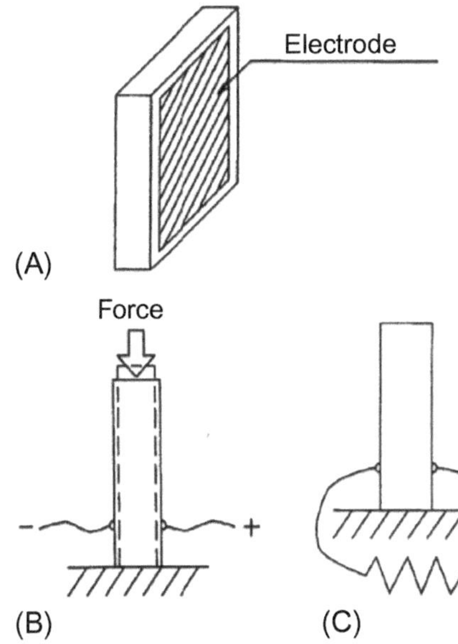

Fig. 5.11 Original idea of the mechanial damping with a piezoelectric. (A) structure of a single plate piezoelectric sample; (B) direct piezoelectric effect; and (C) electric energy dissipation through a resistance.

is produced by the direct piezoelectricity (Fig. 5.11B). Accordingly, the vibration remaining after the removal of the external force induces an alternating voltage, which corresponds to the intensity of that vibration, across the terminals of the single plate. The electric charge produced is allowed to flow and is dissipated as Joule heat, when a resistor is put between the terminals (see Fig. 5.11C). As we discussed in Eq. (5.14), when the external resistance is too large or small, the vibration intensity is not readily reduced, and we need to tune the resistance to match exactly to the piezo-plate impedance, that is, $1/j\omega C$, where ω is the cyclic frequency (i.e., the fundamental mechanical resonance of the piezo-plate, and C is the piezo-plate capacitance).

The bimorph piezoelectric element shown in Fig. 5.1A, an elastic beam sandwiched with two sheets of piezoelectric ceramic plates, is a typical example of a combination of a vibration object and piezoelectric ceramics. The bimorph edge was hit by an impulse force, and the transient vibration displacement decay was monitored by an eddy current-type noncontact displacement sensor (SDP-2300 manufactured by Kaman). Fig. 5.1B shows the measured displacement data which vibrates at the bimorph resonance frequency (295 Hz), and Fig. 5.12 shows the relationship between the damping time constant and an external resistance. It can be seen in the figure that the damping time constant was minimized in the vicinity of 8 kΩ, which is close to the impedance $1/\omega C$.

Let us evaluate the damping constant theoretically. The electric energy U_E generated can be expressed by using the electromechanical coupling factor k and the mechanical energy U_M:

$$U_E = U_M \times k^2. \tag{5.22}$$

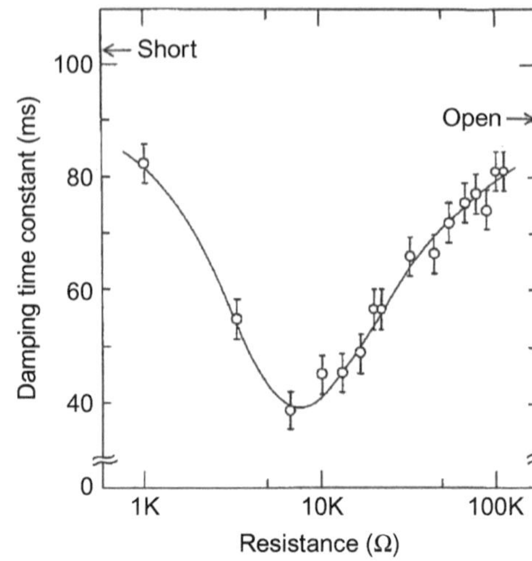

Fig. 5.12 Relationship between the damping time constant and the external resistance.

The piezoelectric damper transforms electric energy into heat energy when the resistor R is connected, and transforming rate of the damper can be raised to a level of up to 50% when the electrical impedance is matched. Accordingly, the vibration energy is transformed at a rate of $(1-k^2/2)$ times with energy vibration repeated, since $k^2/2$ multiplied by the amount of mechanical vibration energy is dissipated as heat energy. As the square of the amplitude is equivalent to the amount of energy, the amplitude decreases at a rate of $(1-k^2/2)^{1/2}$ times with every vibration repeated. If the resonance period is taken to be T_0, the number of vibrations for *t sec* is $2t/T_0$. Consequently, the amplitude in t sec is $(1-k^2/2)^{t/T_0}$. If the residual vibration period is taken to be T_0, the damping in the amplitude of vibration is t sec, which can be expressed as follows:

$$\left(1-k^2/2\right)^{t/T_0}=e^{-t/\tau}. \tag{5.23}$$

Thus, the following relationship for the time constant of the vibration damping is obtained.

$$\tau=-\frac{T_0}{\ln(1-k^2/2)} \tag{5.24}$$

Now let us examine the time constant of the damping using the results for the bimorph in our study. Substitution of $k=0.28$ and $T_0=3.4$ msec in Eq. (5.24) produces $\tau=85$ msec, which seems to be considerably larger than the value of approximately 40 msec obtained experimentally for τ. This is because the theoretical derivation Eq. (5.24) was conducted under the assumption of a loss-free (high Q_M) bimorph. In practice, however, it originally involved mechanical loss, the time constant of which can be obtained as the damping time constant under a short-circuited condition, that is, $\tau_s = 102$ msec. The total vibration displacement can then be expressed as $e^{-t/\tau_{total}} = e^{-t/\tau_s} \times e^{-t/\tau}$. Accordingly,

$$\frac{1}{\tau_{total}} = \frac{1}{\tau_s} + \frac{1}{\tau} \tag{5.25}$$

Substitution of $\tau_s = 102$ msec and $\tau = 85$ msec in Eq. (5.25) produces $\tau_{total} = 46$ msec. This conforms to the result shown in Fig. 5.1B and agrees with the experiment.

Because we used a bimorph structure in this pioneering paper merely due to the simplest geometry, the author feels a sort of guilt on providing a strong influence on subsequent researchers on the bimorph design. The bimorph is not an ideal piezo-device design at all, as we discuss in the following parts.

5.1.5 Three phases in energy harvesting process

After getting into the 1990s, we decided to collect the electric energy without just dissipating the energy via Joule heat, which is the beginning of the piezoelectric energy harvesting; "Chasing two hares to obtain Both." There are three major phases/steps associated with piezoelectric energy harvesting (see Fig. 5.13): (i) *mechanical-mechanical energy transfer*, including mechanical stability of the piezoelectric transducer under large stresses, and mechanical impedance matching, (ii) *mechanical-electrical energy transduction*, relating with the electromechanical coupling factor in the composite transducer structure, and (iii) *electrical-electrical energy transfer*, including electrical impedance matching. A suitable DC/DC

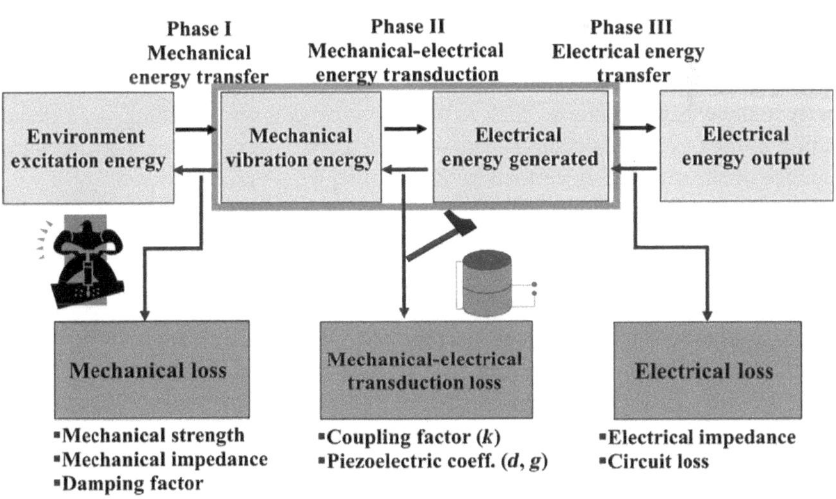

Fig. 5.13 Three major phases associated with piezoelectric energy harvesting): (i) mechanical-mechanical energy transfer, (ii) mechanical-electrical energy transduction, and (iii) electrical-electrical energy transfer.

converter is required to accumulate the electrical energy from a high-impedance piezo-device into a rechargeable battery (low impedance).

This article mainly deals with detailed energy flow analysis in piezoelectric energy harvesting systems with typical stiff "Cymbals" (~100 mW) and flexible piezoelectric transducers Smart Material's "Macro-Fiber Composites" (~1 mW) under cyclic mechanical load (off-resonance), in order to provide comprehensive strategies on how to improve the efficiency of the harvesting system. Energy transfer rates are practically evaluated for all the three steps above. Our application target of the "Cymbal" was set to hybrid vehicles with both an engine and an electromagnetic motor; reducing the engine vibration and harvesting electric energy to car batteries to increase the mileage, while, the target of the flexible piezo-composites was "wearable electric chargers" for portable electronic equipment such as mobile phones. The readers are requested to read this article, taking into account these practical applications.

We should also point out here that there is another research school of piezo-energy harvesting; that is, small energy harvesting (mW or lower) for signal transfer applications, where the efficiency is not a primary objective. This school usually treats an impulse/snap action load to generate instantaneous electric energy for transmitting signals for a short period (100 ms–10 s), without accumulating the electricity in a rechargeable battery. Successful products (million sellers) in the commercial market belong mostly to this category at present, including "Lightning Switch" [9] (remote switch for room lights, with using a unimorph piezoelectric component) by PulseSwitch Systems, VA, and the 25-mm caliber "Programmable Ammunition" [16] (electricity generation with a multilayer piezo-actuator under shot impact) by ATK Integrated Weapon Systems, AZ, as introduced in Section 5.1.3.

5.2 Mechanical-to-mechanical energy transfer

Of all, wasted or unused mechanical energy (vibration source) should be transferred properly to the energy converter such as piezoelectric devices. Mechanical impedance matching is one of the important factors we have to take into account. The mechanical impedance of the material is defined by $Z=(\rho c)^{1/2}$, where ρ is the density and c is the elastic stiffness, or effective parameter values in a composite structure (see Eq. (5.20)). The receiving part of the mechanical energy in the piezo-device should be designed to match the mechanical/acoustic impedance with the vibration source. Otherwise, most of the vibration energy will be reflected at the interface between the vibration source and the harvesting piezoelectric device. Note a proverb, "*pushing a curtain and pushing a wall* [useless task!]," both cases will not transfer mechanical energy efficiently. Fig. 5.14A and B exhibits two extreme examples we will treat in this article: (a) high-energy harvesting from a "hard" machine, such as an engine, and (b) low-energy harvesting from a "soft" machine, such as human motion. Not only mechanical impedance matching, but also mechanical strength and damping factor (i.e., loss) of the device are important.

The "Cymbal" transducer is a preferable device for the high-power purpose [22]. A Cymbal transducer consists of a piezoelectric ceramic disk and a pair of metal endcaps. The metal endcaps play an important role as displacement-direction

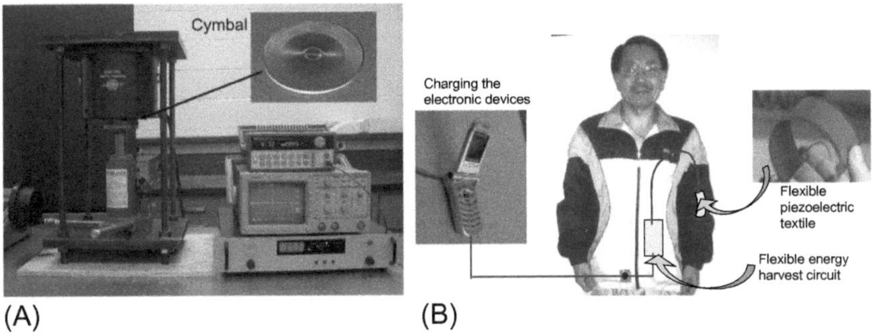

Fig. 5.14 Two extreme vibration source examples: (A) high energy harvesting from a "hard" machine, such as an engine, and (B) low energy harvesting from a "soft" machine, such as human motion.

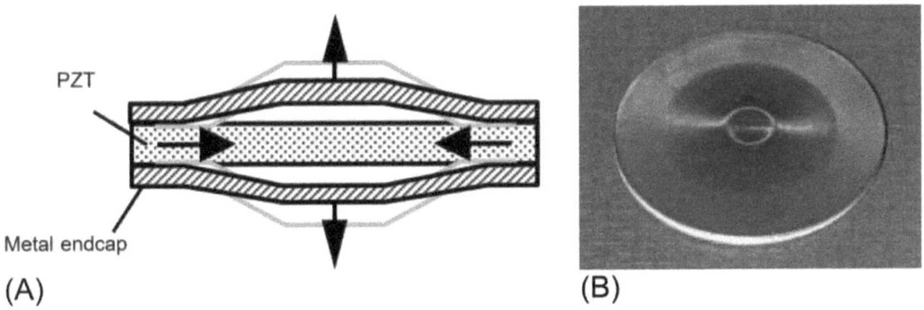

Fig. 5.15 (A) Operation principle of the Cymbal, and (B) photo of the Cymbal.

convertor and displacement amplifier. (See Fig. 5.15.) The Cymbal transducer has a relatively high coupling factor (k_{eff}) and a high stiffness, in comparison with unimorphs/bimorphs, which makes it suitable for a high-force mechanical source. Energy transfer or reflection rate from the hard electromagnetic shaker to a Cymbal was analyzed by changing the rigidity of the Cymbal endcap (i.e., by changing the endcap thickness, 0.3 and 0.4 mm). As shown in Fig. 5.14A, a mechanical shaker)top part in Fig. 5.14A) was employed to apply a maximum of 50 N force in a wide frequency range from pseudo-DC to 1 kHz (i.e., off-resonance usage). The force generated is proportional to the payload mass and the generated acceleration which is controlled by the applied voltage. The acceleration of the payload was computed by performing the real-time differentiation of measured vibration velocity. The vibration velocity was measured by using Polytec Vibrometer (Tustin, CA). A payload mass (cylinder made from aluminum) of 85 g was used in the experiment. The Cymbal transducer was bonded at the top of payload by using a silicone rubber sealer. A bias DC force was applied on the transducer using a homebuilt hydraulic system to avoid the separation problem, that is, bang-bang shock (the lower part in Fig. 5.14A). The mechanical energy transferred to the Cymbal was evaluated from the Cymbal deformation and its effective stiffness. The mechanical-to-mechanical transfer rate of 46% for the 0.4-mm thick sample was dramatically improved to 83%–87% by thinning the endcap down to 0.3 mm; that is, 0.4 mm seems to be too rigid for this shaker's elasticity, which will be discussed again in Table 5.3.

Note, however, that when we increased the bias force level up to 70 N by using a payload mass of 820 g, the sample with 0.3-mm thick endcaps was damaged, and only the 0.4-mm thick sample could endure. Therefore, we should point out that the mechanical strength is another important factor, compromising with the efficiency.

To the contrary, the flexible transducer is a preferable device for "soft" application [23]. The macrofiber composite (MFC) is an actuator that offers reasonably high performance and flexibility in a cost-competitive manner (Smart Material Corp.) (Fig. 5.16) [24]. The MFC consists of rectangular piezo-ceramic rods sandwiched between the layers of adhesive and electroded polyimide film. This film contains interdigitated electrodes that transfer the applied voltage directly to and from the ribbon-shaped rods. This assembly enables in-plane poling, actuation, and sensing in a sealed, durable, ready-to-use package. Uchino et al. are developing intelligent clothing (IC) with piezoelectric energy harvesting system of flexible piezoelectric textiles [23] aiming at a power source for charging up portable equipment such as cellular phones, health-monitoring units, or medical drug delivery devices.

5.3 Mechanical-electrical energy transduction

5.3.1 *Figure of merit (FOM)*

Mechanical energy transferred to the transducer, such as a Cymbal and a flexible transducer, will be converted into electrical energy through the piezoelectric effect. Voltage induced in the transducer can be explained by Eqs. (5.26), (5.27) for a bulk piezoelectric:

$$V = g \times \frac{F(\mathrm{N}) \times t(\mathrm{m})}{A(\mathrm{m}^2)} \tag{5.26}$$

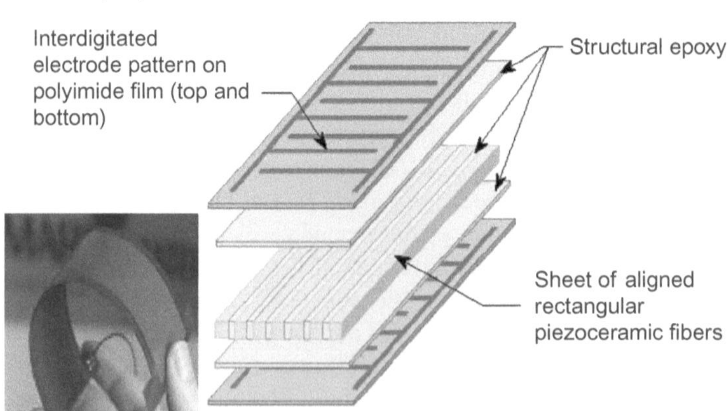

Fig. 5.16 Structure of macro fiber composite (MFC), produced by Smart Material Corp. [24]. Courtesy by Smart Materials.

$$g = d/\varepsilon_0 \varepsilon_r \tag{5.27}$$

Here, g, F, t, and A are piezoelectric voltage constant, applied force, thickness, and area of the piezo-material, respectively. Piezoelectric voltage constant g can be expressed by piezoelectric charge constant d, dielectric constant in vacuum ε_0, and relative dielectric constant ε_r as shown in Eq. (5.27). The output electric power can be calculated as follows:

$$\begin{aligned} P &= \frac{1}{2} C V^2 \cdot f \\ &= \frac{1}{2} \cdot g_{33} \cdot d_{33} \cdot F^2 \cdot \frac{t}{A} \cdot f \end{aligned} \tag{5.28}$$

where C and f are capacitance of the transducer and frequency of the vibration, respectively. Hence output power can be evaluated by the product of g and d as "*Figure of merit*" of the transducer:

$$P \propto g \cdot d \tag{5.29}$$

Note that when the piezo-component is a composite or a hybrid structure, the above constants g and d should be replaced by the "effective" values. It is well known, for example, that the effective d coefficient for a cymbal [25] is expressed by

$$d_{\text{eff}} = d_{33} + A|d_{31}|, \tag{5.30}$$

where the amplification factor A is given by

$$A \propto (1/2)[\text{Cavity diameter}/\text{Cavity depth}]. \tag{5.31}$$

5.3.2 Piezoelectric materials selection

Popularly used piezoelectric materials are metal-oxide materials, in particular, with the perovskite structure. Lead zirconate titanate (PZT)-based ceramics dominate the market in these 60 years. We summarized electromechanical properties of three commercially available piezoelectric PZT-based ceramics in Table 5.1: Hard APC 841, Soft APC 850 (American Piezo Ceramics, Mackeyville, PA), and High g D210 (Dong Il Technology, Korea). We can find the best material (D210) for the energy harvesting application easily by comparing $g \cdot d$ product values.

5.3.3 Design optimization

5.3.3.1 Cymbal

Cymbal design was optimized from the generated electrical energy level [26]. Fig. 5.17 shows a rectification circuit with a full wave rectifier and a capacitor for storing generated electrical energy of the Cymbal transducer in the case of off-

Table 5.1 Electromechanical properties of the commercially available piezoelectric ceramics

Parameter	Hard (APC 841)	Soft (APC 850)	High g (D210)
ε_r	1350	1750	681
k_p (%)	60	63	58
$d_{31}(10^{-12}$ C/N)	109	175	120
g_{31} (10^{-3} Vm/N)	10.5	12.4	20
Q_m	1400	80	89.7
T_c (deg.)	320	360	340
$g_{31} \cdot d_{31}$	1.14×10^{-12}	2.17×10^{-12}	2.40×10^{-12}

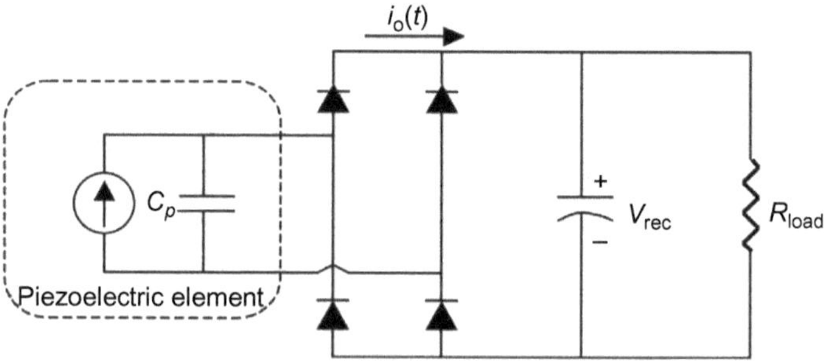

Fig. 5.17 A full-bridge rectifier with a resistive load for piezoelectric energy harvesting.

resonance. Using this circuit, the output voltage and the power were initially measured across the resistive load directly without any amplification circuit to characterize the performance of different transducers. The maximum rectified voltage V_{rec} of a capacitor C_{rec} (10 μF) was charged up to 248 V after saturation. Fig. 5.18 shows the output electrical power from various Cymbal transducers under AC and DC mechanical loads as a function of external load resistance. The thickness of endcap made of steel was changed from 0.3 to 0.5 mm. Prestress (DC bias load) to the Cymbal was 66 N and applied AC (100 Hz) force was varied experimentally from 44 to 70 N. For a small force drive (40 and 55 N), the power level increased with decreasing endcap thickness (from 0.5, to 0.4, and finally to 0.3 mm). With increasing force level up to 70 N, the maximum power of 53 mW was obtained at 400 kΩ with a 0.4-mm steel endcap, because the Cymbal sample with a 0.3-mm thick endcap could not endure under this high force drive (i.e., the cavity depth is collapsed). Note that the maximum power was obtained merely due to large input mechanical vibration level. The 0.3-mm thick samples were better from the energy efficiency viewpoint, which will be discussed again in Table 5.3.

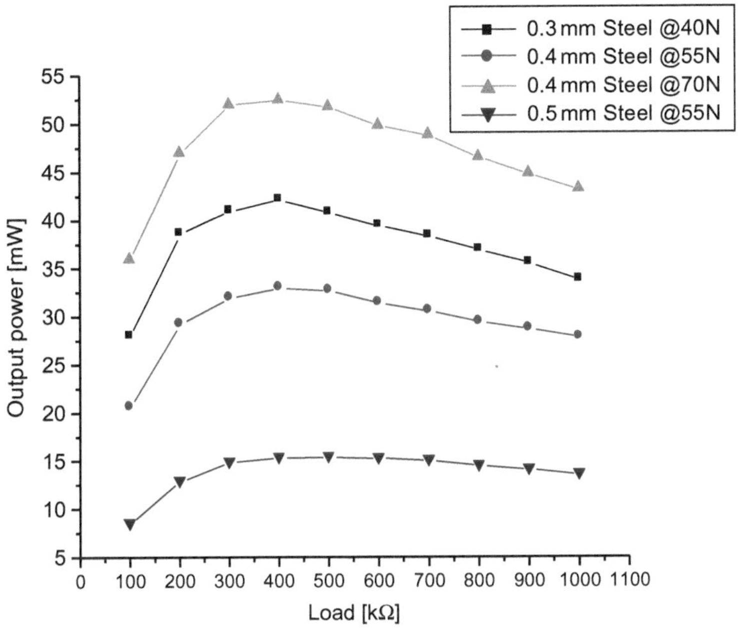

Fig. 5.18 Output electrical power from the various Cymbal transducers under different 100Hz AC mechanical load (shown as @ xx N in the figure under 66N constant DC bias) by changing external electrical load resistance (100–1000 kΩ).

In order to evaluate the mechanical-to-electrical converted energy in the Cymbal, the rectified voltage was used for charging the capacitor C_{rec} of 10 μF in open condition during the charging time (t), for which the capacitor was charged up to 200 V. Note that the measurement results in Table 5.3 include the energy Joule (not the power W, in this case) during a certain charging period.

5.3.3.2 Flexible transducer

Bending mode is usually used for flexible transducers. However, bending vibration induces extension and compression stresses at the same time in the material, and induced electric energy cannot be obtained efficiently because of the cancellation. In order to avoid this problem, a unimorph design was adopted by bonding an elastic substrate to the piezoelectric plate [27].

The d_{31} mode type of Macro-Fiber Composite [MFC (M8528 P2)] from Smart Material Corporation is shown in Fig. 5.19. The piezo-ceramic fibers in the MFC are cut by 350 μm width and 170 μm thickness from a piezoelectric wafer by computer-controlled dicing saw. The total dimensions of the MFC are length 85 mm, width 28 mm, and thickness 0.3 mm. Fig. 5.20 shows the stress distribution of MFC when the mechanical force is applied. As shown in Fig. 5.20A, without

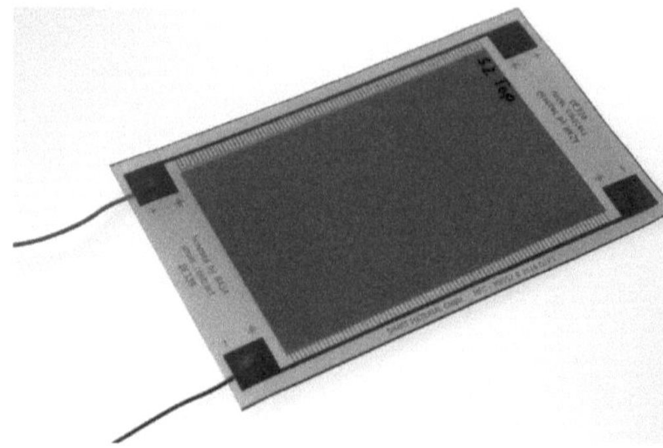

Fig. 5.19 d_{31} mode type of macro fiber composite [MFC (M8528 P2)] from Smart Material Corporation. Courtesy from Smart Material Corp.

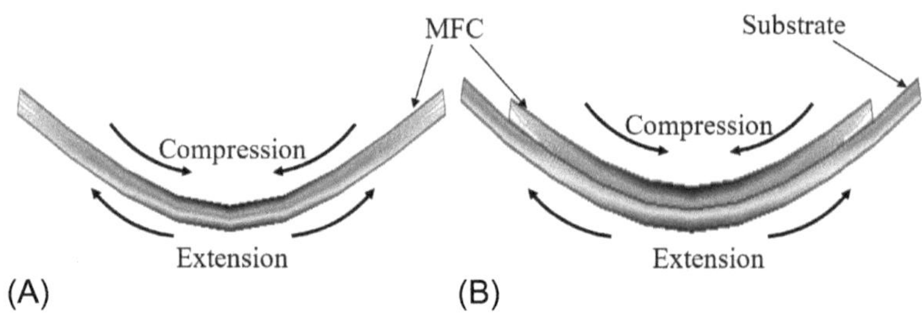

Fig. 5.20 Stress distribution of the macro fiber composite in (A) single plate, and in (B) with an elastic substrate.

any substrate, the extensive stress and compressive stress occur on the top and bottom of the MFC. In this case, the neutral line of the stress distribution is in the middle of the MFC. For this reason the electrical output is very small. However, in Fig. 5.20B, where an additional substrate is bonded on the bottom of the MFC, the neutral line is shifted down in the substrate. Therefore, the electrical output is increased because the MFC has only a compressive stress or extensive stress in the whole volume. Although the uniformity of the compressive stress increases with elastic substrate thickness, the mechanical energy distribution to the MFC decreases, leading to the harvesting electrical energy. In addition, the thickness and material of the substrate should be carefully considered to keep the flexibility of the MFC.

The minimum thickness of the substrate was calculated by the FEM software code ATILA (distributed from Micromechatronics Inc., PA) depending on the material to shift the neutral line away from the MFC. Table 5.2 illustrates the neutral line position and displacement calculated [27]. Typical materials are considered to calculate the neutral line and flexural bending magnitude. For this calculation, the dimensions of MFC are $85\,\text{mm} \times 28\,\text{mm} \times 0.17\,\text{mm}$, and the thickness of the substrate is 0.17 mm. Note that the signs of "+" and "−" indicate the distance from the adjacent line between the MFC and the substrate. In the case of steel, brass, and copper, the

Table 5.2 Neutral line position and the bending displacement calculated by FEM for various substrate materials ($t = 170\,\mu m$) under a fixed mechanical force [27]

	None	Steel	Brass	Copper	Aluminum	Polymer
Neutral line	0.085	−0.03	−0.0025	−0.0125	0.005	0.0825
Displacement	901 μm	79 μm	114 μm	101 μm	129 μm	766 μm

neutral line is located in the substrate below the adjacent line. For the polymer, the neutral line is not changed from the center of MFC same as MFC without any substrate. The displacement in the table is a measure of the flexibility of MFC with a substrate. This displacement is calculated under a fixed mechanical force condition. The aluminum substrate shows the best flexibility than any other metal substrates in Table 5.2. Therefore, aluminum was selected to be used for the substrate in the experiment, even though the neutral line is located slightly above the interface line.

The method to apply the stress on the sample was based on the real application. The mechanical shaker to make vibration at high frequency is not close approach for the piezoelectric energy harvesting with a small mechanical source such as human motion. The MFC was excited to generate a big bending motion by small force at a frequency of around 1–5 Hz. Fig. 5.21 shows an experimental setup for measuring electrical output from a flexible piezo-component under a fixed bending displacement. Note that the small mechanical force used in this experiment means the minimum force which can generate maximum strain in the flexible element without crack.

The voltage signal of the MFC is shown in Fig. 5.22A [27]. The voltage of MFC is considerably increased by bonding an aluminum substrate as shown in Fig. 5.22A. This signal was generated by a small mechanical force with frequency of 5 Hz, and monitored by an oscilloscope (Tektronix, TDS 420A). The addition of the aluminum substrate gave a lower flexibility to the MFC, but the output voltage from MFC with small bending showed much higher voltage signal. The output voltage

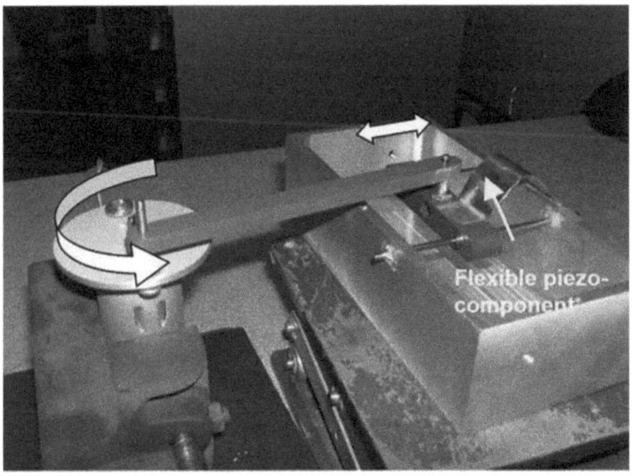

Fig. 5.21 Experimental setup for measuring electrical output from a flexible piezo-component under a fixed bending displacement.

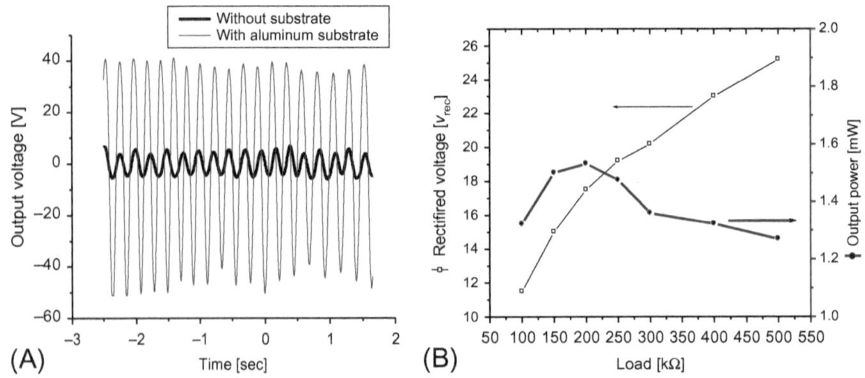

Fig. 5.22 Output of the MFC by around 5 Hz. (A) Output voltage signal of the MFC, and (B) rectified voltage and output power plotted as a function of resistive load.

signal from the MFC was passed through the rectifier and charged a capacitor, and successively discharged through a resistive load. The rectified voltage and output power are shown in Fig. 5.22B. The generated electric power from a small mechanical force at 5 Hz was around 1.5 mW at 200 kΩ. This 200 kΩ electric power corresponds roughly to the matched impedance value of $1/\omega C$ at 5 Hz.

5.3.4 *Energy flow analysis*

Table 5.3 summarized the energy flow analysis on three types of Cymbal transducers: endcap thickness of 0.3 and 0.4 mm with and without bias force under various cyclic vibration levels and drive durations [28, 29]. Note first that the mechanical-to-mechanical energy transfer rate is good for 0.3-mm thick cymbals (83%–87%), while it is rather low for 0.4-mm thick cymbal (46%). This is related to the mechanical impedance matching; 0.4 mm endcap seems to be too rigid (effective stiffness is too high) to match the vibration source shaker.

Second, the received mechanical energy to electrical energy transduction rate can be evaluated from the effective electromechanical coupling factor. Because the value of k_{eff} of the cymbal is around 25%–30%, the energy transduction rate k_{eff}^2 can be evaluated around 6.25%–9%, which agrees well with the experimental results illustrated in Table 5.3. Thus, we obtained the conversion rate from the vibration source energy to the stored electric energy in the cymbal transducer as 7.5%–7.8% for the 0.3-mm thick endcap cymbals, and 2.9% for the 0.4-mm thick endcap cymbal. The reduction of the conversion rate is primarily originated from the mechanical/acoustic impedance mismatch.

5.4 Electrical-to-electrical energy transfer

Piezoelectric materials generally convert mechanical energy to electrical energy with relatively high voltage which means output impedance is relatively high at an off-resonance frequency. On the other hand, energy storage devices such as a rechargeable

Table 5.3 Energy flow analysis on three types of Cymbal transducers: endcap thickness of 0.3 mm and 0.4 mm with and without bias force under various cyclic vibration levels and drive durations [29]

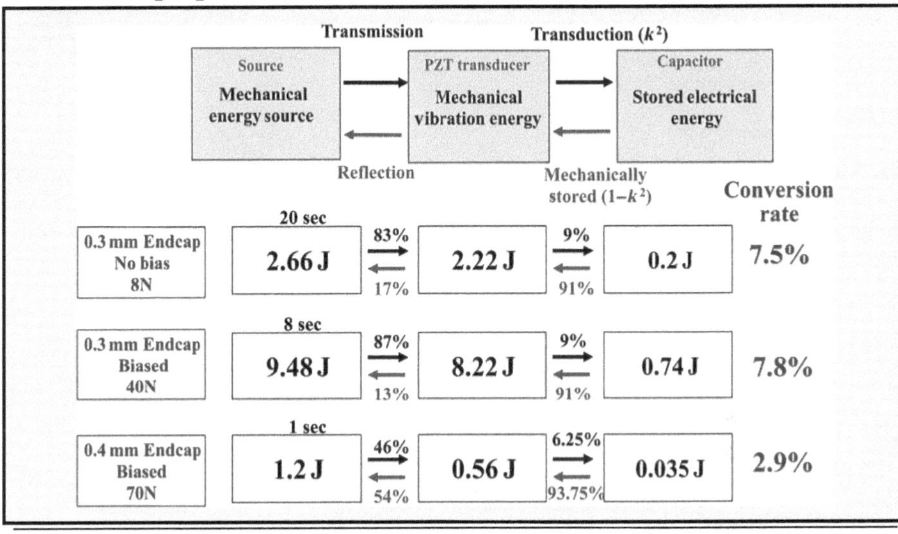

battery have low input impedance (10–100 Ω). Thus, large portion of the excited electrical energy is reflected back, if we connect the battery immediately after the rectified voltage. In order to improve energy transfer efficiency, electrical impedance matching is required.

5.4.1 DC-DC converter

5.4.1.1 Basic of switching regulator

Fig. 5.23 illustrates the basic principle of a switching regulator using a power metal-oxide-semiconductor field-effect transistor (MOSFET). The output voltage v_0 through a load R switches the maximum E (some 100s V) and minimum 0 according to the gate/source voltage (rectangular voltage with the duty ratio of D), so that the average voltage is estimated by

$$v_0 = \frac{T_{on}}{T_{on} + T_{off}} E = dE. (D : \text{duty ratio}) \tag{5.32}$$

If the loss of the electronic component is negligibly small, input power = output power. Thus, using the duty ratio d, the voltage step-up ratio and current step-down ratio should be d, leading to the impedance change by the factor of d^2. If we take $d = 2\%$, the impedance can be reduced by 2500 theoretically.

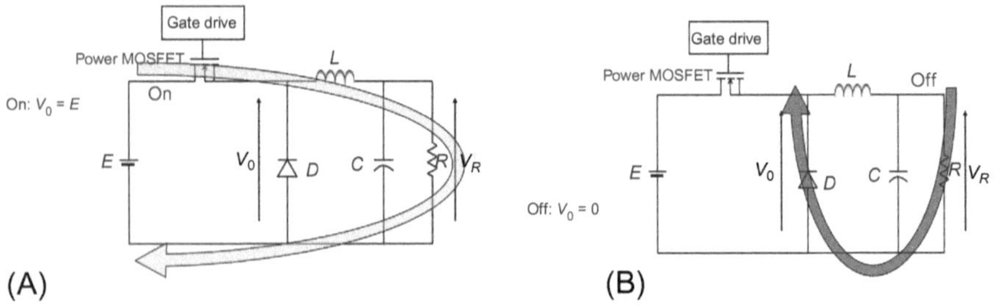

Fig. 5.23 Basic principle of a switching regulator using a power MOSFET. (A) On state and (B) off state. The output voltage through a load R switches the maximum E (some 100s V) and minimum 0 according to the Gate/Source voltage (rectangular On/Off voltage with the duty ratio of D).

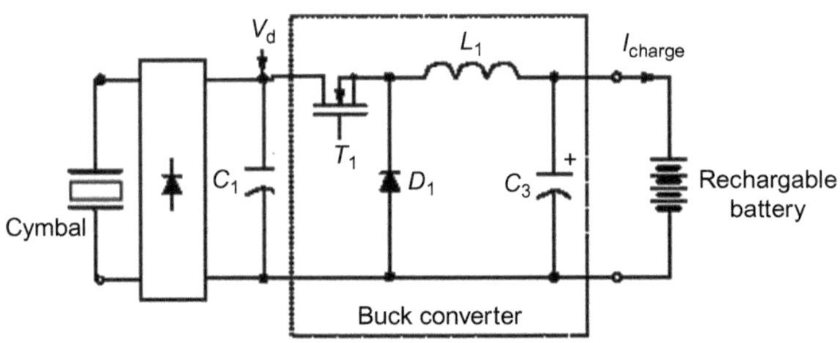

Fig. 5.24 A DC-DC Buck-converter designed to allow transfer of 43 mW power out of 53 mW from the Cymbal (81% efficiency) by converting the original impedance 300 kΩ down to 5 kΩ with a 2% duty cycle and at a switching frequency of 1 kHz.

5.4.1.2 Buck converter

Various converters can step down the voltage to adapt the electrical impedance: Forward converter, Buck Converter, Buck-Boost Converter, Flyback Converter, etc. These converters have low output impedance and low loss characteristics. A DC-DC Buck Converter shown in Fig. 5.24 is the simplest topology using a single MOSFET, which allows transfer of 43 mW power out of 53 mW from the Cymbal (81% efficiency) in practice by converting the original impedance 300 kΩ down to 5 kΩ with a 2% duty cycle and at a switching frequency of 1 kHz [28, 30]. The impedance modification 60 times differs from the above expectation (2500 times), which may be explained by the components' losses in the circuit.

5.4.2 Multilayered cymbal

Since the Buck Converter introduced in the previous section cannot reduce the output impedance sufficiently to match with the rechargeable battery (~50 Ω), we should further reduce the output impedance. Output impedance of the transducer can be changed

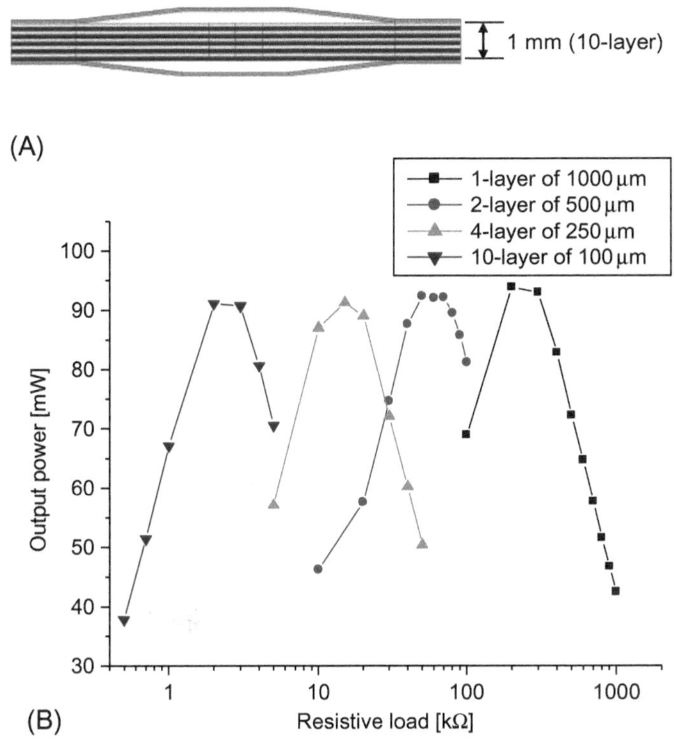

Fig. 5.25 (A) Multilayered Cymbal transducer [28]. (B) Output power characteristics of the multilayered Cymbal transducers.

by changing the transducer structure. Multilayered (ML) transducers have lower impedance [28]. Fig. 5.25A shows cross-sectional view of the ML Cymbal transducer. With increasing the number n of the layers, output impedance decreases by a factor of $1/n^2$. As shown in Fig. 5.25B, the maximum output power shifts to the low-resistance direction which means output impedance can be controlled by the transducer structure. You can find that the matching impedance 300 kΩ in the single layer cymbal is reduced down to 3 kΩ by using a 10-layer cymbal.

It is worth noting that the performance becomes much superior by combining the ML cymbal structure and the DC-DC converter. See Fig. 5.26. When we compare the output power of the 10-layer cymbal with or without the DC-DC converter, using the DC-DC converter reduces the maximum power level from 100 to 80 mW with the matching load shift from 5 to 2 kΩ. However, the output power around 50 Ω (matching to the rechargeable battery) differs significantly; 50 mW with the converter and $<$10 mW without the converter. The reason for this load-insensitive broadening effect is to be clarified in the future.

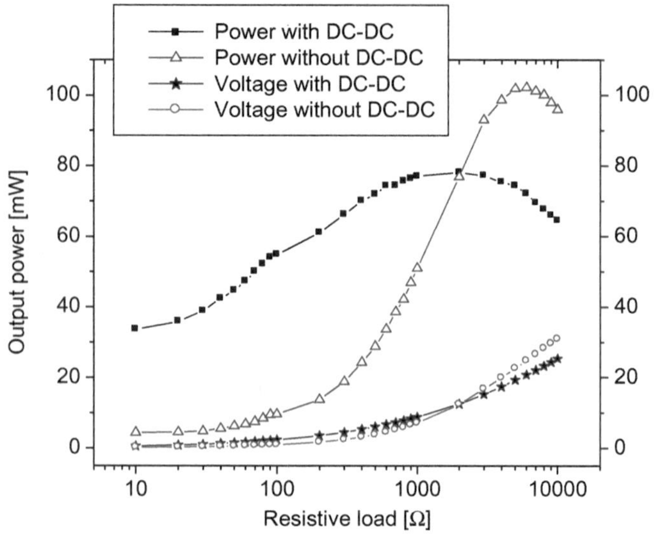

Fig. 5.26 Load resistance dependence of the efficiency of the piezoelectric transformer.

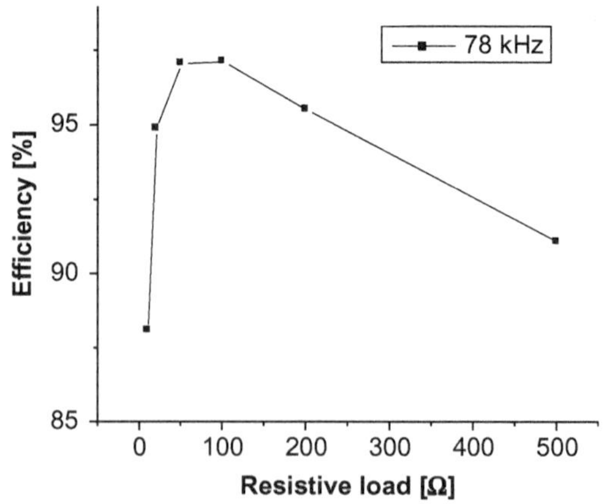

Fig. 5.27 Load resistance dependence of the efficiency of the piezoelectric transformer [31].

5.4.3 Usage of a piezoelectric transformer—Further impedance matching

Another unique circuit design is with a piezoelectric transformer [31]. The piezoelectric transformer used in the circuit has low output impedance around 50 Ω (i.e., resistive at the transformer resonance frequency), and the efficiency of the piezoelectric transformer at its resonance can reach above 97%, as shown in Fig. 5.27. This low output impedance is suitable for an impedance matching to the energy storage devices. Fig. 5.28A illustrates a block diagram of a piezoelectric energy harvesting circuit with a piezoelectric transformer, and Fig. 5.28B is an actual piezoelectric energy harvesting circuit with a ring-dot-type multilayer transformer (see the disk-shaped PZT transformer) [32]. Input and output voltage signals of the piezo-transformer and the final

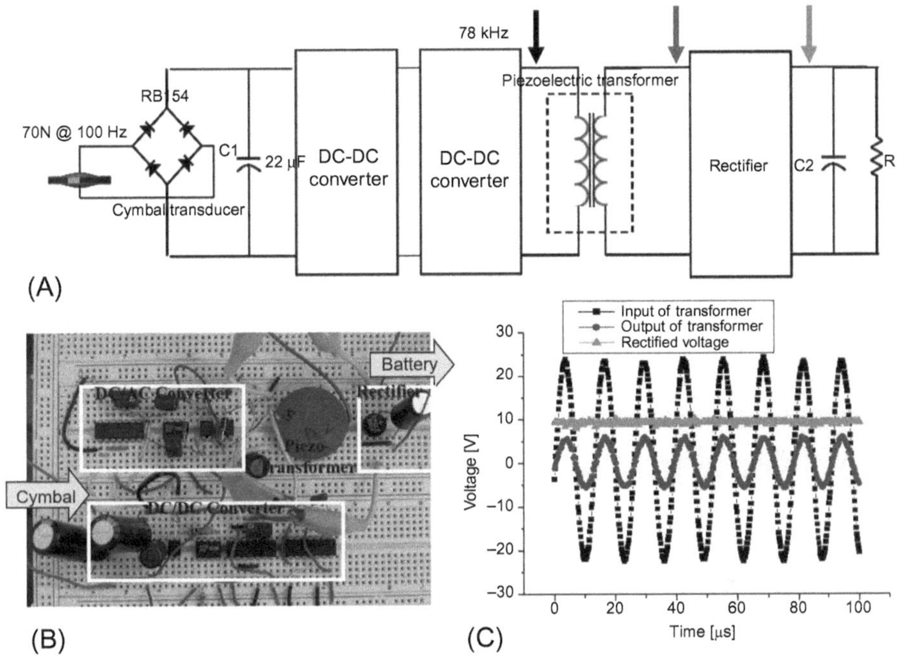

Fig. 5.28 (A) Block diagram of a piezoelectric energy harvesting circuit with a piezoelectric transformer. (B) Actual piezoelectric energy harvesting circuit with a ring-dot type multilayer transformer. (C) Input and output voltage signals of the piezo-transformer and the final rectified voltage, which correspond to the three *arrow* parts in (A).

rectified voltage, which correspond to the three arrow parts in (a), are plotted in Fig. 5.28C. The step-down voltage ratio seems to be 5.

5.5 Summary of the total energy flow

Let us summarize the energy flow/conversion from the input mechanical energy source to the output rechargeable battery. The sample was a cymbal with 0.3-mm thick stainless steel endcaps, inserted below a 4-kg engine weight (40 N bias force). The electromagnetic shaker shook at 100 Hz for 8 s, and the accumulated energy during that time period was measured at each point. The energy levels at all states are summarized in Table 5.4.

Let us discuss why the energy amount decreases with transmitting and transducing successively by numerical analyses:

(1) *Source to Transducer: Mechanical impedance mismatch*
8.22 J/9.48 J = 87% of the mechanical energy is transmitted from the source to piezo-cymbal transducer. If the mechanical impedance is not seriously considered in the transducer (too mechanical-soft or hard), the mechanical energy will not transfer efficiently, and large amount will be reflected at the contact/interface point

(2) *Transduction in the Transducer: Electromechanical coupling*

Table 5.4 Energy flow/conversion analysis in the cymbal energy harvesting process

Source Mechanical energy		Transducer Mechanical energy		Transducer Electrical energy		Circuit-in Electrical energy		Battery Electrical energy
9.48 J	→ ←	8.22 J	→ ←	0.74 J	→ ←	0.42 J	→ ←	0.34 J
	87 %		9 %		Rectifier- 57%		Converter- 81%	

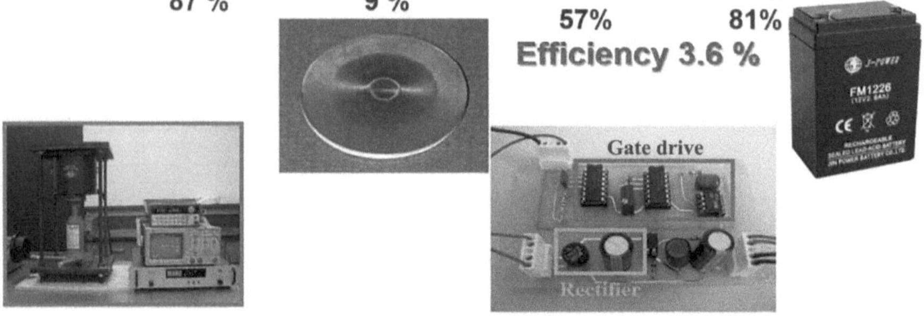

Transduction rate is evaluated by k^2. Since $k_{eff}=0.25$–0.30 in the cymbal, the energy conversion rate will be 9% maximum. The remaining portion will remain as the original mechanical vibration energy. 0.74 J/8.22 J = 9.0%. Since k_{eff} of the bimorph or unimorph is much smaller (10%–15%), the energy conversion rate is smaller than 2%. Using modes with higher k values is highly recommended.

(3) *Transducer to Harvesting Circuit: Electrical impedance matching*

0.42 J/0.74 J = 57% is related with the electrical impedance mismatch between the cymbal output and the circuit input. Unfortunately, a half of the stored electrical energy in the piezo-device cannot be used effectively

(4) *Harvesting Circuit to Rechargeable Battery: DC-DC converter*

0.34 J/0.42 J = 81%. This reduction is partially originated from the energy consumption in the circuit, and partially originated from the electrical impedance mismatch between the circuit output (still around 5 kΩ) and the rechargeable battery impedance (around 10–50 Ω)

In summary, 0.34 J/9.48 J = 3.6% is the current energy harvesting rate from the vibration source to the storing battery. Taking into account the efficiency of pupular amorphous silicon solar cells around 5%–9%, our prototype piezoelectric energy harvesing device with 3.6% efficiency seems to be rather promissing.

5.6 Hybrid energy harvesting—Magnetoelectric devices

The sustainable society requires various sensors for monitoring hazardous environmental factors such as viruses and magnetic fields, which humans cannot sense biologically. Similar to nuclear radiation, magnetic irradiation cannot be easily felt by human, but it may increase the brain cancer probability. We cannot even purchase a magnetic field detector for a low frequency (50 or 60 Hz). The Penn State, in

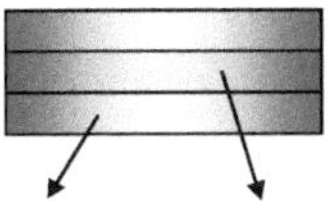

Fig. 5.29 Magnetic noise sensor consisting of a laminated composite of a PZT and two Terfenol-D disks [33].

collaboration with Seoul National University, Korea, developed a simple and handy magnetic noise sensor. Fig. 5.29 shows a schematic structure of the device, in which a PZT disk is sandwiched by two Terfenol-D (magnetostrictor) disks. When a magnetic field is applied on this composite, Terfenol-D expands, which is mechanically transferred to PZT, leading to the generation of electric charge from the PZT. By monitoring the voltage generated in the PZT, we can detect the magnetic field. The key of this device is high effectiveness for a low frequency [33]. As long as the magnetostrictor-piezoelectric laminated composite can generate electric signal under magnetic noise, we can consider a hybrid energy harvesting device on which either magnetic and/or mechanical noises can work.

Recently Finkel et al. [33] demonstrated the above idea by using superior piezoelectric single crystal lead indium niobate-lead magnesium niobate-lead titanate (PIN-PMN-PT) and magnetostrictive single crystal Galfenol. An AC magnetic field ±250 G under a bias field of 250 G applied to the coupled device causes the magnetostrictive Galfenol element to expand, and the resulting stress forces the phase change in the relaxor ferroelectric PIN-PMN-PT single crystal. They have demonstrated high-energy conversion (2 mW at the matching impedance 1 MΩ) in this magnetoelectric device by triggering the F_R-F_O transition in the single crystal by a small AC magnetic field in a broad frequency (off-resonance) range that is important for multi-domain hybrid energy harvesting devices.

5.7 Summary and outlook

1. Three phases/steps for piezoelectric energy harvesting are: (i) mechanical impedance matching, (ii) electromechanical transduction, and (iii) electrical impedance matching.
2. The Cymbal transducer is used for energy harvesting from a high-power mechanical vibration, while the MFC is suitable for a small flexible energy vibration.
3. A Buck Converter is used as the DC/DC converter for realizing the better electrical impedance matching. Further improvement can be realized by using a highly capacitive multilayer design.
4. A piezoelectric transformer is a promising alternative component for the DC/DC converter.
5. As indicated in Table 5.4, the key to dramatic enhancement in the efficiency is to use a high k mode, such as k_{33}, k_t, or k_{15}, rather than flex-tensional modes. Fig. 5.30 summarizes promising piezoelectric device designs for energy harvesting applications. The hinge-lever mechanism is an ideal "mechanical transformer" without losing mechanical energy under an off-resonance condition, and it is easily manufactured using micromachining technologies,

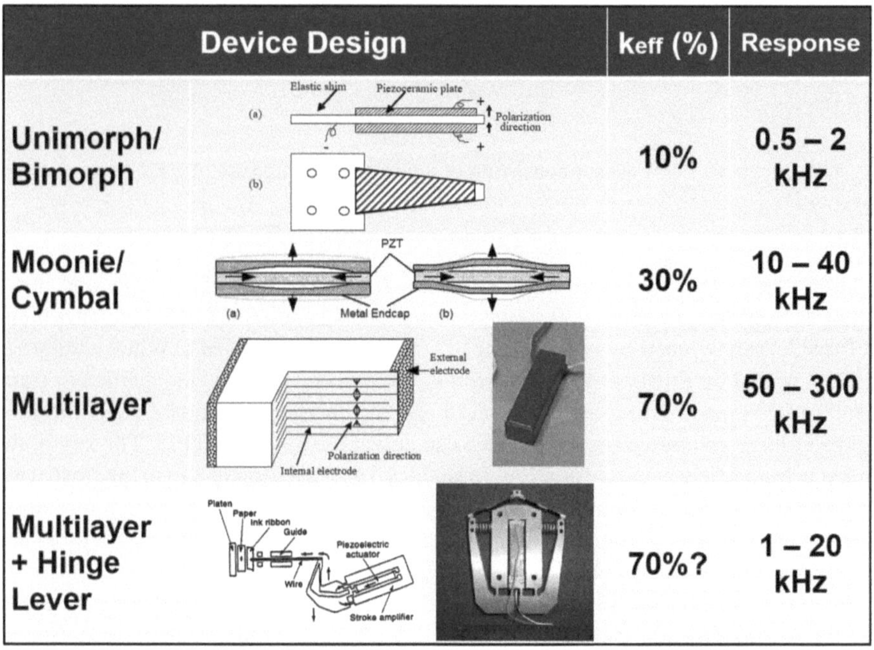

Fig. 5.30 Promising piezoelectric device designs for energy harvesting applications.

which may expand the tunability of the mechanical impedance by keeping a high electromechanical coupling factor k in the range of $k_{33} \approx 70\%$.

6. A hybrid energy harvesting device which operates under either magnetic and/or mechanical noises was introduced, by coupling magnetostrictive and piezoelectric materials.
7. Saving energy is very important in recent years, and the analyses and results in this article can be applied to recover and store the wasted or unused mechanical (and/or magnetic) energy efficiently.

There will be two development directions in piezoelectric energy harvesting areas: (i) remote signal transmission (such as structure health monitoring) in (*mW*) power level, and (ii) energy accumulation in rechargeable batteries in (*W*) power level. Energy harvesting less than *W* does not attract the clients in home appliance and automobile manufacturers. The former will use impulse/snap action mechanisms or low-frequency resonance methods, where the efficiency is not a significant issue, but the minimum spec on the energy should be obtained; while the latter requires dramatic research strategy changes. The researchers are requested to forget the current "biased" knowledge, or not to take a strategy, "because the previous researchers did so." Remembering the fundamental principles explained in this article, challenges a totally unique design and/or idea in the future. Though relatively large investments and research efforts are being put on MEMS/NEMS and "nano harvesting" devices, Uchino, unfortunately, does not have a positive comment on this future. Medical applications seem to be a desirable dream, but obtained/reported energy level

$pW \sim nW$ is a useless level, which is originated from inevitable small volume of the used piezoelectric material (i.e., thin films). Without discovering a genius idea how to combine thousands of these nano-devices in parallel and synchronously in phase, these efforts will be in vain just as a dream, similar to a research fever on metal-oxide "superconductors" in the 1990s.

References

[1] K. Uchino, Ferroelectric Devices, second ed., CRC Press, New York, 2009.
[2] Uchino Estimated the Revenue Based on the Resources, iRAP (Innovative Research & Products), Market Publishers, and IDTechEx.
[3] N. Muensit (Ed.), Energy Harvesting with Piezoelectric and Pyroelectric Materials, Materials Science Foundations, vol. 72, Trans Tech Pub., Stafa-Zuerich, 2011.
[4] J. Wallaschek, M. Neubauer, J. Twiefel, Electromechanical models for energy harvesting systems, in: Energy Harvesting with Piezoelectric and Pyroelectric Materials, Materials Science Foundations, vol. 72, Trans Tech Pub., Stafa-Zuerich, 2011 (Chapter 2).
[5] P. Smithmaitrie, Vibration theory and design of piezoelectric energy harvesting structures, in: Energy Harvesting with Piezoelectric and Pyroelectric Materials, Materials Science Foundations, vol. 72, Trans Tech Pub, Stafa-Zuerich, 2011 (Chapter 3).
[6] S.P. Beeby, D. Zhu, Energy harvesting products and forecast, in: Energy Harvesting with Piezoelectric and Pyroelectric Materials, Materials Science Foundations, vol. 72, Trans Tech Pub., Stafa-Zuerich, 2011 (Chapter 9).
[7] S.W. Arms, C.P. Townsend, J.H. Galbreth, L.C. David, P. Nam, Synchronized System for Wireless Sensing, RFID, Data Aggregation & Remot Reporting, 2009.
[8] K. Uchino, in: Piezoelectric actuators 2004—materials, design, drive/control, modeling and applications, Proceedings of 9th International Conference on New Actuators, A1.0, Bremen, Germany, 14–16 June, 2004, pp. 38–48.
[9] K. Uchino, in: Piezoelectric actuators 2010—piezoelectric devices in the sustainable society, Proceedings of 12th International Conference on New Actuators, A3.0, Bremen, Germany, 14–16 June, 2010. http://www.lightningswitch.com/.
[10] H. Kim, Y. Tadesse, S. Priya, in: S. Priya, D. Inman (Eds.), Energy Harvesting Technologies, Springer Science+Business Media LLC, New York, 2009.
[11] M. Renaud, P. Fiorini, R. van Schaijk, C. van Hoof, Smart Mater. Struct. 18 (2009) 035001.
[12] D. Guyomar, M. Lallart, N. Muensit, C. Lucat, Conversion enhancement for energy harvesting, in: Energy Harvesting with Piezoelectric and Pyroelectric Materials, Materials Science Foundations, vol. 72, Trans Tech Pub., Stafa-Zuerich, 2011 (Chapter 5).
[13] R.P. Lopes, A. Kholkin, Energy harvesting for smart miniaturized systems, in: Energy Harvesting with Piezoelectric and Pyroelectric Materials, Materials Science Foundations, vol. 72, Trans Tech Pub., Stafa-Zuerich, 2011 (Chapter 6).
[14] C. Chang, V.H. Tran, J. Wang, Y.-K. Fuh, L. Lin, Nano Lett. 10 (2010) 726.
[15] Z.L. Wang, J. Song, Science 321 (2006) 242.
[16] http://www.atk.com/MediaCenter/mediacenter_videogallery.asp.
[17] K. Uchino, Ferroelectric Devices, second ed., CRC Press, Boca Raton, FL, 2010.
[18] T. Ikeda, Fundamentals of Piezoelectric Materials Science, Ohm Publishing Co., Tokyo, 1984.
[19] K. Uchino, J. Giniewicz, Micromechatronics, Marcel Dekker, New York, NY, 2003.

[20] L.E. Kinsler, A.R. Frey, A.B. Coppens, J.V. Sanders, Fundamentals of Acoustics, John Wiley & Sons, New York, 1982.
[21] K. Uchino, T. Ishii, J. Ceram. Soc. Jpn. 96 (1988) 863.
[22] H.-W. Kim, A. Batra, S. Priya, K. Uchino, D. Markley, R.E. Newnham, H.F. Hofmann, J. Appl. Phys. 43 (2004) 6178.
[23] K. Uchino, J. Jpn. Soc. Appl. Electromag. 4 (4 (20071210)) (2008) 399.
[24] http://www.smart-material.com/Smart-choice.php?from=MFC.
[25] B. Koc, A. Dogan, J.F. Fernandez, R.E. Newnham, K. Uchino, Jpn. J. Appl. Phys. 35 (1996) 4547.
[26] H.-W. Kim, S. Priya, K. Uchino, R.E. Newnham, J. Electroceram. 15 (2005) 27.
[27] H.-W. Kim, K. Uchino, T. Daue, Proceedings of CD 9th Japan International SAMPE Symposium & Exhibition, SIT Session 05, 29 November–02 December, 2005.
[28] H.-W. Kim, S. Priya, K. Uchino, Jpn. J. Appl. Phys. 45 (2006) 5836.
[29] K. Uchino, Proceedings of 5th International Workshop on Piezoelectric Mater. Appl., State College, PA, 6–10 October, 2008.
[30] H.-W. Kim, S. Priya, H. Stephanau, K. Uchino, IEEE Trans. UFFC 54 (2007) 1851.
[31] K. Uchino, A. Vazquez Carazo, Proceedings of 11th International Conference on New Actuators, Bremen, Germany, 9–11 June, A3.7, 2008, p. 137.
[32] H.-W. Kim, S. Priya, K. Uchino, in: A new method of impedance adaptation for piezoelectric energy harvesting using a cymbal transducer, Proceedings of 10th International Conference on New Actuators, Bremen, Germany, 14–16 June, 2006, A5.4, 2006, pp. 189–192.
[33] J. Ryu, A. Vazquez Carazo, K. Uchino, H.E. Kim, Jpn. J. Appl. Phys. 40 (2001) 4948.

Further reading

[34] P. Finkel, R.P. Moyet, M. Wun-Fogle, J. Restorff, J. Kosior, M. Staruch, J. Stace, A. Amin, Non-resonant magnetoelectric energy harvesting utilizing phase transformation in relaxor ferroelectric single crystals, Actuators 5 (2016) 2, https://doi.org/10.3390/act5010002.

Metal oxides in batteries

6

Faxing Wang, Yi Zhang*, Nengfei Yu*, Lijun Fu*, Yusong Zhu*, Yuping Wu*,‡, Teunis van Ree†*
*State Key Laboratory of Materials-Oriented Chemical Engineering, School of Energy Science and Engineering, Institute for Electrochemical Energy Storage, Nanjing Tech University, Nanjing, People's Republic of China, †Department of Chemistry, University of Venda, Thohoyandou, South Africa, ‡College of Science, Hunan Agricultural University, Changsha, China

6.1 Metal oxides in Ni//MH batteries

6.1.1 Principles and applications of Ni//MH batteries

A rechargeable battery consists of at least one electrochemical cell in which chemical energy is converted directly into electrical energy and vice versa by reversible electrochemical oxidation/reduction reactions. Well-known rechargeable battery systems include lead (Pb)-acid, nickel-cadmium (Ni//Cd), nickel-metal hydride (Ni//MH), metal (Li, Na, Al, etc.)-ion batteries, metal (Li, Na)//S batteries, and metal (Li, Na, Zn, etc.)//air batteries.

A Ni//MH battery uses a hydrogen storage alloy as the negative electrode and nickel hydroxide ($Ni(OH)_2$) as the positive electrode together with an aqueous solution of mainly potassium hydroxide (KOH) as the electrolyte. Unlike the traditional Ni//Cd and Pb//acid batteries based on a dissolution-precipitation mechanism, with the possibility of dendrite formation during charge and discharge, the functioning of a Ni//MH battery can be described as the movement of hydroxide ions between a MH electrode and a nickel hydroxide electrode. During charging, the hydrogen storage alloy reacts with water at the negative electrode to form MH and hydroxyl ions, and $Ni(OH)_2$ is oxidized to NiOOH at the positive electrode, while the reactions proceed in the reverse directions on discharging [1]. When charging takes place, hydrogen moves from the positive electrode to the negative electrode, and during discharge the hydrogen moves in the reverse direction, while the electrolyte does not participate in the reactions. The electrochemical equilibria governing the operation of a Ni//MH battery can be formulated as follows:

$$\text{Positive electrode reaction}: \mathrm{Ni(OH)_2 + OH^- \rightleftharpoons NiOOH + H_2O + e^-} \tag{6.1}$$

$$E_p = 0.49\ \mathrm{V\ (vs.SHE)}$$

$$\text{Negative electrode reaction}: \mathrm{M + H_2O + e^- \rightleftharpoons MH + OH^-} \tag{6.2}$$

$$E_n = -0.83\ \mathrm{V\ (vs.SHE)}$$

Metal Oxides in Energy Technologies. https://doi.org/10.1016/B978-0-12-811167-3.00006-7

$$\text{The cell reaction}: \text{Ni(OH)}_2 + \text{M} \rightleftharpoons \text{NiOOH} + \text{MH} \tag{6.3}$$

$$\Delta E_{\text{cell}} = 1.32\,\text{V}$$

The first electric vehicle (EV) in the world powered by sealed Ni//MH batteries was made by Matsushita Battery Industrial Corp., Japan, in 1992. The power of this EV was equal to that of gasoline vehicles with a cylinder volume of 1500 mL. The weight of the battery was about 400 kg and the output voltage 216 V. The capacity of each battery was 130 Ah with an energy density of 70 Wh kg^{-1}. The driving range was 140 km per charge and the accelerating time from 0 to 40 km h^{-1} was 7 s [2].

At present, the wider application of Ni//MH batteries experiences strong competition from rechargeable Li-ion batteries (LIBs), which have a considerable advantage in terms of specific energy. Nonetheless, they have been used mainly for those applications where LIBs exhibit weak behavior, for example, high power and low cost. Therefore, Ni//MH batteries are still considered to be one of the most likely choices for HEV applications. At the moment, Ni//MH batteries dominate the HEV market because of several major benefits: (1) high specific power density, (2) flexible cell size (capacities may vary from 30 mAh to 250 Ah), (3) safety while charging and discharging (including tolerance to abusive overcharging and overdischarging), (4) easy application to series and series/parallel strings, (5) very good thermal properties, (6) simple and inexpensive charging and electronic control circuits, and (7) environmental compatibility without using heavy metals (such as Pb and Cd) [1]. More than two million hybrid cars (e.g., Toyota Prius, Honda Insight, and Ford Escape Hybrid) have been using Ni//MH technology since 2008.

The challenge for the Ni//MH battery is the large self-discharge rate, which may be related to the hydrogen release rate from the MH electrode. Hydrogen evolution occurs due to the pressure difference between the partial hydrogen pressure in the cell and the equilibrium hydrogen pressure in the hydride alloy electrode. In a sealed cell, the self-discharge rate is lower because the hydrogen partial pressure build-up depresses the hydrogen release from the MH alloys [3].

6.1.2 Positive electrodes for Ni//MH batteries based on metal oxides

The capacity of a Ni//MH battery depends on the specific capacity of the positive electrode and the utilization of active material because of the capacity limitation of the positive electrode. There are two kinds of positive electrodes for Ni//MH batteries, sintered type and paste type. During the initial development of the Ni//MH battery, sintered-type electrodes were used, but they have now been replaced mostly by paste-type electrodes. This type of electrode is manufactured by loading paste on substrate followed by drying and pressing, and facilitates high capacity by reducing substrate weight.

Two nickel hydroxide polymorphs exist: α-phase $Ni(OH)_2$ and β-phase $Ni(OH)_2$, which are transformed into γ-phase NiOOH and β-phase NiOOH, respectively, after full charging. Although α-$Ni(OH)_2$ has a higher theoretical electrochemical capacity

(433 mAh g^{-1} with an average oxidation number of 3.5 or higher for nickel in the γ-phase), it is very unstable in highly alkaline medium and is easily transformed to the β-phase after a few cycles. Mainly α-$Ni(OH)_2$ is slowly dissolved in the aqueous solution, and then re-nucleated to form β-$Ni(OH)_2$. When the positive electrode is overcharged, the active material, that is, β-$Ni(OH)_2$, can be transformed to γ-NiOOH, which is accompanied with a large volumetric change. The α-$Ni(OH)_2$ is formed by discharging γ-NiOOH [4]. This phase is very unstable in the presence of H_2O and alkali and is transformed quickly to β-$Ni(OH)_2$ (Fig. 6.1). Beta-phase spherical nickel hydroxide powders are usually used as active materials since the β-phase has a high tap density (2.1–2.2 g cm^{-3}) and good stability in an alkaline electrolyte. The β-phase nickel hydroxide has its theoretical capacity limitation at 289 m Ahg^{-1}. All the phases are based on NiO_2 layers, built up of edge-sharing NiO_6 octahedra. Both $Ni(OH)_2$ forms crystallize in the hexagonal symmetry with the brucite-type structure in which $Ni(OH)_2$ layers are stacked along the *c*-axis.

In the positive electrode, the less oxidizing state $Ni(OH)_2$ and the more oxidizing state NiOOH are both semiconductors, which are different from the conductive active materials (hydrogen storage alloys) of the negative electrode. Enhancing the utilization of active materials in the battery is very important for improvement of the conductivity and collecting ability of the positive electrode. During the charge and discharge processes occurring at the positive electrode, the electrode is expanding and shrinking, continuously decreasing its mechanical strength, increasing electric resistance, and degenerating the capacity. Even worse, the charge process of the $Ni(OH)_2$ electrode usually occurs in competition with an oxygen evolution reaction (OER), due to the accompanying oxidation of hydroxide ions:

$$4OH^- \rightleftharpoons O_2 + 2H_2O + 4e^- \quad (6.4)$$

Effective charging of the $Ni(OH)_2$ electrode to obtain the maximum capacity depends on particle size, morphology, additives, and electrode preparation. It is therefore of

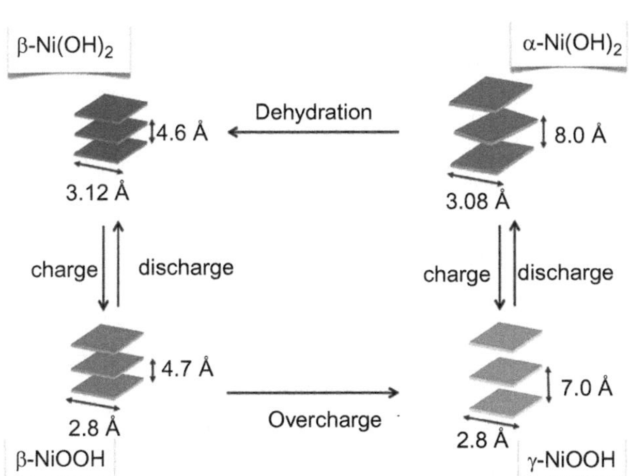

Fig. 6.1 Bode diagram of nickel hydroxides. Reproduced with permission from N. Watanabe, T. Arakawa, Y. Sasaki, T. Yamashita, I. Koiwa, J. Electrochem. Soc. 159 (2012) A1949–A1953, The Electrochemical Society.

great importance to improve the performance of the nickel hydroxide powder used in positive electrodes, to improve the overall performance of Ni//MH batteries.

6.1.3 Modification of the positive electrode for Ni//MH batteries

6.1.3.1 Coating and additives

The most commonly used modification approach is to add a cobalt additive [CoO, $Co(OH)_2$, etc.] directly to the positive electrode material. Cobalt hydroxide is not a competitor to nickel hydroxide as an active material, because its reversibility in energy storage is far from being as good as that of nickel hydroxide. However, when added in small amounts to the positive electrode material, cobalt has several positive effects on the nickel oxide electrode. The presence of cobalt in nickel hydroxide reportedly increases the conductivity and prevents swelling of the nickel hydroxide during cycling [5]. Pure cobalt derivatives can also be added to nickel hydroxide or coated on the surface of spherical nickel hydroxide particles. In this case, cobalt phases form a conducting subnetwork, connecting electronically the low conductive active material particles to the current collector. More specifically, CoO has been dissolved in alkaline electrolyte to form blue Co(II) complex ions, and then added to positive electrode materials to be precipitated on the β-$Ni(OH)_2$ particles. On charging, the β-$Co(OH)_2$ is converted to conductive β-CoOOH. The conductive network of β-CoOOH shortens the 'ring path' and the exchange current density is greatly improved [6]. However, the particle size of the CoO normally used ranges from 0.1 to 0.5 μm. These uneven-sized particles are not dispersed well when mixed with the active β-Ni $(OH)_2$ active material, which may handicap the formation of an integrated β-CoOOH conductive network. To counter this, nanoscale CoO particles were synthesized by decomposition of $CoCO_3$ nanorods [7]. Adding 8 wt% nanoscale CoO was the optimal amount to modify the battery performance at high discharge rates, but adding more nanoscale CoO had no further effect on the overall performance of the Ni//MH batteries.

Various cobalt spinel phases close to Co_3O_4 have also been studied as electrode additives for Ni//MH batteries [5]. A pure phase ($H_{0.60}Li_{0.19}Co_{2.62}O_4$) has been isolated from 2.5 M LiOH, which has a spinel structure close to that of Co_3O_4 but also containing H and Li. The partial compensation of the cobalt vacancies by lithium and hydrogen, thus resulting in the formation of Co^{4+} ions in the octahedral sites, leads to the high electronic conductivity of this phase.

Because oxygen is formed readily at the positive electrode at temperatures above 50°C, the charge efficiency declines significantly, leading to the poor performance of Ni//MH batteries at elevated temperatures [8]. However, a $Ca_3(PO_4)_2/Co(OH)_2$ coating is quite an effective way to improve the high-temperature performance of nickel hydroxide for Ni//MH batteries. The $Ca_3(PO_4)_2$ coating increases the oxygen evolution voltage while $Co(OH)_2$ coating improves the conductivity of the active material. The optimum coating content to obtain a higher discharge capacity at 60°C was around 2%. Cobalt/ytterbium hydroxide has also been used for surface coating of spherical $Ni(OH)_2$ particles to improve their high-temperature performance [9].

The coated spherical $Ni(OH)_2$ was prepared by coprecipitation of Co/Yb hydroxide on the surface of spherical $Ni(OH)_2$. The optimum coating content of ytterbium was around 0.75% atomic concentration to obtain high discharge capacities at both 25° C and 60°C.

Besides cobalt compounds, carbon materials (such as graphite, acetylene black, CNT, and graphene) can be added to the positive electrode material of Ni//MH batteries. Carbon materials are good conductors with little effect on the gravimetric specific capacity while discharging, due to their low density [10]. However, since excess carbon material does not improve the electrochemical properties any further, appropriate carbon material content should be used to suitably modify the performance of the batteries. Highly conductive carbon materials, known for their highly accessible surface, low resistance and good chemical stability, are indeed excellent as conductive agent in Ni//MH batteries. The main way of utilization is still physical mixture of active material and carbon, but this physical mixture of $Ni(OH)_2$ and additives is not effective to reduce the inner resistance of the particles. The utilization efficiency of carbon-containing mixed materials obtained by physical mixing is also relatively low because it is difficult to achieve sufficient contact between the conductor and active material. As a result, the high-rate dischargeability of the positive electrode cannot be improved satisfactorily. In an effort to improve this aspect, a $Ni(OH)_2/C$ composite has been synthesized by a simple coprecipitation method [11]. Compared with pure $Ni(OH)_2$, the synthesized composite had an improved high-rate discharge performance (281 mAh g^{-1} at 1C-rate) and also superior cycling stability and a higher discharge potential. The $Ni(OH)_2/C$ composite not only avoids the formation of γ-NiOOH during high-rate charge-discharge cycling processes, but also helps to reduce the ohmic polarization and charge-transfer polarization.

Graphene is a flat monolayer consisting of sp^2-hybridized carbon atoms tightly packed into a two-dimensional (2D) honeycomb lattice, which has extraordinary electronic transport properties. A β-$Ni(OH)_2$@reduced-graphene-oxide (rGO) composite has been made with a sandwich structure of rGO and β-$Ni(OH)_2$, facilitating diffusion and oxidation/reduction, preventing the aggregation of β-$Ni(OH)_2$ nanoparticles, and stabilizing the structure in the charging and discharging processes [12].

6.1.3.2 Doping

It should be noted that additives that increase the conductivity generally also shift the oxidation peak potential of the nickel hydroxide due to a subsequent reduction in the ohmic drop. However, the core of spherical nickel hydroxide powder is still inactive because of the diffusion barrier. Therefore, another important method has been introduced to improve the performance of the nickel hydroxide—a small fraction of the nickel atoms in the crystal lattice is replaced by other atoms [13], which ideally causes a synergistic effect. Coprecipitated cobalt, calcium, magnesium, copper, and zinc can produce the synergistic effect and improve the electrochemical properties of the active materials [14]. Addition of cobalt can improve electron conductivity and restrict the OER, thus promoting high-rate charge/discharge and high-temperature performance. The doped cobalt can occupy Ni positions in the nickel hydroxide lattice and may exist partly as Co^{4+} during the charge process. Calcium and magnesium can increase the oxygen evolution potential during the charging process. The addition of appropriate amounts of Zn, Ca, and Cu could lower the ionization energy and increase the transition energy, thus accelerating the transport of electrons in electrode materials and improving the electrochemical performance of the nickel hydroxide electrode. However, excessive doping may change the structure of the nickel hydroxide, which could deteriorate its performance.

6.1.3.3 Morphology control

Nanoparticles and microparticles of nickel hydroxide have been synthesized by two different procedures: sonication and stirring [15]. Microparticles of β-$Ni(OH)_2$ were synthesized by a similar procedure using either magnetic stirring or ultrasonication. Nanoparticles of β-$Ni(OH)_2$ with an average particle size of 18 nm were synthesized by chemical precipitation from a $NiCl_2{\cdot}6H_2O$ solution under ultrasonic irradiation in the presence of the surfactant cetyl trimethylammonium bromide (CTAB). CTAB provides advantageous sites for growth of particles, but also stimulates the formation process, comprising nucleation, growth, coagulation, and aggregation. Furthermore, the addition of CTAB can affect the nucleation before and during the precipitation process. The advantages of the preparation of nanocrystalline material by the ultrasonication method are: more uniform distribution/dispersion of the nanoparticles, a marginally higher surface area, and better thermal stability and phase purity. The cyclic voltammograms (CVs) of microparticles and also nanoparticles show pairs of well-defined peaks. With increasing scan rate, the anodic peak current and the cathodic peak current increase but the peak potentials are almost identical. The linear relationship between i and $v^{0.5}$ suggests that the oxidation of nickel hydroxide is a diffusion-limited process. In the case of semi-infinite diffusion, it has been calculated that the proton diffusion coefficients in the microparticles and nanoparticles are 1.1×10^{-11} and 5.1×10^{-11} $cm^2\ s^{-1}$, respectively.

6.1.3.4 Other methods

In most $Ni(OH)_2$ electrodes, poly(tetrafluoroethylene) (PTFE) has been used as a binder. However, the adhesion strength of PTFE is too weak to prevent dropout of active materials from the substrates. For sealed batteries using epoxy resin, styrene butadiene rubber (SBR), polystyrene acrylic acid ester (PSA), and poly(vinyl acetate) (PVA) as positive electrode binders, electrodes using PSA showed better stability than those with other binders [16]. Epoxy resin, SBR, and PVA have a tendency to be oxidized in electrodes. In sealed batteries using these binders, leakage occurred during charging. Actually, oxygen was consumed for binder oxidation at the positive electrode, the negative electrode was excessively charged instead of H_2O formation, and hydrogen evolved from the fully charged negative electrode.

The utilization efficiency of quasi-three-dimensional (Q-3D) substrate electrodes is higher than that of the corresponding punched metal electrodes [17]. The utilization efficiency of the Q-3D-substrate electrode using CoOOH-coated $Ni(OH)_2$ as active material reached 98%. This could be due to improvement in packing efficiency and the high conductivity of the positive electrode.

Generally, the preparation procedures for spherical nickel hydroxide particles are complex. The precipitated nickel hydroxide has to age for a long time, the product quality cannot be guaranteed, and the production cost is high. The conventional coprecipitation procedures for nonspherical nickel hydroxide powders are much simpler than those for spherical powders because there is no hydroxide-forming step. However, nonspherical nickel hydroxide powders always have a lower tap density (1.5–1.8 g cm^{-3}) because of their irregular particle morphology and wide particle size distribution. The low tap density of the nickel hydroxide electrode results in a low volumetric specific capacity, thus seriously limiting the full cell energy density. Recently, a new polyacrylamide (PAM) assisted two-step drying method was proposed to prepare high-density nonspherical nickel hydroxide positive electrode materials [18]. The tap density of these powders was 2.32 g cm^{-3}, which is significantly higher than that of nickel hydroxide powders obtained by the conventional coprecipitation method.

6.2 Metal oxides in LIBs

6.2.1 Principles and applications of LIBs

LIBs are composed of three major components: a positive electrode, a negative electrode, and an electrolyte. Two electrodes with different lithium chemical potentials are in contact through an electrolyte (ionic conductor and electronic insulator). Three steps can be distinguished concerning electrochemical reactions in lithium-ion batteries: Li ion diffusion within the solid electrode material, charge transfer at the interface between the electrodes and electrolyte, and Li ion transport through the electrolyte. Due to these highly reversible processes, lithium-ion batteries can be recharged. The electrolyte of a cell may be a crystalline solid, a glass, a polymer, or a liquid. Compared with solid electrodes, a liquid or polymer electrolyte is preferred because it is difficult to maintain a close solid-solid electrode-electrolyte interface over many charge/discharge cycles when solid electrolytes are utilized in LIBs [19].

Since the first commercial LIBs were introduced by SONY in 1991, the world has seen considerable advancement in the performance of LIBs. Lithium-ion batteries are currently leading the market of portable electronic devices (such as cell phones, laptop computers, iPads, and digital video recorders) and are also competitive power sources for vehicular transport, such as pure electric (EV), plug-in hybrid (PHEV), and hybrid (HEV) EVs. EVs and PHEVs both use electric motors to move, but the PHEV has an electric motor and internal combustion (IC) engine arrangement, combined in a serial configuration. The IC engine is activated only after the battery power has been used up, to run a generator and produce electricity for running the electric motor. The HEV also has an electric motor and IC engine combination, but in the so-called parallel configuration. While the electric motor is used for starting the vehicle and low-speed propulsion, the IC engine takes over for higher speed and long-distance propulsion.

Thus, the HEV needs a high-power LIB pack, whereas the EV and PHEV require a high-energy-density LIB pack [20].

Other uses of LIBs are in medical care in defibrillators, pacemakers, etc., and in the aerospace industry for storing photovoltaic-generated dc electricity. LIBs are normally charged and discharged at a current rate of 0.2–1C, so that full capacity of the cell is stored or consumed in 5–1 h, respectively. The usual operating temperature range of the LIBs is 15–60°C. Governments in Asia, Europe, and North America and their funding agencies are now interested in supporting accelerated R&D activities for a new generation of advanced batteries and onboard energy storage technologies [21].

6.2.2 *Metal oxide-based positive electrodes for LIBs*

There are two kinds of metal oxide-based positive electrode materials: lithium mixed metal oxides ($LiCoO_2$, $LiNiO_2$, $LiNi_{1/3}Co_{1/3}Mn_{1/3}$, $LiMn_2O_4$, and Li_2MnO_3) and transition metal oxides (V_2O_5 and MoO_3).

The commercial positive electrode material, $LiCoO_2$, has been the most successful material for LIBs over the past two decades due to its high-energy density, simple preparation, and long cycling life. The theoretical capacity of $LiCoO_2$ is 274 mAh g^{-1}, but in practice the capacity is only about 140 mAh g^{-1} because the number of reversible Li ions in $LiCoO_2$ is typically <0.5 per unit. Excessive delithiation therefore results in a phase transition from the hexagonal to the monoclinic phase, leading to sudden contraction along the *c*-axis (9% volume change). Furthermore, highly oxidized Co^{4+} ions are likely to decompose electrolytes in the particle surface, resulting in substantial amounts of gas generation. In addition, Co dissolution at 4.5 V at elevated temperatures induces the simultaneous Li dissolution from its surface [22].

$LiNiO_2$ has a layered structure like $LiCoO_2$ but it is cheaper than $LiCoO_2$. Its reversible capacity ($>$180 mAh g^{-1}) is also higher than that of $LiCoO_2$. However, it is not easy to prepare on a large scale because of the difficult oxidation of Ni^{2+} to Ni^{3+} and the easy decomposition of $LiNiO_2$ into Li-deficient compounds Li_dNiO_{2-d} ($0<d<1$) during high-temperature synthesis. Actually, Ni-rich $LiNi_{0.8}Co_{0.2}O_2$ and layered $LiNi_{1/3}Co_{1/3}Mn_{1/3}O_2$ are more often applied as a possible replacement for $LiNiO_2$ because of their higher energy density, stable structure, and easy preparation [23].

$LiNi_{1/3}Co_{1/3}Mn_{1/3}O_2$, with valences of the nickel, cobalt, and manganese ions of +2, +3, and +4, respectively, was first introduced as an alternative positive electrode material in 2001. In this material, only the divalent nickel and trivalent cobalt ions are electroactive, comprising Ni^{2+}/Ni^{4+} and Co^{3+}/Co^{4+} redox couples [24]. Because the tetravalent manganese ions are not electrochemically active, the material is structurally stable, thus effectively preventing the capacity fading usually induced by Mn dissolution and Jahn-Teller distortion.

Because it is relatively inexpensive, environmentally friendly, and manganese resources are abundant, spinel $LiMn_2O_4$ is considered a prime candidate for use as high-power positive electrode materials in lithium-ion batteries. Li ion deintercalation occurs at about 4 V (vs Li^+/Li) during a two-stage process ($LiMn_2O_4/Li_{0.5}Mn_2O_4$ and $Li_{0.5}Mn_2O_4/\gamma\text{-}MnO_2$) with a theoretical capacity of 148 mAh g^{-1}. Further lithiation of

$LiMn_2O_4$ ensues at about 3 V (vs. Li^+/Li), leading to formation of a rock salt phase, $Li_2Mn_2O_4$. However, there are still some problems that should be solved prior to its commercial use. Spinel $LiMn_2O_4$ undergoes extreme capacity fading at high temperatures (>50°C). The poor cyclability is caused by several problems, one of which is Mn dissolution. Manganese dissolution occurs by two mechanisms [25]:

(i) the formation of soluble Mn^{2+} by the disproportionation of $LiMn_2O_4$:

$$2Mn^{3+} \rightarrow Mn^{2+} + Mn^{4+} \tag{6.5}$$

(ii) acid dissolution by hydrogen fluoride (HF):

$$Li_{1-x}Mn_2O_4 + HF \rightarrow \lambda - Mn_2O_4 + LiF + Mn^{2+} + H_2O \tag{6.6}$$

Li_2MnO_3, which can be rewritten as $Li[Li_{1/3}Mn_{2/3}]O_2$, has $C2/m$ space group symmetry and is known to be electrochemically inactive because of the presence of Mn in its +4 valence state. However, it can be made into an electrochemically active positive electrode material by extracting Li and oxygen (or lithium oxide) from the structure by either chemical or electrochemical means, that is:

$$Li_2MnO_3 \rightarrow Li_2O + MnO_2 \tag{6.7}$$

In 1991, inspired by the discovery that acid treatment of the spinel $LiMn_2O_4$ yielded λ-MnO_2 with a Mn_2O_4 spinel framework, researchers synthesized the layered lithium manganese oxide compound $Li_{2-x}MnO_{3-x/2}$ ($0<x<2$) with a cubic-close-packed oxygen anion array by chemical leaching of Li_2O from the rock salt phase Li_2MnO_3 (or $Li_2O \cdot MnO_2$) with acid at 25°C, and got the compound $Li_{1.09}Mn_{0.91}O_2$ (or $0.2Li_2MnO_3 \cdot 0.8LiMnO_2$) after relithiation. The ideal capacity of Li_2MnO_3 is 459 mAh g^{-1} upon total Li deintercalation. However, it suffers from a huge irreversible capacity loss of >130 mAh g^{-1}. The irreversible capacity stems from the fact that instead of re-forming Li_2MnO_3, only half of the Li^+ extracted from Li_2MnO_3 is re-inserted into the MnO_2 structure through the reaction:

$$Li + MnO_2 \rightleftharpoons LiMnO_2 \tag{6.8}$$

More seriously, the emission of O_2 in the organic electrolyte is very dangerous and causes safety issues. In fact, nowadays the Li-rich layered oxide materials xLi_2MnO_3-$(1-x)LiMO_2$ (M = Co, Ni, Fe, Cr, etc.) are becoming promising candidates for positive electrode material because they have a discharge capacity of 280 and 1000 Wh kg^{-1} when operated above 4.7 V (vs. Li^+/Li), which is about twice the capacity of current commercial positive electrode materials for LIBs [26].

For many years, molybdenum trioxide (MoO_3) has been considered an interesting positive electrode material in rechargeable LIBs because of its Mo^{6+}/Mo^{4+} redox ability with an exchange of 2 electrons per metal atom. This implies a theoretical specific charge of up to 372 Ah kg^{-1} and a theoretical energy density of 930 Wh kg^{-1} considering the average discharge potential of 2.5 V (vs. Li/Li^+). V_2O_5 has been intensively studied in recent years for application in LIBs due to its high-energy density, ease of synthesis, and low cost. Specifically, a high discharge capacity of 294 mA h g^{-1} can be delivered when two Li^+ ions are intercalated. In the CV curves, two intense reduction peaks located at 3.30 and 3.1 V (vs. Li^+/Li) correspond to phase changes from α-V_2O_5 to ε-$Li_{0.5}V_2O_5$ and then to δ-LiV_2O_5. The third reduction peak at 2.2 V (vs. Li^+/Li) is attributed to the intercalation of the second Li^+ ion, leading to the formation of γ-$Li_2V_2O_5$. Three obvious oxidation peaks appear during the anodic scanning, at 2.6, 3.4, and 3.5 V (vs. Li^+/Li), respectively. However, the practical application of V_2O_5 as a positive electrode material has been hindered by the poor kinetics, including the slow Li^+ ion diffusion (10^{-12} cm^2 s^{-1}) and low electric conductivity (10^{-2}–10^{-3} S cm^{-1}) [27].

6.2.3 Modification of metal oxide-based positive electrodes for LIBs

6.2.3.1 Coating

Surface coating is a controllable approach that can stabilize the electrode-electrolyte interface, and yields an artificial protective layer that serves as a substitute for the SEI film. Surface coatings of metal oxides (like Al_2O_3 or MgO) can essentially shield the positive electrode material from direct exposure to the electrolyte solution, improving the structural integrity of the active material. Another application of oxide coatings is for the removal of HF from the electrolyte solution. HF is a by-product resulting from the decomposition of $LiPF_6$ by trace amounts of moisture. The HF produced by hydrolysis of $LiPF_6$ would start dissolving the active material and so erode its surface, leading to further capacity decay during prolonged cycling.

The thermal stability of $AlPO_4$-coated $LiCoO_2$ has been examined using overcharge safety tests. Batteries with $LiCoO_2$ and $AlPO_4$-coated $LiCoO_2$ electrodes, respectively, were charged to 12 V and were then kept at that voltage for 2 h. In the case of uncoated $LiCoO_2$, the electrolyte decomposition during overcharge occurred at about 5 V (vs. Li^+/Li). Upon reaching a cell voltage of 12 V, the surface temperature rose to 500°C with the separator melting at c.120°C, causing an internal short circuit followed by fire and explosion of the cell. In the case of the $AlPO_4$-coated $LiCoO_2$, although the batteries swelled because of electrolyte decomposition, the electrodes did not experience short circuits because the surface temperature reached only 60°C. This improved thermal stability can be ascribed to the strong P=O bond and the strong electrostatic attraction between the PO_4^{3-} oxyanion and the Al^{3+} cation in $AlPO_4$ [28].

$LiCoPO_4$ is another electrochemically active compound with a redox potential of about 5 V (vs. Li^+/Li); it is a highly stable material because of the strong P=O bond. An olivine $LiCoPO_4$ phase grown on $LiCoO_2$ has been prepared by heating precipitated $Co_3(PO_4)_2$ nanoparticles and $LiCoO_2$ powder [29]. The swelling property of cells containing $LiCoPO_4$-coated $LiCoO_2$ electrodes was compared with those of cells with bare $LiCoO_2$ and $AlPO_4$-coated $LiCoO_2$ electrodes (Fig. 6.2). Electrolyte decomposition and gas evolution caused an increase in the cell thickness when the charged

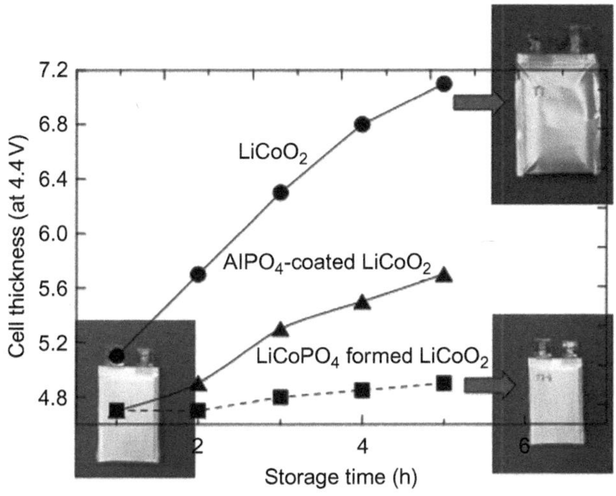

Fig. 6.2 Plot of battery surface temperatures after nail penetration of the Li-ion battery containing bare $LiCoO_2$ (sample A), and $LiCoPO_4$ phase formed in $LiCoO_2$ (sample B). Reproduced with permission from H. Lee, M.G. Kim, J. Cho, Electrochem. Commun. 9 (2007) 149–154, Elsevier.

cells containing bare $LiCoO_2$ and $AlPO_4$-coated $LiCoO_2$ were stored at high temperature. However, the $LiCoPO_4$-coated $LiCoO_2$ material produced very little electrolyte decomposition and insignificant swelling. This indicates that the $LiCoPO_4$ phase reduced exothermic reactions with the electrolyte. $AlPO_4$ coating was also implemented in other positive electrode materials such as $LiNi_{0.8}Co_{0.1}Mn_{0.1}O_2$ [30] and $Li[Li_{0.2}Mn_{0.54}Ni_{0.13}Co_{0.13}]O_2$ [31]. The coating layer on the surface of the layered oxide lattice results in a higher reversible capacity and lower irreversible capacity because it suppresses diffusion and eliminates oxygen vacancies.

The beneficial effect of metal oxide coating is also observed with the spinel $LiMn_2O_4$ material. The coating metal oxides (MO) are treated with HF and form MF_y and H_2O. Several metal oxides MO_x (M = Mg, Zr, Zn, Al, Co) have been investigated as coating layers, and these MO_x-coated $LiMn_2O_4$ electrodes exhibited greatly improved high-temperature cycling performance [32]. For example, colloidal nanoparticles of ZrO_2 were deposited on the surface of the spinel $LiMn_2O_4$ electrode. They suppressed the dissolution of Mn^{3+} and the strong Zr—O bonds at the electrode surface lower the oxygen activity at high potentials. ZrO_2 on the surface also scavenges the acidic HF species released by the electrolyte solution by reaction with HF. As a result, excellent capacity retention on cycling was achieved [33]. To prevent the dissolution of Mn^{2+}, the surface of $LiMn_2O_4$ can be coated with a thin layer of inorganic $Li_2O{\cdot}2B_2O_3$ glass. $Li_2O{\cdot}2B_2O_3$-coated $LiMn_2O_4$ exhibits good wetting properties. The combination of its good wetting properties and relatively low viscosity in the molten state allows easy processing and results in well-distributed coverage with the use of a minimal amount of material [32].

Compared with metal oxide coating, some Li-containing oxides and gel polymer electrolyte (GPE) coatings are more effective for enhancing the electrochemical performance of positive electrodes, because such coating layers have high Li^+ conductivity and can provide tunnels for Li^+ transport during the charge/discharge processes. For example, polyimide (PI) GPE has been introduced onto the $LiNi_{1/3}Co_{1/3}Mn_{1/3}O_2$ surface via thermal imidation of a polymeric acid copolymer comprising pyromellitic dianhydride (PMDA)/oxydianiline (ODA) [34]. The GPE coating is effective in preventing the formation of an undesired resistive layer hindering charge transport at the $LiNi_{1/3}Co_{1/3}Mn_{1/3}O_2$/liquid electrolyte interface.

Carbon coating has remarkable advantages for many transition metal oxides. However, the lithium transition metal oxides are usually synthesized by heating in air at temperatures >700°C. Under these conditions, the carbon shell formed by carbonization of added organic precursors will probably burn away, leading to very little residual carbon. To avoid the loss of carbon, the calcination may be carried out in an inert atmosphere, but the thermal decomposition of the carbon source would probably produce a strongly reducing environment such as H_2, CO, or a combination of the two, which could easily lead to reduction of the transition metal oxides and change the valences of the elements and even their crystal structure. Thus, carbon coating still faces a high hurdle for application on lithium transition metal oxides, and almost all the examples investigated involved a postaddition (mechanical mixing) method, or low temperature and short annealing time when applying the carbon coating [35]. For example, carbon-coated $LiCoO_2$ was obtained by calcining a mixture of $LiCoO_2$ and sugar at 600°C for 30 min in air [36]. As the calcination was conducted for a short time, the carbon coating did not destroy the structure of $LiCoO_2$ or reduce the $LiCoO_2$. After coating the charge capacity of the $LiCoO_2$ increased from 128 to 141 mAh g^{-1}, and the discharge capacity increased from about 110 to 130 mAh g^{-1}. The conductive carbon causes a decrease in the charge transfer resistance and an increase in the diffusion coefficient for lithium ions. Carbon coating has also been employed as a way to improve the electronic conductivity of V_2O_5 [37]. The electronic conductivity of V_2O_5 nanofibers increases with increasing carbon coating. Moreover, carbon-coated V_2O_5 nanofibers showed a slightly lower initial discharge capacity of 300 mAh g^{-1} with improved capacity retention of 65% after 50 cycles at 0.1C rate, whereas native V_2O_5 nanofibers exhibited only 40% capacity retention under the same testing conditions.

An Al_2O_3 coating layer on Li-enriched $Li[Li_{0.12}Mn_{0.54}Ni_{0.13}Co_{0.13}]O_2$ greatly improved the electrochemical performance as well as the thermal stability. The improved electrochemical performance was tentatively attributed to the "buffer" layer provided by the Al_2O_3 coating, through which the activity of the extracted oxygen was reduced and electrolyte decomposition was suppressed at voltages above 4.5 V [38]. More recently, it has been reported that AlF_3 chemically leached extra Li ions from the Li-rich layer materials and triggered a layer-to-spinel phase transformation [39]. The characteristic spinel ($LiMn_2O_4$) plateau at 2.8 V was noticeable and became progressively more distinct with increasing AlF_3 coating. The spinel phase on the Li-rich electrodes was able to negate the previous shortcomings. Different from earlier reports describing surface modification of solid solutions with electrochemically inactive materials (such as Li-Ni-PO_4, AlF_3, and ZnO_2), $FePO_4$ nanoparticles have been uniformly coated on the entire surface of Li_2MnO_3 nanoplates [40]. The coated $FePO_4$ served as a host for some Li^+ ions, which had been leached from the Li_2MnO_3 matrix after the initial charge and did not intercalate back into the MnO_2 during the discharging process, thereby producing electrochemically active Li_xFePO_4. As a result, the coulombic efficiency in the first cycle was near 100%.

Normally, materials composed of spherical particles have higher tap densities than those composed of irregular particles. Li_2MnO_3 microspheres have been synthesized by a simple solid-state reaction. Then, a core-shell structure of porous Co_3O_4 grown on the Li_2MnO_3 microspheres was obtained by a simple hydrothermal method followed by heat treatment [41]. Since the Co_3O_4 outer shell is porous, the Li^+ can easily pass through the outer shell and intercalate/deintercalate into/from the core. The composite exhibited an attractive discharge capacity of 178 mAh g^{-1} with little irreversible capacity loss over the voltage range 2.0–4.8 V (vs. Li^+/Li). Interestingly, the Co_3O_4 in the outer shell successfully solves the irreversible capacity loss and oxygen emission problems of Li_2MnO_3 since the Co_3O_4 shell in the composite can serve as a host to react with "Li_2O" and produce electrochemically active Li_xCoO_y, which can provide available Li^+ for intercalation. Thus, this 'synergistic effect' between the core (Li_2MnO_3 microspheres) and shell (porous Co_3O_4) not only decreases the initial irreversible capacity but also eliminates oxygen emission.

6.2.3.2 Doping

Many studies have utilized more of the reversible capacity of $LiCoO_2$ through the enhanced structural stability of the doped oxides $LiM_xCo_{1-x}O_2$ (M=Al, Cr, Mg). It has been suggested that the coated MO_x (M=Zr, Al, Ti, B) layer reacts with $LiCoO_2$ during heating to form a thin layer consisting of a $LiM_xCo_{1-x}O_2$ solid solution [42]. The improved cyclability was attributed to the suppression of the *c*-axis expansion of the Li_xCoO_2 and dissolution of cobalt in the electrolyte when cycling between 2.75 and 4.4 V (vs. Li^+/Li).

Much research has been carried out to improve the structural stability of Ni-rich positive electrode materials by doping cations (Mg^{2+}, Zr^{4+}, Ti^{4+}) [43]. When Mg^{2+} ions (with ionic radius 0.72 Å) were inserted in Li sites in $LiNi_{0.86}Co_{0.09}Mg_{0.05}O_2$, the cycling stability was enhanced by 70% after 200 cycles between 3.4 and 4.3 V at 60°C [44]. Due to the difference in ionic radius between Mg^{2+} and Ni^+ ions, the Mg^{2+} ions are destabilized within the slabs and then migrate to the interslab space during the charging process. This change in cationic distribution limits the de-cohesion crystallites, which leads to a smaller volume change. Magnesium ions also play an important role in screening the O^{2-}–O^{2-} repulsions during charge and discharge, thus preventing interslab collapse.

Fluorination has been introduced to modify the surface of $LiNi_{0.8}Co_{0.15}Al_{0.05}O_2$ by a one-step facile and dry method [45]. The electronegativity of F is the highest among all the elements, leading to the highest electrophilicity. Metal-fluorine (M-F) bonds replace partial metal-oxygen (M-O) bonds at the surface, improving the bond energy as well as the structural stability. Moreover, the introduction of F can protect the M-O bond from side reaction with the electrolyte and the atmosphere, reducing the release of O from the space lattice and enhancing the safety performance. In addition, the fluoride anion has a greater interaction with the Li^+ ion than the oxide anion, which can lead to faster transference of Li^+ ions between the active material and the electrolyte. As a result, the surface-fluorinated $LiNi_{0.8}Co_{0.15}Al_{0.05}O_2$ had a reversible capacitance of 220 mAh g^{-1} at 0.1C and an excellent long-term cycling stability with 93.6% capacity retention after 80 cycles, which is much better than that of pristine commercial $LiNi_{0.8}Co_{0.15}Al_{0.05}O_2$.

Partially doping of $LiNi_{1/3}Co_{1/3}Mn_{1/3}O_2$ with foreign atoms has also been reported. For example, some Zr^{4+} diffuses into the $LiNi_{1/3}Co_{1/3}Mn_{1/3}O_2$ during high-temperature treatment [46]. Because the ion radius of Zr^{4+} (0.079 nm) is more than that of Co^{3+} (0.0545 nm), Ni^{2+} (0.069 nm), and Mn^{4+} (0.053 nm), the lattice parameters a and c increase as the ratio Zr/(Ni + Co + Mn) increases to 1%. The expansion in the lattice caused by a small amount of Zr^{4+} doping can benefit the transportation of Li^+ ions and enhance the rate capability of $LiNi_{1/3}Co_{1/3}Mn_{1/3}O_2$. The fluorine doping of the $LiNi_{1/3}Co_{1/3}Mn_{1/3}O_2$ promotes the grain size growth and improves its crystallinity [47]. The substitution of oxygen by fluorine should reduce the lattice parameters, given that the radius of F^- (0.134 nm) is less than that of O^{2-} (0.140). The substitution also stabilizes the interface between the electrolyte and the surface layer of the particles when the fluorine content is low.

The advantageous influence of doping has also been observed with spinel $LiMn_2O_4$. Al-doped $LiMn_2O_4$ nanotubes were synthesized using β-MnO_2 nanotubes as self-templates [48]. The obtained products exhibited excellent cycling performance under high current rates at both room and elevated temperatures. The tubular structure, single-crystalline nature, and high crystallinity of the Al-doped spinel nanotubes were well preserved after aging tests for 30 days at 55°C, implying high stability of the Mn—O bond and low occurrence of locally distorted $Mn^{3+}O_6$ groups, due to the Al^{3+} doping. However, the radius of trivalent Al^{3+} is 0.0535 nm, smaller than that of trivalent Mn^{3+}. Consequently, Al^{3+} may substitute Li^+ in the 8a tetrahedral sites, and may cause the original Li^+ to move to 16c octahedral sites. The Li^+ at the octahedral sites cannot de-intercalate at 4 V, so that its electrochemical performance declines. When Cr^{3+} ions are investigated as dopants in $LiMn_2O_4$, it remains a single phase of spinel because the radius of Cr^{3+} is 0.0615 nm, close to that of Mn^{3+} [49]. Its capacity will decrease with increase in the doping amount of chromium, which can be partly attributed to lattice contraction. The $LiCr_xMn_{2-x}O_4$ exhibits better cycle behavior than pristine $LiMn_2O_4$ due to an increase in the stability of the spinel structure. The bond energies of the M-O bonds in MnO_2 (α-type) and CrO_2 are 946 and 1029 kJ mol^{-1}, respectively.

Recently, a novel $TiSi_2/V_2O_5$ heteronanostructure based on a unique nanonet platform was realized [50]. The active material was Ti-doped V_2O_5. $TiSi_2$ nanonets act as structural support and charge transporter. Titanium doping stabilizes V_2O_5, and a thermal annealing induced SiO_2 layer shields the $TiSi_2$ from the electrolyte. The protective SiO_2 coating prevents Li^+ from reacting with $TiSi_2$, which otherwise would lead to the destruction of the nanostructures. Nanobelts of α-MoO_3 can be modified to $H_xMo(O,N)_3$ by a moderate ammonolysis method [51]. Between 200°C and 300°C,

NH_3 gas diffuses between the layers of MoO_3 and reacts with terminal oxygen, forming $H_xMo(O,N)_3$. Hydrogen is introduced into the layers and binds to terminal oxygen, together with the effect of nitridation. Thus, the unit cell volume shrinks significantly, mostly along the *b*-axis. Over the voltage range 1.5–3.5 V (vs. Li^+/Li), the specific capacity of $H_xMo(O,N)_3$ increased to >250 mAh g^{-1} and was cycled 300 times without fading. Nitrogen has a lower electronegativity than oxygen and in this material has a negative charge of −3 instead of −2. The substitution of O by N therefore narrows the band gap of the oxide and results in an increased concentration of charge carriers. Consequently, the electronic conductivity is enhanced by nitrogen doping and may well be increased for a large number of oxides.

6.2.3.3 *Morphology control*

Recent research has focused on nanosizing electrode materials, holding the promise of higher (dis)charge rates. The intended advantage of nanostructured insertion electrodes is the shortening of the diffusion distance through the solid state, since solid-state diffusion is relatively sluggish compared with diffusion through the liquid electrolyte. According to the diffusion formula $t = L^2/2D$, where t is the diffusion time, L is the diffusion distance, and D is the diffusion coefficient, reducing the particle size can significantly shorten the diffusion time of Li^+ ions in active materials, resulting in a much enhanced power performance. In addition, the use of nanomaterials can also enhance the cycling performance of the material by alleviating some of the lattice strain experienced during cycling due to the (de)intercalation of Li^+ ions.

Nanocrystalline $LiCoO_2$ has been synthesized by hydrothermal reaction between CoOOH and an aqueous solution of LiOH [52]. By varying the reaction temperature, the LiOH concentration, and the reaction time, a size-controlled synthesis of $LiCoO_2$ was achieved. Fig. 6.3A shows the discharge curves for various particle sizes of

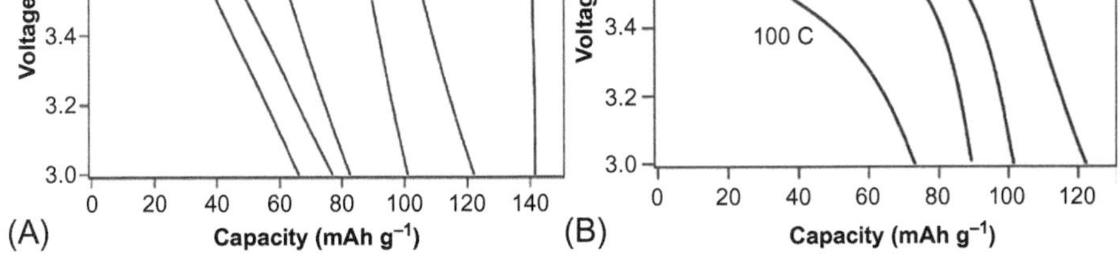

Fig. 6.3 (A) The discharge curves for $LiCoO_2$ in various sizes. (B) Discharge curves of nanocrystalline $LiCoO_2$ (17 nm) at various discharge rates (1–100C rate).
Reproduced with permission from M. Okubo, E. Hosono, J. Kim, M. Enomoto, N. Kojima, T. Kudo, H. Zhou, I. Honma, J. Am. Chem. Soc. 129 (2007) 7444–7452, the American Chemical Society.

nanocrystalline $LiCoO_2$ (bulk to 6 nm) at 1C rate. Extreme size reduction below 15 nm was not favorable for specific capacity. Nanocrystalline $LiCoO_2$ with a size of 17 nm had a high-rate capability even at 100C rate (Fig. 6.3B). Capacitor behavior became more dominant with decreasing crystallite size, while the plateau region at 3.9 V (the capacity for the inner layers) decreased. The increasing capacitor behavior is ascribed to the nanosize effect on the intercalation of Li^+ ions.

Another impressive role of morphology control of electrode materials is noticeable in the spinel $LiMn_2O_4$ material. One-dimensional (1D) $LiMn_2O_4$ nanorods [53], nanowires [54], and nanotubes [55] have been prepared using the correspondingly formed MnO_2 as a self-sacrificial template. The electrode performance can be improved by using these forms of single-crystalline 1D $LiMn_2O_4$. For example, even at very high current rates such as 100°C and 150°C, the discharge curves of $LiMn_2O_4$ nanowires feature an almost flat potential plateau, delivering 100 and 78 mAh g^{-1} at very high rates (60°C and 150°C, respectively) [54]. The $LiMn_2O_4$ nanowires had diameters around 10 nm and were several micrometers in length. Such morphology improves the kinetic properties at very high current rates and is capable of the facile structural transformation of the cubic and tetragonal phase over a large compositional range. The single-crystalline $LiMn_2O_4$ nanowire has also been synthesized by a novel reaction using long single-crystalline $Na_4Mn_9O_{18}$ ($Na_{0.44}MnO_2$) nanowires as a self-template [56]. $Na_{0.44}MnO_2$ nanowire was first synthesized by a simple hydrothermal process. The resulting $Na_{0.44}MnO_2$ is a single-crystalline ultralong length nanowire in a tunnel structure. The formation of $LiMn_2O_4$ nanowire involves the ion-exchange between Li^+ and Na^+ and reaction between $Li_{0.44}MnO_2$ and Li_2MnO_3. These spinel $LiMn_2O_4$ nanowires also displayed high reversible capacities even at remarkably high charge-discharge rates.

The reversible capacity of Li_2MnO_3 is highly dependent on its morphology. For example, Li_2MnO_3 microparticles showed a discharge capacity of <50 mAh g^{-1}, while Li_2MnO_3 nanoparticles delivered a discharge capacity of 250 mAh g^{-1} [57]. Layered $Li_{0.88}[Li_{0.18}Co_{0.33}Mn_{0.49}]O_2$ nanowires have been prepared using $Co_{0.4}Mn_{0.6}O_2$ nanowires and lithium nitrate as precursors at 200°C by a hydrothermal method for fast and high capacity Li ion storage material [58]. The obtained nanowires exhibited a reversible capacity of 230 mAh g^{-1} between 2 and 4.8 V (vs. Li^+/Li), even at the high current rate of 3.6 A g^{-1}.

Morphology control of V_2O_5 materials for LIBs has been widely studied. To obtain V_2O_5 nanosheets, V_2O_5 powder was reacted with H_2O_2 in combination with ultrasonic treatment to generate V_2O_5 gel. Then, the V_2O_5 gel was diluted, freeze dried, and further treated at 450°C in air [59]. The inexpensive raw materials (only commercial V_2O_5 powder and H_2O_2 without using any templates) and facile experimental procedures favor large-scale production of the V_2O_5 nanosheets. These 2D leaf-like V_2O_5 nanosheets exhibited excellent Li^+ storage properties, including high reversible capacity (303 mAh g^{-1}), high-rate capability (104 mAh g^{-1} at 5 A g^{-1}), and good capacity retention upon cycling (capacity fading rate about 0.22% per cycle). The large specific area of the 2D leaf-like nanosheets facilitates electrolyte transport for intercalation and deintercalation of the Li^+ ions; the hierarchical porous structure relaxes the mechanical strain generated upon the charge/discharge cycling.

Hierarchical vanadium-based hollow microflowers have been prepared by a facile solvothermal approach [60]. This novel vanadium-based nanostructure is readily transformed to its V_2O_5 counterpart by calcination in air. The derived V_2O_5 nanosheet-assembled hollow microflowers exhibited excellent lithium storage properties with high capacity and superior cycling stability. Crystalline V_2O_5 thin films have also been obtained by a new atomic layer deposition (ALD) process [61]. Such V_2O_5 films obtained using O_3-based ALD showed superior electrochemical performance as a positive electrode. ALD can greatly benefit energy storage technology, although significant technical and manufacturing challenges, especially cost and complexity, remain.

6.2.4 Metal oxide-based negative electrodes for LIBs

Negative electrode materials based on metal oxides for use in LIBs can be categorized into three different classes:

(1) transition metal oxides obtained by conversion reactions, such as NiO, MoO_2, Co_3O_4, and Fe_2O_3;
(2) metal oxides based on alloy reactions, such as the mixed oxides of Sn, Si, and Ge; and
(3) intercalation compounds, such as TiO_2, $Li_4Ti_5O_{12}$, and Li_3VO_4.

In 2000, researchers reported unexpected electrochemical performance of nanosized transition metal oxides including CoO, FeO, and NiO as negative electrodes for LIBs [62]. These transition metal oxides with various oxidation states can be reversibly reduced and oxidized by conversion routes with the following general Eq. (6.9):

$$M_xO_y + 2yLi^+ + 2ye^- \rightleftharpoons xM + yLi_2O \tag{6.9}$$

These metal oxides usually showed high specific capacities but frequently had low cycling ability associated with the extreme volume changes occurring during the charge/discharge processes. Their low conductivities also resulted in poor rate behavior and further performance degradation.

The lithium storage mechanism in GeO_2 and SnO_2 is described by an initial conversion reaction (Eq. 6.10) followed by an alloying reaction (Eq. 6.11) [63]. The Li^+ ions, however, are not fully removed in the subsequent charging cycle due to irreversible Li_2O formation. This limits the theoretical lithium storage capacity to 4.4 Li^+ ions per GeO_2 unit (1126 mAh g^{-1}) compared with 8.4 Li^+ ions per SnO_2 unit (2152 mAh g^{-1}):

$$MO_2 + 4Li^+ \rightarrow M + 2Li_2O \tag{6.10}$$

$$M + 4.4Li^+ \rightleftharpoons Li_{0.44}M \tag{6.11}$$

Spinel $Li_4Ti_5O_{12}$ is an excellent and safe negative electrode material because (1) it does not undergo volume change during intercalation and de-intercalation of Li^+ ions, and (2) no SEI film is formed. It has a level discharge plateau at approximately 1.55 V (vs. Li/Li^+), which is higher than the reduction potential of most electrolyte solvents so the solvent cannot be reduced and form a solid electrolyte interface. The $Li_4Ti_5O_{12}$ structure can accommodate up to three Li^+ ions with a theoretical specific capacity of 175 mAh g^{-1}; this corresponds to the chemical formula $Li_7Ti_5O_{12}$. However, $Li_4Ti_5O_{12}$ has a quite low electronic conductivity (c.10^{-13} S cm^{-1}) and Li^+ ion diffusion coefficient (10^{-9}–10^{-13} cm^2 s^{-1}) [63].

Titanium dioxide, TiO_2, is a typical safe negative electrode material for use in LIBs, and has an operating voltage above 1 V, which is higher than the potential at which most electrolytes and solvents are reduced. TiO_2 exists in several polymorphic modifications: anatase, rutile, brookite, TiO_2-B (bronze), TiO_2-R (ramsdellite), TiO_2-H (hollandite), TiO_2-II (columbite), and TiO_2-III (baddeyite). Furthermore, the volume expansion of TiO_2 during Li^+ ion intercalation/deintercalation is almost negligible (<4%), endowing it with good structural stability and extended cycle life. Generally, the Li storage and cycling performance of TiO_2 polymorphs depends on the method of preparation, particle size, and morphology. As a general principle, smaller particle sizes ($\leq$200 nm) with a large surface area and porous morphology with interconnected particles can be expected to deliver stable and near theoretical capacity [63].

Li_3VO_4 intercalates Li ions mainly in the voltage range between 0.5 and 1.0 V (vs. Li^+/Li), which is lower than the potential of $Li_4Ti_5O_{12}$ and higher than that of graphite. Its theoretical capacity is 394 mAh g^{-1}, in accordance with $x=2$ in $Li_{3+x}VO_4$. It might act as a new intercalation negative electrode for LIBs. The structure of Li_3VO_4 is built up of oxygen atoms in approximately hexagonal close packing; the cations occupy ordered tetrahedral sites. The octahedral sites are vacancies and connect to each other establishing a 3D channel for lithium ion diffusion [21]. However, Li_3VO_4 requires further modifications to overcome certain limitations, such as its low electronic conductivity, which is lower than its ionic conductivity, and this may cause large resistance polarization and poor rate capability. Despite, Li_3VO_4 has been studied as a fast Li-ion conductor for many years.

6.2.5 Modification of metal oxide-based negative electrodes for LIBs

6.2.5.1 Coating

Coating the surface of the electrode material is a common technique used to tailor the functionality or performance of the electro-active material. Carbon coating is one of the most popular means of easily enhancing the electronic conductivity of negative electrode materials, and can also stabilize SEI films. Importantly, the power ability of a battery ($P=U\times I$, where U is the cell voltage and I is the current density) is determined by its internal resistance (R): higher resistance leads to a more rapid decline of the cell voltage at high current densities due to the ohmic drop (which is the deviation between practical and theoretical voltage, $\Delta U=I\times R$). Furthermore, the temperature increase in the battery is also related to the square of the internal resistance, according to the ohmic law ($Q=I\times R^2$, where Q is the heat generated in the charge/discharge process) [35].

Nanowire arrays of single-crystal SnO_2 coated with amorphous carbon nanotubes have been fabricated by drying and annealing an anodic alumina membrane filled with SnO_2 sol supported by citric acid chelating agent [64]. The nanocomposite exhibited a reversible capacity in the first cycle of 418 mAh g^{-1} based on the weight of SnO_2. After 30 cycles, the discharge capacity was still 353 mAh g^{-1}, while the specific capacity of only SnO_2 powder was 511 mAh g^{-1} on the first discharge, dropping rapidly to 62 mAh g^{-1} after 23 cycles. The large volume expansion of the SnO_2 powder during cycling led to pulverization of the electrode. In the case of the coated material, dimensional confinement of the SnO_2 nanowires by the surrounding CNT limits the volume expansion upon Li^+ ion insertion. Recently, $Mo_3O_{10}(C_6H_8N)_2{\cdot}2H_2O$ nanowires were synthesized as precursors by reaction between ammonium heptamolybdate and aniline in acid solution. This organic-inorganic hybrid precursor provides a simple approach to obtaining MoO_2/C nanowires with a unique sub-nanometer-contacting organic-inorganic hybrid structure [65]. The organic species in the precursors are carbonized to form MoO_2/C hybrid structures during calcination. The MoO_2/C hybrid nanowires showed good rate capability even at a high current density of 1 A g^{-1}, which is much improved compared with that of the bulky micron-sized MoO_2.

Nanosized GeO_2/Ge particles coated by a thin layer of carbon have been synthesized using a simple method involving simultaneous carbon coating and reduction by acetylene gas [66]. The conversion reactions in GeO_2/C and $GeO_2/Ge/C$ samples exhibited better reversibility than nano-GeO_2 and bulk-GeO_2. During the first cycle, the bulk-GeO_2 exhibited discharge and charge capacities of 1587 and 511 mAh g^{-1}, respectively, corresponding to a coulombic efficiency of 32%. The first discharge and charge capacities of nano-GeO_2 were 2444 and 1348 mAh g^{-1}, respectively, with a coulombic efficiency of 55%. The first discharge cycle entailed the reaction of 9.5 Li^+ ions, which is more than the theoretical 8.4 Li^+ ions. The additional 1.1 Li^+ could be due to the formation of the solid electrolyte interphase (SEI) layer on the surface of the material. In the case of the GeO_2/C sample, the first discharge and charge capacities at the same rate were 2682 and 2000 mAh g^{-1}, respectively, corresponding to 75% coulombic efficiency. Thus, carbon coating improved the coulombic efficiency by 20%, and it also enhanced the reversibility of germanium reoxidation during charging. As for the $GeO_2/Ge/C$ sample, the first discharge and charge capacities were 2293 and 1872 mAh g^{-1}, respectively, with a coulombic efficiency of 82%. In general, the alloying reaction shows good reversibility for both the GeO_2/C and the GeO_2-nano samples. In the case of the $GeO_2/Ge/C$ sample, both the alloying and the conversion reaction with lithium are highly reversible. The carbon coating and nanosized particles played an important role in increasing the reversibility of the conversion reaction. More interestingly, the elemental germanium in $GeO_2/Ge/C$ plays a crucial role as a catalyst in improving the reversibility of the conversion reaction of GeO_2. The reaction of Li^+ ions with nano-GeO_2, GeO_2/C, and $GeO_2/Ge/C$ is presented in Fig. 6.4. The interconnected carbon shells of the GeO_2/C and $GeO_2/Ge/C$ samples provide an efficient network for electron transfer, which in turn accelerates the kinetics of the lithium reactions. Moreover, the carbon shells act as a buffer matrix to limit the volume variation during charge and discharge cycles.

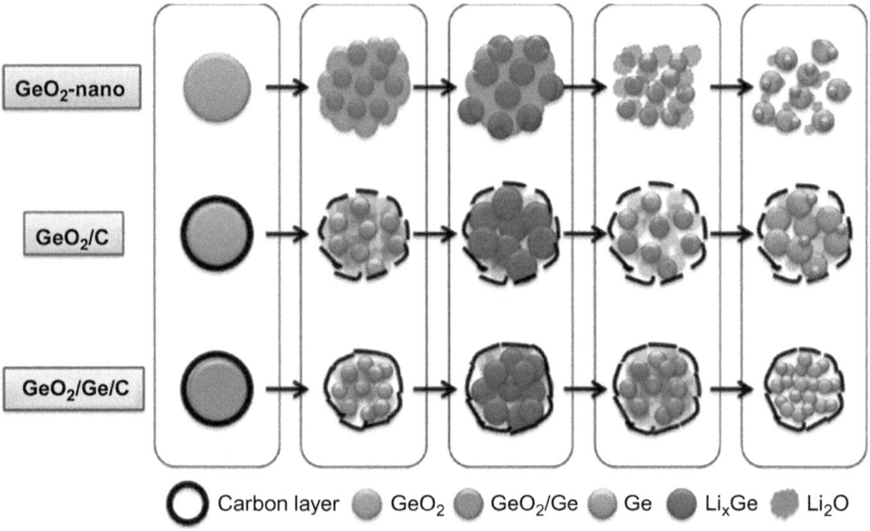

Fig. 6.4 Schematic representation of the lithium reaction mechanism in nano-GeO_2, GeO_2/C, and GeO_2/Ge/C. The elemental germanium in GeO_2/Ge/C plays a crucial role as a catalyst in improving the reversibility of the conversion reaction of GeO_2.
Reproduced with permission from K.H. Seng, M. Park, Z.P. Guo, H.K. Liu, J. Cho, Nano Lett. 13 (2013) 1230–1236, the American Chemical Society.

Magnetite (Fe_3O_4) nanospheres have been wrapped in graphene sheets and further restrained by continuous graphene networks [67]. These hierarchical Fe_3O_4/graphene hybrids provide double protection against the volume changes of Fe_3O_4 during electrochemical processes. Consequently, it delivered a high reversible capacity of 1059 mAh g^{-1} over 150 cycles. $Li_4Ti_5O_{12}$ microspheres composed of nanoflakes wrapped in graphene nanosheets were synthesized by an advanced microwave-hydrothermal method followed by an annealing step. The $Li_4Ti_5O_{12}$/graphene electrode had highly desirable properties: a specific capacity approaching the theoretical value, stable cycling, and exceptional rate capability [68]. Nanosized TiO_2 with a polyaniline (PANi) coating was synthesized by concurrent hydrolysis of $Ti(OC_3H_7)_4$ and polymerization of aniline; subsequent coupling of the PANi-coated TiO_2 with a lithium salt and calcination at 800°C for 24 h produced crystalline $Li_4Ti_5O_{12}$ and simultaneously carbonized the PANi [69]. The PANi carbonization, in turn, led to reduction of some Ti(IV) on the surface to Ti(III). This surface modification enhances the electronic conductivity on the surface and also extends the single-phase insertion/extraction ($Li_{4+\delta}Ti_5O_{12}$ or $Li_{7-\gamma}Ti_5O_{12}$) region outside the two-phase region.

Well-defined $Li_4Ti_5O_{12}$ nanosheets terminated with rutile TiO_2 at the edges have been synthesized by a facile solution-based method [70]. The well-defined structures were visualized directly at atomic resolution by a newly developed aberration-corrected scanning transmission electron microscopy (STEM) technique equipped with annular-bright-field (ABF) and high-angle annular dark-field (HAADF) detectors.

The rutile TiO_2 was grown along the [001] direction, with a thickness of ~1 nm. This direction shows the highest Li^+ diffusion coefficient (10^{-6} cm^2 s^{-1}).

Compared with their oxides, the transition metal nitrides (such as Mo_2N, WN, and TiN) have superior properties in some respects, such as high conductivity, high chemical resistance, hardness, mechanical strength, and high melting point. A drastic improvement in the electrochemical properties of the MoO_2 negative electrode was reported by applying a direct coating of Mo_2N nanolayers [71]. This carbon-free coating strategy only involves MoO_2 hollow nanostructures directly reacted with NH_3 gas, resulting in a uniform Mo_2N nanolayer coating. The resistance of the Mo_2N nanolayer coated MoO_2 hollow nanostructure cells (50 Ω) was less than the 165 Ω obtained for the bare MoO_2 hollow nanostructure cells. Besides a high electric conductivity, the Mo_2N nanocoating layer protects the MoO_2 against oxidation by the air.

6.2.5.2 Doping

The addition of other elements such as B, Al, P, Si, Ge, Ti, Mn, and Fe to SnO_2-based negative electrodes produces amorphous $SnA_{1\text{-}x}O_2$ oxides. This amorphous structure consisting of electrochemically active centers of Sn—O and the other added oxides in an amorphous network of delocalized Sn(II) active centers enables effective storage of Li ions. Moreover, the amorphous glass formed by the other added oxides enhances diffusion of Li^+ ions in contrast with crystalline tin and stannic oxides, so that reversible intercalation and de-intercalation are favored [72].

Anatase TiO_2 nanocrystals doped with both N and S, where the N and S substitutions for lattice O are located, respectively, in the bulk and the surface layer of the crystals, has been obtained by a three-step preparation route. After nitriding for 18 h in gaseous NH_3 at 700°C, the white titania nanoparticles were converted into black particles of cubic TiN. The black TiN_x nanoparticles were then oxidized in air at 340°C to form yellow nitrogen-doped anatase TiO_2 nanocrystals (denoted N-TiO_2). Further S-doping was effected by heating the N-TiO_2 at 400°C in a H_2S atmosphere, which led to formation of brown N/S doped TiO_2 consisting of anatase as major phase and rutile as a minor phase [73]. The effects of nitrogen/sulfur doping on the electronic structure of the TiO_2 crystals can be seen clearly when the UV-visible absorption spectra are compared. A further visible light absorption band with its edge extending to 550 nm can be observed in the N-TiO_2 spectrum (Fig. 6.5A). Moreover, the edge of this new absorption band is nearly parallel to that of the parent TiO_2 because of the narrow band gap of the bulk material. Introduction of S_{sub} in the surface layer of the N/S-doped TiO_2 also changed the electronic structure of the surface layer by the formation of S_{3p} states in the gap, as shown by the apparent red shift of the absorption edge up to around 700 nm. The Li^+ ion storage capacity of the N/S-doped TiO_2 was improved significantly compared with that of the other three samples (N-TiO_2, S-TiO_2, and TiO_2), as shown in Fig. 6.5B. The charge-transfer resistance at the interface between the electrode and the electrolyte was lowered considerably by the co-doping of N and S. Therefore, the highly modified electronic structure apparently causes the much lower resistance, which is the result of the homogeneous distribution of substituted N/S dopants throughout the whole TiO_2 particle.

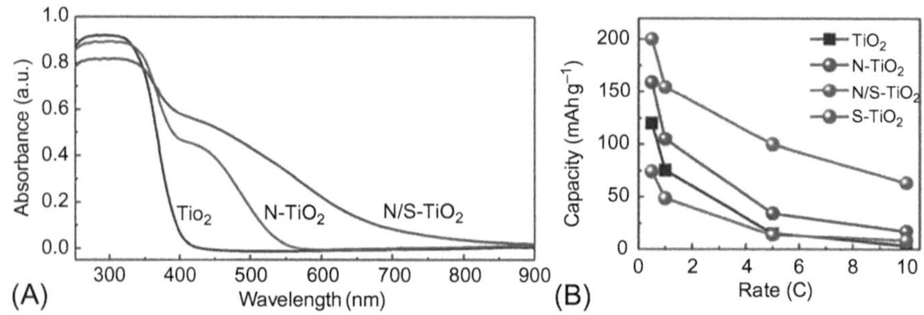

Fig. 6.5 (A) UV-visible absorption spectra of the TiO_2, N-TiO_2, and N/S-TiO_2 crystals; (B) comparison of the specific capacity of TiO_2-based electrodes at different rates. Reproduced with permission from W. Jiao, N. Li, L. Wang, L. Wen, F. Li, G. Liu, H.M. Cheng, Chem. Commun. 49 (2013) 3461–3463, the Royal Society of Chemistry.

6.2.5.3 Morphology control

As for the negative LIB electrodes, nanostructured materials have been designed and synthesized with a variety of morphologies, configurations and porosities, and combinations of microstructures and nanostructures. These materials are foreseen to support the electrochemical process at the negative electrode and retain good structural integration. Designed nanofilms or nanoporous structures can accommodate the volume changes, leading to improved cycling performance. Hollow structures, such as tubes, hollow spheres, and cubes, have been identified as a distinctive class of (hierarchical) micro-/nanostructured materials [74]. Over the years, many different strategies were developed to produce these hollow structures.

The hard-templating route is one of the most common methods to synthesize hollow micro-/nanostructured electrode materials; this methodology comprises the controlled deposition of the designed materials on a variety of disposable templates, followed by removal of the template. For example, hollow TiO_{2-x} porous microspheres consisting of numerous well-crystalline nanocrystals with superior structural integrity and robust hollow interior have been synthesized by a sol-gel template-assisted approach and two-step carbon protected calcination together with hydrogenation [75]. Colloidal mono-dispersed resorcinol/formaldehyde (RF) polymer microspheres were used as a removable template. Amorphous TiO_2 was first deposited on the surface of the polymer resin microspheres by kinetics-controlled coating; then, the dried as-made core-shell structured TiO_2@RF microspheres were heated at 500°C in a N_2 atmosphere to obtain TiO_2@carbon microspheres. After removing the carbon by thermal annealing in air, the porous hollow TiO_2 microspheres were obtained. These hollow microspheres demonstrated excellent lithium storage performance with stable capacity retention for over 300 cycles (a capacity of 151 mAh g^{-1} was still obtained after 300 cycles at 1C, retaining about 82% of the initial capacity of 185 mAh g^{-1}) and enhanced rate capability even up to 10C.

Uniform TiO_2 shells were deposited on Cu_2O polyhedra to form discrete TiO_2@Cu_2O core/shell structures. After subsequent removal of the Cu_2O cores, high quality TiO_2 nanocages with nonspherical shapes, homogeneous shells, and tunable sizes of 100–1000 nm were obtained [76]. The TiO_2 nanocages showed significantly improved electrochemical performance compared with the bulk materials, benefitting from their unique structural characteristics. Specifically, their relatively large surface area endows them with high capacity and favorable response to high-rate cycling relative to solid particles because of the dramatic increase of reactive sites and electrolyte-electrode interface. The void space in the interior allows the volume variation upon lithium storage to be better accommodated.

Considering production simplicity, one-pot synthesis of hollow particles is more attractive. Hollow structured Li_3VO_4 wrapped with graphene nanosheets has been prepared from V_2O_3 by a one-pot and template-free in situ hydrothermal method [77]. The trace oxygen dissolved in the solution gradually oxidized V_2O_3 to VO_4^{3-}. Li_3VO_4 precipitated on the V_2O_3 surfaces, ensuring that the surfaces of the V_2O_3 cubes were effectively covered, making those surfaces less reactive than the freshly exposed interior once the pitting process had started. The smaller particles have a higher Gibbs free energy, which makes them easy to dissolve into the solution. The bigger particles absorbed the surrounding supersaturated solution and kept growing. The composite showed a reversible capacity of 223 mAh g^{-1} even at 20C (1C=400 mAh g^{-1}). After 500 cycles at 10C there was no evident capacity fading.

Hierarchical α-Fe_2O_3 hollow spheres consisting of sheet-like subunits were obtained using a simple quasi-emulsion templating method [78]. First, glycerol was mixed with water, resulting in a quasi-emulsion consisting of oil-in-water microdroplets, which served as soft templates for the ensuing deposition and crystal growth of the shell structure. Sacrificial templating approaches have also been explored extensively. This approach exploits the differences in physicochemical and chemical properties between the sacrificial templates and the shell materials, and is a very effective method to form hollow particles with a variety of shapes. α-Fe_2O_3 nanotubes have been prepared by combining the controlled hydrolysis of Fe^{3+} ions in aqueous solution with etching of preformed Cu nanowires by Fe^{3+} and Cl^- ions [79]. In this process, copper nanowires are oxidized by Fe^{3+} to water-soluble Cu(I)-based complexes, and so they are gradually consumed. In parallel with the oxidation of Cu, Fe^{3+} is reduced to Fe^{2+} and precipitated on the surfaces of the dissolving Cu nanowires as $Fe(OH)_x$, resulting in the formation of a $Fe(OH)_x$ nanotube structure. Controlled annealing of the $Fe(OH)_x$ nanotubes in air produced polycrystalline α-Fe_2O_3 nanotubes, which manifested stable capacity retention at over 1000 mAh g^{-1} for 50 cycles over the voltage range 0.01–3.0 V (vs. Li^+/Li), far superior to that of α-Fe_2O_3 particles tested as reference under identical conditions. The tubular structure nicely accommodates the change in volume during lithium insertion and extraction and also protects the active materials from extreme aggregation which is bound to occur during repeated discharge/charge cycling. A scalable synthesis of Fe_2O_3 microboxes was recently reported; the microboxes consist of hierarchically structured shells manufactured by simple annealing of preformed Prussian blue (PB) microcubes [80]. The preformed metal-organic frameworks (MOFs) served as a template for the synthesis, but also as precursor for the formation of hollow structures of the desired materials. The hierarchical Fe_2O_3 microboxes obtained from these MOFs exhibited a reversible capacity of 945 mAh g^{-1} in the 30th cycle at 200 mA g^{-1}.

Multilayered CoO platelets and nanoparticles have been prepared and the Li cycling behavior (range 0.01–3.0 V vs. Li^+/Li) of the oxides with different surface areas ranging from 20 to 9 $m^2 g^{-1}$ were compared. The CoO platelets with 10 $m^2 g^{-1}$ surface area delivered a high and stable capacity of 800 mAh g^{-1} for up to 100 cycles at a current of 100 mA g^{-1} [81].

NiO nanowalls prepared on Ni foil had a stable and reversible capacity of 635 mAh g^{-1} (at 1.25C rate) over 2–80 cycles. At 1.9C (1343 mA g^{-1}), a stable capacity of 490 mAh g^{-1} was observed over 2–50 cycles [82].

A nanocrystalline MnO thin film with a discharge capacity of 425 mAh g^{-1} at 0.125C after 25 cycles has been prepared [83]. Porous MnO microspheres delivered a capacity of 800 mAh g^{-1} at 50 mA g^{-1} along with a capacity retention of 88% after 50 cycles [84].

Ultrathin SnO_2 nanosheets oriented along the [110] direction have been assembled by oriented attachment of SnO_2 nanoparticles, assisted by additives such as ethanol, NH_3, H_2O, and urea [85]. Apparently the functional groups of these additives cause variations of the surface energy of the SnO_2 nanocrystals because of their different abilities to bond to Sn ions or form hydrogen bonds. After 50 discharge and charge cycles, a reversible capacity of 534 mAh g^{-1} was delivered by the SnO_2 nanosheets. This greatly increased storage capacity is mainly due to the ultrathin thickness and the unique porous structures. The nanometer-sized networks offer negligible ion diffusion times and probably faster phase transitions, and the "breathable" interior spaces can buffer drastic volume changes effectively during the electrode reactions. NiO crystals have been obtained with mostly exposed (110) reactive facets by the thermal conversion of hexagonal $Ni(OH)_2$ nanoplatelets [86]. The NiO crystals obtained have the same crystal features and hexagonal character as the precursor $Ni(OH)_2$ nanoplatelets. Owing to their relatively high surface energy, the (110) crystal planes offer reactive sites for reaction with Li^+ ions, which can enable fast conversion reactions during the charge and discharge processes. Thus, the NiO still had a good lithium storage capacity of 468 mAh g^{-1} at 20C rate and 322 mAh g^{-1} at 40C rate. The well-defined planar structure of Co_3O_4 was controlled easily utilizing a one-step hydrothermal method by simply changing the NaOH and $Co(NO_3)_2 \cdot 6H_2O$ contents without use of a capping agent [87]. The Co_3O_4 octahedra with (111) planes had the highest charge/discharge capacity and the best cycling behavior. This is because the (111) plane contains 3.75 Co(III) ions whereas the (001) plane holds only 2 Co(II) ions.

6.3 Metal oxides in aqueous rechargeable lithium batteries

6.3.1 *Principles and applications of aqueous rechargeable lithium batteries*

The organic electrolytes used in LIBs are combustible and highly toxic, and could therefore become involved in thermal run-away and safety accidents when the battery is fully charged or short circuited. Furthermore, lithium-ion batteries are relatively expensive because organic electrolytes are used and special cell assembly technology is required (e.g., a strictly anhydrous environment has to be maintained during the manufacturing process). To address these issues, aqueous rechargeable lithium batteries (ARLBs) were proposed in 1994 [88]. The mediocre capacity retention and low-energy densities obtained have limited their application in practice, but since 2007, ARLBs based on several different electrode materials subjected to redox reactions have been made by our groups and are now gaining importance as encouraging new battery systems [89].

ARLBs can overcome some of the drawbacks of LIBs. For example, they are essentially safe because no flammable organic electrolytes are used. Moreover, expensive salts are replaced by less costly salts in the aqueous electrolyte. Furthermore, the ionic conductivity is about 100 times higher than that of organic electrolytes, ensuring a high-rate capability and therefore high specific power [90]. Lithium metal oxides with intercalation potentials between 3 and 4 V (vs. Li^+/Li) can be used as positive electrode materials, and electrode materials with a redox potential of about 2–3 V (vs. Li^+/Li) are a viable option for negative ARLB electrode materials.

In contrast to the key factors, such as energy density and cyclability, for a battery used in portable electronic devices, the cost and safety are relatively more important in large-scale stationary energy storage systems, such as smart grids. With this in mind, aqueous rechargeable batteries have great potential for application in stationary energy storage systems [91].

6.3.2 Metal oxide-based positive electrodes for ARLBs

Lithium metal oxides, such as $LiCoO_2$, $LiNi_{1/3}Co_{1/3}Mn_{1/3}O_2$, and $LiMn_2O_4$, are useful positive electrode materials for ARLBs. As in the organic electrolytes, lithium intercalation and de-intercalation also takes place into and from these materials in aqueous electrolytes.

In the aqueous electrolyte, the main redox peaks for $LiCoO_2$ are situated at 0.71 and 0.87 V (vs. SCE), agreeing well with the intercalation/de-intercalation reactions, respectively, of $LiCoO_2$ in organic electrolyte. The particle size of the active materials is very important to achieve good electrochemical performance at high current density in aqueous electrolytes. Crystalline nanoparticles of $LiCoO_2$ have been prepared by a sol-gel method at 550°C [92]. The maximum reversible capacity of the nano-$LiCoO_2$ was 143 mAh g^{-1} at 1 A g^{-1} and still 133 mAh g^{-1} at 10 A g^{-1} (about 70C) in 0.5 M Li_2SO_4 aqueous electrolyte. In addition, its cycling behavior was also very satisfactory, with no evident capacity fading during the initial 40cycles.

Crystalline $LiMn_2O_4$ nanorods have been prepared by a facile hydrothermal method in combination with a traditional solid-state reaction [93]. The diameters of the $LiMn_2O_4$ nanorods were about 80nm, with lengths ranging from 0.2 to 0.8 μm. The CV of the $LiMn_2O_4$ nanorods in aqueous solution had a satisfactory shape, even with a scan rate of up to 50 mV s^{-1}, and two redox peaks could be clearly identified; this is a better result than is obtained in organic electrolytes.

The sol-gel method generally refers to the hydrolysis and condensation of metal alkoxides or alkoxide precursors, leading to dispersions of oxide particles in a sol.

The sol is then dried or gelled by solvent removal or by chemical reaction. The rates of hydrolysis and condensation are important parameters that affect the properties of the final products [94]. Nanochain $LiMn_2O_4$ was prepared by a starch-assisted sol-gel method [95]. The starch carrier makes a major contribution to the shape. The spinel structure has a nanochain morphology and consists of beads of about 100 nm diameter (Fig. 6.6A), presenting a reversible capacity of 110 mAh g^{-1} at 4.5C and 95 mAh g^{-1} even at 91C in 0.5 M Li_2SO_4 aqueous electrolyte. When charged at 136C, 84% capacity was retained (Fig. 6.6B).

$LiMn_2O_4$ nanotubes with a preferred orientation of (400) planes have been prepared using multiwall carbon nanotubes as sacrificial template [96]. First, acid-treated MWCNTs were immersed in an aqueous solution of $KMnO_4$ and coated with MnO_2. Then, the MnO_2-coated MWCNTs were mixed with LiOH and heated to 700°C to remove the MWCNTs. The nanotubes haves a length of >2 μm and a tetragonal open end with an edge length of about 80–100 nm (Fig. 6.6C). MWCNTs consist of graphene sheets rolled into concentric cylinders. The graphene plane coordinates with Mn atoms due to π-3d orbital interaction. As a result, the orientated MnO_2 was deposited on the MWCNTs using the surface graphene sheets as templates. The prepared $LiMn_2O_4$ also retains its typical crystal orientation. At a charging potential of 1.4 V, capacities of 97.3 mAh g^{-1} (88.5% of the capacity), 80.2 mAh g^{-1} (72.9% of the capacity), and 59.3 mAh g^{-1} (53.9% of the capacity) were obtained at charging rates of 120C (30s), 300C (12s), and 600C (6s), respectively (Fig. 6.6D). Both the unique nanotube characteristics and the tailored facets play important roles in achieving superfast second-level charge capability. There are more (001) or (010) planes on the edges of these planes or vertical to the (400) planes, and the 8a sites for lithium intercalation and deintercalation are situated at the (001) or (010) planes. Therefore, more Li sites are exposed to the aqueous electrolyte due to the preferred growth of (400) planes.

Using polystyrene (PS) colloids as a template, a porous $LiMn_2O_4$ consisting of nanograins with high crystallinity was prepared [97]. The diameter of the pores was in the range 200–250 nm, close to the average diameter of the original PS spheres (Fig. 6.6E). The surrounding walls are composed of unique nano grains with a diameter of <50 nm. The porous $LiMn_2O_4$ presented excellent capacity retention, with no >7% capacity loss after 10,000 full cycles at 9C (Fig. 6.6F). This means that, if the ARLB is charged and discharged once a day, its lifetime can be at least 30 years, which is very attractive. Whereas the discharge capacity of the solid $LiMn_2O_4$ electrode decreases rapidly over the first 2000 cycles, the porous structure can accommodate strain and stress during the charge and discharge processes.

One potential disadvantage of nanomaterials is the lower packing density leading to lower volumetric energy densities. In this regard, hierarchical microstructures and nanostructures have become an effective strategy. The tradeoff between using nanostructured electrodes and designing dense architectures may give rise to new scientific questions at the mesoscale. The best way to improve both rate capability and tap density would be to use micro-sized particles that consist of primary nanoparticles. $LiNi_{1/3}Co_{1/3}Mn_{1/3}O_2$ agglomerates composed of many primary nanoparticles have been

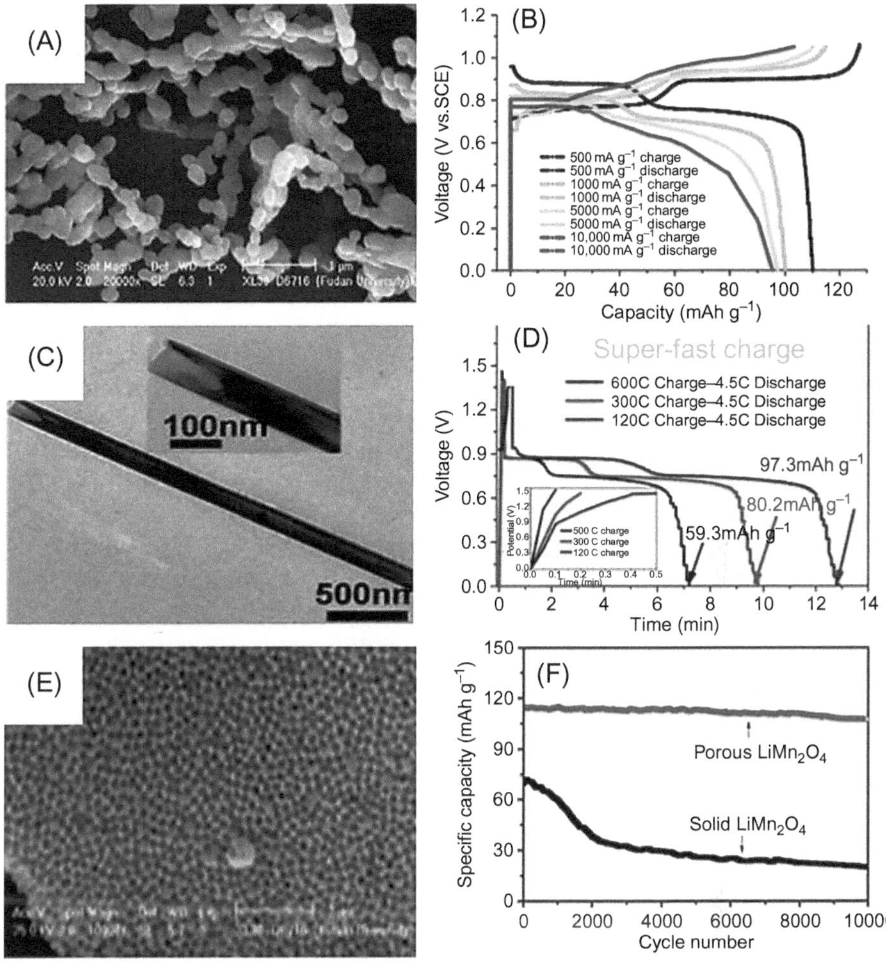

Fig. 6.6 (A) SEM micrographs of $LiMn_2O_4$ nanochains; (B) $LiMn_2O_4$ nanochain charge/discharge curves; (C) TEM micrographs of the prepared $LiMn_2O_4$ nanotubes and (d) their second-level charge capability; and (e) SEM micrographs of the as-prepared porous $LiMn_2O_4$ and (f) its cycling behavior.
Parts A and B: Reproduced with permission from W. Tang, S. Tian, L.L. Liu, L. Li, H.P. Zhang, Y.B. Yue, Y. Bai, Y.P. Wu, K. Zhu, Electrochem. Commun. 13 (2011) 205–208, Elsevier. Parts C and D: Reproduced with permission from W. Tang, Y.Y. Hou, F.X. Wang, L.L. Liu, Y.P. Wu, K. Zhu, Nano Lett. 13 (2013) 2036–2040, the American Chemical Society. Parts E and F: Reproduced with permission from Q. Qu, L. Fu, X. Zhan, D. Samuelis, J. Maier, L. Li, S. Tian, Z. Lia, Y. Wu, Energy Environ. Sci. 4 (2011) 3985–3990, the Royal Society of Chemistry.

synthesized by using vapor grown carbon fibers (VGCFs) as templates [98]. On being charged at 180C and discharged at 3C, the discharge capacity reached 72 mAh g^{-1} and the charge time needed was only 20 s. After 50 cycles, a discharge capacity of 65 mAh g^{-1} was attained with a charge time of 15 s.

6.3.3 Metal oxide-based negative electrodes for ARLBs

Vanadium oxides, molybdenum oxides, and some lithium intercalation compounds have been reported as negative electrode materials for ARLBs. Research on negative electrode materials is quite limited owing to the challenges concerning dissolution of the active cations. In the case of metal oxides, dissolution of the active cation will become worse when the electrochemical activity is increased by nanostructuring. Therefore, some conductive polymers or carbon are coated on the surface of negative electrode materials.

A hybrid of polypyrrolidine (PPy) coated V_2O_5 nanowires and MWCNTs has been prepared as negative electrode material for ARLBs [99]. PPy not only serves as a conducting additive reducing the weight ratio of inert materials in the electrodes, but also acts as a coating preventing the dissolution of active material and the associated release of metal ions during the charge and discharge processes. The PPy coating markedly improved the cycling performance of the ARLB system, with no evident capacity loss after 500 cycles, except during the initial cycles. Similarly, an ARLB consisting of a nanocomposite of PPy@MoO_3 and $LiMn_2O_4$ displayed excellent rate capability and good cycling performance [100]. The average charge and discharge voltages were 1.40 and 1.22 V, respectively. The energy density of this ARLB was 45 Wh kg^{-1} at 350 W kg^{-1} and even reached 38 Wh kg^{-1} at 6 kW kg^{-1}. This excellent rate capability compared with that of supercapacitors is ascribed to the conductive PPy coating decreasing the charge transfer resistance.

Hierarchical LiV_3O_8 nanofibers consisting of nanosheets with exposed (100) facets, have been made using electrospinning followed by calcination [101]. At electrolyte pH values <5, LiV_3O_8 nanosheets with exposed (100) facets lessened proton co-intercalation into the LiV_3O_8 electrode. The (010) and (001) facets of the of LiV_3O_8 crystal structure provide many channels while the (100) facet has fewer and smaller channels, so that it is more difficult for H^+ ions to intercalate into the electrode.

However, the energy density of these ARLB systems is still much lower than that of conventional LIBs due to the narrow electrochemical window of water. In generation-II ARLB systems, Li or Mg metal coated with LISICON film in a GPE are used as the negative electrodes (Fig. 6.7). The LISICON film consisting of

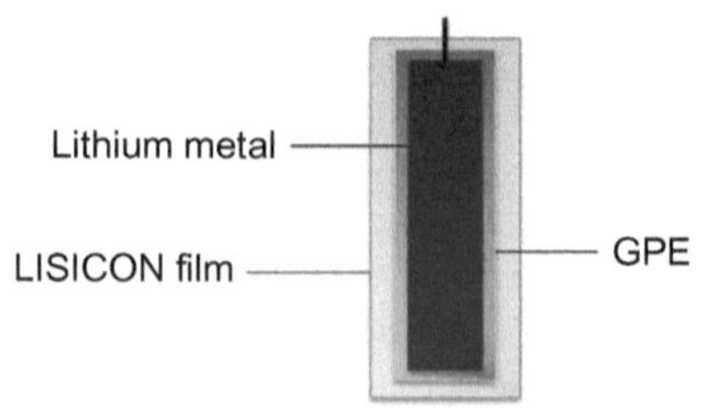

Fig. 6.7 Schematic illustration of the coated lithium metal electrode for aqueous rechargeable lithium batteries.
Reproduced with permission from X. Wang, Y. Hou, Y. Zhu, Y. Wu, R. Holze, Sci. Rep. 3 (2013) 1401 – 1405, Nature.

Li_2O-Al_2O_3-SiO_2-P_2O_5-TiO_2-GeO_2 is a solid electrolyte which can also act as a separator. Its ionic conductivity of Li^+ ions is about 0.1 mS cm^{-2} at room temperature. Owing to its solid structure, protons, water, hydrated and solvated ions cannot pass through [102]. These battery systems deliver output voltages of about 3–4 V, about three times higher than that of traditional ARLBs, and their energy density is around 300 Wh kg^{-1} [103,104].

6.4 Metal oxides in Na-ion batteries

6.4.1 Principles and applications of Na-ion batteries

Sodium-ion batteries (NIBs) were originally developed in the early 1980s, approximately over the same time period as LIBs [105]. In recent years, NIBs have drawn increasing attention for large-scale energy storage, because of the natural abundance, low cost, and environmental friendliness of sodium [106]. NIBs are considered to be one of the most appealing alternative rechargeable batteries to LIBs.

Since both sodium and lithium are alkali metals, they share similar chemical properties including ionicity, electronegativity, and electrochemical reactivity. Accordingly, they have comparable synthetic protocols and electrochemical performance, which shows that Na ion batteries could be successfully developed based on previously applied approaches or methods used for the lithium counterpart.

Nonaqueous NIBs, such as the "18,650"-type battery developed within the French framework of RS2E, are characterized by high-energy density (90 Wh kg^{-1}) and excellent cycling behavior for up to 2000 cycles [107]. The electrode materials in LIBs provide the best library for research on NIBs because many Na-ion insertion hosts have their roots in Li-ion insertion hosts.

6.4.2 Metal oxide-based positive electrodes for Na-ion batteries

Vanadium oxides have been the focus of many recent studies regarding high capacity positive electrode materials for NIBs. The hydrothermal or solvothermal method has been widely used for the synthesis of various 1D nanostructured V_2O_5 positive electrode materials for NIBs [108–110]. The main advantages of the hydrothermal and solvothermal processes include fast reaction kinetics, short processing time, good phase purity, high crystallinity, and low cost. Single-crystalline bilayered V_2O_5 nanobelts with large interlayer spacing of the (001) crystal planes exhibit a maximum capacity of 231 mAh g^{-1} [110]. The as-prepared V_2O_5 presents a noticeable distance between the closest approach bilayer stacks (11.53 Å) along the *c*-axis. The superior electrochemical performance is ascribed to the unique bilayered vanadium oxide nanobelts with dominantly exposed (100) crystal planes, which provide large interlayer spacing for facile Na-ion intercalation/deintercalation. However, VO_x nanotubes exhibited faster capacity-fading when applied as the positive electrode for NIBs [108,109]. This is probably because of the lower valence of vanadium in the material and the presence of the amine templates that were used for the large-scale fabrication of the nanotubes.

A simple and versatile method for the preparation of V_6O_{13} microflowers was developed using a simple hydrothermal route with the aid of an alkali metal nitrate salt [111]. The obtained V_6O_{13} microflowers containing a small amount of impurity displayed a high capacity of up to 225 mAh g^{-1} when used in a NIB. The impurity phase in the V_6O_{13} micro-flowers may lead to less phase transition upon sodium intercalation. As a member of the vanadium oxide family, NaV_3O_8 has been shown to be a promising positive electrode material for Na ion storage [112]. The PPy@NaV_3O_8 nanosheets core-shell composites showed two distinct pairs of redox peaks located at 3.35 and 2.35 V (vs. Na^+/Na) (cathodic, reduction), and at 3.47 and 2.61 V (vs. Na^+/Na) (anodic, oxidation), and all the redox peaks almost overlap in the first three cycles.

Layered Na_xCoO_2 is known as the oldest layered oxide studied as sodium insertion host. The early NIBs were demonstrated with P2- and P3-type Na_xCoO_2 as positive electrode materials in 1988 [113]. Maximum reversible capacities of 140 mAh g^{-1} for O_3, 130 mAh g^{-1} for a P3 phase sample, and 120 mAh g^{-1} for P2 phases were observed. The voltage profile, operating voltage, and phase transition behavior for the different layered polymorphs are very similar. Although a monoclinic phase transition is observed for Li_xCoO_2 ($x \approx 0.5$), accompanied by a short voltage plateau, Na_xCoO_2 shows even more complicated phase transitions on sodiation/desodiation [114].

Na_xMnO_2 can be categorized into two different types of structures: tunnel structures ($x = 0.4$ and 0.44) and layered structures ($x = 1.0$ and 0.7). Among all the positive materials identified as possible candidates to be used in NIB applications, $Na_{0.44}MnO_2$ is particularly attractive because its crystal structure has suitable large-sized tunnels for sodium incorporation. Single-crystalline $Na_{0.44}MnO_2$ nanowire showed an initial charge capacity of 50 mAh g^{-1} corresponding to 0.18 Na-ion deintercalation from $Na_{0.44}MnO_2$, in which the potential plateaus were observed at 3.2 and 3.45 V (vs. Na^+/Na) [115]. The next discharge/charge cycle displayed a reversible capacity of 120 mAh g^{-1}, corresponding to 0.43 Na-ion intercalation/deintercalation in Na_xMnO_2 ($0.26 < x < 0.69$). In addition to the abovementioned two potential plateaus at 3.45 and 3.2 V (vs. Na^+/Na), four potential plateaus at 3.0, 2.6, 2.5, and 2.2 V (vs. Na^+/Na) were also observed during the charging and discharging processes [115,116]. For P2-phase $Na_{0.7}MnO_2$, Mn^{3+}/Mn^{4+} is responsible for the conduction in Na_xMnO_2. The $Na_{0.7}MnO_2$ nanoplates demonstrated a near theoretical capacity of 164 mAh g^{-1} (its theoretical capacity is 170 mAh g^{-1}) [117]. The improvement is ascribed to the predominantly exposed active (100) facet, which could facilitate fast Na^+ ion intercalation/deintercalation during the discharge and charge processes. However, its cycling decay due to the Jahn-Teller effect of Mn^{3+}/Mn^{4+} is still obvious.

Recently, researchers prepared layered O_3-type $NaCrO_2$ with good capacity retention and high-rate capability using an emulsion drying method [118]. The 3.4% carbon-modified $NaCrO_2$ used had a good capacity retention of about 90% after 300 cycles, and delivered a maximum capacity of 106 mAh g^{-1} at a current density of 5.5 A g^{-1}. A high-rate capacity of 99 mAh g^{-1} was obtained at 150C, corresponding to a very short discharge time of 27 s. Carbon-modified $NaCrO_2$ has much higher electrical conductivity (10^{-1} S cm^{-1}) than that of bare $NaCrO_2$ (8×10^{-5} S cm^{-1}), which can accelerate the reversible insertion/extraction of Na ions and the facile complementary redox reactions of the Cr^{4+}/Cr^{3+} couple.

Motivated by the search for other structures similar to $Na_{0.44}MnO_2$, some new Na ion intercalation metal oxides, $Na_{2.7}Ru_4O_9$ and $Na_{0.54}Mn_{0.5}Ti_{0.51}O_2$, were introduced [119,120]. The unique network structure of $[RuO_6]$ octahedra provided a 1D chain that is large enough to accommodate three Na sites. $Na_{2.7}Ru_4O_9$ shows three redox peaks between 2.8 and 3.9 V (vs. Na^+/Na), but once Na^+ ions were removed from the $Na_{2.7}Ru_4O_9$ structure, only a portion (<60%) could be re-intercalated [119]. The $Na_{0.54}Mn_{0.5}Ti_{0.51}O_2$ nanorods delivered a discharge capacity of 100 mAh g^{-1} after 150 cycles at 0.028 A g^{-1}, and 67 mAh g^{-1} after 400 cycles at 0.14 A g^{-1} [120]. These nanorods were formed perpendicular to the Na^+ ion tunnels, thereby significantly shortening the diffusion paths of the sodium ions and benefiting transfer kinetics. This result once again demonstrates the importance of tailored facets in electrode materials.

α-$NaFeO_2$ is known as a typical example of an O_3-type layered structure and is easily prepared by a conventional solid-state reaction. Reversibility in $NaFeO_2$ electrode materials is significantly influenced by cutoff conditions when charging [121]. Although charging capacity increases as a function of cutoff voltage, the reversible capacity noticeably decreases when charged beyond 3.5 V (vs. Na^+/Na). The reversible capacity at a cutoff voltage of 3.4 V is 80 mAh g^{-1}, indicating that approximately 0.3 mol of Na^+ ions are reversibly extracted from $NaFeO_2$ and inserted into the $Na_{0.7}FeO_2$ host structures.

6.4.3 Metal oxide-based negative electrodes for Na-ion batteries

One of the first transition metal oxides used as negative electrode for NIBs was spinel $NiCo_2O_4$, prepared by precipitation of a mixed oxalate precursor, followed by thermal decomposition in air at 320°C [122]. The spinel oxide exhibited poor electrochemical performance with high polarization and low reversible capacity (c.200 mAh g^{-1}), considering its high theoretical specific capacity of c.890 mAh g^{-1}.

A simple and scalable method comprising in situ engraving of Cu foil with Na_2O has been realized to prepare aligned porous CuO arrays [123]. During the first discharge process, three reduction peaks were observed, located at 1.6, 0.46, and 0.05 V (vs. Na^+/Na), corresponding to multistep electrochemical reactions that involve the generation of an intermediate copper oxide phase ($Cu^{II}_{1-x}Cu^{I}_{x}O_{1-x/2}$), the formation of a Cu_2O phase, and the decomposition of Cu_2O by Na into Cu and Na_2O, respectively. Subsequently, four oxidation peaks are found at about 0.5, 1.3, 1.85, and 2.3 V (vs. Na^+/Na) during the charging process. The small peaks positioned at 0.5 and 1.3 V (vs. Na^+/Na) were assigned to the oxidation of a small amount of residual Cu_2O to CuO, while the 1.85 V (vs. Na^+/Na) peak corresponds to the process:

$$2Cu + Na_2O \rightarrow Cu_2O + 2Na \quad (6.12)$$

The fourth peak, at 2.3 V (vs. Na^+/Na), may result from the oxidation of the Cu_2O into CuO.

Sodium transition metal (Ti, V) oxides have been utilized also as negative materials for NIBs by an insertion-type mechanism. For example, up to 0.5 Na^+ ion was reversibly inserted into a layered Na_xVO_2 host at 1.5 V (vs. Na^+/Na), which corresponds to a capacity of 120 mAh g^{-1}. A layered $NaTiO_2$ has been prepared, which can accommodate about 0.5 Na atoms per formula unit at 1 V (vs. Na^+/Na). The relatively low redox potential of titanium oxides compared with other transition metals is ascribed to the high reducing activity of trivalent titanium cations [124].

A high reversible capacity of >150 mAh g^{-1} was obtained with anatase-type TiO_2, (corresponding to approximately 0.5 mol sodium insertion) over the voltage range of 0–2.0 V (vs. Na^+/Na), even though the large surface area of nanosized anatase inevitably results in low coulombic efficiency (42%) [125]. The voltage profile of nanosized anatase TiO_2 is a sloping curve, and voltage plateaus are not observed. $Na_2Ti_3O_7$ consists of zigzag layers of titanium-oxygen octahedra with Na^+ ions in the interlayer spaces that can be exchanged easily, and it also exhibits a particularly low potential [126]. The insertion of two additional Na^+ ions (amounting to 180 mAh g^{-1}) in the lattice and the reduction of two of the three Ti(IV) species to Ti(III) occur at a reversible plateau around 0.3 V (vs. Na^+/Na) at a low cycling rate (C/25). Moreover, it is possible to increase the rate capability by decreasing the particle size.

The effect of oxygen vacancies has been investigated when using MoO_{3-x} nanosheets as negative electrode in a NIB. The MoO_{3-x} nanosheets exhibited improved electrochemical performance compared with the MoO_3 counterpart due to the enhanced electric conductivity and Na^+ diffusion coefficient [127]. Al_2O_3 coating reduced the cycling-induced SEI formation, promoting the effect of oxygen vacancies under deep-discharge conditions.

6.5 Metal oxides in Mg-ion batteries

6.5.1 *Principles and applications of Mg-ion batteries*

Mg is an environmentally benign element and the fifth most abundant element in the earth's crust. The divalent nature of Mg^{2+} ions has been a hurdle for fast intercalation kinetics, but can also offer an opportunity for realizing very high capacities. Magnesium has a low reduction potential of −2.37 V (vs. SHE), a volumetric capacity (3832 mAh mL^{-1}) nearly twice that of lithium (2026 mAh mL^{-1}) derived from its divalent nature, and relatively low cost due to the abundance of magnesium resources. More importantly, Mg does not form dendrites during reversible electrochemical plating/stripping, which eliminates crucial safety concerns when used as a negative electrode material [128].

At present, there are a few companies trying to develop rechargeable Mg batteries, including Sony, LG, Toyota, and Pellion Technologies (a Massachusetts Institute of Technology spin-off company). The last decade has witnessed a large amount of work dedicated to Mg-ion batteries.

6.5.2 Metal oxide-based positive electrodes for Mg-ion batteries

Reversible intercalation of Mg ions has been reported for transition metal oxides including V_2O_5, MnO_2, MoO_3, and TiO_2.

V_2O_5 is a very important metal oxide with the fascinating possibility of hosting different cations. The electrochemical reactivity of cations such as Ca^{2+} and Mg^{2+} in crystalline V_2O_5 materials has been investigated [129]. V_2O_5 nanotubes were obtained as the main product from V_2O_5 precursors and primary amines in a sol-gel reaction followed by hydrothermal treatment [130]. AC impedance measurement demonstrated that Mg^{2+} ion diffusion is faster into the nanotubes than into polycrystalline V_2O_5. It should be noted that the diffusion of Mg^{2+} ions can be facilitated by water molecules, because water can solvate Mg^{2+} ions and promote the insertion/extraction process [131]. However, water in the electrolyte also accelerates the formation of a passivation film on the surface of the Mg anode, which is unfavorable for the electrochemical performance.

The electrochemistry of the nanosized α-MnO_2 structure (hollandite 2×2 tunneled) was studied recently [132]. It presented a specific capacity of 280 mAh g^{-1} in the first discharge using Mg salt/THF as electrolyte. However, the discharge/charge capacity faded during cycling due to the partial collapse of the α-MnO_2 tunnel structure associated with Mg^{2+} ion intercalation/deintercalation. The Mg^{2+} ion deintercalation/intercalation out of/into δ-MnO_2 showed a lower discharge capacity (70 mAh g^{-1}) than the α-phase, but the δ-MnO_2 retained more of its original capacity over more cycles [133].

In another study, MoO_3 films obtained by electrodeposition were reversibly cycled at a relatively high capacity of 220 mAh g^{-1} at a potential around 1.8 V (vs. Mg^{2+}/Mg) [134]. XRD and Raman spectra clearly showed that a two-phase reaction occurs during the magnesiation of MoO_3. When TiO_2 is used as positive electrode material for Mg-ion batteries, the electrolyte is the key factor in achieving electrochemical activity. TiO_2 showed poor Mg^{2+} ion intercalation/deintercalation performance in the organic magnesium organohaloaluminate electrolyte, that is, $(PhMgCl)_2$-$AlCl_3$/THF [135]. However, with the highly compatible $Mg(BH_4)_2$/$LiBH_4$/tetraglyme electrolyte, commercial TiO_2 (25–30 nm in size) delivered a high specific capacity (155 mAh g^{-1}) and excellent cycling stability (90 cycles). It also had well-defined charging and discharging plateaus at approximately 0.87 and 1.16 V (vs. Mg^{2+}/Mg), respectively. The co-intercalation of Li^+ and Mg^{2+} ions may contribute to the exceptional performance of the Mg ion batteries.

6.5.3 Metal oxide-based negative electrodes for Mg-ion batteries

Negative electrode materials for Mg-ion batteries have been investigated to a limited extent. Mg metal itself can serve as negative electrode material, but in this case the choice of possible electrolyte is narrowed down to mainly Grignard reagents and ethereal solvents. The recent development of new Mg alloys and low-strain compound negative electrodes (such as $Li_4Ti_5O_{12}$) provides opportunities for the use of more conventional electrolytes.

On cycling, the reversible capacity of the $Li_4Ti_5O_{12}$ electrode [136] gradually increases during the initial cycles and reaches a maximum value of 175 mAh g^{-1} in the 15th cycle (Fig. 6.8A). Its typical electrochemical behavior with a significant plateau, gradually changes to a sloped discharge-charge voltage curve. It exhibits good cycling stability and coulombic efficiency at a current density of 300 mA g^{-1} for 500 cycles (Fig. 6.8B), which can be attributed to the zero-strain characteristic of $Li_4Ti_5O_{12}$. The in situ synchrotron XRD (Fig. 6.8C) shows that the peak intensity of the cycled electrode increases slightly with cycling, but no obvious peak shifts or new Bragg reflections are observed, indicating small volume changes during the charging/discharging processes. This result may be attributed to the comparable effective ionic radii of Mg^{2+} (0.72 Å) and Li^+ (0.76 Å). The excellent cycling stability of $Li_4Ti_5O_{12}$ in Mg-ion batteries could benefit from the small volume changes during the chargedischarge process [136].

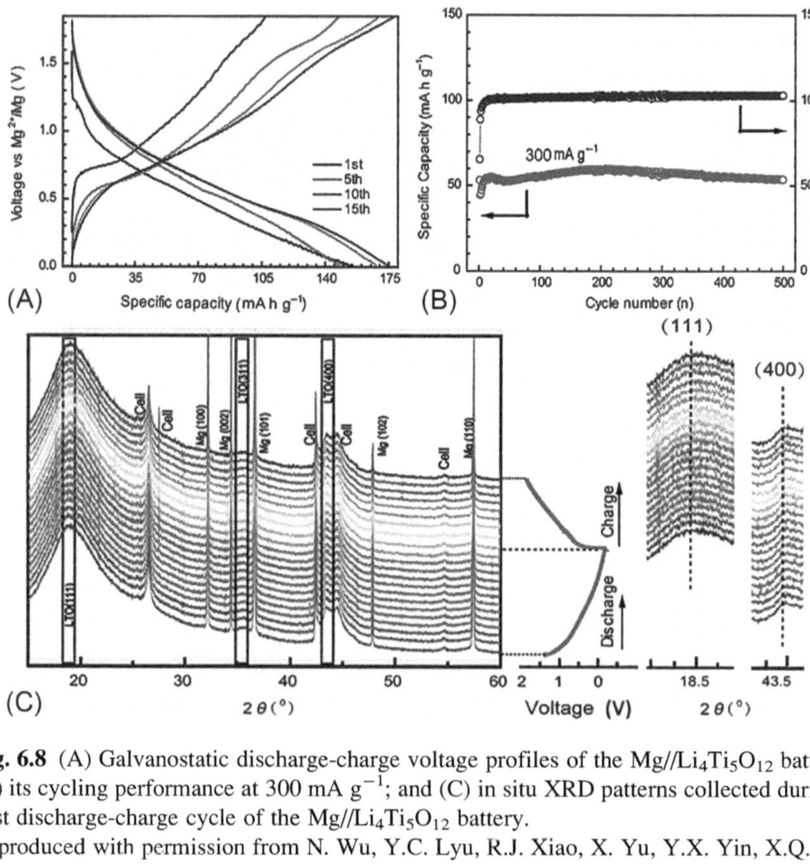

Fig. 6.8 (A) Galvanostatic discharge-charge voltage profiles of the Mg//$Li_4Ti_5O_{12}$ battery; (B) its cycling performance at 300 mA g^{-1}; and (C) in situ XRD patterns collected during the first discharge-charge cycle of the Mg//$Li_4Ti_5O_{12}$ battery.

Reproduced with permission from N. Wu, Y.C. Lyu, R.J. Xiao, X. Yu, Y.X. Yin, X.Q. Yang, H. Li, L. Gu, Y.G. Guo, NPG Asia Mater. 6 (2014) e120, Nature.

6.6 Metal oxides in Al-ion batteries

The initial work on Al-based batteries mainly focused on primary batteries. The first rechargeable Al-ion battery was described in 2011 [137]. This battery used V_2O_5 nanowires as the positive electrode against aluminum metal as the negative electrode in an ionic liquid-based electrolyte. The open circuit voltage of the Al-ion battery was 1.8 V (vs. Al^{3+}/Al). It demonstrated a well-defined and very stable Al^{3+} intercalation plateau at 0.55 V (vs. Al^{3+}/Al). In the first cycle, the battery exhibited a capacity of 305 mAh g^{-1}, against 273 mAh g^{-1} after the 20th cycle. An Al-ion battery based on VO_2 nanorods as the positive electrode has also been described [138]. First-principles calculations were employed to investigate theoretically the crystal structure change and the intercalation-deintercalation mechanism of the Al ions. In the initial cycle, the Al-ion battery delivered a discharge capacity of 165 mAh g^{-1}. Clear charge and discharge voltage plateaus could be seen around 0.7 and 0.5 V (vs. Al^{3+}/Al).

Electrochemical aluminum storage of anatase TiO_2 nanotube arrays in aqueous $AlCl_3$ solution was investigated recently [139]. A pair of large reversible redox peaks was observed at 1.26 and 0.84 V (vs. SCE), corresponding to the reversible aluminum ion insertion/extraction reactions in anatase TiO_2. In the cathodic process, the hydrogen evolution reaction occurred at a potential of −1.43 V (vs. SCE) after the aluminum ion insertion, which ensures aluminum insertion in preference to hydrogen evolution. The maximum discharge capacity of 75 mAh g^{-1} was obtained after activation at a capacity 4 mA cm^{-2}, slightly higher than that of anatase TiO_2 for the Li^+ ion insertion in alkaline solution. The reversible aluminum storage is about 0.074 mol per mol anatase calculated from the maximum discharge capacity, implying that there is still room for increasing the aluminum insertion capacity. Black mesoporous anatase TiO_2 nanoleaves have also been synthesized by a solution plasma process [140]. It not only exhibited a sustained high aluminum storage performance of 278 mAh g^{-1} (corresponding to $Al_{0.27}TiO_2$) at a current density of 0.05 A g^{-1}, but also exhibited an excellent rate capability (141 mAh g^{-1} at a current density of 2.0 A g^{-1}), with excellent cycling stability.

6.7 Outlook

Metal oxides with high specific capacity have been considered the most promising electrode materials for various rechargeable batteries. The further development of metal oxide electrode materials still faces several challenges, such as poor electronic conductivity and the huge volume change during the charge/discharge processes. Poor conductivity always leads to poor rate performance and low-power density, while large volume changes will induce pulverization which results in insufficient cycling stability. Numerous approaches have attempted to address these challenges. Appropriate heteroatom doping can help to achieve higher electronic conductivity due to increased free electron concentrations. Surface modification and/or coating with various conductive materials are also used to relieve these severe problems. Various carbon materials and conductive polymers are extensively studied as buffering and conductive agents for metal oxide electrodes. The design and fabrication of metal oxide electrodes with different morphologies and architectures, such as 1D nanostructures (e.g., nanowires, nanorods, and nanotubes), 2D nanostructures (e.g., nanosheets), and 3D hierarchical nanostructure arrays, has resulted in enhanced cyclability and rate capability. In general, the rational design and fabrication of metal oxide electrodes with desirable architectures and optimized properties has been shown to be an effective approach to produce high-performance rechargeable batteries. Undoubtedly, research on metal oxide electrode materials for rechargeable batteries has a very bright future.

Although research has led to considerable breakthroughs, some challenges for using metal oxides in rechargeable batteries still remain:

(1) A good understanding of the surface chemistry between electrode materials and electrolytes is needed for further design and exploration of high-performance metal oxide electrode materials.
(2) It is well known that unique morphologies can influence the electrochemical performance significantly. Controllable fabrication of the structure and morphology of metal oxides is still required. It should be noted that little emphasis has been placed so far on the low-cost synthesis of high-performance nanostructured metal oxides. Developing large-scale fabrication techniques for less-expensive nanostructured metal oxides, with sustainable and minimal environmental impact on resources is crucial for widespread commercial application.
(3) Current rechargeable battery systems are not only limited by the performance of the electrode materials. Metal oxide-based electrode materials should be systematically developed in parallel with the progress made in electrolytes and separators.
(4) Metal oxides can be used for further exploration of new battery chemistries such as Zn//NiO, Zn//Co_3O_4, and metal/air.

Acknowledgments

Financial support from National Materials Genome Project (2016YFB0700600), China National Funds for Distinguished Youth Scientists (51026004), and NSFC (U1601214, 51502137, and 21603103) is gratefully acknowledged.

References

[1] Y. Liu, H. Pan, M. Gao, Q. Wang, J. Mater. Chem. 21 (2011) 4743–4755.
[2] F. Zhan, L.J. Jiang, B.R. Wu, Z.H. Xia, X.Y. Wei, G.R. Qin, J. Alloys Compd. 293–295 (1999) 804–808.
[3] W.H. Zhu, Y. Zhu, B.J. Tatarchuk, Int. J. Hydrog. Energy 39 (2014) 19789–19798.
[4] N. Watanabe, T. Arakawa, Y. Sasaki, T. Yamashita, I. Koiwa, J. Electrochem. Soc. 159 (2012) A1949–A1953.
[5] F. Tronel, L.G. Demourgues, M. Menetrier, L. Croguennec, L. Goubault, P. Bernard, C. Delmas, Chem. Mater. 18 (2006) 5840–5851.
[6] J.B. Wu, J.P. Tu, W.K. Zhang, H. Huang, J. Alloys Compd. 431 (2007) 321–325.
[7] J.B. Wu, J.P. Tu, T.A. Han, Y.Z. Yang, W.K. Zhang, X.B. Zhao, J. Power Sources 156 (2006) 667–672.
[8] X. He, J. Ren, W. Li, C. Jiang, C. Wan, Electrochim. Acta 51 (2006) 4533–4536.

[9] X.M. He, C.Y. Jiang, W. Li, C.R. Wan, J. Electrochem. Soc. 153 (2006) A566–A569.
[10] J.B. Wu, J.P. Tu, Z. Yu, X.B. Zhang, J. Electrochem. Soc. 153 (2006) A1847–A1851.
[11] W.K. Zhang, X.H. Xia, H. Huang, Y.P. Gan, J.B. Wu, J.P. Tu, J. Power Sources 184 (2008) 646–651.
[12] B. Li, H. Cao, J. Shao, H. Zheng, Y. Lu, J. Yin, M. Qu, Chem. Commun. 47 (2011) 3159–3161.
[13] C. Fierro, A. Zallen, J. Koch, M.A. Fetcenko, J. Electrochem. Soc. 153 (2006) A492–A496.
[14] W. Zhang, W. Jiang, L. Yu, Z. Fu, W. Xia, M. Yang, Int. J. Hydrog. Energy 34 (2009) 473–480.
[15] M.A. Kiani, M.F. Mousavi, S. Ghasemi, J. Power Sources 195 (2010) 5794–5800.
[16] H. Fukunaga, M. Kishimi, N. Igarashi, T. Ozaki, T. Sakai, J. Electrochem. Soc. 152 (2005) A42–A46.
[17] H. Fukunaga, M. Kishimi, T. Ozaki, T. Sakai, J. Electrochem. Soc. 152 (2005) A126–A131.
[18] E. Shangguan, Z. Chang, H. Tang, X.Z. Yuan, H. Wang, Int. J. Hydrog. Energy 35 (2010) 9716–9724.
[19] J.B. Goodenough, Acc. Chem. Res. 46 (2013) 1053–1061.
[20] M.V. Reddy, G.V. Subba Rao, B.V.R. Chowdari, Chem. Rev. 113 (2013) 5364–5457.
[21] W.H. Zhu, Y. Zhu, Z. Davis, B.J. Tatarchuk, Appl. Energy 106 (2013) 307–313.
[22] F. Wang, X. Wang, Z. Chang, Y. Zhu, L. Fu, X. Liu, Y. Wu, Nanoscale Horiz. 1 (2016) 272–289.
[23] L.J. Fu, H. Liu, C. Li, Y.P. Wu, E. Rahm, R. Holze, H.Q. Wu, Solid State Sci. 8 (2006) 113–128.
[24] F. Wang, Y. Liu, X. Wang, Z. Chang, Y. Wu, R. Holze, ChemElectroChem 2 (2015) 1024–1030.
[25] K.T. Lee, S. Jeong, J. Cho, Acc. Chem. Res. 46 (2013) 1161–1170.
[26] H. Yu, H.S. Zhou, J. Phys. Chem. Lett. 4 (2013) 1268–1280.
[27] A. Pan, H.B. Wu, L. Yu, X.W. Lou, Angew. Chem. Int. Ed. 52 (2013) 2226–2230.
[28] J. Cho, Y.W. Kim, B. Kim, J.G. Lee, B. Park, Angew. Chem. Int. Ed. 42 (2003) 1618–1621.
[29] H. Lee, M.G. Kim, J. Cho, Electrochem. Commun. 9 (2007) 149–154.
[30] J. Cho, H. Kim, B. Park, J. Electrochem. Soc. 151 (2004) A1707–A1711.
[31] Y. Wu, A. Vadivel Murugan, A. Manthiram, J. Electrochem. Soc. 155 (2008) A635–A641.
[32] C. Li, H.P. Zhang, L.J. Fu, H. Liu, Y.P. Wu, E. Rahm, R. Holze, H.Q. Wu, Electrochim. Acta 51 (2006) 3872–3883.
[33] M.M. Thackeray, C.S. Johnson, J.S. Kim, Electrochem. Commun. 5 (2003) 752–758.
[34] J.H. Park, J.H. Cho, S.B. Kim, W.S. Kim, S.Y. Lee, S.Y. Lee, J. Mater. Chem. 22 (2012) 12574–12581.
[35] H. Li, H. Zhou, Chem. Commun. 48 (2012) 1201–1217.
[36] Q. Cao, H.P. Zhang, G.J. Wang, Q. Xia, Y.P. Wu, H.Q. Wu, Electrochem. Commun. 9 (2007) 1228–1232.
[37] Y.L. Cheah, R. Hagen, V. Aravindan, R. Fiz, S. Mathur, S. Madhavi, Nano Energy 2 (2013) 57–64.
[38] Y. Wu, A. Manthiram, J. Electrochem. Solid-State Lett. 9 (2006) A221–A224.
[39] Y.K. Sun, M.J. Lee, C.S. Yoon, J. Hassoun, K. Amine, B. Scrosati, Adv. Mater. 24 (2012) 1192–1196.

[40] F. Wang, S. Xiao, M. Li, X. Wang, Y. Zhu, Y. Wu, A. Shirakawa, J. Peng, J. Power Sources 287 (2015) 416–421.
[41] F. Wang, Z. Chang, X. Wang, Y. Wang, B. Chen, Y. Zhu, Y. Wu, J. Mater. Chem. A 3 (2015) 4840–4845.
[42] B. Park, J. Cho, Y.J. Kim, T.J. Kim, Angew. Chem. Int. Ed. 40 (2001) 3367–3369.
[43] Y. Cho, P. Oh, J. Cho, Nano Lett. 13 (2013) 1145–1152.
[44] C. Pouillerie, F. Perton, P. Biensan, J.P. Peres, M. Broussely, C. Delmas, J. Power Sources 96 (2001) 293–302.
[45] L. Zhu, Y. Liu, W. Wu, X. Wu, W. Tang, Y. Wu, J. Mater. Chem. A 3 (2015) 15156–15162.
[46] Y. Huang, J. Chen, J. Ni, H. Zhou, X. Zhang, J. Power Sources 188 (2009) 538–545.
[47] M. Kageyama, D. Li, K. Kobayakawa, Y. Sato, Y.S. Lee, J. Power Sources 157 (2006) 494–500.
[48] Y.L. Ding, J. Xie, G.S. Cao, T.J. Zhu, H.M. Yu, X.B. Zhao, J. Phys. Chem. C 115 (2011) 9821–9825.
[49] B. Wang, Y. Xia, L. Feng, D. Zhao, J. Power Sources 44 (1993) 539–546.
[50] S. Zhou, X. Yang, Y. Lin, J. Xie, D. Wang, ACS Nano 6 (2012) 919–924.
[51] X.J. Wang, R. Nesper, C. Villevieille, P. Novák, Adv. Energy Mater. 3 (2013) 606–614.
[52] M. Okubo, E. Hosono, J. Kim, M. Enomoto, N. Kojima, T. Kudo, H. Zhou, I. Honma, J. Am. Chem. Soc. 129 (2007) 7444–7452.
[53] D.K. Kim, P. Muralidharan, H.W. Lee, R. Ruffo, Y. Yang, C.K. Chan, H. Peng, R. A. Huggins, Y. Cui, Nano Lett. 8 (2008) 3948–3952.
[54] H.W. Lee, P. Muralidharan, R. Ruffo, C.M. Mari, Y. Cui, D.K. Kim, Nano Lett. 10 (2010) 3852–3856.
[55] Y.L. Ding, J. Xie, G.S. Cao, T.J. Zhu, H.M. Yu, X.B. Zhao, Adv. Funct. Mater. 21 (2011) 348–355.
[56] E. Hosono, T. Kudo, I. Honma, H. Matsuda, H. Zhou, Nano Lett. 9 (2009) 1045–1051.
[57] J. Lim, J. Moon, J. Gim, S. Kim, K. Kim, J. Song, J. Kang, W.B. Im, J. Kim, J. Mater. Chem. 22 (2012) 11772–11777.
[58] Y. Lee, M.G. Kim, J. Cho, Nano Lett. 8 (2008) 957–961.
[59] Y. Li, J. Yao, E. Uchaker, J. Yang, Y. Huang, M. Zhang, G. Cao, Adv. Energy Mater. 3 (2013) 1171–1175.
[60] A.Q. Pan, H.B. Wu, L. Zhang, X.W. Lou, Energy Environ. Sci. 6 (2013) 1476–1479.
[61] X. Chen, E. Pomerantseva, P. Banerjee, K. Gregorczyk, R. Ghodssi, G. Rubloff, Chem. Mater. 24 (2012) 1255–1261.
[62] P. Poizot, S. Laruelle, S. Grugeon, L. Dupont, J.M. Tarascon, Nature 407 (2000) 496–499.
[63] H. Li, X. Liu, T. Zhai, D. Li, H. Zhou, Adv. Energy Mater. 3 (2013) 428–432.
[64] N. Zhao, G. Wang, Y. Huang, B. Wang, B. Yao, Y. Wu, Chem. Mater. 20 (2008) 2612–2614.
[65] Q. Gao, L. Yang, X. Lu, J. Mao, Y. Zhang, Y. Wu, Y. Tang, J. Mater. Chem. 20 (2010) 2807–2812.
[66] K.H. Seng, M. Park, Z.P. Guo, H.K. Liu, J. Cho, Nano Lett. 13 (2013) 1230–1236.
[67] W. Wei, S. Yang, H. Zhou, I. Lieberwirth, X. Feng, K. Müllen, Adv. Mater. 25 (2013) 2909–2914.
[68] Y. Shi, J. Gaoa, H.D. Abruñaa, H. Liu, H. Li, J. Wang, Y. Wu, Nano Energy 8 (2014) 297–304.
[69] Y. Wang, H. Liu, K. Wang, H. Eiji, Y. Wang, H. Zhou, J. Mater. Chem. 19 (2009) 6789–6795.

[70] Y.Q. Wang, L. Gu, Y.G. Guo, H. Li, X.Q. He, S. Tsukimoto, Y. Ikuhara, L.J. Wan, J. Am. Chem. Soc. 134 (2012) 7874–7879.
[71] J. Liu, S. Tang, Y. Lu, G. Cai, S. Liang, W. Wang, X. Chen, Energy Environ. Sci. 6 (2013) 2691–2697.
[72] Y.P. Wu, E. Rahm, R. Holze, Electrochim. Acta 47 (2002) 3491–3507.
[73] W. Jiao, N. Li, L. Wang, L. Wen, F. Li, G. Liu, H.M. Cheng, Chem. Commun. 49 (2013) 3461–3463.
[74] L. Zhang, H.B. Wu, X.W. Lou, Adv. Energy Mater. 4 (2014) 1300958–1300969.
[75] C. Wang, F. Wang, Y. Zhao, Y. Li, Q. Yue, Y. Liu, Y. Liu, A.A. Elzatahry, A. Al-Enizi, Y. Wu, Y. Deng, D. Zhao, Nano Res. 9 (2016) 165–173.
[76] Z. Wang, X.W. Lou, Adv. Mater. 24 (2012) 4124–4129.
[77] Y. Shi, J.Z. Wang, S.L. Chou, D. Wexler, H.J. Li, K. Ozawa, H.K. Liu, Y.P. Wu, Nano Lett. 13 (2013) 4715–4720.
[78] B. Wang, J.S. Chen, H.B. Wu, Z.Y. Wang, X.W. Lou, J. Am. Chem. Soc. 133 (2011) 17146–17148.
[79] Z.Y. Wang, D.Y. Luan, S. Madhavi, C.M. Li, X.W. Lou, Chem. Commun. 47 (2011) 8061–8063.
[80] L. Zhang, H.B. Wu, S. Madhavi, H.H. Hng, X.W. Lou, J. Am. Chem. Soc. 134 (2012) 17388–17391.
[81] W.L. Yao, J. Yang, J.L. Wang, Y. Nuli, J. Electrochem. Soc. 155 (2008) A903–A908.
[82] B. Varghese, M.V. Reddy, Z. Yanwu, C.S. Lit, T.C. Hoong, G.V. Subba Rao, B.V. R. Chowdari, A.T.S. Wee, C.T. Lim, C.H. Sow, Chem. Mater. 20 (2008) 3360–3367.
[83] X. Yu, Y. He, J. Sun, K. Tang, H. Li, L. Chen, X. Huang, Electrochem. Commun. 11 (2009) 791–794.
[84] K. Zhong, B. Zhang, S. Luo, W. Wen, H. Li, X. Huang, L. Chen, J. Power Sources 196 (2011) 6802–6808.
[85] C. Wang, G. Du, K. Stahl, H. Huang, Y. Zhong, J.Z. Jiang, J. Phys. Chem. C 116 (2012) 4000–4011.
[86] D. Su, M. Ford, G. Wang, Sci. Rep. 2 (2012) 924–930.
[87] X. Xiao, X. Liu, H. Zhao, D. Chen, F. Liu, J. Xiang, Z. Hu, Y. Li, Adv. Mater. 24 (2012) 5762–5766.
[88] W. Li, J.R. Dahn, D.S. Wainwright, Science 264 (1994) 1115–1118.
[89] G. Wang, L. Fu, N. Zhao, L. Yang, Y. Wu, H. Wu, Angew. Chem. Int. Ed. 46 (2007) 295–297.
[90] Z. Chang, Y. Yang, M. Li, X. Wang, Y. Wu, J. Mater. Chem. A 2 (2014) 10739–10755.
[91] W. Tang, Y. Zhu, Y. Hou, L. Liu, Y. Wu, K.P. Loh, H. Zhang, K. Zhu, Energy Environ. Sci. 6 (2013) 2093–2104.
[92] W. Tang, L.L. Liu, S. Tian, L. Li, Y.B. Yue, Y.P. Wu, S.Y. Guan, K. Zhu, Electrochem. Commun. 12 (2010) 1524–1526.
[93] W. Tang, L.L. Liu, S. Tian, L. Li, L.L. Li, Y.B. Yue, Y. Bai, Y.P. Wu, K. Zhu, R. Holze, Electrochem. Commun. 13 (2011) 1159–1162.
[94] Y. Liu, B. Zhang, F. Wang, Z. Wen, Y. Wu, Pure Appl. Chem. 86 (2014) 593–609.
[95] W. Tang, S. Tian, L.L. Liu, L. Li, H.P. Zhang, Y.B. Yue, Y. Bai, Y.P. Wu, K. Zhu, Electrochem. Commun. 13 (2011) 205–208.
[96] W. Tang, Y.Y. Hou, F.X. Wang, L.L. Liu, Y.P. Wu, K. Zhu, Nano Lett. 13 (2013) 2036–2040.
[97] Q. Qu, L. Fu, X. Zhan, D. Samuelis, J. Maier, L. Li, S. Tian, Z. Lia, Y. Wu, Energy Environ. Sci. 4 (2011) 3985–3990.
[98] F. Wang, S. Xiao, Z. Chang, Y. Yang, Y. Wu, Chem. Commun. 49 (2013) 9209–9211.

[99] W. Tang, X.W. Gao, Y.S. Zhu, Y.B. Yue, Y. Shi, Y.P. Wu, K. Zhu, J. Mater. Chem. 22 (2012) 20143–20145.
[100] W. Tang, L. Liu, Y. Zhu, H. Sun, Y. Wu, K. Zhu, Energy Environ. Sci. 5 (2012) 6909–6913.
[101] L. Liang, M. Zhou, Y. Xie, Chem. Asian. J. 7 (2012) 565–571.
[102] X. Wang, Y. Hou, Y. Zhu, Y. Wu, R. Holze, Sci. Rep. 3 (2013) 1401–1405.
[103] X. Wang, Q. Qu, Y. Hou, F. Wang, Y. Wu, Chem. Commun. 49 (2013) 6179–6181.
[104] Z. Chang, Y. Yang, X. Wang, M. Li, Z. Fu, Y. Wu, R. Holze, Sci. Rep. 5 (2015) 11931–11938.
[105] G.H. Newman, L.P. Klemann, J. Electrochem. Soc. 127 (1980) 2097–2099.
[106] D. Larcher, J.M. Tarascon, Nat. Chem. 7 (2015) 19–29.
[107] French research network on electrochemical energy storage (RS2E), http://www.energie-rs2e.com/en/news/na-ion-batteries-promising-prototype (accessed: 16.05.17).
[108] H. Kim, R.H. Kim, S.S. Lee, Y. Kim, D.Y. Kim, K. Park, ACS Appl. Mater. Interfaces 6 (2014) 11692–11697.
[109] H. Kim, D.Y. Kim, Y. Kim, S.S. Lee, K. Park, ACS Appl. Mater. Interfaces 7 (2015) 1477–1485.
[110] D. Su, G. Wang, ACS Nano 7 (2013) 11218–11226.
[111] H. Fei, Y. Lin, M. Wei, J. Colloid Interface Sci. 425 (2014) 1–4.
[112] H. Kang, Y. Liu, M. Shang, T. Lu, Y. Wang, L. Jiao, Nano 7 (2015) 9261–9267.
[113] L.W. Shacklette, T.R. Jow, L. Townsend, J. Electrochem. Soc. 135 (1988) 2669–2674.
[114] N. Yabuuchi, K. Kubota, M. Dahbi, S. Komaba, Chem. Rev. 114 (2014) 1636–11682.
[115] E. Hosono, T. Saito, J. Hoshino, M. Okubo, Y. Saito, D.N. Hamane, T. Kudo, H. Zhou, J. Power Sources 217 (2012) 43–46.
[116] Y. Cao, L. Xiao, W. Wang, D. Choi, Z. Nie, J. Yu, L.V. Saraf, Z. Yang, J. Liu, Adv. Mater. 23 (2011) 3155–3160.
[117] D. Su, C. Wang, H. Ahn, G. Wang, Chem. Eur. J. 19 (2013) 10884–10889.
[118] C.Y. Yu, J.S. Park, H.G. Jung, K.Y. Chung, D. Aurbach, Y.K. Sun, S.T. Myung, Energy Environ. Sci. 8 (2015) 2019–2026.
[119] Y.H. Jung, S.T. Hong, D.K. Kim, J. Electrochem. Soc. 160 (2013) A897–A900.
[120] X. Jiang, S. Liu, H. Xu, L. Chen, J. Yang, Y. Qian, Chem. Commun. 51 (2015) 8480–8483.
[121] N. Yabuuchi, H. Yoshida, S. Komaba, Electrochemistry 80 (2012) 716–719.
[122] R. Alcantara, M. Jaraba, P. Lavela, J.L. Tirado, Chem. Mater. 14 (2002) 2847–2848.
[123] S. Yuan, X. Huang, D. Ma, H. Wang, F. Meng, X. Zhang, Adv. Mater. 26 (2014) 2273–2279.
[124] S.Y. Hong, Y. Kim, Y. Park, A. Choi, N.S. Choi, K.T. Lee, Energy Environ. Sci. 6 (2013) 2067–2081.
[125] Y. Xu, E.M. Lotfabad, H.L. Wang, B. Farbod, Z.W. Xu, A. Kohandehghan, D. Mitlin, Chem. Commun. 49 (2013) 8973–8975.
[126] P. Senguttuvan, G. Rousse, V. Seznec, J.M. Tarascon, M.R. Palacín, Chem. Mater. 23 (2011) 4109–4111.
[127] Y. Xu, M. Zhou, X. Wang, C. Wang, L. Liang, F. Grote, M. Wu, Y. Mi, Y. Lei, Angew. Chem. 54 (2015) 1–5.
[128] J. Muldoon, C.B. Bucur, T. Gregory, Chem. Rev. 114 (2014) 11683–11720.
[129] G.G. Amatucci, F. Badway, A. Singhal, B. Beaudoin, G. Skandan, T. Bowmer, I. Plitz, N. Pereira, T. Chapman, R. Jaworski, J. Electrochem. Soc. 148 (2001) A940–A950.
[130] L. Jiao, H. Yuan, Y. Wang, J. Cao, Y. Wang, Electrochem. Commun. 7 (2005) 431–436.
[131] M. Okoshi, Y. Yamada, A. Yamada, H. Nakai, J. Electrochem. Soc. 160 (2013) A2160–A2165.

[132] R. Zhang, X. Yu, K.W. Nam, C. Ling, T.S. Arthur, W. Song, A.M. Knapp, S.N. Ehrlich, X.Q. Yang, M. Matsui, Electrochem. Commun. 23 (2012) 110–113.
[133] S. Rasul, S. Suzuki, S. Yamaguchi, M. Miyayama, Electrochim. Acta 82 (2012) 243–249.
[134] G. Gershinsky, H.D. Yoo, Y. Gofer, D. Aurbach, Langmuir 29 (2013) 10964–10972.
[135] S. Su, Z. Huang, Y. NuLi, F. Tuerxun, J. Yang, J. Wang, Chem. Commun. 51 (2015) 2641–2644.
[136] N. Wu, Y.C. Lyu, R.J. Xiao, X. Yu, Y.X. Yin, X.Q. Yang, H. Li, L. Gu, Y.G. Guo, NPG Asia Mater. 6 (2014) e120.
[137] N. Jayaprakash, S.K. Das, L.A. Archer, Chem. Commun. 47 (2011) 12610–12612.
[138] W. Wang, B. Jiang, W. Xiong, H. Sun, Z. Lin, L. Hu, J. Tu, J. Hou, H. Zhu, S. Jiao, Sci. Rep. 3 (2013) 3383–3388.
[139] S. Liu, J.J. Hu, N.F. Yan, G.L. Pan, G.R. Li, X.P. Gao, Energy Environ. Sci. 5 (2012) 9743–9746.
[140] Y.J. He, J.F. Peng, W. Chu, Y.Z. Li, D.G. Tong, J. Mater. Chem. A 2 (2014) 1721–1731.

Metal oxides in supercapacitors

7

Lei Zhou, Chunyang Li*, Xiang Liu*, Yusong Zhu*, Yuping Wu*, Teunis van Ree*[†]
*State Key Laboratory of Materials-Oriented Chemical Engineering, School of Energy Science and Engineering, Institute for Electrochemical Energy Storage, Nanjing Tech University, Nanjing, People's Republic of China, [†]Department of Chemistry, University of Venda, Thohoyandou, South Africa

7.1 Introduction

Supercapacitors, also known as electrochemical capacitors, emerged around the globe as new energy storage systems in the 1970s [1, 2]. They are promising new energy storage devices, which can play a bridging role between batteries and conventional capacitors. In the past several decades, supercapacitors have attracted a vast amount of attention due to the higher power density than that of rechargeable batteries, higher energy density compared with conventional dielectric capacitors, and long cycle life [3–11]. A huge amount of scientific and technological research has been committed to develop the promise of superior power and energy of the supercapacitors as a hybrid system for electric vehicles, owing to the steadily growing consumption of petroleum and environmental concerns. Supercapacitors could provide higher power compensation to meet the requirements of moving, accelerating, and climbing power for electric vehicles. They have already been used successfully in electronic products, energy storage, and other fields. Their market size is expanding gradually with the rapid development of modern society.

Consequently, the US Department of Energy has accepted a development plan in which supercapacitors will play an important role, with batteries, in upcoming energy storage systems [12]. Other countries and companies have also invested huge amounts in researching and developing supercapacitor technologies. Although the electrochemical capacitor is becoming a promising energy storage device due to its advantages in respect of power density and cycling life, it has an unsatisfactory low-energy density [13, 14]. To overcome this obstacle, one of the most intensive approaches is the development of new materials. Electrode materials mainly include carbon materials, metal oxides, conducting polymer materials, and battery-type materials [15–24]. Each kind of electrode material has its merits and drawbacks. In this chapter, we focus on introducing the progress of research on metal oxides in supercapacitors.

Metal Oxides in Energy Technologies. https://doi.org/10.1016/B978-0-12-811167-3.00007-9

7.2 Outline of electrochemical capacitors

7.2.1 Principle and classification of supercapacitors

Supercapacitors are known as power compensation and energy storage devices. Three types of capacitors can be distinguished, depending on the energy storage principle: electric double-layer capacitors (EDLCs), pseudocapacitors, and hybrid supercapacitors [25, 26]. They are described in detail in the following sections.

7.2.1.1 Electric double-layer capacitors

EDLCs are charge storage devices, which are similar to lithium ion batteries in design and assembly. In general, EDLCs are composed of two electrodes, an electrolyte and a separator [27]. The separator electrically insulates the positive electrode and negative electrode in an organic electrolyte system.

As shown schematically in Fig. 7.1, the electrical double-layer capacitance derives from the separation of electric charges, which is generated by directional alignment of ions and electrons at the interface between the electrode material and electrolyte [28]. In particular, the excess electric charges aggregate on the positive or negative electrode surfaces, while electrolyte ions with compensating charges accumulate in the electrolyte solution to maintain electrical neutrality. When the system is given an

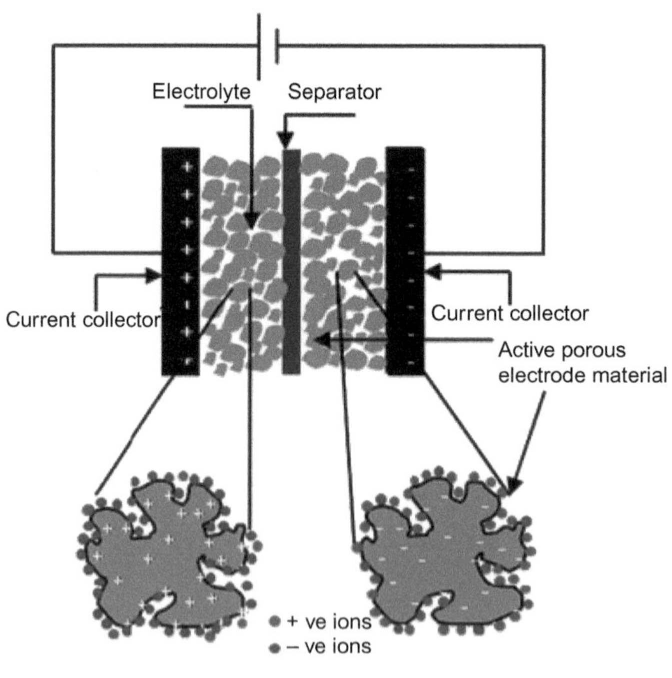

Fig. 7.1 Schematic representation of electric double layer capacitors (EDLCs). Reproduced by permission from the Royal Society of Chemistry.

external load or is charged, the cations in the electrolyte migrate toward the negative electrode and the anions migrate to the positive electrode, while the electrons travel through the external circuit from the negative to the positive electrode [29]. In this way, the double layer is formed at the interfaces with the electrodes. When an external load is removed or charging finishes, positive electric charges on the electrode and anions in the electrolyte attract each other while negative electric charges on the electrode and cations in the electrolyte attract each other in order to stabilize the double layers on the electrodes [29]. Thus, a relatively stable voltage is generated. During the process of discharge, the reverse processes happen. The mechanism of electric charge generation on the surface of the electrode implies that the concentration of electrolyte remains constant regardless of the process of charging/discharging. As a result, energy accumulates in the double-layer interface.

The above mechanism comprises surface dissociation and ion adsorption facilitated not only by the electrolyte but also by crystal lattice defects [30]. During the charging/discharging processes, the electrode materials do not react electrochemically, and physical charge accumulation takes place at the solid/electrolyte interface. Therefore, electric double-layer charge storage is a surface process, and the surface characteristics of the electrode enormously affect the capacitance of the system. Based on this mechanism, the selection of electrode material directly determines the electrical performances of EDLCs. High surface area carbon materials have been used to develop high-density capacitors due to availability and abundance. Moreover, carbon electrodes can assume a vast amount of manufactured shapes and forms such as foams, fibers, porous carbon, carbon nanotubes (CNTs), graphene, etc. [18, 31–40] Although carbon materials have been utilized as high surface area electrode material ever since the beginning of EDLC development, they are still among the most popular electrode materials because of their availability and low cost.

7.2.1.2 Pseudocapacitors

Pseudocapacitors, also called faradaic supercapacitors, are different from EDLCs [41] in that pseudocapacitance arises from fast and reversible faradaic reactions (redox reactions) at or near the electrode surface as shown in Fig. 7.2 [26]. The faradaic electrochemical processes [42] comprise the passage of charge across the double layer, resulting in faradaic current passing through the supercapacitor cell. These processes occurring in the supercapacitor enhance the energy density and also increase the specific capacitance; on the other hand, pseudocapacitors suffer from a relatively low power density and lack of cycling stability when compared with EDLCs [43, 44]. The most commonly used pseudocapacitive materials are metal oxides (such as RuO_2 and MnO_2) [45, 46] and electronically conducting polymers (such as polyanilines, polypyrroles, and polythiophenes) [47–50].

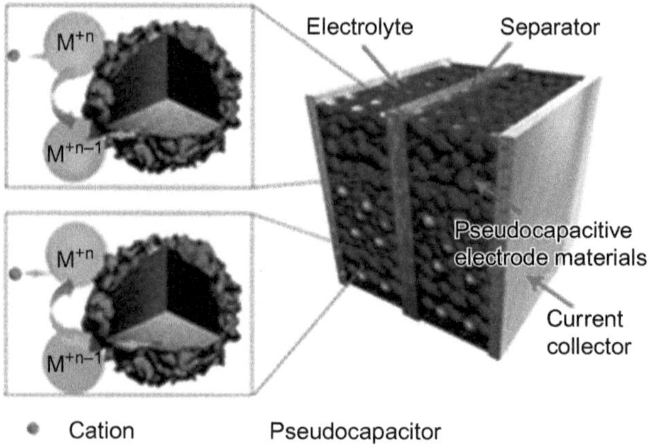

Fig. 7.2 Schematic representation of pseudocapacitors.
Reproduced by permission from Elsevier.

7.2.1.3 Hybrid supercapacitors

Hybrid supercapacitors have recently come into focus and have progressed vastly [51–60]. They consist of two different types of electrode materials, a separator to isolate the two electrodes electrically, and an electrolyte. The structure of hybrid supercapacitors is illustrated schematically in Fig. 7.3A [61]. To further enhance the energy density, hybrid supercapacitors usually comprise a redox (faradaic) reaction electrode or a battery-type electrode, with an electric double-layer electrode (typically carbon materials) in an organic or aqueous electrolyte [62, 63]. In such a hybrid supercapacitor, both mechanisms, electrical double-layer capacitance and faradaic or battery capacitance, happen simultaneously. The redox or battery-type electrodes have a large energy density, while the nonfaradaic capacitive electrode has high power density and excellent cycling stability [62, 64]. Thus, the electrode assembly can improve markedly the energy density of the device. In the field of hybrid capacitors, scientific and technical workers have developed both high voltage and high-energy density lithium and sodium ion capacitors [57, 58, 62]. The structure of lithium ion capacitors is illustrated schematically in Fig. 7.3B [26].

7.2.2 Characteristics of supercapacitors

In Fig. 7.4 the plots of power density versus energy density, also called Ragone plots, are shown for the most important modern storage devices [12]. As shown in Fig. 7.4, it is clear that supercapacitors attain the highest power density (1–10 kW kg^{-1}) when compared with other energy storage systems such as Li-ion batteries, Ni-MH batteries, and lead acid battery cells. In other words, they can be fully charged or discharged in shorter times. Therefore, they have an important role in complementing or replacing batteries in terms of power density. The main characteristics of supercapacitors include the following:

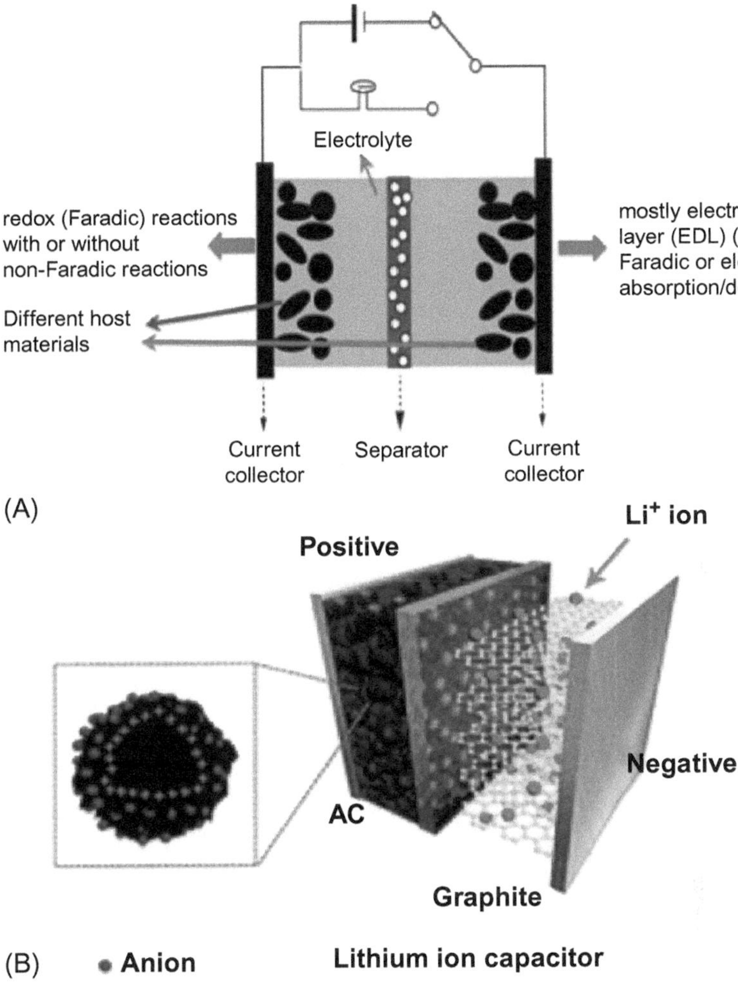

Fig. 7.3 Schematic illustration of (A) hybrid supercapacitors and (B) lithium ion capacitors. (A): Reproduced by permission from Royal Society of Chemistry. (B): Reproduced by permission from Elsevier.

(1) High power density. Supercapacitors can be completely charged or discharged in seconds, whereas batteries normally require several hours [12, 65].

(2) Excellent cycling stability. General speaking, EDLCs can run over 10,000 cycles without any noticeable performance degradation. Even though fast redox reactions occur in the charging and recharging processes, they also enjoy a much longer lifetime than batteries [66, 67].

(3) Wide range of operating temperatures, typically between −40°C and +70°C [29].

(4) High efficiency. Supercapacitors charge and discharge reversibly over the full-operating voltage range, with a relatively low-energy loss during each cycle [68].

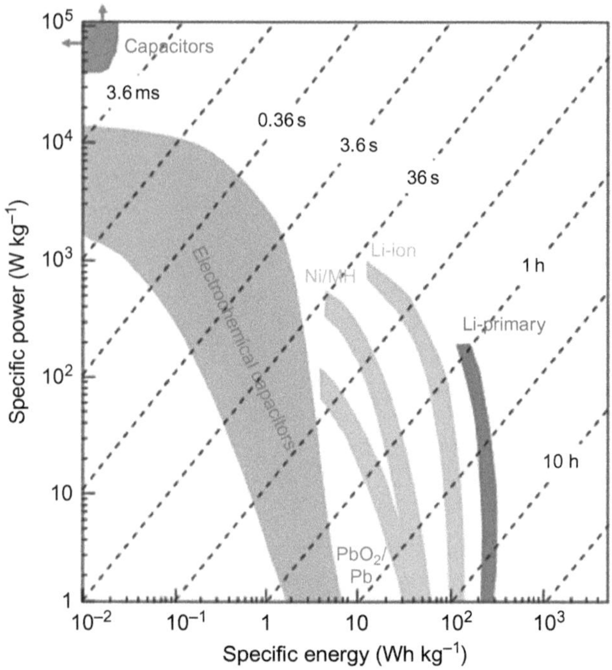

Fig. 7.4 Ragone plots of various electrical energy storage devices.
Reproduced by permission from Macmillan.

(5) Environmentally friendly. Toxic materials are normally not involved and waste materials are easily disposed of [69, 70].
(6) Low cost. The raw materials are abundant and widely available [71].
(7) Safety. Supercapacitors are much safer than lithium-ion batteries due to their using aqueous electrolytes [72].

In summary, supercapacitors are excellent in the areas of power density, cycling stability, operating temperature, efficiency, environmental friendliness, cost, and safety.

7.2.3 Electrolytes for supercapacitors

The choice of electrolyte is very important for supercapacitor performance. On the one hand, the system voltage of a supercapacitor depends on the stability of the electrolyte at the applied potential. On the other hand, the energy density is limited by the breakdown voltage of the electrolyte, and the power density also relies on electrolyte conductivity. Thus, a high-operating voltage system can provide both high-energy density and power density as long as a suitable electrolyte is used.

The requirements for electrolytes include high breakdown voltage, superior electrochemical stability, high ionic concentration, low viscosity, low resistivity, low volatility, low toxicity, and low cost [29]. There are currently three types of electrolyte employed in supercapacitors: aqueous electrolyte, organic electrolyte, and ionic liquids [26].

The aqueous electrolytes currently employed can be categorized into three types: acidic solution (such as H_2SO_4 solution), alkaline solution (such as KOH solution), and neutral solution (such as Li_2SO_4, Na_2SO_4, or KCl solution) [73–81]. Aqueous electrolytes provide both higher ionic concentration and lower resistance compared with organic electrolytes. Therefore, supercapacitors using an aqueous electrolyte solution may well have a higher capacitance and more power than capacitors containing organic electrolytes. In addition, aqueous electrolytes have a big advantage over organic electrolytes: they can be prepared and employed without too tight control of the production process, whereas organic electrolytes require strict preparation procedures to get pure electrolytes. However, aqueous electrolytes have a narrow voltage window of 1.23 V (due to the thermodynamic decomposition of water) [82], much smaller than those of organic electrolytes. Because of the narrow voltage window the energy and power densities in aqueous electrolytes have only a limited chance of improvement, resulting in the typically lower energy densities of supercapacitors with aqueous electrolytes.

Organic electrolytes allow a much wider voltage window of about 3.5 V compared with aqueous electrolytes [83]. This results in a large advantage with respect to higher energy density when using an organic electrolyte rather than aqueous electrolyte. Among the organic electrolytes, propylene carbonate (PC) is the most commonly used solvent, because PC-based electrolytes are environmentally friendly and provide a wide voltage window as well as good conductivity [26]. From the foregoing, it is clear that the water content of organic electrolytes should be as low as possible, preferably below 3–5 ppm [26], to avoid a significant voltage reduction. There are also other considerations when using organic electrolytes. Compared with aqueous electrolyte, for one thing, their complex purification and preparation procedures may cause safety problems due to the flammability and toxicity of the organic solvent. Furthermore, their low conductivity could lead to lower power and smaller capacitance.

Ionic liquids, which are known as room-temperature molten salts, are attractive candidates for electrolytes due to desirable properties [84], such as their high thermal and chemical stability, extensive electrochemical stability window between 2 and 6 V, low flammability, nontoxicity, the wide variety of cation and anion combinations, etc. [85–87] Ionic liquids are composed entirely of ions, they are solvent-free and liquids at room temperature, making them attractive "green electrolytes" [88, 89]. However, they are usually highly viscous liquids and have low ionic conductivity at ambient temperatures, seriously influencing their electrochemical performance. As a result, designing an ionic liquid with higher conductivity for practical applications remains a challenging prospect [90]. Although ionic liquids have not so far met the requirements for practical applications, their future development will be of great significance in achieving high performance, safety and "green" energy storage systems.

7.3 Metal oxides in supercapacitors

7.3.1 Preparation methods for metal oxides in supercapacitors

In the following, we introduce the main preparation methods among the many published methods for preparation of metal oxides.

(1) High-temperature solid-state method. This is a convenient and simple method for the preparation of metal oxides. Generally speaking, the solid precursors are mixed intimately and then reacted at a high temperature. For example, Co_3O_4 can be prepared by the high-temperature solid-state method [90].
(2) Hydrothermal method. In this system, the precursor is dissolved in deionized water or organic solvent and then heated in an airtight reaction autoclave. The resulting high temperature and pressure reaction produces a crystalline product. Some metal oxides (such as MnO_2 and NiO) can be prepared by the hydrothermal method [91, 92].
(3) Templating method. This is an important method for the synthesis of porous materials. Organic materials are usually prepared to form a template with a distinct morphology and then the metal sample grows at their surface. The target product is obtained by removing the organic template. In this process, it is also important to choose a suitable organic template to control the open spaces of the porous material. For instance, hollow-structured Co_3O_4 and high surface area MnO_2 can be fabricated by a templating derived method [93, 94].
(4) Sol-gel method. In this process the precursors, which usually are chemically reactive compounds, firstly form a sol by undergoing hydrolysis and condensation reactions. The sol slowly polymerizes to form a gel, and then the gel generates nanomaterial by drying and curing. For instance, the sol-gel method was used to prepare MnO_2 nanowire arrays and $NiO/NiCo_2O_4/Co_3O_4$ composite, resulting in improved electrochemical behavior for electrochemical capacitors [95, 96].
(5) Electrodeposition method. Here, electrochemical principles are used to deposit metal ions and insoluble particles in suspension to obtain the desired product on the cathode surface. The process is convenient and simple, and the product's thickness and morphology can be controlled by choosing an appropriate oxidation voltage, solution temperature, and deposition time. For example, RuO_2 and MnO_2 have been prepared by the electrodeposition method [97, 98].
(6) Chemical vapor deposition (CVD). Some compounds or gases generate a nonvolatile solid membrane layer or nanomaterial on a suitable substrate surface. The basic requirement is that there must be a heterogeneous reaction on the substrate surface. This method combines low deposition temperature with a high deposition rate. Moreover, the composition and crystallinity of the solid membrane can be controlled accurately by the sedimentation conditions. Metal oxides such as ZnO and TiO_2 can be prepared by this method [99, 100].
(7) Chemical precipitation method. In this method, a chemical reaction between different components in solution is used to form an insoluble product. When the precipitation reagent solution is added dropwise to an inorganic solution, the product is formed rapidly and uniformly dispersed. The target product is obtained after filtration and cleaning steps followed by heat treatment. For instance, $Ni(OH)_2$ and SnO_2 are usually prepared by a chemical precipitation method, especially for large-scale synthesis of carbon material/metal oxide composites [101, 102].

7.3.2 Electrochemical performance of selected metal oxides

7.3.2.1 RuO_2

Ruthenium oxide (RuO_2) is a widely studied pseudocapacitive material due to its highly reversible redox reactions, three distinct oxidation states accessible within the voltage window of 1.2 V, high theoretical specific capacitance (1360 F g^{-1} for $RuO_2 \cdot 0.5H_2O$), good thermal stability, high electronic conductivity (300 S cm^{-1}), superior cycle lifespan, and high rate capability [45, 103–105]. The electrochemical behavior of RuO_2 in acidic solution has been largely researched over the last 30 years [12]. The pseudocapacitive behavior is generally ascribed to a fast and reversible electron transfer reaction coupled with adsorption of protons at or near the surface of RuO_2 particles. These processes are shown schematically in Fig. 7.5, where the Ru oxidation state changes from (III) to (IV) [42]. Furthermore, some factors have an important influence on the electrochemical behavior of RuO_2, such as specific surface area, combined water, degree of particle crystallinity, particle size, electrode architecture, and electrolytes used, resulting in a specific capacitance of 200 up to 1200 F g^{-1} [106–113]. For instance, ruthenium oxide thin-film electrodes, which were electrodeposited on titanium substrates, showed highly reversible reactivity and superior cycle stability, with their specific capacitance reaching 786 F g^{-1} [114]. In another example, an arrayed porous RuO_2 structure displayed a specific capacitance of up

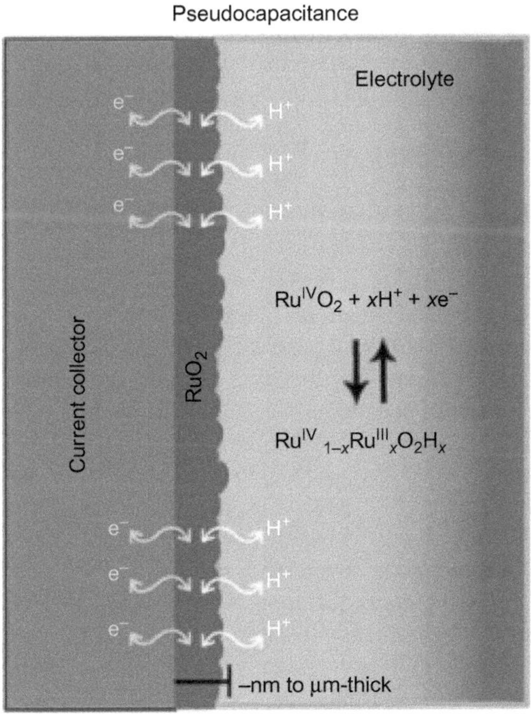

Fig. 7.5 Schematic illustration of ruthenium oxide (RuO_2) in acidic solution.
Reproduced by permission from Elsevier.

to 1300 F g^{-1} because the special structure increases the surface area and then enhances pathways for diffusion, completing the redox reaction with a high pseudocapacitance [45]. A similar trend based on specific surface area was also observed by Patake and Bi [115, 116]. In addition, the capacitance values of RuO_2 can be modified by using different electrolytes and electrolyte concentrations in the devices [117–119].

Although RuO_2 can provide outstanding charge storage performance, the high cost coupled with the precious metal oxide and the difficulties in large-scale production restrict its practical application. Therefore, there has been a continued focus on reducing the use of Ru and improving its utilization [120–122]. Two feasible methods are being developed: fabricating ruthenium-based composite materials in which RuO_2 is deposited on other low-cost materials (such as activated carbon, graphene, CNTs) and preparing composites by doping RuO_2 with base metal oxides [106, 123–128].

7.3.2.2 MnO_2

Among the metal oxides, manganese dioxide (MnO_2) is often considered to be the most promising electrode material for electrochemical capacitors owing to its low cost, nontoxicity, abundance, environmental friendliness, and good theoretical specific capacitance between 1100 and 1300 F g^{-1} [129–131]. In 1999, Goodenough and Lee [80] were the first to report the pseudocapacitive behavior of manganese oxide in an aqueous solution (2 M KCl). This was followed by a great deal of exploring to establish the charge storage mechanism in MnO_2 electrodes. The charge storage mechanism is based on surface adsorption of electrolyte cations M^+ (e.g., K^+, Na^+, Li^+) as well as proton incorporation according to the following reaction [12, 130]:

$$MnO_2 + xM^+ + yH^+ + (x+y)e^- \rightleftharpoons MnOOM_xH_y \quad (7.1)$$

In general, MnO_2 micro-powders or micrometer-thick films display a specific capacitance of about 150 F g^{-1} in neutral aqueous electrolytes within a voltage window of 1 V due to its poor electronic conductivity, which is much lower than that of RuO_2 [12]. However, crystalline MnO_2 has several crystal structures, including α-, β-, γ-, and δ-MnO_2, obtained by different preparation conditions, with different specific capacitances as a result. For instance, α-MnO_2 electrodes display a specific capacitance ranging from 265 to 320 F g^{-1} when measured in aqueous Na_2SO_4 electrolyte solution within a voltage window of 1 V [132]. In general, δ-MnO_2 exhibits a much higher specific capacitance than β-MnO_2 and γ-MnO_2 in Na_2SO_4 electrolyte solution due to its two-dimensional layered structure [133, 134]. On the other hand, the pseudocapacitive performance of MnO_2 is also affected by several physical factors such as morphology, thickness of the electrode layer, and specific surface area [135–138]. Manganese dioxide can exist in many different morphologies such as microspheres, nanowires, nanorods, nanobelts, nanowhiskers, and nanotubes, with specific surface area values typically between 20 and 150 m^2 g^{-1} [139–146]. Generally speaking, a large specific surface area shortens the transport/diffusion path lengths for both ions and electrons, leading to faster kinetics and thus higher charge/discharge capacities [147]. This is probably why MnO_2 nanowires on smaller diameter electrodes have a higher capacitance than nanorods (350 F g^{-1} opposed to 243 F g^{-1}) [148].

To promote the electrical conductivity and so obtain a high specific capacitance for MnO_2, researchers have put significant efforts into exploring some efficient and feasible approaches, including depositing MnO_2 onto carbon materials and fabricating metal-doped nanostructured MnO_2 [149, 150]. The use of carbon-based supports increases the electrical conductivity of MnO_2 electrodes and also enhances their surface area and ordered mesopores. For instance, MnO_2 composites on CNT support display a specific capacitance of 700 F g^{-1} between −0.1 and 0.9 V (vs. SCE) at a scan rate of 20 mV s^{-1} in aqueous Na_2SO_4 solution due to the CNT providing electron-conducting paths and forming a superior conductive network [151]. In another example, nanostructured Co-doped MnO_2 electrodes exhibited a specific capacitance of 621 F g^{-1}; even at a high scan rate of 200 mV s^{-1} they still yielded a specific capacitance of 377 F g^{-1} [152]. In conclusion, to obtain the best electrochemical performance of supercapacitors, MnO_2 electrode materials can be modified by preparing composites with carbon materials and doping metal into the manganese oxide, resulting in optimizing parameters such as electrical conductivity, surface area, particle size, and pore structure.

7.3.2.3 Nickel oxide and hydroxide

(1) Nickel oxide (NiO) is a well-known electrode material in the field of supercapacitors thanks to its high theoretical specific capacitance (about 3750 F g^{-1}), low cost, environmental friendliness, and ease of preparation [153–155]. The pseudocapacitance of NiO originates from the following redox reaction [156, 157]:

$$NiO + OH^- \rightleftharpoons NiOOH + e^- \tag{7.2}$$

Although NiO has enormous advantages in theory, its relatively poor electrical conductivity and low-specific surface area hinder its practical application. One of the solutions to these problems is to synthesize nanostructures with large surface areas, which are associated with more faradaic active sites and higher pseudocapacitance [158]. Therefore, different nanostructured nickel oxides (such as nanowires, nanoflakes, nanocolumns, nanosheets, porous nanoflowers, and hollow nanospheres) have been successfully fabricated in the past few years by a variety of methods [159–164]. For instance, Cao et al. [165] reported the synthesis of flower-like NiO hollow nanospheres with a specific surface area of 176 m^2 g^{-1} based on an efficient gas/liquid interfacial microwave-assisted process, which exhibited a maximum specific capacitance of 585 F g^{-1} and had excellent cycling stability at the current density of 5 A g^{-1} in alkaline electrolyte solution, as shown in Fig. 7.6. Ding et al. prepared NiO hollow nanospheres via a template method and obtained a maximum specific capacitance of 570 F g^{-1} even at the high current density of 0.5 A g^{-1} [158]. The porous structure of the hollow nanospheres enhances the spatial dispersion, resulting in both higher surface area and better mass transport, guaranteeing a steady supply of ions for faradaic reactions [35].

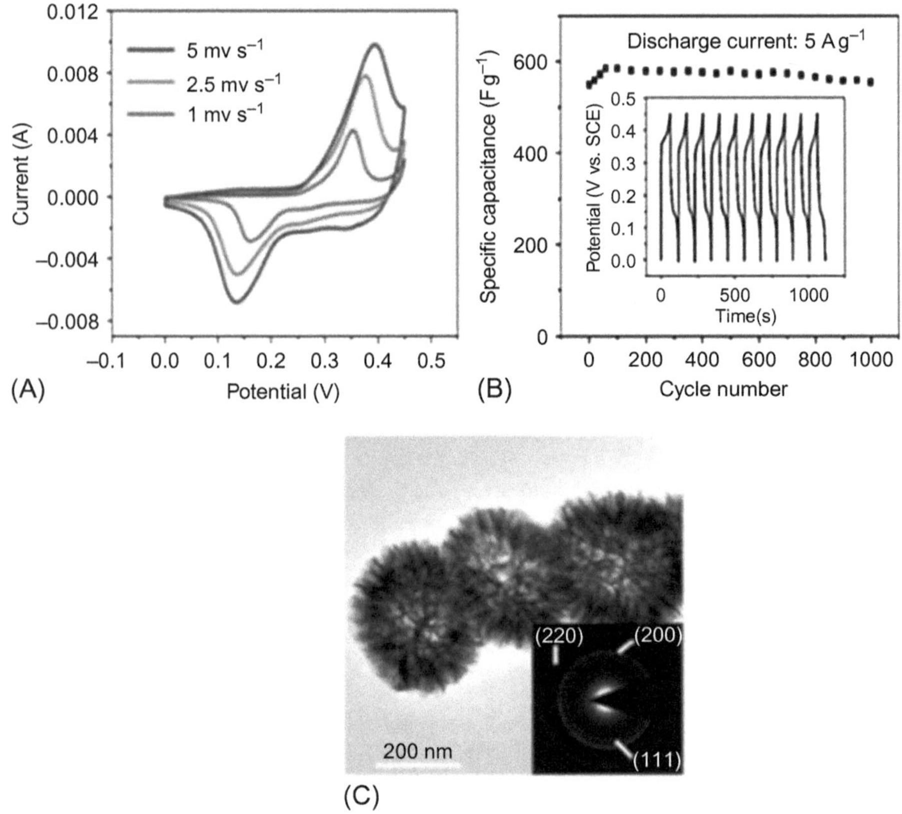

Fig. 7.6 Electrochemical properties of flower-like NiO hollow nanospheres in 2 M KOH solution: (A) CV curves at different scan rates; (B) at a discharge current of 5 A g^{-1} (the inset is the charge/discharge curve of the NiO electrode); characterization: (C) TEM image of the flower-like NiO hollow nanospheres (inset: SAED pattern).
Reproduced by permission from the Royal Society of Chemistry.

For further improvement, a modern trend in NiO materials is the development of nanostructured carbon-based composite materials. For example, Gund et al. [166] prepared a NiO/MWCNT nanohybrid electrode, which exhibited a maximum specific capacitance of 1727 F g^{-1} at 5 mA cm^{-2} of current density with 91% capacitance retention after 2000 cycles, confirming the high specific capacitance and better stability attainable with such materials. The CNT-based nanocomposite electrodes provide a higher surface area with conducting channels, leading to a fast electrochemical kinetic process and outstanding chemical stability to NiO nanomaterials through the $\pi-\pi$ bonding between the side wall of the MWCNTs and NiO [167]. What is more, this nanohybrid also provides a one-dimensional nanostructure, which not only maximizes the utilization ratio of active materials but also decreases the diffusion path length significantly [167]. In addition, Zhu et al. [168] synthesized a rGO/NiO electrode with a high specific capacitance of 770 F g^{-1} and improved rate capability. Zhao et al. studied the electrochemical properties of monolayer graphene/NiO nanosheets, which show a high specific capacitance of 525 F g^{-1}. The electrochemical performance of these rGO-based composites is significantly enhanced compared with those of pure NiO [169]. Bai et al. [170] made a rGO/CNT/NiO composite by a simple method under mild conditions, with specific capacitance as high as 1000 F g^{-1} at a current density of 4 A g^{-1} with a loss rate of 5% after 2000 cycles, demonstrating excellent cycling stability. Apparently the enhanced reduction degree of rGO and tunnels for electron transport provided by the CNTs greatly enhance the electron transport properties [170]. Therefore, the conductivity of nickel oxide can be improved by designing carbon-based composite materials; these can be used to manufacture outstanding electrodes for supercapacitor applications with high specific capacitance, rate capability and cycling stability, which is a favorable development for future large-capacity energy storage.

(2) Nickel hydroxide [$Ni(OH)_2$] is an attractive electrode material because of its high theoretical capacity, superior redox behavior, and potential applications in alkaline batteries and supercapacitors [171]. Its main reaction mechanism as positive electrode material for supercapacitors is described by Refs. [67, 172]

$$Ni(OH)_2 + OH^- \rightleftharpoons NiOOH + H_2O + e^- \tag{7.3}$$

However, $Ni(OH)_2$ usually suffers from poor stability, low conductivity, and large volume changes during the charge/discharge processes [173]. $Ni(OH)_2$-carbon composites involving high surface-area conductive materials such as CNTs, activated carbon, or graphene, show remarkably enhanced electrochemical performance due to improvement of the conductivity of the composites and shortening of the electron and ion diffusion pathways; the latter aspect results in more efficient charge and mass exchange [172, 174–178]. For instance, the flower-like $Ni(OH)_2$/GO/CNT composites exhibit a high specific capacitance of 1815 F g^{-1} (96% of the theoretical pseudocapacitance) at a current density of 2 A g^{-1} and still show 97% capacitance retention after 2000 cycles at 10 A g^{-1}, indicating good cycling performance [179]. Sui et al. [180] reported that $Ni(OH)_2$/AC/CNT composites present a high specific capacitance of 1038 F g^{-1} at a current density of 1 A g^{-1}. Wang and coworkers [177] reported $Ni(OH)_2$-graphene composites exhibiting a specific capacitance of 1335 F g^{-1} at a charge and discharge current density of 2.8 A g^{-1}.

Heteroatom doping has also been introduced to optimize the poor electrical conductivity of $Ni(OH)_2$. For example, Co-doped $Ni(OH)_2$ shows a specific capacitance of 684 F g^{-1} in 1 mol L^{-1} NaOH electrolyte solution, which is much higher than that of pure $Ni(OH)_2$ and/or Co_3O_4 [181]. The simultaneous presence of zinc and cobalt ions in the $Ni(OH)_2$ layer enhances the efficacy of the active material and reduces the resistance of the unit cell, enhancing the capacitor's specific energy and power densities [180]. This is because cobalt significantly improves the electronic conductivity, and zinc increases the utilization of active material and improves the ion migration path [180].

7.3.2.4 Cobalt oxide and hydroxide

(1) Cobalt oxide (Co_3O_4) is generally considered one of the best candidates for electrode material in the field of supercapacitors owing to its superior reversible redox behavior, excellent cycle stability, large surface area, and outstanding corrosion stability [182, 183]. The redox reactions of Co_3O_4 in alkaline electrolyte solution can be expressed as [184, 185]

$$Co_3O_4 + OH^- + H_2O \rightleftharpoons 3CoOOH + e^- \quad (7.4)$$

$$CoOOH + OH^- \rightleftharpoons CoO_2 + H_2O + e^- \quad (7.5)$$

Many groups have studied Co_3O_4-carbon composites including CNTs and graphene. Our group has also reported that Co_3O_4 with preferred orientation along the (220) plane could be prepared in situ and coated on MWCNT via a simple hydrothermal procedure. The fabricated Co3O4@MWCNT nanocable shows excellent electrochemical performance due to a unique-orientated nanocable structure in 0.5 M KOH solution as shown in Fig. 7.7 [185]. In detail, its specific capacitance is as

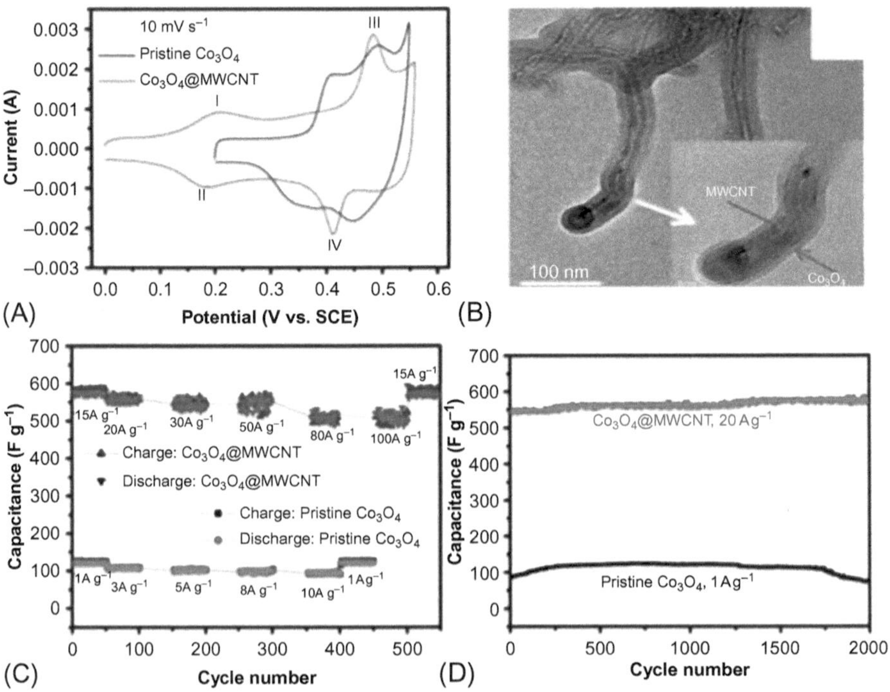

Fig. 7.7 Electrochemical characterization of pristine Co_3O_4 and the manufactured Co_3O_4@MWCNT nanocable: (A) CV curves at 10 mV s^{-1}; (B) TEM micrographs of the Co_3O_4@MWCNT nanocables; (C) capacitance at different current densities; and (D) cycling behavior.

Reproduced by permission from the American Chemical Society.

high as 590 F g^{-1} at 15 A g^{-1} and 510 F g^{-1} even at 100 A g^{-1} in the potential range −0.2 to 0.58 V (vs. SCE), confirming its superior rate capability; there is no capacitance fading up to 2000 cycles. In addition, Xiang et al. [186] have reported the in-situ growth of 20 nm-sized Co_3O_4 nanoparticles on chemically reduced graphene oxide (rGO) sheets to form a rGO-Co_3O_4 composite during hydrothermal processing. The rGO-Co_3O_4 composite originally had a specific capacitance of 416 F g^{-1} and retained 95.6% capacity after 1000 cycles at a current density of 2 A g^{-1} in 2 M KOH electrolyte solution. These carbon-based composite electrodes all display high rate capability and excellent long-term stability.

Some studies have concentrated on a variety of morphologies and microstructures for Co_3O_4, such as nanowires, nanorods, nanotubes, nanoboxes, and mesoporous materials. For instance, Gao et al. [187] prepared Co_3O_4 nanowire arrays on nickel foam via a template-free growth method, which shows a maximum specific capacitance of 746 F g^{-1} at a current density of 5 mA cm^{-2}, and maintaining 85% capacity retention after 500 charge-discharge cycles. Co_3O_4 nanorods synthesized by a simple and efficient hydrothermal method, had a specific capacitance of 280 F g^{-1} at 5 mV s^{-1} in 2 M KOH, while single-crystal Co_3O_4 nanorods provided a capacitance of 456 F g^{-1} at a current density of 1 A g^{-1} [188, 189]. Using a chemical deposition method, Co_3O_4 nanotubes with an outer wall diameter of 300 nm and a wall thickness of 50 nm were made, with a specific capacitance of 574 F g^{-1} at a current density of 0.1 A g^{-1} [190]. In addition, mesoporous Co_3O_4 nanoparticles were synthesized having a surface area of 184.8 m^2 g^{-1} and a specific capacitance of 370 F g^{-1} with a cycling efficiency of 95% after 1000 cycles; mesoporous Co_3O_4 nanoboxes with a surface area of 235 m^2 g^{-1} had a specific capacitance of 623 F g^{-1} with a negligible decay of 4% over 1000 cycles [191, 192]. These studies show that the electrochemical behavior of Co_3O_4 is strongly dependent on morphology, specific surface area, and electrical conductivity. Therefore, high-performance supercapacitors based on Co_3O_4 electrodes can be made using appropriate morphology and microstructure.

(2) Cobalt hydroxide [$Co(OH)_2$]

Cobalt hydroxide [$Co(OH)_2$] has the potential to become a high-performance electrode material owing to its layered structure with a large interlayer spacing, providing a large surface area and a high ion insertion/extraction rate [193, 194]. The pseudo-faradaic reaction at a low potential of $Co(OH)_2$ and the faradaic reaction at a higher potential are expressed in the following equations [195, 196]:

$$Co(OH)_2 + OH^- \rightleftharpoons CoOOH + H_2O + e^- \tag{7.6}$$

$$CoOOH + OH^- \rightleftharpoons CoO_2 + H_2O + e^- \tag{7.7}$$

The application of $Co(OH)_2$-based materials in supercapacitors has been reported by many groups. For instance, mesoporous $Co(OH)_2$ nanowires with a surface area of 101 m^2 g^{-1} were synthesized by a lysine-assisted hydrothermal method and had a specific capacitance of 421 F g^{-1} at 10 mA cm^{-2} (1.33 A g^{-1}) [197]. Li et al. [197] have reported nanostructured $Co(OH)_2$ flakes deposited on three-dimensional ordered nickel/silicon microchannel plates by electrochemical deposition; this material had a specific capacitance of 690 F cm^{-2} in the potential range −0.1 to 0.4 V at a current density of 10 mA cm^{-2}. A flexible nanoflake film of porous $Co(OH)_2$ deposited by galvanostatic electrodeposition on a stainless steel mesh, had a capacitance of 609 F g^{-1} [198]. Apart from designing the morphology of $Co(OH)_2$, many efforts have been focused on combining $Co(OH)_2$ and carbon materials to improve the conductivity. For example, Ghosh et al. prepared a $Co(OH)_2$/graphene hybrid composite consisting of platelets by means of a simple hydrothermal process. This material has a specific capacitance of 532 F g^{-1} at 2 A g^{-1} current density and 94% specific capacitance retention over 500 charge-discharge cycles in 6 M KOH electrolyte [199]. In addition, Zhao et al. [200] made $Co(OH)_2$/graphene/Ni electrodes by a combination of radio-frequency plasma-enhanced CVD and electrodeposition, which had a specific capacitance up to 693.8 F g^{-1} at a current density of 2 A g^{-1} with 75.5% specific capacitance retention over 2000 consecutive charge-discharge cycles.

From the above discussion it follows that $Co(OH)_2$-based electrodes can provide higher specific capacitances by means of variations in morphology, specific surface area, and electrical conductivity. However, these high specific capacitances are found over narrow potential ranges, thereby limiting the practical application of these materials in supercapacitors.

7.3.2.5 Iron oxides

The main iron oxides are Fe_3O_4 (magnetite) and Fe_2O_3 (hematite). Fe_3O_4 is a naturally abundant pseudocapacitive material with a reasonable metallic electrical conductivity of 10^2–10^3 S cm^{-1}. However, its low specific capacitance limits practical applications. For instance, Zhu et al. [201] report that a Fe_3O_4 electrode provides a specific capacitance of only 5.3 F g^{-1} in 1 M Na_2SO_4 electrolyte solution. However, Chung et al. [202] made porous Fe_3O_4 films by electrochemical deposition that had a specific capacitance of 105 F g^{-1} in 1 M Na_2SO_4 solution, apparently because the optimized porosity and crystallinity result in a large available surface area providing better electrolyte penetration and improved charge transfer reactions. In addition, a hydrothermally synthesized Fe_3O_4 film exhibited typical pseudocapacitive behavior in 1 M Na_2SO_3 electrolyte solution, showing a specific capacitance of 118.2 F g^{-1} between −1 and 0.1 V with a capacity retention of 88.75% after 500 cycles [203]. Apparently, the performance of Fe_3O_4 electrode materials for supercapacitors depends strongly on oxide crystallites and the electrolyte anion species.

Hematite, the main polymorph of Fe_2O_3, also exhibits pseudocapacitance and has the advantages of low cost and environmental friendliness. However, its electronic conductivity is a limitation for practical applications. To address the issue, the strategy of composites based on Fe_2O_3 was introduced to improve its conductivity, resulting in better ion diffusion. For example, Cheng et al. [204] manufactured a CNT@Fe_2O_3 sponge composite electrode with a specific capacitance of up to 296.3 F g^{-1} at a scan rate of 5 mV s^{-1}, which results from the combination of the electric double-layer capacitance of the porous CNT network and the enhanced pseudocapacitance of α-Fe_2O_3. Our group synthesized unique core-shell MnO_2@Fe_2O_3 nanospindles by a simple and cost-effective hydrothermal method; this nanostructure plays an important role in increasing the surface area and active sites for faradaic reactions. These composite electrodes display a maximum specific capacity of 159 F g^{-1} at a current density of 0.1 A g^{-1}, and its capacitance retention remains at 97.4% after 5000 cycles in a 0.5 mol L^{-1} K_2SO_4 neutral aqueous electrolyte as shown in Fig. 7.8 [205]. It seems that Fe_2O_3-based composite materials may be developed as promising electrodes for practical supercapacitors.

7.3.2.6 TiO_2

TiO_2 is a very important material for energy storage systems because of the good intercalation/deintercalation behavior of metal ions (such as Li^+ and Na^+) without the formation of solid electrolyte interface by-products and electrode collapse caused by volume changes; these characteristics contribute to its high power capacity and long lifespan [206–211]. In terms of properties, TiO_2 is suitable for use as negative electrode material in organic electrolyte for hybrid supercapacitors. For instance, Wang reported the fabrication of a hybrid Li-ion capacitor consisting of a TiO_2 nanobelt

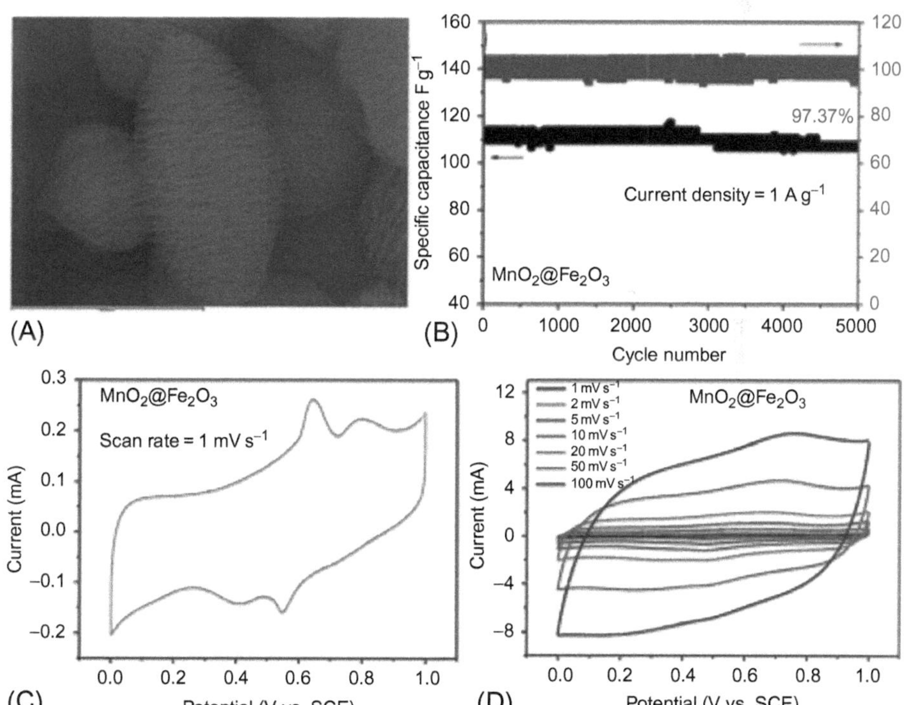

Fig. 7.8 (A) SEM micrographs of the $MnO_2@Fe_2O_3$ nanospindles; (B) cycling performance of the nanospindles at a current density of 1 A g^{-1}; (C) and (D) cyclic voltammetric curves of nanospindles at different scan rates in a 0.5 mol L^{-1} K_2SO_4 solution.
Reproduced by permission from the Royal Society of Chemistry.

array anode and graphene hydrogel cathode in 1 M $LiPF_6$ electrolyte solution; this cell delivers a maximum energy density of 82 Wh kg^{-1} and a power density of 19 kW kg^{-1} (achieved at 21 Wh kg^{-1}) and has a reasonable cycle stability (73% retention after 600 cycles) within the voltage range of 0–3.8 V [212]. Furthermore, anode materials made of aligned or randomly oriented anatase and brookite crystals have been studied, showing that three-dimensional TiO_2 nanotube arrays and networks promote redox reactions of many oxides [such as RuO_2 and $Co(OH)_2$] in an aqueous electrolyte. These effects are obtained by promoting an initial effective dispersion of the oxide nanoparticles, and possibly by catalyzing charge-transfer reactions [213, 214]. For example, flower-like RuO_2-TiO_2 nanocomposites with a size of 180–200 nm exhibit excellent capacitive performance in 0.5 M H_2SO_4 electrolyte [214]. In another example, $Co(OH)_2$-TiO_2 nanotube composites prepared by Tao et al. [214] presented a maximum specific capacitance of 229 F g^{-1} in 6 M KOH electrolyte. Therefore, these kinds of composites show great potential for use in electrochemical supercapacitors.

7.3.2.7 SnO_2

SnO_2 has also been investigated as an alternative electrode material for use in supercapacitors. However, it has a much lower specific capacitance than other metal oxides. For instance, a SnO_2 composite electrode prepared by impregnation of a wet resorcinol formaldehyde gel with tin oxide, had a specific capacitance of only 66 F g^{-1} in acidic aqueous electrolyte [215]. An SnO_2-RuO_2 composite electrode had a specific capacitance of only 33 F g^{-1} at a scan rate of 50 mV s^{-1} in 1M H_2SO_4 solution [126]. On the other hand, an amorphous SnO_2 material has been prepared by electrochemical deposition; this material had a specific capacitance of 285 F g^{-1} at a scan rate of 10 mV s^{-1} [216]. The introduction of Al_2O_3 into SnO_2 can also markedly improve its specific capacitance in sodium chloride electrolyte solutions, which is much greater than that of pure SnO_2 [217]. It appears that, although the specific capacitance of SnO_2 can been improved by modification, its low potential limits further application in supercapacitors.

7.3.2.8 V_2O_5

The vanadium in V_2O_5 can attain several oxidation states, resulting in both surface and bulk redox reactions. It has therefore been explored for its potential application in energy storage devices [218]. Generally speaking, V_2O_5 has a higher capacitance in KCl electrolyte than in other electrolyte solutions. For instance, Lee and Goodenough [219] reported that amorphous vanadium oxide, which was prepared by quenching a fine V_2O_5 powder heated at 950°C for 30 min, resulted in a specific capacitance of 350 F g^{-1} in KCl electrolyte over the potential range of 0–0.8 V (vs. SCE). Furthermore,Wee et al. [220] investigated the specific capacitance of V_2O_5 nanofibers in different electrolytes and found that the electrode presented its highest specific capacitance of 190 F g^{-1} in 2 M KCl electrolyte, compared with 1 M H_2SO_4 (106 F g^{-1}) and 2 M KOH (8 F g^{-1}). Thus, its specific capacitance strongly depends on the electrolyte solution used; 2 M KCl seems to be the most suitable electrolyte. In our group, $V_2O_5{\cdot}0.6H_2O$ nanobelts were synthesized by a hydrothermal method and presented four distinct couples of redox peaks over the potential range of 0–1.0 V (vs. SCE) in 0.5 M K_2SO_4 [221]. The redox peaks in K_2SO_4 solution can be attributed to the intercalation/deintercalation of K^+ ions into/from the $V_2O_5{\cdot}0.6H_2O$ lattice accompanied with the electrochemical reduction of V^{5+} to different valence states. A specific capacitance as high as 180.7 F g^{-1} was attained at a current rate of 2C [221].

Unfortunately, the further development of V_2O_5 electrodes is hamstrung by poor electronic conductivity and high dissolution in aqueous solution [222, 223]. More specifically, the dissolution of vanadium will lead to capacity fading and poor cycling performance, which is an urgent problem to be solved. To address electronic conductivity, CNTs have been investigated as a conducting material to form V_2O_5-CNT composites, which have larger specific capacitances than pure V_2O_5 [224]. It has been shown that the addition of CNTs can increase the specific capacitance in KCl solution to almost three times that of vanadium oxide alone [224, 225]. A $V_2O_5{\cdot}xH_2O$-CNT film electrode also had a high specific capacitance of 910 F g^{-1}, three times higher than that of the $V_2O_5{\cdot}xH_2O$ film electrode (300 F g^{-1}) [226]. Although the combination of V_2O_5 with CNTs is an effective strategy to improve electronic transport, this kind of composite cannot prevent vanadium dissolution [227, 228]. In our group, polypyrrole (PPy) was grown uniformly on the surface of V_2O_5 nanoribbon using dodecyl benzenesulfonate anion (DBS^-) as a surfactant. The obtained core-shell-structured PPy@V_2O_5 nanocomposite is expected to resolve the above two problems simultaneously, utilizing the electronic conductivity and polymeric coating effect of PPy. The PPy@V_2O_5 electrode exhibits no obvious capacitance loss after 10,000 cycles and is a promising candidate as anode material for supercapacitors [229].

In summary, the electrochemical performance of V_2O_5 can be improved to some extent by employing a suitable strategy.

7.3.2.9 Composite oxides

Composite oxides have received enormous attention in recent years owing to their great potential for use in energy storage systems. They usually form an $A_xB_{3-x}O_4$ spinel structure, where A and B represent two different transition metals, including Ni, Co, Fe, Mo, and Mn [230]. Two different metal cations can offer a broad range of redox reactions and enhance electronic conductivity, which contribute to electrochemical performance. For instance, the charge storage mechanism of $NiCo_2O_4$ in alkaline electrolyte during the redox process corresponds to the reversible reactions of Co^{2+}/Co^{3+} and Ni^{2+}/Ni^{3+} transitions as given in the following equations [231]:

$$NiCo_2O_4 + OH^- + H_2O \rightleftharpoons NiOOH + 2CoOOH + e^- \quad (7.8)$$

$$NiOOH + OH^- \rightleftharpoons NiO_2 + H_2O + e^- \quad (7.9)$$

$$CoOOH + OH^- \rightleftharpoons CoO_2 + H_2O + e^- \quad (7.10)$$

During the charge/discharge process, the valence state changes from Co^{3+} to Co^{4+}, from Co^{2+} to Co^{3+}, and from Ni^{2+} to Ni^{3+} on the surface of bulk electrode materials, where fast and reversible faradaic reactions take place. The electrochemical redox potentials of Co^{2+}/Co^{3+}, Ni^{2+}/Ni^{3+}, and Co^{3+}/Co^{4+} transitions are so close that the redox peaks overlap each other [232].

Among many kinds of composite oxides, $NiCo_2O_4$ has been explored extensively in the field of supercapacitors due to its low cost, nontoxicity, natural abundance as well as good conductivity (ranging from 0.05 to 10^{-6} S cm^{-1}) [231, 233]. Furthermore, it also exhibits a wide diversity of nanostructures, which are feasible for superior capacitive properties. Thus, this section mainly describes the electrochemical performance of different $NiCo_2O_4$ morphologies. For instance, Pu et al. [234] reported that porous nanoplate $NiCo_2O_4$ with a high surface area exhibited a capacitance of 294 F g^{-1} at a current density of 1 A g^{-1} with cycling stability of 89.8% after 2200 cycles. Urchin-like $NiCo_2O_4$ nanostructures with a surface area of 99.3 m^2 g^{-1} have been found to display a superior specific capacitance of 1650 F g^{-1} at current densities of 1 A g^{-1} [235]. Three-dimensional $NiCo_2O_4$ microspheres with radial chain-like nanowires have also been reported, providing capacitances of 1284 and 986 F g^{-1} at current densities of 2 and 20 A g^{-1}, respectively [236]. These examples prove that the porosity and high specific surface area result in excellent electrochemical performance because these characteristics provide smooth paths for electrolyte penetration and ion diffusion into inner active sites.

The application of $NiCo_2O_4$ as an electrode material is a current hotspot in energy storage research. Given the above considerations, we believe that it is critical for enhancing electrochemical properties to continue to develop $NiCo_2O_4$ materials with special morphologies and structures. In particular, well-designed three-dimensional architectures are significant for obtaining both high power density and high-energy density in supercapacitors by providing an effective path for fast electron transport and shortened paths for electron transport as well as metal cation species diffusion [235].

7.4 Summary and outlook

In this chapter, we introduced the fundamental principles, characteristics, and applications of supercapacitors, especially the recent advances in research on supercapacitors using metal oxides as electrode materials. From what we have described above, we can see that metal oxides have developed rapidly in the fields of theory and practice. On the whole, higher energy densities are offered by metal oxides than by conventional carbon materials, and they also have better electrochemical stability than polymer materials. Metal oxides store energy similar to carbon materials and also undergo faradaic reactions between the electrode materials and ions within appropriate potential windows. The most studied metal oxides in supercapacitors include RuO_2, MnO_2, NiO, Co_3O_4, V_2O_5, TiO_2, Fe_3O_4, SnO_2, and $NiCo_2O_4$ due to their environmental benefit, ultrahigh theoretical capacitance, and excellent chemical stability.

However, the limited energy density of supercapacitors remains a key challenge hampering their wider use in energy storage applications.

A possible strategy to overcome this barrier is to fabricate materials with diverse nanostructures or topographies, with high capacitances and wide potential windows. Furthermore, the progress of nanoscience and nanotechnology has provided an impetus for developing novel electrode structures. Generally speaking, well-designed electrode materials should meet several requirements, such as (1) large specific surface area for more active sites, (2) low internal electrical resistance for highly efficient charge transport, (3) suitable pore-size distribution for high-speed ion diffusion, and (4) a metal existing in two or more oxidation states for faradaic reactions to take place in the composite electrode. These factors will contribute to improving the utilization of electrode materials and shorten the transport path of both ions and electrons, resulting in higher specific capacitance, better cycling performance, and wider potential ranges.

Unfortunately, metal oxides have a key weakness in their poor conductivity. Therefore, inspired research is needed to develop novel electrode materials by the coupling of metal oxides with carbon materials to provide effective electrodes for supercapacitors. It has been shown that it is essential to incorporate metal oxide nanoparticles in nanostructured carbon materials, such as carbon fiber, mesoporous carbons, CNTs, graphene, etc. The metal oxide/carbon composites can not only enhance the utilization of the active materials, but also improve the electrical conductivity and mechanical strength of the composite materials.

In recent years, flexible and deformable energy storage devices have also arisen due to their potential applications in wearable and portable electronic products, such as rollup displays, electronic paper, and wearable systems for personal multimedia. Consequently, designing smart flexible supercapacitors by integrating some of the properties such as ultralight, ultrathin, transparent, stretchable, small-size, and even compression-tolerant properties will continue to be a promising field and attract more and more attention in the future. In addition, the operational safety and application of "green chemistry" in flexible supercapacitors should also be considered for practical applications.

In summary, we believe that the development of metal oxide nanomaterials will pave the way for the fabrication of the next-generation flexible supercapacitors with satisfying energy densities at high power densities.

Acknowledgments

Financial support from the National Materials Genome Project (2016YFB0700600), China National Funds for Distinguished Youth Scientists (51026004), and NSFC (U1601214 and 51502137) is gratefully acknowledged.

References

[1] B.E. Conway, Electrochemical Supercapacitors: Scientific Fundamentals and Technological Applications, Kluwer Academic/Plenum Publishers, New York, 1999.

[2] B.E. Conway, Transition from supercapacitor to battery behavior in electrochemical energy storage, J. Electrochem. Soc. 138 (1991) 1539–1548.
[3] J.R. Miller, P. Simon, Electrochemical capacitors for energy management, Science 321 (2008) 651–652.
[4] L.Q. Mai, A.M. Khan, X.C. Tian, K.M. Hercule, Y.L. Zhao, Synergistic interaction between redox-active electrolyte and binder-free functionalized carbon for ultrahigh supercapacitor performance. Nat. Commun. 4 (2013)2923https://doi.org/10.1038/ncomms3923.
[5] Q. Lu, M.W. Lattanzi, Y.P. Chen, X.M. Kou, W.F. Li, X. Fan, K.M. Unruh, J.G. Chen, J. Q. Xiao, Supercapacitor electrodes with high energy and power densities prepared from monolithic NiO/Ni nanocomposites, Angew. Chem. Int. Ed. 50 (2011) 6847–6850.
[6] S. Boukhalfa, K. Evanoff, G. Yushin, Atomic layer deposition of vanadium oxide on carbon nanotubes for high-power supercapacitor electrodes, Energy Environ. Sci. 5 (2012) 6872–6879.
[7] L.B. Liu, Y. Yu, C. Yan, K. Li, Z.J. Zheng, Wearable energy-dense and power-dense supercapacitor yarns enabled by scalable graphene-metallic textile composite electrodes. Nat. Commun. 6 (2015)7260https://doi.org/10.1038/ncomms8260.
[8] H. Wang, H.S. Casalongue, Y. Liang, H.G. Dai, $Ni(OH)_2$ nanoplates grown on graphene as advanced electrochemical pseudocapacitor materials, J. Am. Chem. Soc. 132 (2010) 7472–7477.
[9] C.G. Liu, Z.N. Yu, D. Neff, A. Zhamu, B.Z. Jang, Graphene-based supercapacitor with an ultrahigh energy density, Nano Lett. 10 (2010) 4863–4868.
[10] Y. He, W. Chen, X. Li, Z. Zhang, J. Fu, C. Zhao, E. Xie, Freestanding three-dimensional graphene/MnO_2 composite networks as ultralight and flexible supercapacitor electrodes, ACS Nano 7 (2013) 174–182.
[11] X.H. Lu, M.H. Yu, T. Zhai, G.M. Wang, S.L. Xie, T.Y. Liu, High energy density asymmetric quasi-solid-state supercapacitor based on porous vanadium nitride nanowire anode, Nano Lett. 13 (2013) 2628–2633.
[12] P. Simon, Y. Gogotsi, Materials for electrochemical capacitors, Nat. Mater. 7 (2008) 845–854.
[13] Q.T. Qu, Y. Shi, S. Tian, Y.H. Chen, Y.P. Wu, A new cheap asymmetric aqueous supercapacitor: activated carbon/$NaMnO_2$, J. Power Sources 194 (2009) 1222–1225.
[14] Q.T. Qu, L.L. Li, S. Tian, G.X. Wang, R. Holze, A cheap asymmetric supercapacitor with high energy at high power: activated carbon//$K_{0.27}MnO_2 \cdot 0.6H_2O$, J. Power Sources 195 (2010) 2789–2794.
[15] E. Frackowiak, Carbon materials for supercapacitor application, Phys. Chem. Chem. Phys. (15) (2007) 1774–1785.
[16] Y. Zhu, S. Murali, M.D. Stoller, K.J. Ganesh, W.W. Cai, P.J. Ferreira, A. Pirkle, K. A. Cychosz, M. Thomes, O. Su, E.A. Stach, R.S. Ruoff, Carbon-based supercapacitors produced by activation of graphene, Science 332 (2011) 1537–1541.
[17] Y.P. Zhai, Y.Q. Dou, D.Y. Zhao, P.F. Fulvio, S. Dai, Carbon materials for chemical capacitive energy storage, Adv. Mater. 23 (2011) 4828–4850.
[18] D.N. Futaba, K. Hata, T. Yamada, T. Hiraoka, Y. Hayamizu, Y. Kakudate, O. Tanaike, H. Hatori, M. Yumura, S. Iijima, Shape-engineerable and highly densely packed single-walled carbon nanotubes and their application as super-capacitor electrodes, Nat. Mater. 5 (2006) 987–994.
[19] X. Lang, A. Hirata, T. Fujita, M.W. Chen, Nanoporous metal/oxide hybrid electrodes for electrochemical supercapacitors, Nat. Nanotechnol. 6 (2011) 232–236.

[20] X. Xia, J. Tu, Y. Zhang, X. Wang, C. Gu, X. Zhao, H.J. Fan, High-quality metal oxide core/shell nanowire arrays on conductive substrates for electrochemical energy storage, ACS Nano 6 (2012) 5531–5538.
[21] G.A. Snook, P. Kao, A.S. Best, Conducting-polymer-based supercapacitor devices and electrodes, J. Power Sources 196 (2011) 1–12.
[22] M. Mastragostino, C. Arbizzani, F. Soavi, Conducting polymers as electrode materials in supercapacitors, Solid State Ionics 148 (2002) 493–498.
[23] B.H. Zhang, Y. Liu, Z. Chang, Y.Q. Yang, Y.P. Wu, R. Holze, Nanowire $Na_{0.35}MnO_2$ from a hydrothermal method as a cathode material for aqueous asymmetric supercapacitors, J. Power Sources 253 (2014) 98–103.
[24] F.X. Wang, S.Y. Xiao, X.W. Gao, Y.S. Zhu, H.P. Zhang, Y.P. Wu, R. Holze, Nanoporous $LiMn_2O_4$ spinel prepared at low temperature as cathode material for aqueous supercapacitors, J. Power Sources 242 (2013) 560–565.
[25] A. Burke, Ultracapacitors: why, how, and where is the technology, J. Power Sources 91 (2000) 37–50.
[26] C. Zhong, Y. Deng, W.B. Hu, J.L. Qiao, L. Zhang, J.J. Zhang, A review of electrolyte materials and compositions for electrochemical supercapacitors, Chem. Soc. Rev. 44 (2015) 7484–7539.
[27] E. Frackowiak, F. Beguin, Carbon materials for the electrochemical storage of energy in capacitors, Carbon 39 (2001) 937–950.
[28] L.L. Zhang, X.S. Zhao, Carbon-based materials as supercapacitor electrodes, Chem. Soc. Rev. 38 (2009) 2520–2531.
[29] G. Wang, L. Zhang, J. Zhang, A review of electrode materials for electrochemical supercapacitors, Chem. Soc. Rev. 41 (2012) 797–828.
[30] A. Ghosh, Y.H. Lee, Carbon-based electrochemical capacitors, ChemSusChem 5 (2012) 480–499.
[31] J. Lee, K. Sohn, T. Hyeon, Fabrication of novel mesocellular carbon foams with uniform ultralarge mesopores, J. Am. Chem. Soc. 123 (2001) 5146–5147.
[32] H. Cheng, Z. Dong, C. Hu, Y. Zhao, Y. Hu, L. Qu, N. Chen, L. Dai, Textile electrodes woven by carbon nanotube-graphene hybrid fibers for flexible electrochemical capacitors, Nanoscale 5 (2013) 3428–3434.
[33] K. Okajima, K. Ohta, M. Sudoh, Capacitance behavior of activated carbon fibers with oxygen-plasma treatment, Electrochim. Acta 50 (2005) 2227–2231.
[34] L.F. Chen, X.D. Zhang, H.W. Liang, M.G. Kong, Q.F. Guan, P. Chen, Z.Y. Wu, S.K. Yu, Synthesis of nitrogen-doped porous carbon nanofibers as an efficient electrode material for supercapacitors, ACS Nano 6 (2012) 7092–7102.
[35] D.W. Wang, F. Li, M. Liu, G.Q. Lu, H.M. Cheng, 3D aperiodic hierarchical porous graphitic carbon material for high-rate electrochemical capacitive energy storage, Angew. Chem. Int. Ed. 120 (2008) 379–382.
[36] C. Niu, E.K. Sichel, R. Hoch, D. Moy, H. Tennent, High power electrochemical capacitors based on carbon nanotube electrodes, Appl. Phys. Lett. 70 (1997) 1480–1482.
[37] M. Kaempgen, C.K. Chan, J. Ma, Y. Cui, G. Gruner, Printable thin film supercapacitors using single walled carbon nanotubes, Nano Lett. 9 (2009) 1872–1876.
[38] J. Liu, Charging graphene for energy, Nat. Nanotechnol. 9 (2014) 739–741.
[39] M.F. El-Kady, V. Strong, S. Dubin, R.B. Kaner, Laser scribing of high-performance and flexible graphene-based electrochemical capacitors, Science 335 (2012) 1326–1330.
[40] J.R. Miller, R.A. Outlaw, B.C. Holloway, Graphene double-layer capacitor with ac line filtering performance, Science 329 (2010) 1637–1639.

[41] H. Wang, H.S. Casalongue, Y.Y. Liang, H.J. Dai, $Ni(OH)_2$ nanoplates grown on graphene as advanced electrochemical pseudocapacitor materials, J. Am. Chem. Soc. 132 (2010) 7472–7477.

[42] G. Yu, X. Xie, L. Pan, Z. Bao, Y. Cui, Hybrid nanostructured materials for high-performance electrochemical capacitors, Nano Energy 2 (2013) 213–234.

[43] Y. Zhang, H. Feng, X.B. Wu, L.Z. Wang, A.Q. Zhang, T.C. Xia, H.C. Dong, X.F. Li, L. S. Zhang, Progress of electrochemical capacitor electrode materials: a review, Int. J. Hydrog. Energy 34 (2009) 4889–4899.

[44] C.M. Chuang, C.W. Huang, H. Teng, J.M. Ting, Effects of carbon nanotube grafting on the performance of electric double layer capacitors, Energy Fuel 24 (2010) 6476–6482.

[45] C.C. Hu, K.H. Chang, M.C. Lin, Y.T. Wu, Design and tailoring of the nanotubular arrayed architecture of hydrous RuO_2 for next generation supercapacitors, Nano Lett. 6 (2006) 2690–2695.

[46] L. Yuan, X. Lu, X. Xiao, T. Zhai, J. Dai, F. Zhang, B. Hu, X. Wang, L. Gong, J. Chen, C. Hu, Y. Tong, J. Zhou, Z. Wang, Flexible solid-state supercapacitors based on carbon nanoparticles/MnO_2 nanorods hybrid structure, ACS Nano 6 (2011) 656–661.

[47] K. Zhang, L.L. Zhang, X.S. Zhao, J. Wu, Graphene/polyaniline nanofiber composites as supercapacitor electrodes, Chem. Mater. 22 (2010) 1392–1401.

[48] Q. Wu, Y. Xu, Z. Yao, A. Liu, G. Shi, Supercapacitors based on flexible graphene/polyaniline nanofiber composite films, ACS Nano 4 (2010) 1963–1970.

[49] D. Zhang, X. Zhang, Y. Chen, P. Yu, C. Wang, Enhanced capacitance and rate capability of graphene/polypyrrole composite as electrode material for supercapacitors, J. Power Sources 196 (2011) 5990–5996.

[50] H. Zhang, Z. Hu, M. Li, L. Hu, S. Jiao, A high-performance supercapacitor based on a polythiophene/multiwalled carbon nanotube composite by electropolymerization in an ionic liquid microemulsion, J. Mater. Chem. A 2 (2014) 17024–17030.

[51] X. Wang, S. Kajiyama, H. Iinuma, E. Hosono, S. Oro, I. Moriguchi, M. Okubo, A. Yamada, Pseudocapacitance of MXene nanosheets for high-power sodium-ion hybrid capacitors. Nat. Commun. 6 (2015)6544https://doi.org/10.1038/ncomms7544.

[52] J.J. Yang, C.H. Choi, H.B. Seo, S.G. Park, Voltage characteristics and capacitance balancing for $Li_4Ti_5O_{12}$/activated carbon hybrid capacitors, Electrochim. Acta 86 (2012) 277–281.

[53] B.E. Conway, W.G. Pell, Double-layer and pseudocapacitance types of electrochemical capacitors and their applications to the development of hybrid devices, J. Solid State Electrochem. 7 (2003) 637–644.

[54] J.H. Kim, S.H. Kang, K. Zhu, J.Y. Kim, N.R. Neale, A.J. Frank, Ni-NiO core-shell inverse opal electrodes for supercapacitors, Chem. Commun. 47 (2011) 5214–5216.

[55] H. Jiang, L. Yang, C. Li, P.S. Lee, High-rate electrochemical capacitors from highly graphitic carbon-tipped manganese oxide/mesoporous carbon/manganese oxide hybrid nanowires, Energy Environ. Sci. 4 (2011) 1813–1819.

[56] J.B. Jiang, G.R. Tan, S.J. Peng, D. Qian, J.L. Liu, D.M. Luo, Y.C. Liu, Electrochemical performance of carbon-coated $Li_3V_2(PO_4)_3$ as a cathode material for asymmetric hybrid capacitors, Electrochim. Acta 107 (2013) 59–65.

[57] K. Lu, B. Song, X. Gao, H.X. Dai, J.T. Zhang, H.Y. Ma, High-energy cobalt hexacyanoferrate and carbon micro-spheres aqueous sodium-ion capacitors, J. Power Sources 303 (2016) 347–353.

[58] V. Khomenko, E. Raymundo-Piñero, F. Béguin, High-energy density graphite/AC capacitor in organic electrolyte, J. Power Sources 177 (2008) 643–651.

[59] L.M. Chen, Q.Y. Lai, Y.J. Hao, Y. Zhao, X.Y. Ji, Investigations on capacitive properties of the AC/V_2O_5 hybrid supercapacitor in various aqueous electrolytes, J. Alloys Compd. 467 (2009) 465–471.
[60] X.W. Wang, M.X. Li, Z. Chang, Y.Q. Yang, Y.P.W. X Liu, Co_3O_4@ MWCNT nanocable as cathode with superior electrochemical performance for supercapacitors, ACS Appl. Mater. Interfaces 7 (2015) 2280–2285.
[61] F. Wang, S. Xiao, Y. Hou, C. Hu, L. Liu, Y. Wu, Electrode materials for aqueous asymmetric supercapacitors, RSC Adv. 3 (2013) 13059–13084.
[62] F. Wang, X. Wang, Z. Chang, X. Wu, X. Liu, L. Fu, Y. Zhu, Y. Wu, W. Huang, A quasi solid state sodium ion capacitor with high energy density, Adv. Mater. 27 (2015) 6962–6968.
[63] M.S. Hong, S.H. Lee, S.W. Kim, Use of KCl aqueous electrolyte for 2 V manganese oxide/activated carbon hybrid capacitor, Electrochem. Solid-State Lett. 5 (2002) A227–A230.
[64] V. Aravindan, J. Gnanaraj, Y.S. Lee, S. Madhavi, Insertion-type electrodes for nonaqueous Li-ion capacitors, Chem. Rev. 114 (2014) 11619–11635.
[65] K.H. An, W.S. Kim, Y.S. Park, J.M. Moon, D.J. Bae, S.C. Lim, Y.S. Lee, Y.H. Lee, Electrochemical properties of high-power supercapacitors using single-walled carbon nanotube electrodes, Adv. Funct. Mater. 11 (2011) 387–392.
[66] H. Jiang, P.S. Lee, C. Li, 3D carbon based nanostructures for advanced supercapacitors, Energy Environ. Sci. 6 (2013) 41–53.
[67] Z. Wu, X.L. Huang, Z.L. Wang, J.J. Xu, H.G. Wang, X.B. Zhang, Electrostatic induced stretch growth of homogeneous β-$Ni(OH)_2$ on graphene with enhanced high-rate cycling for supercapacitors. Sci. Rep. 4 (2014)3669https://doi.org/10.1038/srep03669.
[68] C.Y. Chen, Z.Y. Shih, Z.S. Yang, H.T. Chang, Carbon nanotubes/cobalt sulfide composites as potential high-rate and high-efficiency supercapacitors, J. Power Sources 215 (2012) 43–47.
[69] D. Zhang, X. Zhang, Y. Chen, C. Wang, Y. Ma, An environment-friendly route to synthesize reduced graphene oxide as a supercapacitor electrode material, Electrochim. Acta 69 (2012) 364–370.
[70] B. Dyatkin, V. Presser, M. Heon, M.R. Lukatskaya, M. Beidaghi, Y. Gogotsi, Development of a green supercapacitor composed entirely of environmentally friendly materials, ChemSusChem 6 (2013) 2269–2280.
[71] P. Yang, Y. Ding, Z. Lin, Z. Chen, Y. Li, P. Qiang, M. Ebrahimi, W. Mai, C. Wang, Z. Wang, Low-cost high-performance solid-state asymmetric supercapacitors based on MnO_2 nanowires and Fe_2O_3 nanotubes, Nano Lett. 14 (2014) 731–736.
[72] C. Arbizzani, M. Biso, D. Cericola, M. Lazzari, F. Soavi, M. Mastragostino, Safe, high-energy supercapacitors based on solvent-free ionic liquid electrolytes, J. Power Sources 185 (2008) 1575–1579.
[73] F. Lufrano, P. Staiti, Performance improvement of Nafion based solid state electrochemical supercapacitor, Electrochim. Acta 49 (2004) 2683–2689.
[74] D. Hulicova-Jurcakova, M. Seredych, G.Q. Lu, T.J. Bandosz, Combined effect of nitrogen- and oxygen-containing functional groups of microporous activated carbon on its electrochemical performance in supercapacitors, Adv. Funct. Mater. 19 (2009) 438–447.
[75] X. Wang, M. Li, Z. Chang, Y. Wang, B. Chen, L. Zhang, Y. Wu, Orientated Co_3O_4 nanocrystals on MWCNTs as superior battery-type positive electrode material for a hybrid capacitor, J. Electrochem. Soc. 162 (2015) A1966–A1971.

[76] L. Zhu, W. Wu, Y. Zhu, W. Tang, Y. Wu, Composite of CoOOH nanoplates with multiwalled carbon nanotubes as superior cathode material for supercapacitors, J. Phys. Chem. C 119 (2015) 7069–7075.

[77] L. Zhou, Z. Yang, C. Li, B. Chen, Y. Wang, L. Fu, Y. Zhu, X. Liu, Y. Wu, Prussian blue as positive electrode material for aqueous sodium-ion capacitor with excellent performance, RSC Adv. 6 (2016) 109340–109345.

[78] F.X. Wang, S.Y. Xiao, Y.S. Zhu, Z. Chang, C.L. Hu, Y.P. Wu, R. Holze, Spinel $LiMn_2O_4$ nanohybrid as high capacitance positive electrode material for supercapacitors, J. Power Sources 246 (2014) 19–23.

[79] J. Shao, X. Li, Q. Qu, Y. Wu, Study on different power and cycling performance of crystalline $K_xMnO_2{\cdot}nH_2O$ as cathode material for supercapacitors in Li_2SO_4, Na_2SO_4, and K_2SO_4 aqueous electrolytes, J. Power Sources 223 (2013) 56–61.

[80] H.Y. Lee, J.B. Goodenough, Supercapacitor behavior with KCl electrolyte, J. Solid State Chem. 144 (1999) 220–223.

[81] L.Z. Fan, J. Maier, High-performance polypyrrole electrode materials for redox supercapacitors, Electrochem. Commun. 8 (2006) 937–940.

[82] G. Wang, L. Fu, N. Zhao, L. Yang, Y. Wu, H. Wu, An aqueous rechargeable lithium battery with good cycling performance, Angew. Chem. Int. Ed. 119 (2007) 299–301.

[83] R. Marom, S.F. Amalraj, N. Leifer, J. David, A. Doron, A review of advanced and practical lithium battery materials, J. Mater. Chem. 21 (2011) 9938–9954.

[84] A. Balducci, U. Bardi, S. Caporali, M. Mastragostinoa, F. Soavia, Ionic liquids for hybrid supercapacitors, Electrochem. Commun. 6 (2004) 566–570.

[85] H. Ohno, K. Fukumoto, Progress in ionic liquids for electrochemical reaction matrices, Electrochemistry 76 (2008) 16–23.

[86] G.A. Snook, P. Kao, A.S. Best, Conducting-polymer-based supercapacitor devices and electrodes, J. Power Sources 196 (2011) 1–12.

[87] M. Galiński, A. Lewandowski, I. Stępniak, Ionic liquids as electrolytes, Electrochim. Acta 51 (2006) 5567–5580.

[88] Y. Chen, X. Zhang, D. Zhang, P. Yu, Y. Ma, High performance supercapacitors based on reduced graphene oxide in aqueous and ionic liquid electrolytes, Carbon 49 (2011) 573–580.

[89] M. Armand, F. Endres, D.R. MacFarlane, H. Ohno, B. Scrosati, Ionic-liquid materials for the electrochemical challenges of the future, Nat. Mater. 8 (2009) 621–629.

[90] F. Meng, Z. Fang, Z. Li, W. Xu, M. Wang, Y. Liu, J. Zhang, W. Wang, D. Zhao, X. Guo, Porous Co_3O_4 materials prepared by solid-state thermolysis of a novel Co-MOF crystal and their superior energy storage performances for supercapacitors, J. Mater. Chem. A 1 (2013) 7235–7241.

[91] D.L. Yan, Z.L. Guo, G.S. Zhu, Z.Z. Yu, H.R. Xu, A.B. Yu, MnO_2 film with three-dimensional structure prepared by hydrothermal process for supercapacitor, J. Power Sources 199 (2012) 409–412.

[92] Y. Jiang, D. Chen, J. Song, Z. Jiao, Q. Ma, H. Zhang, L. Cheng, B. Zhao, Y. Chu, A facile hydrothermal synthesis of graphene porous NiO nanocomposite and its application in electrochemical capacitors, Electrochim. Acta 91 (2013) 173–178.

[93] D. Wang, Y. Yu, H. He, J. Wang, W. Zhao, H. Abruna, Template synthesis of hollow-structured Co_3O_4 nanoparticles as high-performance anodes for lithium-ion batteries, ACS Nano 9 (2015) 1775–1781.

[94] C.M. Ghimbeu, A. Malak-Polaczyk, E. Frackowiak, C. Vix-Guterl, Template-derived high surface area λ-MnO_2 for supercapacitor applications, J. Appl. Electrochem. 44 (2014) 123–132.

[95] X. Wang, X. Wang, W. Huang, P.J. Sebastian, S. Gamboa, Sol-gel template synthesis of highly ordered MnO_2 nanowire arrays, J. Power Sources 140 (2005) 211–215.
[96] M.C. Liu, L.B. Kong, C. Lu, X.M. Li, Y.C. Luo, L. Kang, A sol-gel process for fabrication of $NiO/NiCo_2O_4/Co_3O_4$ composite with improved electrochemical behavior for electrochemical capacitors, ACS Appl. Mater. Interfaces 4 (2012) 4631–4636.
[97] B.O. Park, C.D. Lokhande, H.S. Park, K.D. Jung, O.S. Joo, Performance of supercapacitor with electrodeposited ruthenium oxide film electrodes effect of film thickness, J. Power Sources 134 (2004) 148–152.
[98] S.L. Chou, J.Z. Wang, S.Y. Chew, H.K. Liu, S.X. Dou, Electrodeposition of MnO_2 nanowires on carbon nanotube paper as free-standing, flexible electrode for supercapacitors, Electrochem. Commun. 10 (2008) 1724–1727.
[99] M.A. Azam, N.S.A. Manaf, E. Talib, M.S.A. Bistamam, Aligned carbon nanotube from catalytic chemical vapor deposition technique for energy storage device: a review, Ionics 19 (2013) 1455–1476.
[100] X. Wang, G. Yushin, Chemical vapor deposition and atomic layer deposition for advanced lithium ion batteries and supercapacitors, Energy Environ. Sci. 8 (2015) 1889–1904.
[101] R.R. Salunkhe, J.J. Lin, V. Malgras, S.X. Dou, J.H. Kim, Y. Yamauchi, Large-scale synthesis of coaxial carbon nanotube/$Ni(OH)_2$ composites for asymmetric supercapacitor application, Nano Energy 11 (2015) 211–218.
[102] S.N. Pusawale, P.R. Deshmukh, J.L. Gunjakar, C.D. Lokhande, SnO_2-RuO_2 composite films by chemical deposition for supercapacitor application, Mater. Chem. Phys. 139 (2013) 416–422.
[103] H. Lee, M.S. Cho, I.H. Kim, J.D. Nam, Y. Lee, RuO_x/polypyrrole nanocomposite electrode for electrochemical capacitors, Synth. Met. 160 (2010) 1055–1059.
[104] I.H. Kim, K.B. Kim, Electrochemical characterization of hydrous ruthenium oxide thin-film electrodes for electrochemical capacitor applications, J. Electrochem. Soc. 153 (2006) A383–A389.
[105] K. Sakiyama, S. Onishi, K. Ishihara, K. Orita, T. Kajiyama, N. Hosoda, T. Hara, Deposition and properties of reactively sputtered ruthenium dioxide films, J. Electrochem. Soc. 140 (1993) 834–839.
[106] Y.T. Kim, K. Tadai, T. Mitani, Highly dispersed ruthenium oxide nanoparticles on carboxylated carbon nanotubes for supercapacitor electrode materials, J. Mater. Chem. 15 (2005) 4914–4921.
[107] J.K. Lee, H.M. Pathan, K.D. Jung, O.S. Joo, Electrochemical capacitance of nanocomposite films formed by loading carbon nanotubes with ruthenium oxide, J. Power Sources 159 (2006) 1527–1531.
[108] R. Fu, Z. Ma, J.P. Zheng, Proton NMR and dynamic studies of hydrous ruthenium oxide, J. Phys. Chem. B 106 (2002) 3592–3596.
[109] J.P. Zheng, P.J. Cygan, T.R. Jow, Hydrous ruthenium oxide as an electrode material for electrochemical capacitors, J. Electrochem. Soc. 142 (1995) 2699–2703.
[110] W. Sugimoto, K. Yokoshima, Y. Murakami, Y. Takasu, Charge storage mechanism of nanostructured anhydrous and hydrous ruthenium-based oxides, Electrochim. Acta 52 (2006) 1742–1748.
[111] F. Pico, E. Morales, J.A. Fernandez, T.A. Centeno, J. Ibanez, R.M. Rojas, J.M. Amarilla, J.M. Rojo, Ruthenium oxide/carbon composites with microporous or mesoporous carbon as support and prepared by two procedures. A comparative study as supercapacitor electrode, Electrochim. Acta 54 (2009) 2239–2245.

[112] H. Kim, B.N. Popov, Characterization of hydrous ruthenium oxide/carbon nanocomposite supercapacitors prepared by a colloidal method, J. Power Sources 104 (2002) 52–61.
[113] M. Egashira, Y. Matsuno, N. Yoshimoto, M. Morita, Pseudo-capacitance of composite electrode of ruthenium oxide with porous carbon in non-aqueous electrolyte containing imidazolium salt, J. Power Sources 195 (2010) 3036–3040.
[114] Y.Z. Zheng, H.Y. Ding, M.L. Zhang, Hydrous-ruthenium-oxide thin film electrodes prepared by cathodic electrodeposition for supercapacitors, Thin Solid Films 516 (2008) 7381–7385.
[115] V.D. Patake, C.D. Lokhande, Chemical synthesis of nano-porous ruthenium oxide (RuO_2) thin films for supercapacitor application, Appl. Surf. Sci. 254 (2008) 2820–2824.
[116] R.R. Bi, X.L. Wu, F.F. Cao, L.Y. Jiang, Y.G. Guo, L.J. Wan, Highly dispersed RuO_2 nanoparticles on carbon nanotubes: facile synthesis and enhanced supercapacitance performance, J. Phys. Chem. C 114 (2010) 2448–2451.
[117] M. Egashira, Y. Matsuno, N. Yoshimoto, M. Morita, Pseudo-capacitance of composite electrode of ruthenium oxide with porous carbon in non-aqueous electrolyte containing imidazolium salt, J. Power Sources 195 (2010) 3036–3040.
[118] Y.S. Lin, K.Y. Lee, K.Y. Chen, Y.S. Huang, Superior capacitive characteristics of RuO_2 nanorods grown on carbon nanotubes, Appl. Surf. Sci. 256 (2009) 1042–1045.
[119] P. Chen, H. Chen, J. Qiu, C. Zhou, Inkjet printing of single-walled carbon nanotube/RuO_2 nanowire supercapacitors on cloth fabrics and flexible substrates, Nano Res. 3 (2010) 594–603.
[120] T.R. Jow, J.P. Zheng, Electrochemical capacitors using hydrous ruthenium oxide and hydrogen inserted ruthenium oxide, J. Electrochem. Soc. 145 (1998) 49–52.
[121] Q.L. Fang, D.A. Evans, S.L. Roberson, J.P. Zheng, Ruthenium oxide film electrodes prepared at low temperatures for electrochemical capacitors, J. Electrochem. Soc. 148 (2001) A833–A837.
[122] C.C. Hu, Y.H. Huang, Effects of preparation variables on the deposition rate and physicochemical properties of hydrous ruthenium oxide for electrochemical capacitors, Electrochim. Acta 46 (2001) 3431–3444.
[123] J.M. Miller, B. Dunn, T.D. Tran, R.W. Pekala, Deposition of ruthenium nanoparticles on carbon aerogels for high energy density supercapacitor electrodes, J. Electrochem. Soc. 144 (1997) L309–L311.
[124] Z.S. Wu, D.W. Wang, W. Ren, J. Zhao, G. Zhou, F. Li, H.M. Cheng, Anchoring hydrous RuO_2 on graphene sheets for high-performance electrochemical capacitors, Adv. Funct. Mater. 20 (2010) 3595–3602.
[125] W. Wang, S. Guo, I. Lee, K. Ahmed, J.B. Zhong, Z. Favors, F. Zaera, M. Ozkan, C. S. Ozkan, Hydrous ruthenium oxide nanoparticles anchored to graphene and carbon nanotube hybrid foam for supercapacitors, Sci. Rep. 4 (2014) 4452.
[126] S.L. Kuo, N.L. Wu, Composite supercapacitor containing tin oxide and electroplated ruthenium oxide, Electrochem. Solid-State Lett. 6 (2003) A85–A87.
[127] F. Pico, J. Ibanez, T.A. Centeno, C. Pecharroman, R.M. Rojas, J.M. Amarilla, J.M. Rojo, $RuO_2 \cdot xH_2O$/NiO composites as electrodes for electrochemical capacitors: effect of the RuO_2 content and the thermal treatment on the specific capacitance, Electrochim. Acta 51 (2006) 4693–4700.
[128] Y.G. Wang, Z.D. Wang, Y.Y. Xia, An asymmetric supercapacitor using RuO_2/TiO_2 nanotube composite and activated carbon electrodes, Electrochim. Acta 50 (2005) 5641–5646.

[129] S.C. Pang, M.A. Anderson, T.W. Chapman, Novel electrode materials for thin film ultracapacitors: comparison of electrochemical properties of sol-gel-derived and electrodeposited manganese dioxide, J. Electrochem. Soc. 147 (2000) 444–450.
[130] M. Toupin, T. Brousse, D. Bélanger, Influence of microstucture on the charge storage properties of chemically synthesized manganese dioxide, Chem. Mater. 14 (2002) 3946–3952.
[131] S.L. Kuo, N.L. Wu, Investigation of pseudocapacitive charge-storage reaction of MnO_2 nH_2O supercapacitors in aqueous electrolytes, J. Electrochem. Soc. 153 (2006) A1317–A1324.
[132] C.C. Hu, T.W. Tsou, Ideal capacitive behavior of hydrous manganese oxide prepared by anodic deposition, Electrochem. Commun. 4 (2002) 105–109.
[133] J. Yan, T. Wei, J. Cheng, Z.J. Fan, M.L. Zhang, Preparation and electrochemical properties of lamellar MnO_2 for supercapacitors, Mater. Res. Bull. 45 (2010) 210–215.
[134] S.W. Donne, A.F. Hollenkamp, B.C. Jones, Structure, morphology and electrochemical behaviour of manganese oxides prepared by controlled decomposition of permanganate, J. Power Sources 195 (2010) 367–373.
[135] B. Babakhani, D.G. Ivey, Anodic deposition of manganese oxide electrodes with rod-like structures for application as electrochemical capacitors, J. Power Sources 195 (2010) 2110–2117.
[136] N. Nagarajan, M. Cheong, I. Zhitomirsky, Electrochemical capacitance of MnO_x films, Mater. Chem. Phys. 103 (2007) 47–53.
[137] Y.K. Zhou, M. Toupin, D. Belanger, T. Brousse, F. Favier, Electrochemical preparation and characterization of Birnessite-type layered manganese oxide films, J. Phys. Chem. Solids 67 (2006) 1351–1354.
[138] S. Devaraj, N. Munichandraiah, High capacitance of electrodeposited MnO_2 by the effect of a surface-active agent, Electrochem. Solid-State Lett. 8 (2005) A373–A377.
[139] L. Zhang, L. Kang, H. Lv, Z.K. Su, K. Ooi, Z.H. Liu, Controllable synthesis, characterization, and electrochemical properties of manganese oxide nanoarchitectures, J. Mater. Res. 23 (2008) 780–789.
[140] M.S. Wu, P.C.J. Chiang, J.T. Lee, J.C. Lin, Synthesis of manganese oxide electrodes with interconnected nanowire structure as an anode material for rechargeable lithium ion batteries, J. Phys. Chem. B 109 (2005) 23279–23284.
[141] L.C. Zhang, Z.H. Liu, H. Lv, X.H. Tang, K. Ooi, Shape-controllable synthesis and electrochemical properties of nanostructured manganese oxides, J. Phys. Chem. C 111 (2007) 8418–8423.
[142] L. Yuan, X.H. Lu, X. Xiao, T. Zhai, J.J. Dai, F.C. Zhang, B. Hu, X. Wang, L. Gong, J. Chen, C.G. Hu, Y.X. Tong, J. Zhou, Z.L. Wang, Flexible solid-state supercapacitors based on carbon nanoparticles/MnO_2 nanorods hybrid structure, ACS Nano 6 (2011) 656–661.
[143] Q. Qu, P. Zhang, B. Wang, Y. Chen, T. Shu, Y. Wu, R. Holze, Electrochemical performance of MnO_2 nanorods in neutral aqueous electrolytes as a cathode for asymmetric supercapacitors, J. Phys. Chem. C 113 (2009) 14020–14027.
[144] R. Ma, Y. Bando, L. Zhang, T. Sasaki, Layered MnO_2 nanobelts: hydrothermal synthesis and electrochemical measurements, Adv. Mater. 16 (2004) 918–922.
[145] V. Subramanian, H. Zhu, B. Wei, Nanostructured MnO_2: hydrothermal synthesis and electrochemical properties as a supercapacitor electrode material, J. Power Sources 159 (2006) 361–364.
[146] F. Su, M. Miao, Asymmetric carbon nanotube-MnO_2 supercapacitors for wearable electronics, Nanotechnology 25 (2014)135401.

[147] J. Luo, H.T. Zhu, H.M. Fan, J.K. Liang, H.L. Shi, G.H. Rao, J.B. Li, Z.M. Du, Z.X. Shen, Synthesis of single-crystal tetragonal α-MnO_2 nanotubes, J. Phys. Chem. C 112 (2008) 12594–12598.
[148] M.S. Wu, Electrochemical capacitance from manganese oxide nanowire structure synthesized by cyclic voltammetric electrodeposition, Appl. Phys. Lett. 87 (2005)153102.
[149] G. Yu, L. Hu, N. Liu, H. Wang, M. Vosgueritchian, Y. Yang, Y. Cui, Z. Bao, Enhancing the supercapacitor performance of graphene/MnO_2 nanostructured electrodes by conductive wrapping, Nano Lett. 11 (2011) 4438–4442.
[150] J. Wen, X. Ruan, Z. Zhou, Preparation and electrochemical performance of novel ruthenium-manganese oxide electrode materials for electrochemical capacitors, J. Phys. Chem. Solids 70 (2009) 816–820.
[151] J. Yan, Z. Fan, T. Wei, J. Cheng, B. Shao, K. Wang, L. Song, M. Zhang, Carbon nanotube/MnO_2 composites synthesized by microwave-assisted method for supercapacitors with high power and energy densities, J. Power Sources 194 (2009) 1202–1207.
[152] K.R. Prasad, N. Miura, Electrochemically synthesized MnO_2-based mixed oxides for high performance redox supercapacitors, Electrochem. Commun. 6 (2004) 1004–1008.
[153] M.S. Wu, Y.A. Huang, J.J. Jow, W.D. Yang, C.Y. Hsieh, H.M. Tsai, Anodically potentiostatic deposition of flaky nickel oxide nanostructures and their electrochemical performances, Int. J. Hydrog. Energy 33 (2008) 2921–2926.
[154] R.S. Jayashree, P.V. Kamath, Suppression of the $\alpha \rightarrow \beta$-nickel hydroxide transformation in concentrated alkali: role of dissolved cations, J. Appl. Electrochem. 31 (2001) 1315–1320.
[155] K.W. Nam, K.H. Kim, E.S. Lee, W.S. Yoon, X.Q. Yang, K.B. Ki, Pseudocapacitive properties of electrochemically prepared nickel oxides on 3-dimensional carbon nanotube film substrates, J. Power Sources 182 (2008) 642–652.
[156] K.C. Liu, M.A. Anderson, Porous nickel oxide/nickel films for electrochemical capacitors, J. Electrochem. Soc. 143 (1996) 124–130.
[157] J.W. Lee, T. Ahn, J.H. Kim, J.M. Ko, J.D. Kim, Nanosheets based mesoporous NiO microspherical structures via facile and template-free method for high performance supercapacitors, Electrochim. Acta 56 (2011) 4849–4857.
[158] S. Ding, T. Zhu, J. Chen, Z. Wang, C. Yuan, X. Low, Controlled synthesis of hierarchical NiO nanosheet hollow spheres with enhanced supercapacitive performance, J. Mater. Chem. 21 (2011) 6602–6606.
[159] D. Su, H.S. Kim, W.S. Kim, G.X. Wang, Mesoporous nickel oxide nanowires: hydrothermal synthesis, characterization and applications for lithium-Ion batteries and supercapacitors with superior performance, Chem. Eur. J. 18 (2012) 8224–8229.
[160] J. Liu, J. Jiang, M. Bosman, H.J. Fan, Three-dimensional tubular arrays of MnO_2-NiO nanoflakes with high areal pseudocapacitance, J. Mater. Chem. 22 (2012) 2419–2426.
[161] J.W. Lang, L.B. Kong, W.J. Wu, Y.C. Luo, L. Kang, Facile approach to prepare loose-packed NiO nano-flakes materials for supercapacitors, Chem. Commun. (2008) 4213–4215.
[162] F. Cao, G.X. Pan, X.H. Xia, P.S. Tang, H.F. Chen, Synthesis of hierarchical porous NiO nanotube arrays for supercapacitor application, J. Power Sources 264 (2014) 161–167.
[163] Y. Ren, L. Gao, From three-dimensional flower-like α-$Ni(OH)_2$ nanostructures to hierarchical porous NiO nanoflowers: microwave-assisted fabrication and supercapacitor properties, J. Am. Ceram. Soc. 93 (2010) 3560–3564.
[164] X. Yan, X. Tong, J. Wang, C. Gong, M. Zhang, L. Liang, Rational synthesis of hierarchically porous NiO hollow spheres and their supercapacitor application, Mater. Lett. 95 (2013) 1–4.

[165] C.Y. Cao, W. Guo, Z.M. Cui, W.G. Song, W. Cai, Microwave-assisted gas/liquid interfacial synthesis of flowerlike NiO hollow nanosphere precursors and their application as supercapacitor electrodes, J. Mater. Chem. 21 (2011) 3204–3209.
[166] G.S. Gund, D.P. Dubal, S.S. Shinde, C.D. Lokhande, Architectured morphologies of chemically prepared NiO/MWCNTs nanohybrid thin films for high performance supercapacitors, ACS Appl. Mater. Interfaces 6 (2014) 3176–3188.
[167] M.P. Anantram, F. Leonard, Physics of carbon nanotube electronic devices, Rep. Prog. Phys. 69 (2006) 507.
[168] X. Zhu, H. Dai, J. Hu, L. Ding, J. Li, Reduced graphene oxide-nickel oxide composite as high performance electrode materials for supercapacitor, J. Power Sources 203 (2012) 243–249.
[169] B. Zhao, J. Song, P. Liu, W. Xu, T. Fang, Z. Jiao, H. Zhang, Y. Jiang, Monolayer graphene/NiO nanosheets with two-dimension structure for supercapacitors, J. Mater. Chem. 21 (2011) 18792–18798.
[170] Y. Bai, M. Du, J. Chang, L. Gao, Supercapacitors with high capacitance based on reduced graphene oxide/carbon nanotubes/NiO composite electrodes, J. Mater. Chem. A 2 (2014) 3834–3840.
[171] G.W. Yang, C.L. Xu, H.L. Li, Electrodeposited nickel hydroxide on nickel foam with ultrahigh capacitance, Chem. Commun. (2008) 6537–6539.
[172] L.L. Zhang, Z. Xiong, X.S. Zhao, A composite electrode consisting of nickel hydroxide, carbon nanotubes, and reduced graphene oxide with an ultrahigh electrocapacitance, J. Power Sources 222 (2013) 326–332.
[173] J. Ji, L. Zhang, H. Ji, Y. Li, X. Zhao, X. Bai, X. Fan, F. Zhang, R. Ruoff, Nanoporous $Ni(OH)_2$ thin film on 3D ultrathin-graphite foam for asymmetric supercapacitor, ACS Nano 7 (2013) 6237–6243.
[174] P. Lin, Q. She, B. Hong, X. Liu, Y. Shi, Z. Shi, M. Zheng, Q. Dong, The nickel oxide/CNT composites with high capacitance for supercapacitor, J. Electrochem. Soc. 157 (2010) A818–A823.
[175] J.H. Park, O.O. Park, K.H. Shin, C.S. Jim, J.H. Kim, An electrochemical capacitor based on a $Ni(OH)_2$/activated carbon composite electrode, Electrochem. Solid-State Lett. 5 (2002) H7–H10.
[176] Q. Huang, X. Wang, J. Li, C. Dai, S. Gamboa, P. Sebastian, Nickel hydroxide/activated carbon composite electrodes for electrochemical capacitors, J. Power Sources 164 (2007) 425–429.
[177] S. Yang, X. Wu, C. Chen, H. Dong, W. Hu, X. Wang, Spherical α-$Ni(OH)_2$ nanoarchitecture grown on graphene as advanced electrochemical pseudocapacitor materials, Chem. Commun. 48 (2012) 2773–2775.
[178] H.L. Wang, H.S. Casalongue, Y.Y. Liang, H.J. Dai, $Ni(OH)_2$ nanoplates grown on graphene as advanced electrochemical pseudocapacitor materials, J. Am. Chem. Soc. 132 (2010) 7472–7477.
[179] X. Ma, J. Liu, C. Liang, X. Gong, R. Che, A facile phase transformation method for the preparation of 3D flower-like β-$Ni(OH)_2$/GO/CNTs composite with excellent supercapacitor performance, J. Mater. Chem. A 2 (2014) 12692–12696.
[180] J.H. Park, S. Kim, O.O. Park, J.M. Ko, Improved asymmetric electrochemical capacitor using Zn-Co co-doped $Ni(OH)_2$ positive electrode material, Appl. Phys. A 82 (2006) 593–597.
[181] J.H. Zhong, A.L. Wang, G.R. Li, J.W. Wang, Y.N. Qu, Y.X. Tong, Co_3O_4/$Ni(OH)_2$ composite mesoporous nanosheet networks as a promising electrode for supercapacitor applications, J. Mater. Chem. 22 (2012) 5656–5665.

[182] V. Srinivasan, J.W. Weidner, Capacitance studies of cobalt oxide films formed via electrochemical precipitation, J. Power Sources 108 (2002) 15–20.

[183] Z. Xun, C. Cai, W. Xing, T. Lu, Electrocatalytic oxidation of dopamine at a cobalt hexacyanoferrate modified glassy carbon electrode prepared by a new method, J. Electroanal. Chem. 545 (2003) 19–27.

[184] X. Wang, M. Li, Z. Chang, Y. Yang, Y. Wu, X. Liu, Co_3O_4@MWCNT nanocable as cathode with superior electrochemical performance for supercapacitors, ACS Appl. Mater. Interfaces 7 (2015) 2280–2285.

[185] C. Yuan, L. Yang, L. Hou, J. Li, Y. Sun, X. Zhang, L. Shen, X. Lu, S. Xiong, X. Lou, Flexible hybrid paper made of monolayer Co_3O_4 microsphere arrays on rGO/CNTs and their application in electrochemical capacitors, Adv. Funct. Mater. 22 (2012) 2560–2566.

[186] C.C. Xiang, M. Li, M.J. Zhi, A. Manivannan, N.Q. Wu, A reduced graphene oxide/Co_3O_4 composite for supercapacitor electrode, J. Power Sources 226 (2013) 65–70.

[187] Y. Gao, S. Chen, D. Cao, G. Wang, J. Yin, Electrochemical capacitance of Co_3O_4 nanowire arrays supported on nickel foam, J. Power Sources 195 (2010) 1757–1760.

[188] G. Wang, X. Shen, J. Horvat, B. Wang, H. Liu, D. Wexler, J. Yao, Hydrothermal synthesis and optical, magnetic, and supercapacitance properties of nanoporous cobalt oxide nanorods, J. Phys. Chem. C 113 (2009) 4357–4361.

[189] L. Cui, J. Li, X.G. Zhang, Preparation and properties of Co_3O_4 nanorods as supercapacitor material, J. Appl. Electrochem. 39 (2009) 1871–1876.

[190] J. Xu, L. Gao, J. Cao, W. Wang, Z. Chen, Preparation and electrochemical capacitance of cobalt oxide (Co_3O_4) nanotubes as supercapacitor material, Electrochim. Acta 56 (2010) 732–736.

[191] M. Zheng, J. Cao, S. Liao, J. Liu, H. Chen, Y. Zhao, W. Dai, G. Ji, J. Cao, J. Tao, Preparation of mesoporous Co_3O_4 nanoparticles via solid-liquid route and effects of calcination temperature and textural parameters on their electrochemical capacitive behaviors, J. Phys. Chem. C 113 (2009) 3887–3894.

[192] T.Y. Wei, C.H. Chen, K.H. Chang, S.Y. Lu, C.C. Hu, Cobalt oxide aerogels of ideal supercapacitive properties prepared with an epoxide synthetic route, Chem. Mater. 21 (2009) 3228–3233.

[193] V. Gupta, T. Kusahara, H. Toyama, S. Gupta, N. Miura, Potentiostatically deposited nanostructured α-$Co(OH)_2$: a high performance electrode material for redox-capacitors, Electrochem. Commun. 9 (2007) 2315–2319.

[194] V. Gupta, S. Gupta, N. Miura, Al-substituted α-cobalt hydroxide synthesized by potentiostatic deposition method as an electrode material for redox-supercapacitors, J. Power Sources 177 (2008) 685–689.

[195] L. Cao, F. Xu, Y.Y. Liang, H.L. Li, Preparation of the novel nanocomposite $Co(OH)_2$/ultra-stable Y Zeolite and its application as a supercapacitor with high energy density, Adv. Mater. 16 (2004) 1853–1857.

[196] C. Yuan, X. Zhang, L. Hou, L. Shen, D. Li, F. Zhang, C. Fan, J. Li, Lysine-assisted hydrothermal synthesis of urchin-like ordered arrays of mesoporous $Co(OH)_2$ nanowires and their application in electrochemical capacitors, J. Mater. Chem. 20 (2010) 10809–10816.

[197] M. Li, S. Xu, T. Liu, F. Wang, P. Yang, L. Wang, P. Chu, Electrochemically-deposited nanostructured $Co(OH)_2$ flakes on three-dimensional ordered nickel/silicon microchannel plates for miniature supercapacitors, J. Mater. Chem. A 1 (2013) 532–540.

[198] F. Fusalba, P. Gouerec, D. Villers, D. Belanger, Electrochemical characterization of polyaniline in nonaqueous electrolyte and its evaluation as electrode material for electrochemical super- capacitors, J. Electrochem. Soc. 148 (2001) A1–A6.

[199] D. Ghosh, S. Giri, C.K. Das, Preparation of CTAB-assisted hexagonal platelet $Co(OH)_2$/graphene hybrid composite as efficient supercapacitor electrode material, ACS Sustain. Chem. Eng. 1 (2013) 1135–1142.
[200] C. Zhao, X. Wang, S. Wang, Y. Wang, Y. Zhao, W. Zheng, Synthesis of $Co(OH)_2$/graphene/Ni foam nano-electrodes with excellent pseudocapacitive behavior and high cycling stability for supercapacitors, Int. J. Hydrog. Energy 37 (2012) 11846–11852.
[201] H. Zhu, D. Yang, L. Zhu, Hydrothermal growth and characterization of magnetite (Fe_3O_4) thin films, Surf. Coat. Technol. 201 (2007) 5870–5874.
[202] K.W. Chung, K.B. Kim, S.H. Han, H. Lee, Novel synthesis and electrochemical characterization of nano-sized cellular Fe_3O_4 thin film, Electrochem. Solid-State Lett. 8 (2005) A259–A262.
[203] J. Chen, K. Huang, S. Liu, Hydrothermal preparation of octadecahedron Fe_3O_4 thin film for use in an electrochemical supercapacitor, Electrochim. Acta 55 (2009) 1–5.
[204] X. Cheng, X. Gui, Z. Lin, Y. Zheng, M. Liu, R. Zhan, Y. Zhu, Z. Tang, Three-dimensional α-Fe_2O_3/carbon nanotube sponges as flexible supercapacitor electrodes, J. Mater. Chem. A 3 (2015) 20927–20934.
[205] L. Zhu, Z. Chang, Y. Wang, B. Chen, Y. Zhu, W. Tang, Y. Wu, Core-shell MnO_2@Fe_2O_3 nanospindles as a positive electrode for aqueous supercapacitors, J. Mater. Chem. A 3 (2015) 22066–22072.
[206] D. Deng, M.G. Kim, J.Y. Lee, J. Cho, Green energy storage materials: nanostructured TiO_2 and Sn-based anodes for lithium-ion batteries, Energy Environ. Sci. 2 (2009) 818–837.
[207] N. Li, G. Liu, C. Zhen, F. Li, L. Zhang, H. Cheng, Battery performance and photocatalytic activity of mesoporous anatase TiO_2 nanospheres/graphene composites by template-free self-assembly, Adv. Funct. Mater. 21 (2011) 1717–1722.
[208] J. Xu, C. Jia, B. Cao, W. Zhang, Electrochemical properties of anatase TiO_2 nanotubes as an anode material for lithium-ion batteries, Electrochim. Acta 52 (2007) 8044–8047.
[209] H. Xiong, M.D. Slater, M. Balasubramanian, C.S. Johnson, T. Rajh, Amorphous TiO_2 nanotube anode for rechargeable sodium ion batteries, J. Phys. Chem. Lett. 2 (2011) 2560–2565.
[210] Y. Xu, E.M. Lotfabad, H.L. Wang, B. Farbod, Z.W. Xu, A. Kohandehghan, D. Mitlin, Nanocrystalline anatase TiO_2: a new anode material for rechargeable sodium ion batteries, Chem. Commun. 49 (2013) 8973–8975.
[211] Z.H. Bi, M.P. Paranthaman, P.A. Menchhofer, R.R. Dehoff, C.A. Bridges, M.F. Chi, B. K. Guo, X.G. Sun, S. Dai, Self-organized amorphous TiO_2 nanotube arrays on porous Ti foam for rechargeable lithium and sodium ion batteries, J. Power Sources 222 (2013) 461–466.
[212] H. Wang, C. Guan, X. Wang, H. Fan, A high energy and power Li-ion capacitor based on a TiO_2 nanobelt array anode and a graphene hydrogel cathode, Small 11 (2015) 1470–1477.
[213] C.C. Hu, C.W. Wang, T.H. Wu, K.H. Chang, Anodic composite deposition of hydrous RuO_2-TiO_2 nanocomposites for electrochemical capacitors, Electrochim. Acta 85 (2012) 90–98.
[214] F. Tao, Y. Shen, Y. Liang, H. Li, Synthesis and characterization of $Co(OH)_2$/TiO_2 nanotube composites as supercapacitor materials, J. Solid State Electrochem. 11 (2007) 853–858.
[215] S.W. Hwang, S.H. Hyun, Synthesis and characterization of tin oxide/carbon aerogel composite electrodes for electrochemical supercapacitors, J. Power Sources 172 (2007) 451–459.

[216] K.R. Prasad, N. Miura, Electrochemical synthesis and characterization of nanostructured tin oxide for electrochemical redox supercapacitors, Electrochem. Commun. 6 (2004) 849–852.

[217] A. Jayalakshmi, N. Venugopal, K.P. Raja, M.M. Rao, Nano SnO_2-Al_2O_3 mixed oxide and SnO_2-Al_2O_3-carbon composite oxides as new and novel electrodes for supercapacitor applications, J. Power Sources 158 (2006) 1538–1543.

[218] T. Kudo, Y. Ikeda, T. Watanabe, M. Hibino, M. Miyayama, H. Abe, K. Kajita, Amorphous V_2O_5/carbon composites as electrochemical supercapacitor electrodes, Solid State Ionics 152 (2002) 833–841.

[219] H.Y. Lee, J.B. Goodenough, Ideal supercapacitor behavior of amorphous $V_2O_5 \cdot nH_2O$ in potassium chloride (KCl) aqueous solution, J. Solid State Chem. 148 (1999) 81–84.

[220] G. Wee, H.Z. Soh, Y.L. Cheah, S.G. Mhaisalkar, M. Srinivasan, Synthesis and electrochemical properties of electrospun V_2O_5 nanofibers as supercapacitor electrodes, J. Mater. Chem. 20 (2010) 6720–6725.

[221] Q.T. Qu, Y. Shi, L.L. Li, W.L. Guo, Y.P. Wu, H.P. Zhang, S.Y. Guo, R. Holze, $V_2O_5 \cdot 0.6H_2O$ nanoribbons as cathode material for asymmetric supercapacitor in K_2SO_4 solution, Electrochem. Commun. 11 (2009) 1325–1328.

[222] Z. Chen, Y. Qin, D. Weng, Q. Xiao, Y. Peng, X. Wang, H. Li, F. Wei, Y. Lu, Design and synthesis of hierarchical nanowire composites for electrochemical energy storage, Adv. Funct. Mater. 19 (2009) 3420–3426.

[223] M. Sathiya, A.S. Prakash, K. Ramesha, J.M. Tarascon, A.K. Shukla, V_2O_5-anchored carbon nanotubes for enhanced electrochemical energy storage, J. Am. Chem. Soc. 133 (2011) 16291–16299.

[224] Z. Chen, V. Augustyn, J. Wen, Y. Zhang, M. Shen, B. Dunn, Y. Lu, High performance supercapacitors based on intertwined CNT/V_2O_5 nanowire nanocomposites, Adv. Mater. 23 (2011) 791–795.

[225] M. Jayalakshmi, M.M. Rao, N. Venugopal, K.B. Kim, Hydrothermal synthesis of SnO_2-V_2O_5 mixed oxide and electrochemical screening of carbon nano-tubes (CNT), V_2O_5, V_2O_5-CNT, and SnO_2-V_2O_5-CNT electrodes for supercapacitor applications, J. Power Sources 166 (2007) 578–583.

[226] I.H. Kim, J.H. Kim, B.W. Cho, Y.H. Lee, K.B. Kim, Synthesis and electrochemical characterization of vanadium oxide on carbon nanotube film substrate for pseudocapacitor applications, J. Electrochem. Soc. 153 (2006) A989–A996.

[227] Y.S. Hu, X. Liu, J.O. Muller, R. Schlogl, J. Maier, D.S. Su, Synthesis and electrode performance of nanostructured V_2O_5 by using a carbon tube in tube as a nanoreactor and an efficient mixed conducting network, Angew. Chem. Int. Ed. 48 (2009) 210–214.

[228] Q. Qu, Y. Zhu, X. Gao, Y. Wu, Core-shell structure of polypyrrole grown on V_2O_5 nanoribbon as high performance anode material for supercapacitors, Adv. Energy Mater. 2 (2012) 950–955.

[229] Y. Liang, H. Wang, J. Zhou, Y. Li, J. Wang, T. Regier, H. Dai, Covalent hybrid of spinel manganese-cobalt oxide and graphene as advanced oxygen reduction electrocatalysts, J. Am. Chem. Soc. 134 (2012) 3517–3523.

[230] X. Wang, X. Han, M. Lim, N. Singh, C. Gan, M. Jan, P. Lee, Nickel cobalt oxide-single wall carbon nanotube composite material for superior cycling stability and high-performance supercapacitor application, J. Phys. Chem. C 116 (2012) 12448–12454.

[231] P.S. Pati, L.D. Kadam, C.D. Lokhande, Preparation and characterization of spray pyrolysed cobalt oxide thin films, Thin Solid Films 272 (1996) 29–32.

[232] D.P. Dubal, P. Gomez-Romero, B.R. Sankapal, R. Holze, Nickel cobaltite as an emerging material for supercapacitors: an overview, Nano Energy 11 (2015) 377–399.

[233] F. Cheng, J. Shen, B. Peng, Y. Pan, Z. Tao, J. Chen, Rapid room-temperature synthesis of nanocrystalline spinels as oxygen reduction and evolution electrocatalysts, Nat. Chem. 3 (2011) 79–84.
[234] J. Pu, J. Wang, X. Jin, F. Cui, E. Sheng, Z. Wang, Porous hexagonal $NiCo_2O_4$ nanoplates as electrode materials for supercapacitors, Electrochim. Acta 106 (2013) 226–234.
[235] Q. Wang, B. Liu, X. Wang, S. Ran, L. Wang, D. Chen, G. Shen, Morphology evolution of urchin-like $NiCo_2O_4$ nano-structures and their applications as pseudocapacitors and photoelectrochemical cells, J. Mater. Chem. 22 (2012) 21647–21653.
[236] R. Zou, K. Xu, T. Wang, G. He, Q. Liu, X. Liu, Z. Zhang, J. Hu, Chain-like $NiCo_2O_4$ nanowires with different exposed reactive planes for high-performance supercapacitors, J. Mater. Chem. A 1 (2013) 8560–8566.

Further reading

[237] F.X. Wang, X.W. Wu, X.H. Yuan, Z.C. Liu, L.J. Fu, Y.S. Zhu, Q.M. Zhou, Y.P. Wu, W. Huang, Latest advances on supercapacitors from new electrode materials to novel device designs, Chem. Soc. Rev. 46 (2017) 6816–6854.
[238] Y. Jin, H. Chen, M. Chen, N. Liu, Q. Li, Graphene-patched CNT/MnO_2 nanocomposite papers for the electrode of high-performance flexible asymmetric supercapacitors, ACS Appl. Mater. Interfaces 5 (2013) 3408–3416.
[239] J. Yan, Z. Fan, W. Sun, W. Sun, G. Ning, T. Wei, Q. Zhang, R. Zhang, L. Zhi, F. Wei, Advanced asymmetric supercapacitors based on $Ni(OH)_2$/graphene and porous graphene electrodes with high energy density, Adv. Funct. Mater. 22 (2012) 2632–2641.
[240] L. Sui, S. Tang, Y. Chen, Z. Dai, H. Huangfu, Z. Zhu, X. Qin, Y. Deng, G. Haarberg, An asymmetric supercapacitor with good electrochemical performances based on $Ni(OH)_2$/AC/CNT and AC, Electrochim. Acta 182 (2015) 1159–1165.

Metal oxide semiconductors for solar water splitting

8

Jing Wang, Teunis van Ree†, Yuping Wu‡, Peng Zhang§, Lian Gao§*
*State Key Laboratory of Materials-Oriented Chemical Engineering, College of Chemistry and Chemical Engineering, Nanjing Tech University, Nanjing, China, †Department of Chemistry, University of Venda, Thohoyandou, South Africa, ‡State Key Laboratory of Materials-Oriented Chemical Engineering, School of Energy Science and Engineering, Institute for Electrochemical Energy Storage, Nanjing Tech University, Nanjing, People's Republic of China, §School of Materials Science and Engineering, Shanghai Jiaotong University, Shanghai, China

8.1 Introduction

8.1.1 General introduction

Energy is the driving force for the survival and development of human beings. The development from the utilization of dry cattle dung and straw in prehistoric society and underdeveloped communities, to the intense use of coal, oil, and natural gas in modern society indicates the enormous progress of human society with the improvement in energy utilization. Energy provides not only light and heat, but also the possibility of long distance travel in the sky, on the ground, and even through interplanetary space. Applying friction to wood was an early way to start a fire, followed by flintstone and tinder, while the modern match as we know it came around 1800 CE. Although coal, natural gas, and oil had been discovered, they were mainly used for cooking and lighting rather than industrial applications. People heated boilers using wood until the use of the steam engine during the industrial revolution. Later, it was found that the steam engine can be started more conveniently and effectively using coal for energy, and wood gradually became the most widely used energy source. Then, as a result of the invention and popularization of automobiles, petroleum began to dominate the world energy market due to its convenience and greater effectiveness than coal. The next stage seems to be the use of natural gas as the primary energy source. Currently, coal, oil, and natural gas account for most of the world's energy consumption (Fig. 8.1) [1]. However, the extreme use of fossil fuel has led to environmental issues and the present energy crisis. The ideal solution to the energy crisis must be to reduce the consumption of fossil fuels and remove the element carbon from fuels. Replacing current fuels with hydrogen has received intense attention in the scientific and industrial fields. Hydrogen can be prepared by several techniques, such as water electrolysis and reforming of fossil fuels.

The excessive release of greenhouse gases and rising costs is promoting the exploration of renewable energy sources. Solar energy irradiating the surface of the Earth (1.3×10^5 TW) exceeds the current global human energy consumption

Metal Oxides in Energy Technologies. https://doi.org/10.1016/B978-0-12-811167-3.00008-0

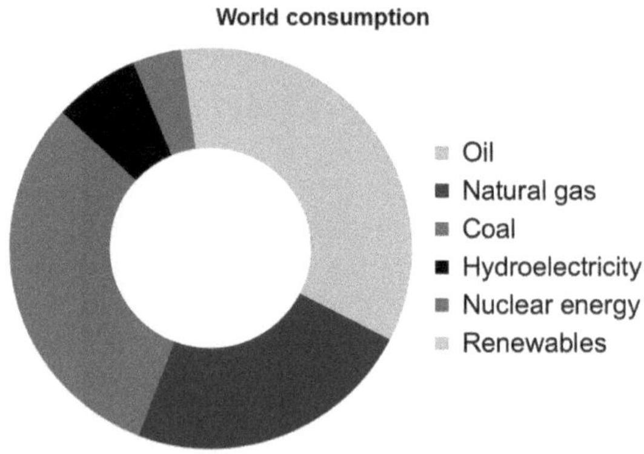

Fig. 8.1 World primary energy consumption situation in recent years.

(16 TW in 2010) [2]. If solar energy can be utilized in appropriate and practical ways, it will have amazing economic and social value. Research on solar energy is mainly divided into solar-to-electricity and solar-to-hydrogen (STH). Research on the former is relatively mature and great progress has been made in industrial applications. To overcome limitations in distribution, space, climate, and cost, further efforts need to be made to explore novel solar cell materials and storage materials. Solar energy can be stored in chemical bonds by photocatalysis of solar-to-chemical fuels, by which peaks of energy utilization and solar irradiation can be avoided. Photocatalysis is considered an effective supplemental approach to photovoltaic cells in solar energy utilization; solar water splitting is promising to be the ideal method to produce hydrogen, and solar hydrogen can be applied as a limitless source of clean fuel in many fields.

8.1.2 Mechanism and thermodynamics of solar water splitting

In 1972, Honda and Fujishima carried out photoelectrochemical (PEC) water oxidation experiments on TiO_2 electrodes [3]. Since then, solar water splitting using semiconductors has been researched extensively by photocatalytic and PEC methods, according to the two basic approaches: a particle system and an electrode system (Fig. 8.2). Photocatalytic water splitting is performed using the semiconductor powder

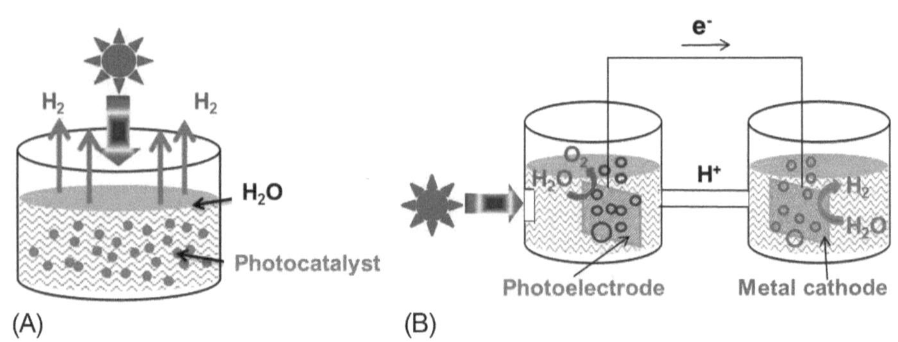

Fig. 8.2 Schematic illustration of solar water splitting: (A) particle system and (B) electrode system.

as photocatalyst (Fig. 8.2A). The photocatalyst is dispersed directly into a quartz container with an aqueous solution irradiated by sunlight, and the formation of hydrogen is readily detected.

The processes involved in a photocatalytic water-splitting reaction generally consist of three steps (Fig. 8.3) [4]: photon absorption, charge separation, and migration to surface reaction sites, and surface redox reaction. When the photocatalyst absorbs photons with energy higher than the bandgap, electrons are excited to the conduction band (CB) while leaving holes in the valence band (VB). Photogenerated electron-hole pairs separate and migrate to the photocatalyst surface. In this step, some recombination of charges (electrons and holes) will take place in bulk materials and on the surface of the photocatalyst, bulk recombination and surface recombination, which is a critical factor that limits the solar-to-chemical energy conversion efficiency for semiconductor photocatalysts. It is necessary to suppress the charge recombination to improve the photocatalytic performance. In a typical water-splitting reaction, the electrons reduce protons into hydrogen and the holes concurrently oxidize OH^- into O_2 on the surface of the photocatalyst. Due to the simplicity of operation, this system is very popular. However, the generated H_2 and O_2 are mixed, which is an obvious disadvantage and limits extensive application in water splitting. This problem can be overcome using a PEC water-splitting system (Fig. 8.2B).

Water splitting is a thermodynamically uphill reaction, and photocatalytic and PEC water-splitting reactions possess similar thermodynamics. According to the Nernst equation, the free energy change for decomposition of 1 mol H_2O into 1 mol H_2 and ½ mol O_2 is 237.178 kJ at 298 K and 1 bar, which corresponds to a thermodynamic potential of 1.23 eV for the water-splitting reaction. Therefore, a suitable band

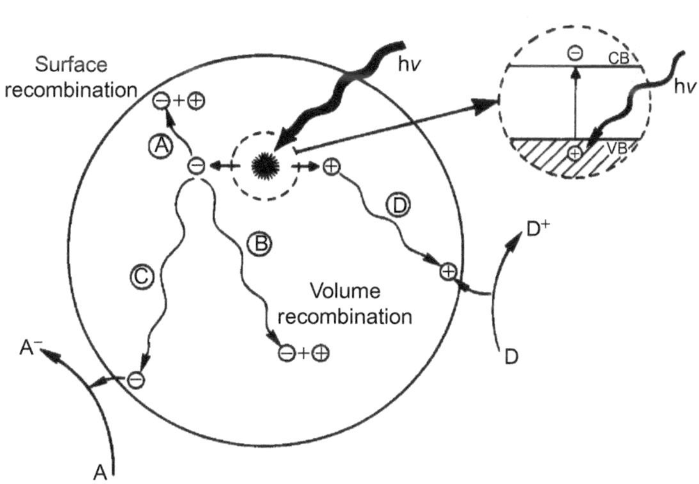

Fig. 8.3 Illustration of the processes involved in photocatalytic hydrogen generation by water splitting.
Reproduced with permission from A.L. Linsebigler, G. Lu, J.T. Yates, Chem. Rev. 95 (1995) 735–758.

position is necessary for a single photocatalyst to perform a photocatalytic water-splitting reaction, and is a thermodynamic requirement rather than a sufficient condition. The bandgap of the semiconductor must straddle the water oxidation and reduction potentials, which are +0 and + 1.23 V versus the normal hydrogen electrode (NHE) (pH = 0), respectively (Fig. 8.4A). In the photocatalytic *Z*-scheme system (Fig. 8.4B), two kinds of photocatalyst can be connected in series, in such a way that water reduction and oxidation will take place on two different photocatalysts, respectively. When an electrode is in contact with an electrolyte, the band bends upward for an n-type photoanode (Fig. 8.4C) [2]. Barriers formed at the bent band can drive the separation of photogenerated electron-hole pairs. The photoexcited electrons will be transferred to a counter electrode through the back contact and an external circuit to reduce H^+ to H_2, while the photoexcited holes accumulate on the surface of the photoanode and oxidize H_2O to O_2. The chemical reaction equations corresponding to these processes under illumination are shown as follows [5]:

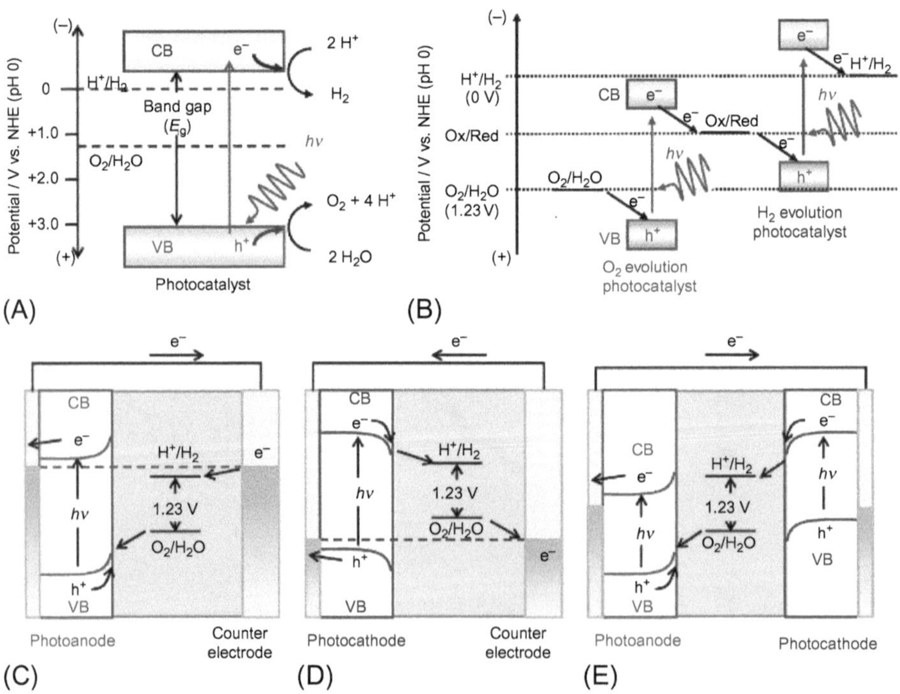

Fig. 8.4 Energy diagrams of photocatalytic water splitting based on (A) one-step excitation and (B) two-step excitation (Z-scheme); photoelectrochemical (PEC) water splitting using (C) photoanode, (D) photocathode, and (E) photoanode and photocathode in a tandem configuration. The bandgaps are depicted smaller in (B) and (E) to emphasize that semiconductors with a narrow bandgap can be employed.
Reproduced with permission from T. Hisatomi, J. Kubota, K. Domen, Chem. Soc. Rev. 43 (2014) 7520–7535.

Photon absorption:

$$4h\nu \rightarrow 4h^+ + 4e^- \quad (8.1)$$

Anode (oxidation):

$$H_2O + 2h^+ \rightarrow 2H^+ + ½O_2 \quad (8.2)$$

Cathode (reduction):

$$2H^+ + 2e^- \rightarrow H_2 \quad (8.3)$$

Overall reaction:

$$2H_2O + 4h\nu \rightarrow 2H_2 + O_2 \quad (8.4)$$

The chemical equations show that the number of absorbed photons is equal to that of the photogenerated electron-hole pairs, which are twice that of the produced hydrogen molecules and four times that of the generated oxygen molecules.

Compared with the n-type photoanode, a p-type photocathode possesses bands bent downwards. The photoexcited electrons will accumulate on the semiconductor surface, the photoexcited holes migrate to the counter electrode, and H_2O will be reduced and oxidized to H_2 and O_2, respectively (Fig. 8.4D). Currently, no single semiconductor can produce hydrogen and oxygen by photocatalytic or PEC reaction with ideal conversion efficiency and prolonged stability. In a Z-scheme water-splitting system, an n-type photoanode and a p-type photocathode with mutually matched structure of bandgaps are connected in tandem (Fig. 8.4E). It is different from the system consisting of a single photoelectrode and a counter electrode, in that highly effective solar water splitting will be affected without a bias.

8.1.3 Overview of the main approach to solar water splitting

Solar water splitting can be achieved by several different approaches using semiconductor materials, but is mainly done by photocatalytic processes, PEC cells, photovoltaic-based cells, and dye-sensitized solar water splitting. As mention before, the most important distinction among these approaches is that some use suspended semiconductor particles, and others a system featuring semiconductor electrodes. In this chapter, photocatalytic and PEC water-splitting systems are discussed in detail. They have similar mechanisms and, to some extent, thermodynamics. Due to the involvement of particles, the mechanism of the photocatalytic system mainly relies on the kinetic competition of charge transfer processes, rather than semiconductor-liquid junctions. PEC cells encounter the opposite situation. A solid-liquid junction is formed at the interface between semiconductor electrodes and the aqueous electrolyte, which plays an important role in promoting effective separation of photogenerated electron-hole pairs. At any rate, semiconductors play an essential role in successful solar water splitting, and are presented in Section 8.2.

8.2 Semiconductors for solar water splitting

8.2.1 General requirements of semiconductors for photocatalysts and photoelectrodes

8.2.1.1 Energy band of semiconductors

An important application of semiconductors is their use as photocatalysts and photoelectrodes for water splitting to generate hydrogen and oxygen [6]. In semiconductors, the lowest energy band, called the VB, is generally fully occupied by electrons, and the highest energy band, which is normally unoccupied, is the CB. The energy difference between the lower energy level of CB (E_c) and the upper energy level of VB (E_v) is the bandgap energy (E_g). The bandgap energy E_g determines the theoretical maximum STH efficiency (Fig. 8.5) [7]. A good example is provided by anatase TiO_2 (A-TiO_2) with a bandgap of 3.2 eV; it can absorb only the UV portion of the solar spectrum, which leads to a low theoretical maximum STH efficiency of 1%. Hematite (α-Fe_2O_3), with a bandgap of 2.2 eV, has a high theoretical maximum STH efficiency of 15%. To achieve a high STH efficiency, semiconductors with a narrow bandgap should be used to absorb sunlight over a wide spectrum.

To accomplish the thermodynamically uphill water-splitting reaction, a Gibbs free energy of 237.18 kJ mol^{-1} is required, which is 1.23 eV under standard conditions in

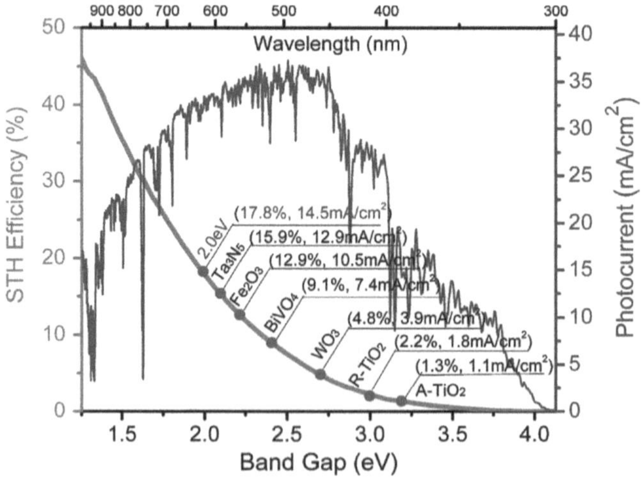

Fig. 8.5 Dependence of the theoretical maximum solar-to-hydrogen (STH) efficiency and the photocurrent density of photoelectrodes on the bandgap under AM 1.5 G irradiation (100 mW cm^{-2}).
Reproduced with permission from J. Li, N. Wu, Catal. Sci. Technol. 4 (2014) 4440.

terms of electrochemical potential. For an overall solar water-splitting process, the CB edge of a semiconductor should be more negative than the potential of hydrogen production (0 V_{NHE}) and its VB edge should be more positive than the oxygen generation potential (+1.23 V_{NHE}), which facilitates the transfer of photogenerated charges and the water redox reaction. Therefore, 1.23 eV is the theoretical minimum bandgap of a semiconductor for water splitting, which corresponds to light with a wavelength of approximately 1100 nm, according to the following equation [8]:

$$E_g\,(\mathrm{eV}) = h\nu = hc/\lambda = 1240/\lambda\,(\mathrm{nm}) \tag{8.5}$$

where h is Planck's constant (6.626×10^{-34} J·s) and c is the speed of light (3.0×10^8 m·s^{-1}); 1 eV equals 1.062×10^{-19} J. Fig. 8.6 shows the band levels of some

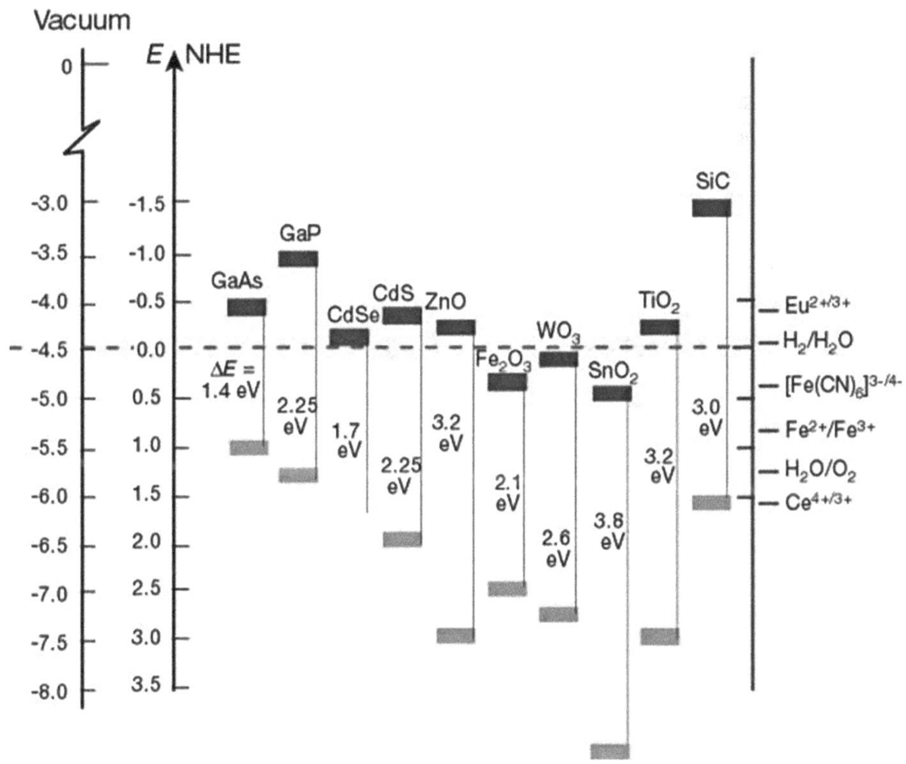

Fig. 8.6 Band positions of several semiconductors measured in aqueous electrolyte at pH 1. The lower edge of the conduction band (CB) (top bar) and upper edge of the valence band (VB) (bottom bar) are exhibited along with the bandgap in electron volts. The energy scale is indicated in electron volts, referencing with the normal hydrogen electrode (NHE) or the vacuum level. The standard potentials of several redox couples are presented against the standard hydrogen electrode potential on the middle right side.
Reproduced with permission from M. Grätzel, Nature 414 (6861) (2001) 338–344.

common semiconductors [6]. Some oxides, such as TiO_2 and ZnO, possess suitable band structures for solar water splitting, but owing to wide bandgaps, they cannot take full advantage of the visible light portion of the sun's spectrum. On the other hand, a semiconductor with a narrow bandgap can improve the light-harvesting efficiency by absorbing visible light. Therefore, appropriate band engineering is important for designing photocatalysts and photoelectrodes.

8.2.1.2 Interface of semiconductor and electrolyte

The semiconductor-aqueous electrolyte interface is similar to a Schottky junction in some respects. The interface behavior can be described by a diffuse ionic double-layer model: an equilibrium will be reached when the electrochemical potentials of the semiconductor and electrolyte are equal, which is expressed as $E_F = E_{F,redox}$. E_F is the Fermi level, while $E_{F,redox}$ is the Fermi level in a redox system. In a PEC system, a semiconductor photoelectrode and a metal electrode are linked and immersed into the electrolyte, which serve as work electrode and counter electrode, respectively [9]. Initially, there is no contact between the two electrodes and the electrolyte, and there is no aqueous equilibrium (Fig. 8.7A). Because of the absence of excess net charge at the interface, the conduction and VBs are all flat, and the semiconductor potential is called the flat band potential (V_{FB}). In this system (Fig. 8.7B), in the dark, the Fermi level of the n-type semiconductor is higher than that of the metal cathode, so that electrons leave the semiconductor via an ohmic contact and are transferred to the metal cathode through an external circuit, while holes are left behind in a space charge region. When the Fermi levels of the photoanode and photocathode are at the same energy, equilibrium is reached. The conduction and VB edges bend upward by an amount E_B to form a potential barrier (e.g., Schottky barrier), which hampers further electron transfer. The Fermi level of the metal cathode turns out to be lower than the potential for water reduction, so that a barrier to electron transfer is established between these two energy levels. It is not conducive to water splitting. When irradiated (Fig. 8.7C), the extent of band bending decreases before reaching a new equilibrium, due to the generation of photo-induced electrons and holes. The Fermi level of the metal cathode is still lower than the water reduction potential at the new equilibrium. When a bias is applied to improve the Fermi level of the metal cathode (Fig. 8.7D), the Fermi level rises above the water reduction potential, electrons are injected into the electrolyte from the cathode to reduce water to hydrogen, while holes are also injected into the electrolyte from the photoanode to complete the oxidation of water to oxygen.

Unlike an n-type semiconductor, the bands of a p-type semiconductor bend downwards in a PEC system under illumination, and photogenerated electrons are injected into the electrolyte from the photocathode to complete the water reduction, and photoexcited holes are transferred to the metal anode and injected into the electrolyte to complete water oxidation. When the Fermi level of the metal electrode is lower than the water reduction potential, application of an external bias can not only sustain current flow, but also increase the extent of band bending, which further drives the separation of photogenerated electron-hole pairs in an electric field.

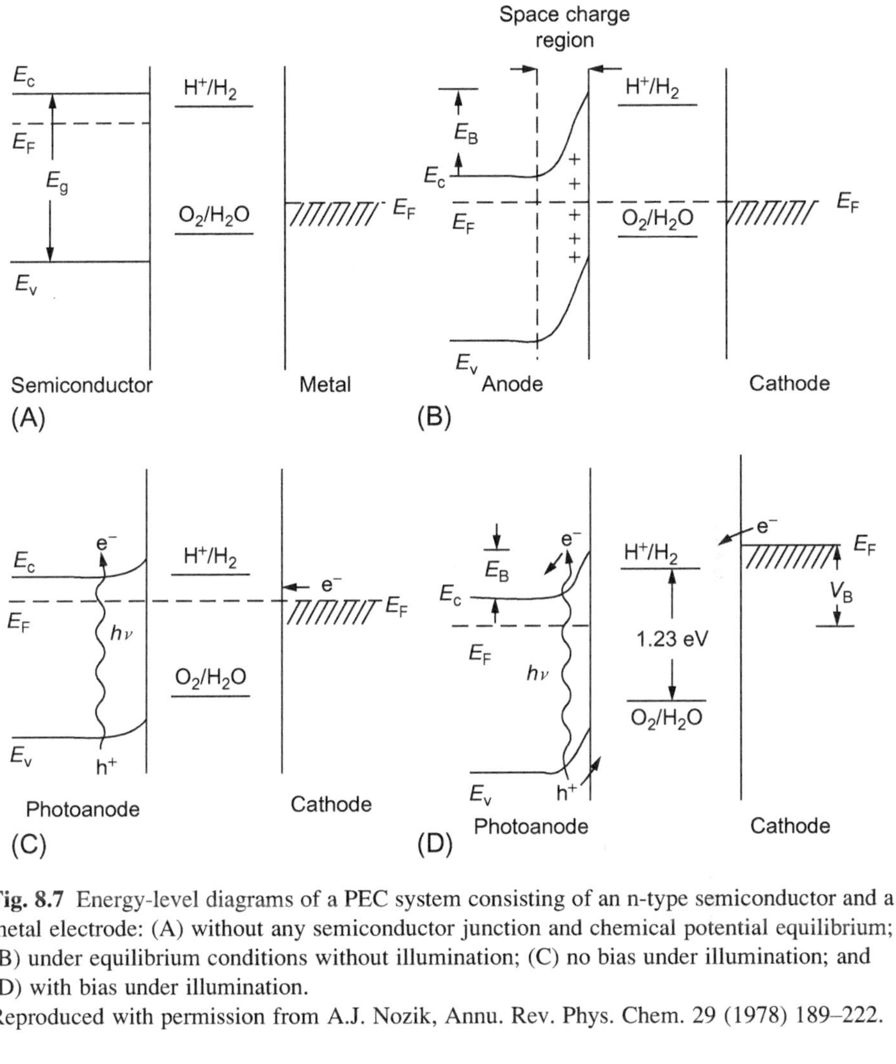

Fig. 8.7 Energy-level diagrams of a PEC system consisting of an n-type semiconductor and a metal electrode: (A) without any semiconductor junction and chemical potential equilibrium; (B) under equilibrium conditions without illumination; (C) no bias under illumination; and (D) with bias under illumination.
Reproduced with permission from A.J. Nozik, Annu. Rev. Phys. Chem. 29 (1978) 189–222.

8.2.1.3 Stability of semiconductors

The stability of semiconductors is an important issue limiting their practical use for solar water splitting. Instability may be the consequence of three major factors, namely the nature of the electrolyte in contact with the semiconductor, the electrical potential applied, and irradiation. In turn, semiconductor instability results in chemical dissolution, electrochemical corrosion, and photocorrosion. For a given combination of pH and potential, the semiconductor will be stable in the dark while it is unstable when irradiated. This contributes to the strong oxidizability and reducibility of photogenerated holes and electrons in bulk materials, respectively, resulting in oxidative or reductive decomposition of semiconductors.

Many metal oxide semiconductors are thermodynamically unstable. In addition, photocorrosion can also be caused by kinetic factors instead of thermodynamics, as a result of the photoexcited holes consuming the semiconductor itself. Take some common semiconductors for example: so far, TiO_2 has been considered the most stable semiconductor photocatalyst, although it is unstable thermodynamically; WO_3 is stable in acidic media while α-Fe_2O_3 is stable under alkaline conditions; Cu_2O has poor stability due to self-photocorrosion in electrolyte solution; and photocorrosion of CdS is also an obstacle for practical applications.

However, there are several possible approaches to overcome these problems. The simplest method is to introduce sacrificial agents. A good redox couple may be found to accomplish the decomposition reaction with favorable kinetics. Typical sacrificial couples are several simple organic compounds (such as alcohols, saccharides, aldehydes, and carboxylic acids), inorganic salts (such as chalcogenides), and multivalent ions (e.g., $Ce^{4+/3+}$). Depositing a stable protective layer on the entire surface of the semiconductor is also an effective approach, with TiO_2 used frequently due to its chemical stability. This kind of design can rapidly remove photogenerated carriers from a semiconductor. In addition, depositing an appropriate cocatalyst on the surface of the semiconductor provides a physical barrier between the unstable semiconductor and electrolyte. Many kinds of cocatalysts have been introduced and are described in the following section.

8.2.2 Typical photocatalysts and photoelectrodes

Since the pioneering work on solar water splitting by Fujishima [3], various semiconductors for photocatalysts and photoelectrodes have attracted serious attention of many scientists. Generally, semiconductors can be classified as metal oxides and nonmetal oxides. In the following section, several representative ones are discussed in detail.

8.2.2.1 Metal oxides

Most metal oxides are n-type semiconductors and have been investigated as photocatalysts and photoanodes for solar water splitting over the last few decades. Typical examples are TiO_2, WO_3, $BiVO_4$, α-Fe_2O_3, and Cu_2O.

8.2.2.1.1 Titanium dioxide (TiO_2)

TiO_2 was initially introduced by Fujishima and Honda [3] as semiconductor for solar water splitting in a PEC system. It is one of the most effective semiconductors for solar-to-chemical energy conversion by far, and has several attractive advantages of other abundant resources, such as nontoxicity, photochemical stability, and low cost. However, its wide bandgap of 3.2 eV means it has a low light absorption efficiency (can only absorb the UV portion of the solar spectrum) and the theoretical maximum STH efficiency is limited to a low level ($\eta = 1.3\%$ for anatase and 2.2% for rutile TiO_2) (Fig. 8.5). Over the past decades, many efforts have been made to decrease the bandgap by introduction of dopants [10–12]. A series of metal ions doped into TiO_2

nanoparticles significantly influenced photoactivity, carrier separation, and transfer rates [13]. Doping with some transition metals (Mo, V, Al, and Pb) decreases the photoactivity of TiO_2 powder, and the absorption spectra will shift to a longer wavelength region with an increasing concentration of dopant [14]. Doping with a metal cation also strongly influences the properties of TiO_2 as a semiconductor. Fe(III)-doped TiO_2 nanoparticles with different Fe(III) concentrations can be prepared at different pH values using a hydrothermal method. A close relationship exists among phase purity, pH values, Fe(III) concentration, distribution of iron ions, and PEC behavior of the modified TiO_2 semiconductors [15]. Compared with pure TiO_2, N-doped TiO_2 powders and films have improved optical absorption and photocatalytic activity under visible light (wavelengths below 500 nm), which is mainly due to the bandgap narrowing caused by doping nitrogen into substitutional sites of the TiO_2 semiconductor [10]. Optical response and photocatalytic activity of TiO_2 can be increased also by co-doping Cl^- and Br^- using a hydrothermal method [16]. A C-doped TiO_2 semiconductor prepared by controlled combustion of Ti metal in a natural gas flame absorbs visible light (wavelength below 500 nm) and has a narrower bandgap than rutile (2.32 eV), which is closely related to the substitution of partial lattice oxygen atoms by carbon [17]. This material has better water-splitting photoactivity than pure TiO_2 (Fig. 8.8). At an applied potential of 0.3 V, C-doped TiO_2 (CM-*n*-TiO_2, flame-made) has a total conversion efficiency of 11% and a maximum photoconversion efficiency of 8.35% with 40 mW·cm^{-2} irradiation, which is much better than the 1% at 0.6 V obtained with pure TiO_2 (oven-made).

Ti^{3+} doping is another direction of research. This self-doping is different from heteroelement doping (e.g., C, N, and F), and should reduce defect formation [18].

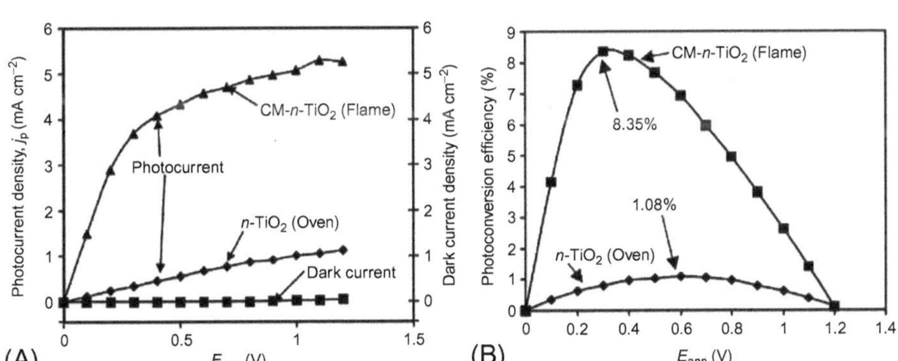

Fig. 8.8 (A) Photocurrent density as a function of applied potential E_{app} of carbon-modified (CM)-*n*-TiO_2 (flame-made) and the reference *n*-TiO_2 photoelectrodes (made by electric tube furnace or oven) under xenon lamp illumination at an intensity of 40 mW cm^{-2}. Dark current density of CM-*n*-TiO_2 (flame-made) as a function of applied potential is also shown. (B) Photoconversion efficiency as a function of applied potential E_{app} of CM-*n*-TiO_2 (flame-made) and the reference *n*-TiO_2 (made by electric tube furnace or oven) photoelectrodes. Reproduced with permission from S.U.M. Khan, M. Al-Shahry, W.B. Ingler, Science 297 (5590) (2002) 2243–2245.

The presence of Ti^{3+} in the bulk material of partially reduced TiO_2 extends the photoresponse region from the UV to the visible light, leading to good visible-light photocatalytic activity for hydrogen evolution from water. The Ti^{3+} sites in TiO_2 are highly stable in air and water under illumination [19]. Ti^{3+} self-doping in TiO_2 contributes to effective enhancement of PEC performance with respect to visible-light absorption and electrical conductivity. Self-doped hierarchical TiO_2 nanotube arrays (NTs) with the top layer serving as TiO_2 photonic crystals (TiO_2 NTPCs) have been prepared by a simple two-step anodization method and then self-doped by ethylene glycol (EG) reduction in a microwave process (MWR-TiO_2 NTPCs); this material shows a 10-fold increase in visible-light photocurrent density compared with the non-doped material (Fig. 8.9). The improved visible-light absorption is attributed to the introduction of interband states of the Ti^{3+} oxygen vacancy, and the increased electrical conductivity is the result of the high donor density [20].

Surface disorder engineering is also an effective approach to narrow the TiO_2 bandgap [21]. Unlike the doping method, surface disorder is introduced using a hydrogenation process to enhance solar absorption. The disorder-engineered TiO_2 nanocrystals (black powders in a macro-perspective) exhibit substantial solar-driven photocatalytic activities, including the photooxidation of organic molecules in water and the production of hydrogen with the use of a sacrificial reagent (methanol) (Fig. 8.10) [12]. In addition, after several testing cycles, the disordered-engineered TiO_2 nanocrystals also exhibited a high stability in the solar water-splitting reaction under simulated sunlight (Fig. 8.10).

Titania photoelectrodes with different nanostructures are also receiving growing attention from more and more researchers. Highly ordered titania NTs of variable wall thickness have been prepared by anodizing a Ti sheet; the nanotube wall thickness was shown to greatly influence the photoanodic performance [22]. The best hydrogen production rate was found for a sample with a pore diameter of 22 nm and a wall thickness of 34 nm; the hydrogen production rate under 100 mW cm^{-2} UV illumination was 960 μmol $h^{-1} W^{-1}$ (24 mL $h^{-1} W^{-1}$) at an overall conversion efficiency of 6.8%. A cross-linked TiO_2 nanowire anode containing mixed anatase and rutile phases has been fabricated by a drop cast method. The random orientation of the wires not only provides a high surface area, but also helps to increase the effective optical path length by scattering light into the plane of the film. Up to 1.05% solar energy conversion efficiency was obtained, and the photocurrent density was up to 2.6 mA cm^{-2} under an illumination of AM 1.5 G [23]. Self-aligned highly ordered TiO_2 NAs with a length of 45 μm have also been formed by potentiostatic anodization of Ti foil; the corresponding photocurrent and conversion efficiency were up to 26 mA cm^{-2} and 16.25%, respectively [24]. Mesoporous TiO_2 thin films synthesized by two approaches (sol-gel method and preformed crystalline TiO_2 nanoparticles) illustrate different morphologies and PEC performance (Fig. 8.11A–G). The sol-gel derived sample has a water-splitting efficiency about 10 times higher than that obtained for TiO_2 nanoparticles, due to its good electronic connectivity [25].

Three-dimensional (3D) TiO_2 networks have been synthesized by atomic layer deposition (ALD) of TiO_2 films on cellulose nanofiber (CNF) templates

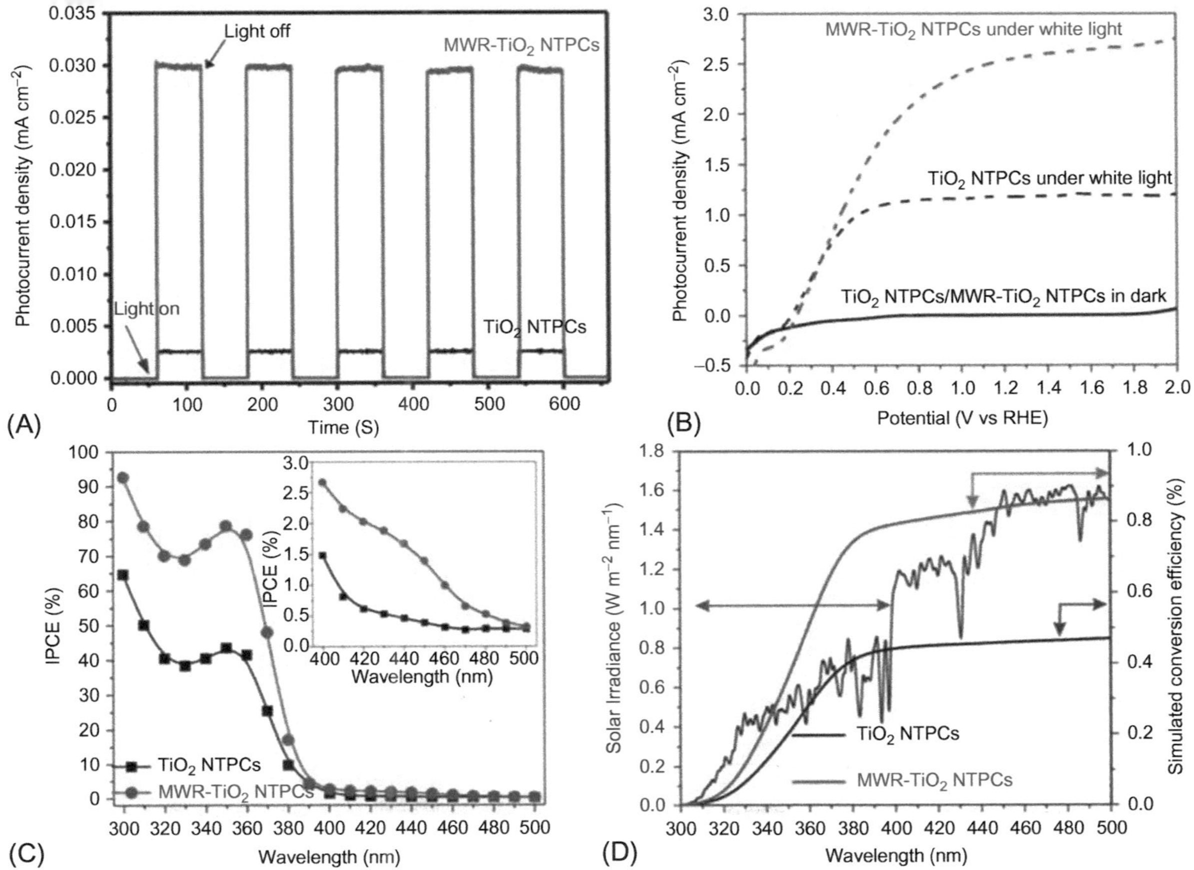

Fig. 8.9 PEC performance of TiO_2 NTPC and MWR-TiO_2 NTPC photoelectrodes: (A) photocurrent density versus time plots at an applied potential of 1.23 V versus RHE; (B) linear sweep voltammogram curves collected under AM 1.5 G; (C) incident photon-to-current efficiency (IPCE) plots; the inset shows magnified IPCE spectra over an incident wavelength range of 400–500 nm; and (D) photoconversion efficiency for the TiO_2 NTPCs and MWR-TiO_2 NTPCs as a function of wavelength.

Reproduced with permission from Z. Zhang, X. Yang, M. N. Hedhili, E. Ahmed, L. Shi, P. Wang, ACS Appl. Mater. Interfaces 6 (2014) 691–696.

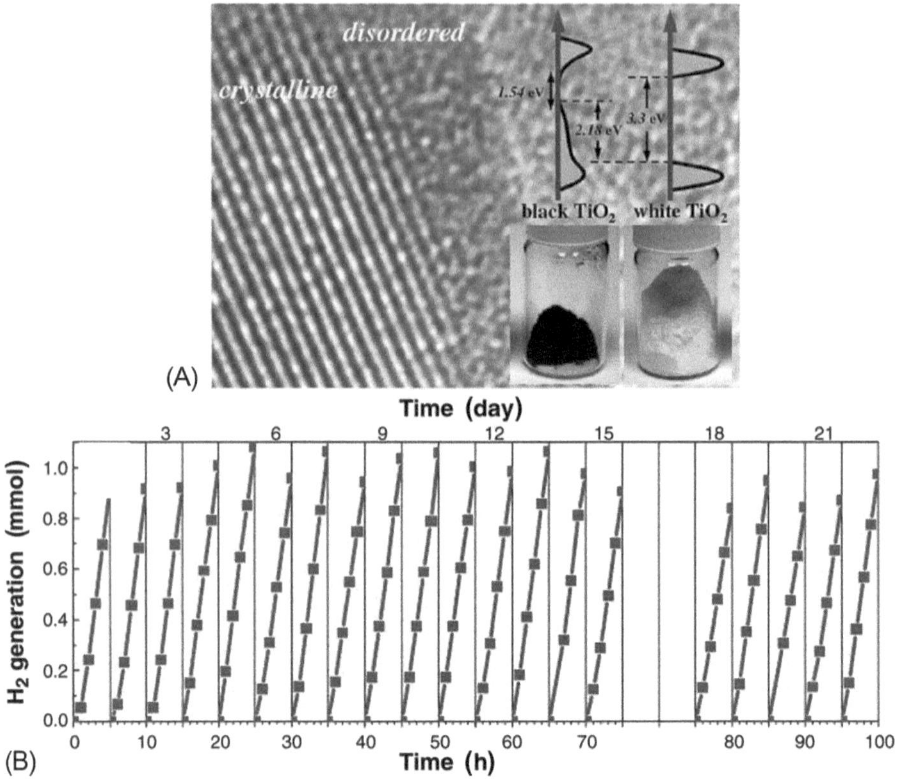

Fig. 8.10 (A) Schematic illustration of the state density of disorder-engineered TiO_2 nanocrystals as compared with that of unmodified TiO_2 nanocrystals and (B) cycling measurement of hydrogen gas generation by direct photocatalytic water splitting with disorder-engineered black TiO_2 nanocrystals under simulated sunlight.
Reproduced with permission from X. Chen, L. Liu, P.Y. Yu, S.S. Mao, Science 331 (6018) (2011) 746–750.

(Fig. 8.12A), which offer a great surface area for PEC water splitting. The maximum photocurrent density is $0.4\,mA\,cm^{-2}$ in a traditional PEC system (Fig. 8.12B), while it reaches $1\,mA\,cm^{-2}$ in a capillary PEC process (where the electrolyte is supplied through capillary force-driven nano/micro-channels in the CNF film) [26]. Hierarchical TiO_2 NTs, prepared by two-step anodization, have optimized photocurrent density and photoconversion efficiency on the nanopore/nanotube TiO_2 NTs of $1.59\,mA\,cm^{-2}$ at 1.23 V versus RHE and 0.84%, respectively [27]. Single-crystalline branched TiO_2 nanorods (B-NRs) have been prepared by a solution method. The B-NRs simultaneously offer a large contact area with the electrolyte, excellent light-trapping characteristics, and a highly conductive pathway for charge carrier collection, and they show excellent PEC water-splitting performance. A photocurrent density of

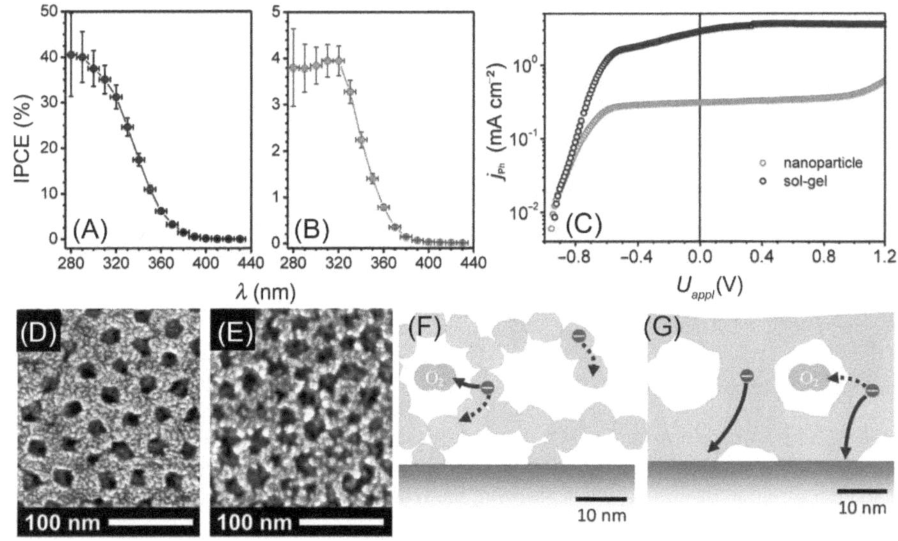

Fig. 8.11 IPCE of (A) sol-gel and (B) nanoparticle TiO_2 films measured at $U_{appl} = 0\,V$; (C) $I - U$ characteristics of mesoporous TiO_2 photoelectrodes; (D and E) HRSEM images of mesoporous TiO_2 thin films prepared by sol-gel route and nanoparticle route, respectively; (F and G) sketches of the electronic transport within the mesoporous photoelectrodes: (F) the small nanoparticles and the template particle-size mismatch result in a short conduction path cross section. Recombination with, for example, dissolved oxygen is facilitated due to the high surface area; and (G) the sol-gel films therefore show high photocurrents due to thick and continuous pore walls and a lower recombination rate.
Reproduced with permission from P. Hartmann, D.-K. Lee, B.M. Smarsly, J. Janek, ACS Nano 4 (2010) 3147–3154.

$0.83\,mA/cm^{-2}$ at 0.8 V versus RHE under AM 1.5 G illumination was obtained, with an incident photon-to-current efficiency (IPCE) of 67% at 380 nm with an applied bias of 0.6 V versus RHE; this is nearly two times higher than the bare NRs without branches (Fig. 8.13) [28].

Due to the synergistic effects of the increased surface area to volume ratio of the nanosized crystals and the raised CB minimum, the photooxidation capability, photoactivity, and hydrogen production rate of nanosized TiO_2 samples prepared by the hydrothermal method are all enhanced compared with micron-sized TiO_2 samples at the same conditions [29].

Combining p-type Si and rutile TiO_2 nanowires has been investigated as the hydrogen-generating photocathode and oxygen-generating photoanode, respectively, comprising a fully integrated nanosystem for direct solar water splitting (Fig. 8.14) [30]. The Si nanowires were fabricated by a reactive-ion etching (RIE) approach on patterned single-crystalline Si wafers, and TiO_2 nanowires were grown from p-Si backbone nanowires under hydrothermal conditions. Indium and Pt nanoparticles

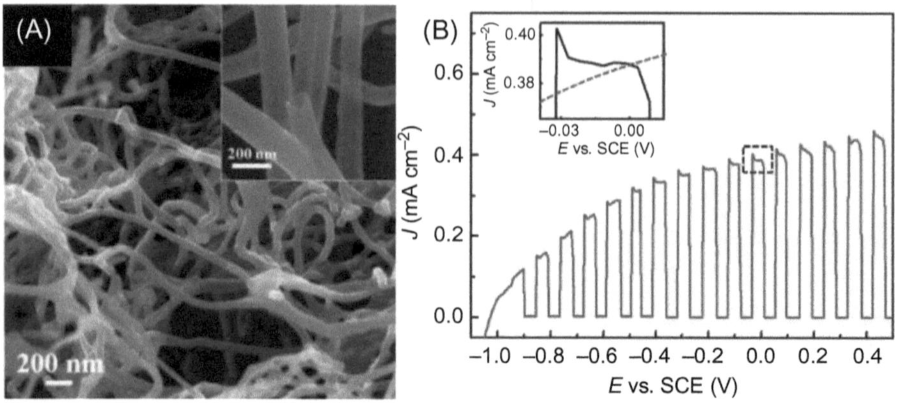

Fig. 8.12 (A) Morphology of 3D TiO_2 networks; inset shows an enlarged broken tip revealing the hollow tubular structure. (B) $J-V$ characteristics of 3D TiO_2 networks; inset shows enlarged top region of one J_{ph} cycle showing the feature of the initial current spike. Reproduced with permission from Z. Li, C. Yao, Y. Yu, Z. Cai, X.D. Wang, Adv. Mater. 26 (2014) 2262–2267.

were loaded on the nanowire surfaces to reduce the reaction overpotential. This system shows stable catalytic performance and a solar-to-fuel conversion efficiency of 0.12% under simulated sunlight illumination, which is comparable with that of photosynthesis in plants.

Some oxide semiconductors with a narrower bandgap than TiO_2 may be unable to generate enough driving force for full water splitting, which has an important relationship with kinetic overpotentials and ohmic resistances [31]. However, they have much higher theoretical STH efficiencies and have been intensively researched for decades, for instance, WO_3, $BiVO_4$, α-Fe_2O_3, and Cu_2O.

8.2.2.1.2 Tungsten trioxide (WO_3)

WO_3, which has a bandgap of 2.7 eV, has been reported as a novel n-type semiconductor for solar water splitting, because it is inexpensive, environmentally benign, and chemically stable in acidic aqueous media. Importantly, the maximum theoretical STH efficiency is 6%, which is higher than that of TiO_2 [32]. However, the main drawback is the formation of peroxo species on the surface competing with O_2 production.

WO_3 nanoparticles have been prepared by a sol-gel method, and then formed into thin films by a doctor blade approach following heat treatment [33]. The IPCE values obtained under front illumination and back irradiation are very different, with the maximum IPCE reaching 70% at 380 nm from back illumination, which is higher than the 50% obtained by front irradiation. The maximum quantum conversion efficiency (QE) of WO_3 is lower than that of TiO_2, which may be due to the low light absorption coefficient. Therefore, the scattering and absorption of light should be enhanced by changing the accumulation mode of the semiconductor particles to further improve the IPCE. More transparent WO_3 thin film can be obtain by a similar method using

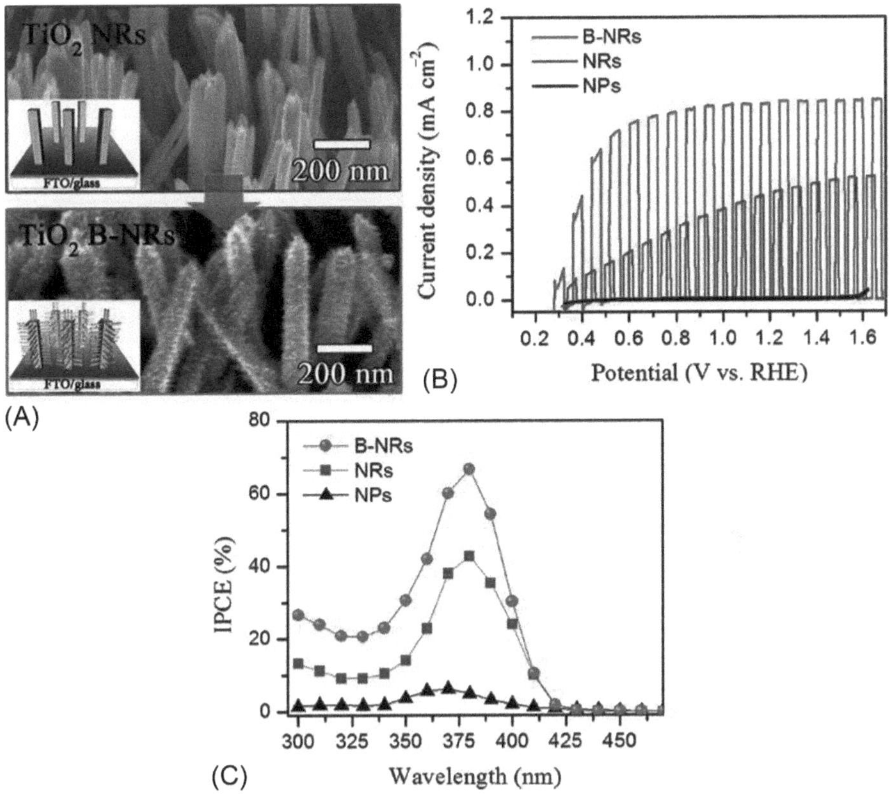

Fig. 8.13 (A) SEM images of TiO_2 nanorods (NRs) and TiO_2 B-NRs. Insets show corresponding schematic descriptions; (b) chopped $I-V$ curves illuminated by a xenon lamp; and (C) IPCE spectra measured at an applied bias of 0.6 V versus RHE.
Reproduced with permission from I.S. Cho, Z. Chen, A.J. Forman, D.R. Kim, P.M. Rao, T.F. Jaramillo, X. Zheng, Nano Lett. 11 (2011) 4978–4984.

polyethylene glycol (PEG)-300 as a stabilizing agent [34, 35]. The IPCE measured in 1 M $HClO_4$ electrolyte is >75% at 400 nm, with an applied bias of 1 V versus RHE.

The PEC performance of particle thin films differs from those of monocrystalline and polycrystalline thin films. The charge separation process of thin film prepared from nanoparticles is dominated by diffusion migration. A mechanism of charge separation and transportation can be determined by IPCE values obtained under front and back irradiation. To improve the performance of the WO_3 photoanode, various methods also have been adopted over several years. A porous structured WO_3 thin film with a pore wall thickness of 10 nm and length of 70 nm has been fabricated using anodic oxidation, which was expected to improve light scattering and current density by its large surface area [36]. However, the experimental result showed an ideal IPCE

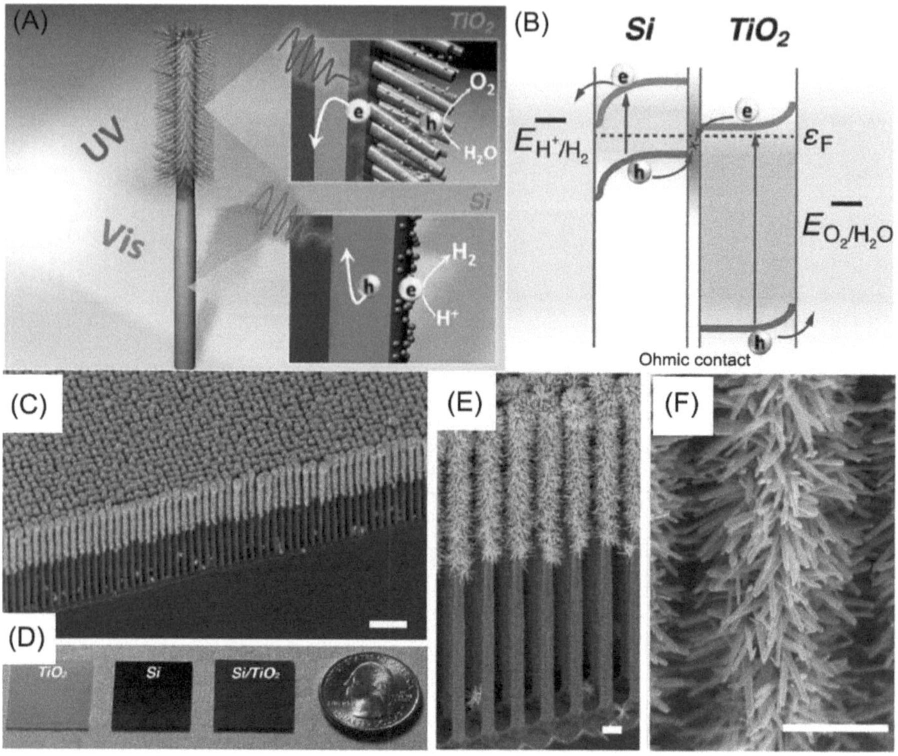

Fig. 8.14 (A) Structural schematics representation of the nanotree heterostructure; (B) energy band diagram of the nanotree heterostructure for solar-driven water splitting; (C) false-colored, large-scale SEM image of a Si/TiO_2 nanotree array; (D) comparison of the optical images of a TiO_2 nanowire (NW) substrate, a Si NW substrate, and a Si/TiO_2 nanotree substrate; (E) SEM image of the details of a nanotree heterostructure; (F) magnified SEM image showing the large surface area of the TiO_2 segment used for water oxidation. The scale bars are 10 μm (C) and 1 μm (F).
Reproduced with permission from C. Liu, J. Tang, H.M. Chen, B. Liu, P. Yang, Nano Lett. 13 (2013) 2989–2992.

of 25% at 350 nm and 1 V versus RHE. Other WO_3 thin films with different morphologies have been fabricated by a simple sol-gel method, with various acids (HCl, $HClO_4$, and H_2SO_4) added simultaneously to obtain different compactness and porous structures [37]. The use of HCl and $HClO_4$ is expected to form semitransparent thin films with porous structure, while a compact transparent thin film will be obtained without any acid. The IPCE of as-synthesized WO_3 prepared in HCl solution is up to 0.3% under irradiation at 86 mW cm^{-2}, which is higher than the 0.1% obtained without acid addition. However, it is lower than that of TiO_2 (0.4%), which is because the CB of WO_3 is too low, and an applied bias higher than that of TiO_2 is needed to achieve solar water splitting. These results show that thin films made from particles not only can change the light scattering and carrier transportation property of photoelectrodes, but also perform better than compact thin films. A similar approach has also been reported with the use of $HClO_4$ to form WO_3 following a heat treatment at 500°C for 30 min, and corresponding IPCE values were measured at different wavelengths [38]. The IPCE values obtained were 39% at 410 nm and 46% at 305 nm, respectively, and the scale of photoresponse can be expanded to 470 nm.

Porous structures and connect mode of particles play important roles in IPCE. Compact and transparent WO_3 photoelectrodes can be fabricated by sputtering at a temperature of 270°C [39]. The photocurrent density of as-synthesized WO_3 is 2.7 mA cm^{-2} at 1.6 V versus SCE under AM 1.5, which is higher than the 1.4 mA cm^{-2} reported almost simultaneously [38]. Mesoporous WO_3 thin films have been prepared using a potentiostatic anodization method, the use of NMF/H_2O/NH_4F mixed electrolyte being helpful to obtain highly efficient WO_3 photoanodes [40]. The synthesized WO_3 photoanodes gave high H_2O electrolysis yields of between 70% and 90% in 1 M H_2SO_4 at a potential bias of 1 V versus SCE and close to 100% in the presence of MeOH, and a photocurrent of 9 mA cm^{-2} under AM 1.5 irradiation (λ = 300–700 nm; 370 mW cm^{-2}). Ordered WO_3 nanowire (hexagonal structure) and nanoflake (monoclinic structure) films have been synthesized on fluorine-doped tin oxide (FTO)-coated glass substrates by a solvothermal method (Fig. 8.15A) [41]. The amounts of water, oxalic acid, and urea in the precursor play important roles in the formation of film morphology. The PEC measurements show a photocurrent of 1.43 mA cm^{-2} under AM 1.5G illumination and IPCE >60% at 400 nm (Fig. 8.15B and C).

A sphere-like WO_3 nanoparticle film with mesoporous nanostructure and high transparency has been synthesized by controlling the weight ratio of the tungsten precursor to PEG, resulting in a photocurrent of 3.7 mA cm^{-2} under AM 1.5 irradiation (λ = 300–700 nm) [42]. 3D WO_3 nanostructures on carbon paper were obtained using a high-temperature reactive vapor deposition process with tungsten powder as the source material in the presence of oxygen; the material exhibited a strong photocurrent response under visible light illumination (λ > 420 nm) and good photocatalytic performance [43].

Coupling tungsten oxide with other semiconductors to form a junction has also attracted increasing attention in the solar water-splitting field. Various WO_3-based junctions have been employed in the past [44]. The WO_3 photoanode incorporated with RGO shows significantly enhanced PEC activity for water splitting [45]. The RGO in the combined WO_3-RGO electrode provides good surface contact and plays a favorable role in decreasing charge carrier recombination at the particle interface of the WO_3. A photocurrent of 1.1 mA cm^{-2} at 1 V versus Ag/AgCl was obtained, but better PEC performance is only possible at a relatively more positive bias (>1 V). Electrodeposited WO_3 thin film with an ALD-deposited Al_2O_3 overlay shows a threefold increase of the photocurrent efficiency and enhanced faradaic efficiency (increasing from 17.7% to 54.1%). Apparently, the Al_2O_3 overlay suppresses the formation of surface peroxo species by decreasing the number of electron trapping sites on the WO_3 surface while promoting hole trapping (Fig. 8.16) [46]. Solar water splitting is facilitated and carrier recombination is delayed.

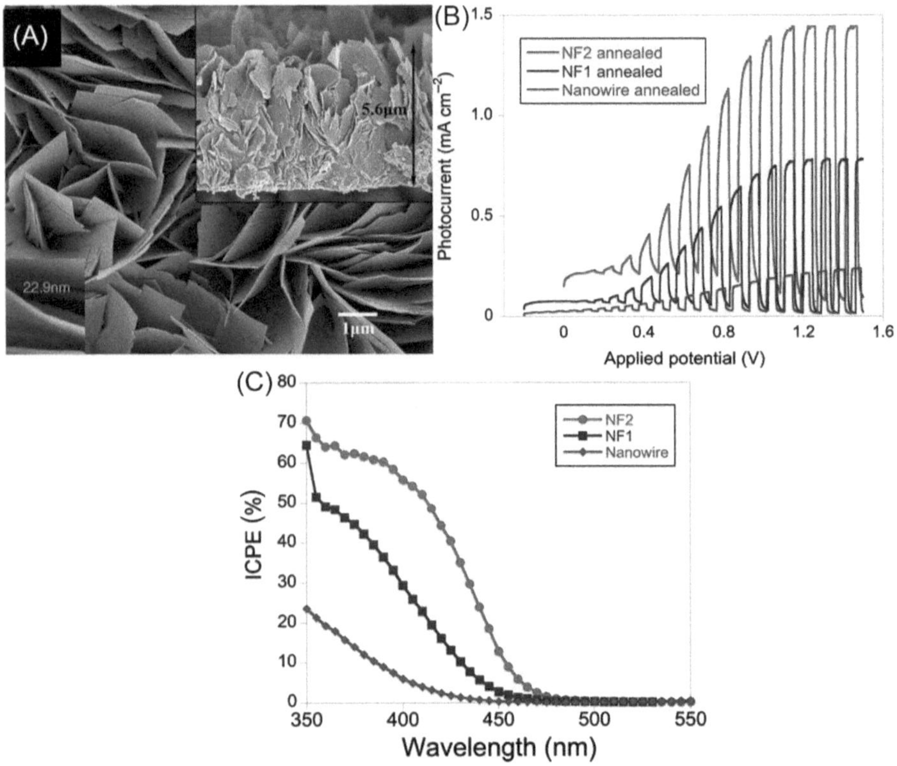

Fig. 8.15 (A) FESEM images of WO_3 nanoflake. Inset shows film cross-section; (B) current-potential plots for annealed NW and two flake samples, under chopped visible light in an aqueous solution of 0.1 mol L^{-1} sodium sulfate (Na_2SO_4); and (C) IPCE of the three samples. Reproduced with permission from J. Su, X. Feng, J.D. Sloppy, L. Guo, C.A. Grimes, Nano Lett. 11 (2011) 203–208.

WO_3 modified with nanoparticulate Pt species (PtO_x/WO_3) in a suspension system can accomplish water oxidation by irradiation with visible light at pH 5.9 using IO_3^- as an electron acceptor, and improved water oxidation activity also can be obtained by further modifying with a small amount of secondary cocatalysts (e.g., MnO_x, CoO_x, RuO_2, or IrO_2). A sample of RuO_2/WO_3 has shown the best apparent quantum yield of 14.4% at 420 nm [47]. Nanocomposite Au:WO_3 films with different gold contents have been synthesized by a sol-gel method [48]. The best PEC performance for a sample containing 1 mol% Au was obtained by a higher interface length between photoanode/electrolyte and lower Au surface accumulation, which is dominated by surface morphology. The highest IPCE and amount of hydrogen produced are about 20% at 360 nm and 2.7 μmol h^{-1}, respectively. 3D branched WO_3 nanosheet arrays with layered C_3N_4 heterojunctions and CoO_x nanoparticles have been explored as flexible photoanode for efficient PEC water oxidation [49]. Owing to enhanced light

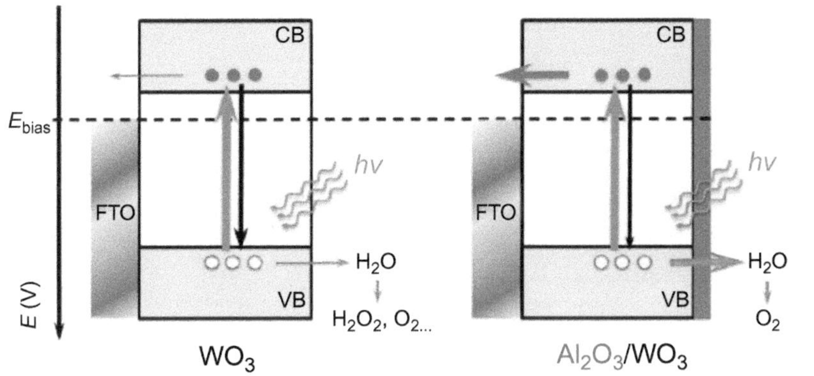

Fig. 8.16 Illustration of hole and electron mediated charge transfer occurring on (left) WO_3 and (right) Al_2O_3/WO_3 illuminated electrodes biased at a potential around the thermodynamic water oxidation potential (0.97 V vs. Ag/AgCl). The purple rectangle on the right side represents the alumina overlayer. For the sake of clarity, electrons (red full circles) and holes (blue open circles) are represented as free on the conduction (CB) and valence (VB) bands, respectively. Likewise, blue and red arrows represent hole and electron-mediated processes, respectively. Green arrows represent the photogeneration of electron-hole pairs. Black arrows represent recombination losses. Relative arrow thicknesses represent the rate of the processes (the thicker the arrow, the faster and more effective the process).
Reproduced with permission from W. Kim, T. Tachikawa, D. Monllor-Satoca, H. Kim, T. Majima, W. Choi, Energy Environ. Sci. 6 (2013) 3732–3739.

harvesting, efficient separation of photogenerated electron-hole pairs, fast charge transfer at the interface, and the unique 3D layered nanostructural feature, the as-synthesized junctions had a maximum IPCE of 57.8% at 350 nm and photocurrent density of 5.76 mA cm^{-2} at 2.1 V versus Ag/AgCl for solar water oxidation. Tungstic acid thin layer-coated WO_3 octahedra bound with {111} basal planes (Fig. 8.17A) have been fabricated using a simple solvothermal route with the assistance of urea; they showed high visible-light-driven photocatalytic reducibility to remove dissolved Ag^+ in photoprocessing wastewater, with a removal efficiency about 11 and 74 times that of commercial tungsten oxide particles and P25, respectively, under the same conditions [50]. Noble metal particles can be grown on the surface of 3D urchin-like WO_3 in situ (Fig. 8.17B), and the as-synthesized metal/WO_3 hybrid composites have better visible-light photocatalytic activity in the degradation reaction of rhodamine B [51]. Titanium(IV) doped WO_3 nanocuboids have been synthesized by a simple approach, showing that the band structure can be changed by doping Ti(IV) into the lattice of WO_3 to improve visible-light-driven photocatalytic performance for organic pollutant decomposition (Fig. 8.17C) [52].

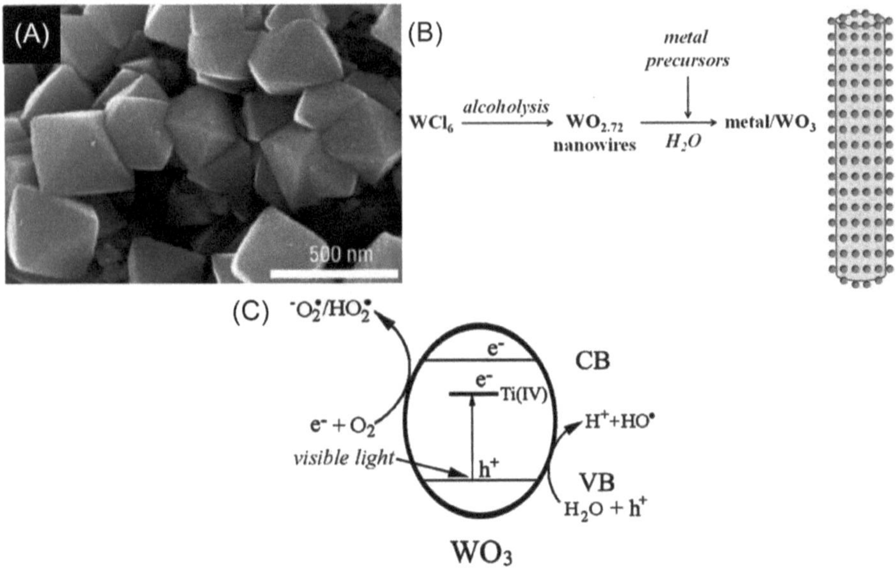

Fig. 8.17 (A) Typical SEM image of the as-prepared octahedral crystals. (B) Schematic procedure for in situ loading of metal particles on WO_3. (C) Schematic diagram of the charge separation and the photocatalytic activity for the Ti(IV)-doped WO_3 photocatalyst.
A: Reproduced with permission from Z.G. Zhao, Z.F. Liu, M. Miyauchi, Chem. Commun. 46 (2010) 3321–3323. B: Reproduced with permission from G. Xi, J. Ye, Q. Ma, N. Su, H. Bai, C. Wang, J. Am. Chem. Soc. 134 (2012) 6508–6511. C: Reproduced with permission from C. Feng, S. Wang, B. Geng, Nanoscale 3 (2011) 3695–3699.

8.2.2.1.3 Bismuth vanadate ($BiVO_4$)

$BiVO_4$ is a very interesting candidate for photoanode materials due to its relatively low cost, the theoretical maximum STH of 9.1%, a maximum photocurrent of $7.5\,mA\,cm^{-2}$ under standard AM 1.5 solar light irradiation, and a bandgap of 2.4 eV. The VB edge lies at approximately 2.4 eV versus RHE, providing enough overpotential for holes to accomplish water oxidation while the CB edge is located just short of the thermodynamic level for H_2. Its bandgap is slightly larger than that of an ideal photoanode (which is c.2.0 eV), but its very negative CB position may compensate for this disadvantage [53]. However, $BiVO_4$ suffers from several disadvantages:

1. Poor photo-induced electron transportation. Before charges (especially electrons) reach the interface between photoelectrode and electrolyte, recombination of most of the electron-hole pairs (approximately 60%–80%) has taken place in the bulk material;
2. Slow kinetics of oxygen evolution. The kinetics for oxygen evolution is very slow;
3. Low CB level. The CB edge of $BiVO_4$ is $<0\ V_{RHE}$, so that an external bias potential needs to be applied for PEC hydrogen evolution.

First research on the $BiVO_4$ photocatalyst for solar water oxidation was reported by Kudo et al. in 1998 [54]. Great progress in preparation and application has been made over the last 20 years. Early work on $BiVO_4$ as semiconductor focused mainly on photocatalysis in a suspension photocatalytic system for water oxidation or photodegradation of organic compounds [55]. Because of the drawback of the low CB level, more recent research has focused on the application of $BiVO_4$ photoelectrodes in a PEC system. Representative preparation methods of $BiVO_4$ photoanodes for solar water splitting mostly entail solution-based synthesis, electrochemical synthesis, and gas-phase synthesis [55].

Bismuth vanadate films have been prepared by a modified method using metal–organic decomposition by spin coating [56]. This method was used to dope Mo^{6+} into $BiVO_4$ by adding molybdenyl acetylacetonate to the precursor solution, and RhO_2 was then loaded on the Mo-doped $BiVO_4$ electrode by a simple impregnation method (Fig. 8.18) [57]. The obtained sample has a porous structure and a photocurrent density of 2.16 mA cm^{-2} at 1.0 V versus RHE in natural seawater under AM 1.5 G sunlight (1000 W m^{-2}) (Fig. 8.19). Doping $BiVO_4$ with W^{6+} (W:$BiVO_4$) is achieved by incorporating tungstic acid in the precursor solution, resulting in slightly smaller feature sizes than those of the $BiVO_4$ photoanode (Fig. 8.19A and B) [58]. PEC experiments have shown that interfacing a cobalt-based water-splitting catalyst (Co-Pi) with these W:$BiVO_4$ photoanodes eliminates most of the losses caused by surface electron-hole recombination (Fig. 8.19C). After surface modification with Co-Pi, a very large cathodic shift (c.440 mV) in the onset potential of the W:$BiVO_4$ photoanode found for sustained PEC water oxidation at pH 8 (Fig. 8.19D).

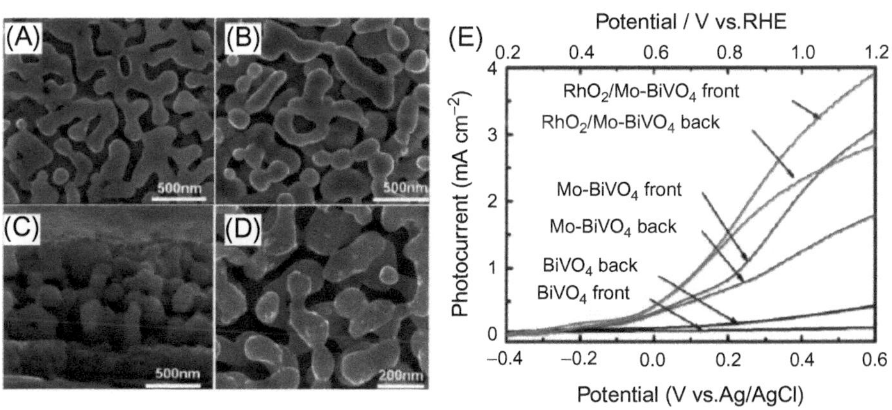

Fig. 8.18 SEM images of typical samples: (A) the surface of pure $BiVO_4$ film; (B) the surface of Mo-doped $BiVO_4$ film; (C) a cross-section of Mo-doped $BiVO_4$ on a fluorine-doped tin oxide (FTO) substrate; (D) RhO_2 particles on the surface of a Mo-doped $BiVO_4$ film; and (E) photocurrent-potential curves for $BiVO_4$, Mo-doped $BiVO_4$, and Mo-doped $BiVO_4$ with RhO_2 surface modification illuminated from the backside and the front side.
Reproduced with permission from W. Luo, Z. Yang, Z. Li, J. Zhang, J. Liu, Z. Zhao, Z. Wang, S. Yan, T. Yu, Z. Zou, Energy Environ. Sci. 4 (2011) 4046–4051.

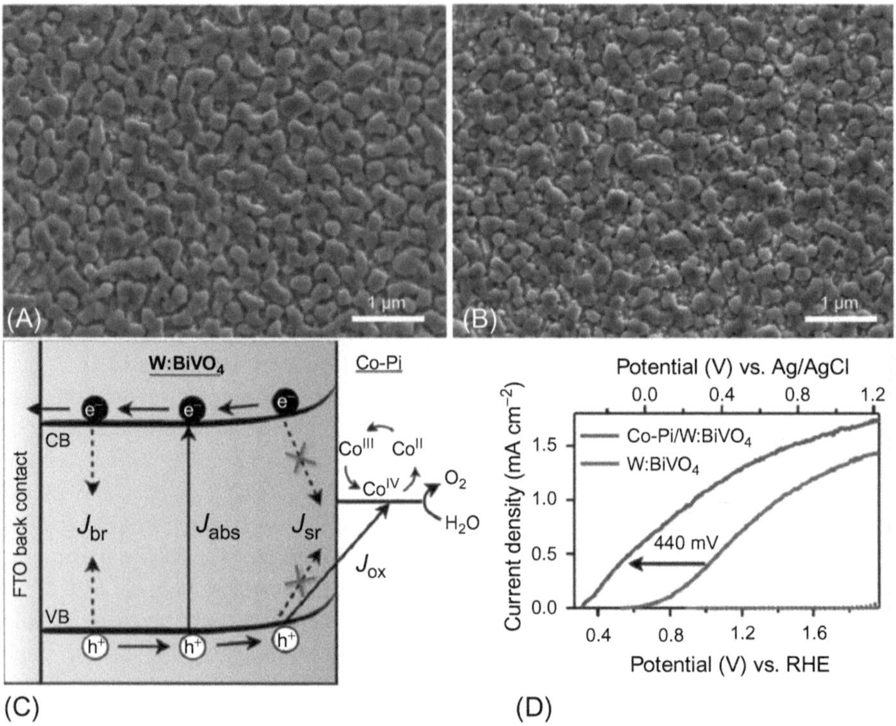

Fig. 8.19 Top-view SEM images of representative (A) $BiVO_4$ and (B) W:$BiVO_4$ photoanodes. (C) Energy diagram showing the kinetic processes active in the Co-Pi/W:$BiVO_4$ PEC photoanodes; and (D) $J-V$ curves measured for a W:$BiVO_4$ photoanode before and after photoassisted electrodeposition of Co-Pi.
Reproduced with permission from D.K. Zhong, S. Choi, D.R. Gamelin, J. Am. Chem. Soc. 133 (2011) 18370–18377.

A porous bismuth vanadate electrode doped with 2 atom% Mo (i.e., $BiV_{0.98}Mo_{0.02}O_4$) has been fabricated by surfactant-assisted metal-organic decomposition at 500°C, and a cobalt-phosphate-based oxygen evolution catalyst (Co-Pi OEC) has been loaded on the surface of a Mo-doped $BiVO_4$ photoanode using electrochemical deposition [59]. The Co-Pi/$BiV_{0.98}Mo_{0.02}O_4$ composite photoelectrode produces a stable photocurrrent density of 1.0 mA cm^{-2} at 1.0 V versus Ag/AgCl under AM 1.5 illumination in an aqueous 0.5 M Na_2SO_4 solution buffered at pH 7 with phosphate. A $BiVO_4$ photoanode has also been formed by a spray pyrolysis metal-organic deposition (MOD) method [60]. Chemical bath deposition (CBD) is also a simple method to prepare $BiVO_4$ photoelectrodes. The deposition time, temperature, pH of the solution, and the buffer solution have important effects on the morphologies of the $BiVO_4$ films, with compact films with a smaller microcrystal exhibiting better performance. The best PEC is produced in a buffer solution at pH ~ 7 at 85°C for a period of 30 h [61]. $BiVO_4$ arrays containing various dopants can also be formed using a simple drop cast method [62–64]. Molybdenum doping seems to be more effective than W doping in promoting photocatalytic oxygen production [65]. Compared with the photocurrent (0.42 mA cm^{-2}) of undoped $BiVO_4$, a photocurrent of c.2.38 mA cm^{-2} was attained by Mo-doped $BiVO_4$, and 1.98 mA cm^{-2} by W doping, under AM 1.5 illumination at 1.23 V versus RHE [65].

Hydrogen-treated $BiVO_4$ (H-$BiVO_4$) photoanodes can be fabricated by annealing $BiVO_4$ films in a hydrogen atmosphere at elevated temperatures (200–400°C) [66]. In this process, hydrogen gas reduces $BiVO_4$ to introduce oxygen vacancies as well as hydrogen impurities. A maximum photocurrent density of 3.5 mA cm^{-2} was reached at 1.0 V versus Ag/AgCl, which is 1 order of magnitude higher than that of air-annealed $BiVO_4$ obtained at the same potential. The enhanced photoactivity is attributed to the increased donor density of H-$BiVO_4$, which facilitates charge transport and collection. Another approach similar to Mo/W doping has been employed using PO_4 oxoanions to replace VO_4 oxoanions in $BiVO_4$. PO_4 oxoanion doping did not change the optical absorption and crystal structure of $BiVO_4$, whereas it did lower the charge transportation resistance [67]. Consequently, PEC and photocatalytic water oxidation were promoted considerably by a factor of about 30 compared with pristine $BiVO_4$. Yb^{3+} and Er^{3+} co-doped $BiVO_4$ also show good NIR photoactivity [68, 69].

Investigation of facet-dependent photocatalytic activity of $BiVO_4$ for water oxidization has shown that efficient charge separation can be realized on different crystal facets ({010} and {110}) separately under photo-irradiation [70, 71]. The photo-induced electrons and holes move to the {010} and {110} facets, respectively, and consequently reduction and oxidation will take place on {010} and {110}, respectively (Fig. 8.20A). Many kinds of cocatalyst can be loaded by photo-deposition on the surface of the $BiVO_4$ crystal; Au, Pt, and Ag particles are deposited selectively on the {010} facets, while MnO_x and PbO_2 particles are deposited only on the {110} facets (Fig. 8.20B). Cocatalyst loading results in much higher activity in both the photocatalytic and photoelectrocatalytic water oxidation reactions, compared with a photocatalyst with randomly distributed cocatalysts. A sample with Pt on the {010} facets and MnO_x on the {110} facets showed much higher activity in both photocatalytic and PEC water oxidation, compared with its counterparts with randomly distributed Pt and PbO_2 cocatalysts (Fig. 8.20C and D). Several other cocatalysts in similar photocatalytic systems also showed enhanced oxygen evolution (Fig. 8.20E and F) [71].

Heterojunctions of WO_3/$BiVO_4$ have also been investigated for PEC water splitting. To form a WO_3/$BiVO_4$ heterojunction, $BiVO_4$ was spin coated on WO_3 NR-array films made by solvothermal deposition on FTO (Fig. 8.21A and B) [72]. Due to the high surface area and improved separation of the photogenerated charge at the WO_3/$BiVO_4$ interface, the heterojunction structure exhibited enhanced PEC activity (Fig. 8.21C). A WO_3/W:$BiVO_4$ core/shell nanowire photoanode has been made by combining flame vapor deposition and drop casting [73]; $BiVO_4$ and WO_3 serve as the primary light absorber and electron conductor, respectively. Light absorption and charge separation are enhanced simultaneously, resulting in a photocurrent of 3.1 mA cm^{-2} and an IPCE of c.60% at 300–450 nm and a potential of 1.23 V versus RHE (Fig. 8.21). Vertically oriented WO_3 NRs capped with very thin $BiVO_4$ absorber layers have been prepared using glancing angle deposition and physical

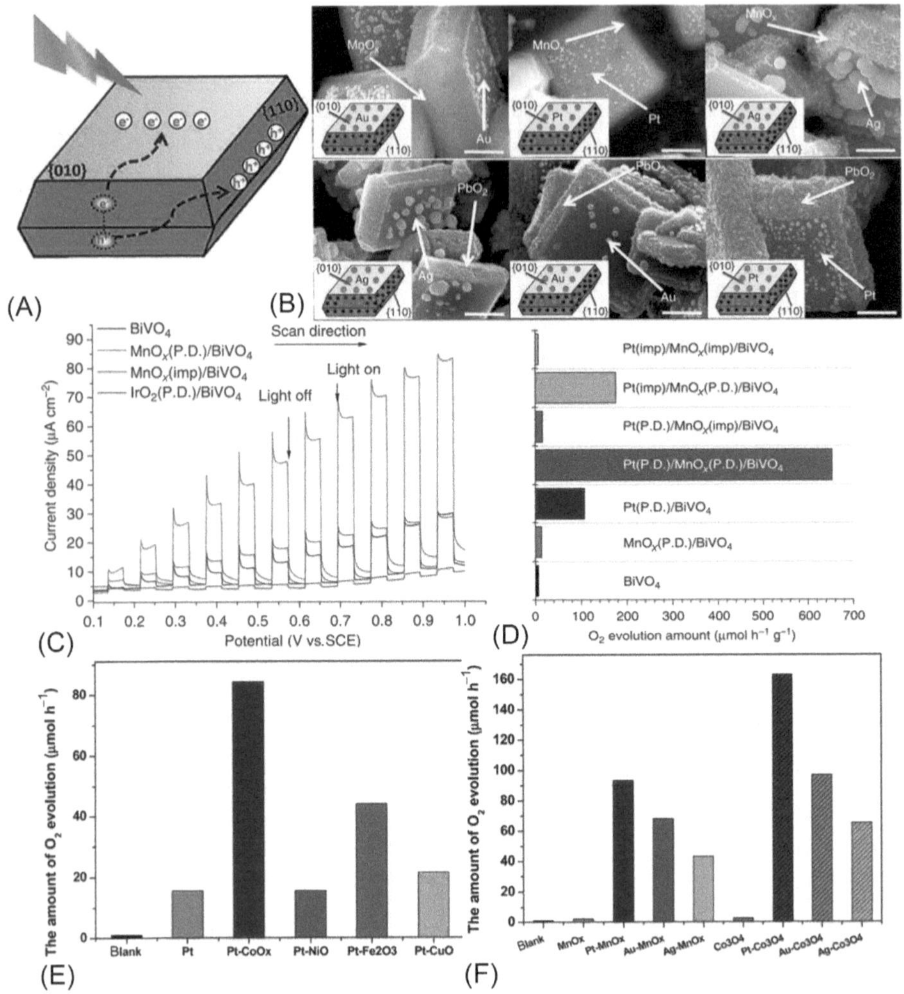

Fig. 8.20 (A) Charge separation between the {010} and {110} facets; (B) SEM images of dual components photodeposited on the surface of $BiVO_4$: Au/MnO_x/$BiVO_4$; Pt/MnO_x/$BiVO_4$; Ag/MnO_x/$BiVO_4$; Ag/PbO_2/$BiVO_4$; Au/PbO_2/$BiVO_4$; and Pt/PbO_2/$BiVO_4$. The contents of the deposited metals/metal oxides are all 5 wt%. Scale bar, 500 nm; (C) photochemical performance of four typical photoelectrodes; (D) photocatalytic water oxidation performance of $BiVO_4$; (E) photocatalytic water oxidation activity of $BiVO_4$ with different oxidation cocatalysts with Pt fixed as the reduction cocatalyst; and (F) photocatalytic water oxidation activity of different reduction cocatalysts with MnO_x or Co_3O_4 fixed as the oxidation cocatalysts.
A, B, C, and D reproduced with permission from R. Li, F. Zhang, D. Wang, J. Yang, M. Li, J. Zhu, X. Zhou, H. Han, C. Li, Nat. Commun. 4 (2013) 1432; E and F reproduced with permission from R. Li, H. Han, F. Zhang, D. Wang, C. Li, Energy Environ. Sci. 7 (2014) 1369–1376.

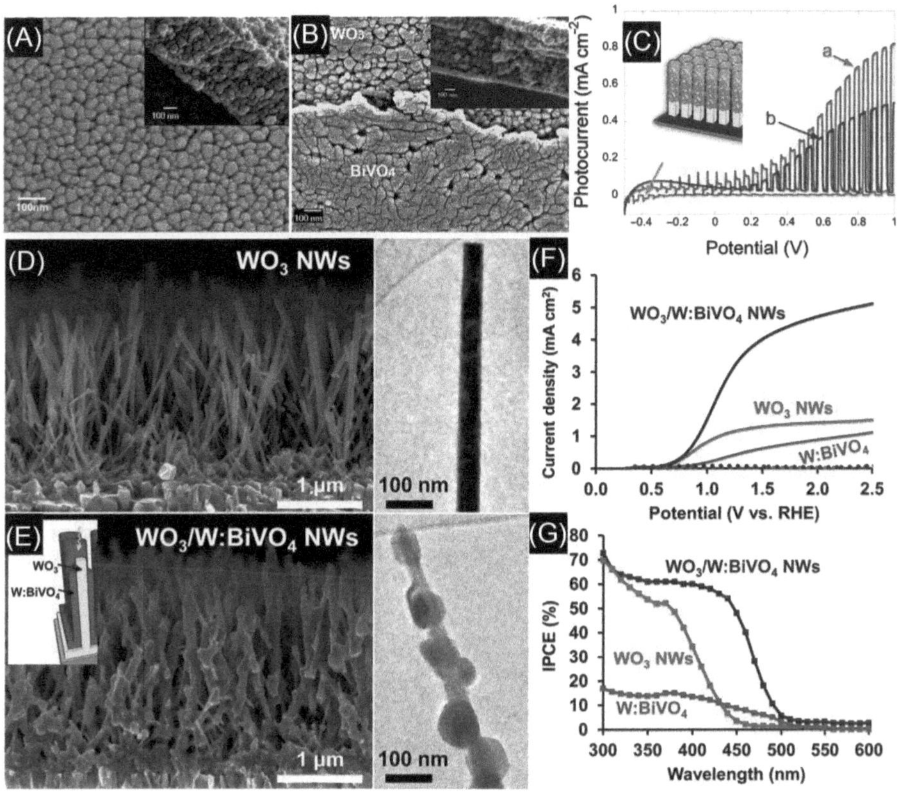

Fig. 8.21 (A, B, C) WO_3-NRs/$BiVO_4$ photoanode synthesized by solvothermal deposition and spin coating. Inset of (C) is a structural schematic representation of WO_3-NRs/$BiVO_4$. (D) SEM and (E) TEM images of the bare WO_3 NW array and WO_3/W:$BiVO_4$ core/shell NWs, respectively. The W:$BiVO_4$ shell consists of a single layer of densely packed nanoparticles. Inset of (E) is a structural schematic representation of the WO_3/W:$BiVO_4$ core/shell NW; (F) $J-V$ curves in 0.5 M potassium phosphate electrolyte buffered at pH 8; and (G) IPCE result measured at 1.23 V versus RHE.
A, B, and C reproduced with permission from J. Su, L. Guo, N. Bao, C.A. Grimes, Nano Lett. 11 (2011) 1928–1933. D, E, F, and G reproduced with permission from P.M. Rao, L. Cai, C. Liu, I.S. Cho, C.H. Lee, J.M. Weisse, P. Yang, X. Zheng, Nano Lett. 14 (2014) 1099–1105.

sputtering techniques. When modified with Co-Pi, the NRs deliver a remarkably stable photocurrent of 3.2 mA cm^{-2} at 1.23 V versus RHE in Na_2SO_4 electrolyte under simulated solar light irradiation [74].

8.2.2.1.4 α-Hematite (α-Fe_2O_3)

α-Fe_2O_3 is an abundant, nontoxic, and stable oxide, and has a narrow bandgap (1.9–2.2 eV) and a large theoretical maximum STH (12.9%), and is therefore considered an ideal semiconductor for solar water splitting. However, its very short hole diffusion length (2–4 nm), low minority charge carrier mobility, and poor water oxidation kinetics hinder practical application [7]. A relatively high applied bias is required for water splitting, and a suspension system is unworkable. To overcome these shortcomings, researchers have tried for decades to improve its preparation and properties.

Translucent Si-doped and undoped hematite thin films have been fabricated by two different methods, ultrasonic spray pyrolysis (USP) and atmospheric pressure chemical vapor deposition (APCVD) [75]. Iron(III) acetylacetonate ($Fe(AcAc)_3$) and iron pentacarbonyl ($Fe(CO)_5$) served as precursors, respectively, and tetraethyl orthosilicate (TEOS) was used as silicon dopant. Silicon doping significantly influenced the morphology and photoresponse of the films. The USP Si-doped hematite thin film has a changed morphology and increased photocurrent compared with USP undoped hematite electrodes (Fig. 8.22A). The APCVD Si-doped sample has a dendritic morphology (Fig. 8.22B) and produces a photocurrent of 1.45 mA cm^{-2} at 1.23 V versus RHE, whereas the undoped hematite electrode shows less developed branches on the surface (Fig. 8.22B, inset) and has a very weak photocurrent density (below 1 μA cm^{-2}) at the same applied potential. The enhanced photocurrent is the result of the increased electrical conductivity, because silicon acts as an electron donor in the hematite lattice.

Doped and undoped α-Fe_2O_3 thin film electrodes for PEC hydrogen production have been prepared using reactive magnetron sputtering (RMS); Si-doped and Ti-doped samples exhibit better PEC properties than the undoped photoanode [76]. The improvement is attributed to an enhanced charge-transfer rate coefficient at the surface and the passivation of the grain boundaries by the dopants. These are caused by the highly disordered surface and higher grain boundary recombination of the Si- and Ti-doped samples. Ti-doped α-Fe_2O_3 thin film has also been fabricated by APCVD using $Fe(CO)_5$ and $TiCl_4$ as precursors (Fig. 8.23A and B) [77]. With 0.8 atom% Ti in hematite, the maximum IPCE for water splitting in alkaline solution was found to be 27.2% under an applied bias of 0.6 V versus Ag/AgCl at 400 nm. Ultrathin Ti-doped α-Fe_2O_3 photoanodes have been prepared by APCVD through pyrolysis of ferrocene at 450°C on Ti foil (Fig. 8.23C and D) [78]. The photocurrent of this ultrathin hematite photoanode was up to 0.9 mA cm^{-2} at 0.6 V versus SCE under AM 1.5 G illumination. This improved performance is ascribed to the diffusion and doping of Ti^{4+} from the Ti substrate during pyrolysis deposition of the hematite on the Ti substrate.

Composites of α-Fe_2O_3 with other metal oxides have also been reported recently. A nanostructured host scaffold (WO_3) has been used to support a thin layer of α-Fe_2O_3 nanoparticles, which improves the light absorption and increases the surface area of the α-Fe_2O_3 [79]. The photocurrent in host-guest electrodes is increased by 20% and the absorbed photon conversion efficiency (APCE) is improved to 8.0% because of the host/guest architecture. A WO_3/Fe_2O_3 heterojunction electrode has been formed via a sol-gel approach [80]. The visible light response and the conversion efficiency improved significantly compared with the WO_3 and Fe_2O_3 electrodes alone, which is because the transfer of photogenerated electrons in WO_3/Fe_2O_3 is easier than that in WO_3 and Fe_2O_3 alone. In addition, NiO/α-$Ni(OH)_2$-hematite electrodes delivered

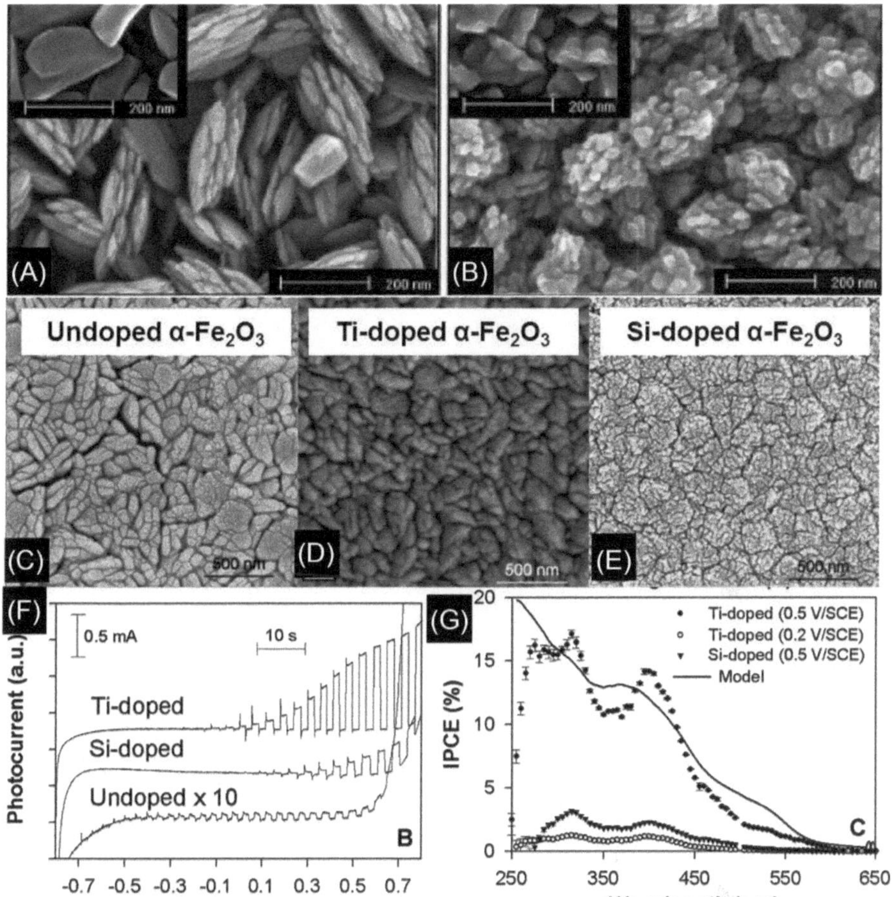

Fig. 8.22 Typical HR-SEM images of Si-doped hematite films prepared from (A) ultrasonic spray pyrolysis (USP) and (B) atmospheric pressure chemical vapor deposition (APCVD); (A, B insets) undoped hematite thin films; SEM images of (C) undoped, (D) Ti-doped, (E) Si-doped α-Fe_2O_3 films; (F) chopped $I-V$ curves of doped and undoped α-Fe_2O_3 films; and (G) IPCE as a function of wavelength of the doped α-Fe_2O_3 films.
A and B reproduced with permission from I. Cesar, A. Kay, J.A. Gonzalez Martinez, M. Grätzel, J. Am. Chem. Soc. 128 (2006) 4582–4583; C, D, E, F, and G reproduced with permission from J.A. Glasscock, P.R.F. Barnes, I.C. Plumb, N. Savvides, J. Phys. Chem. C 111 (2007) 16477–16488.

an excellent photocurrent of 16 mA cm^{-2}, which is 16 times higher than that of the pristine hematite film [81]. Apparently the synergistic effect of Fe and Ni accounts for the strong increase in photocurrent. A p-n junction of NiO/Fe_2O_3 can also promote charge separation, with NiO acting as an efficient hole acceptor and reducing the over-potential for water oxidation [82]. An α-Fe_2O_3 composite on TiO_2 also improves water

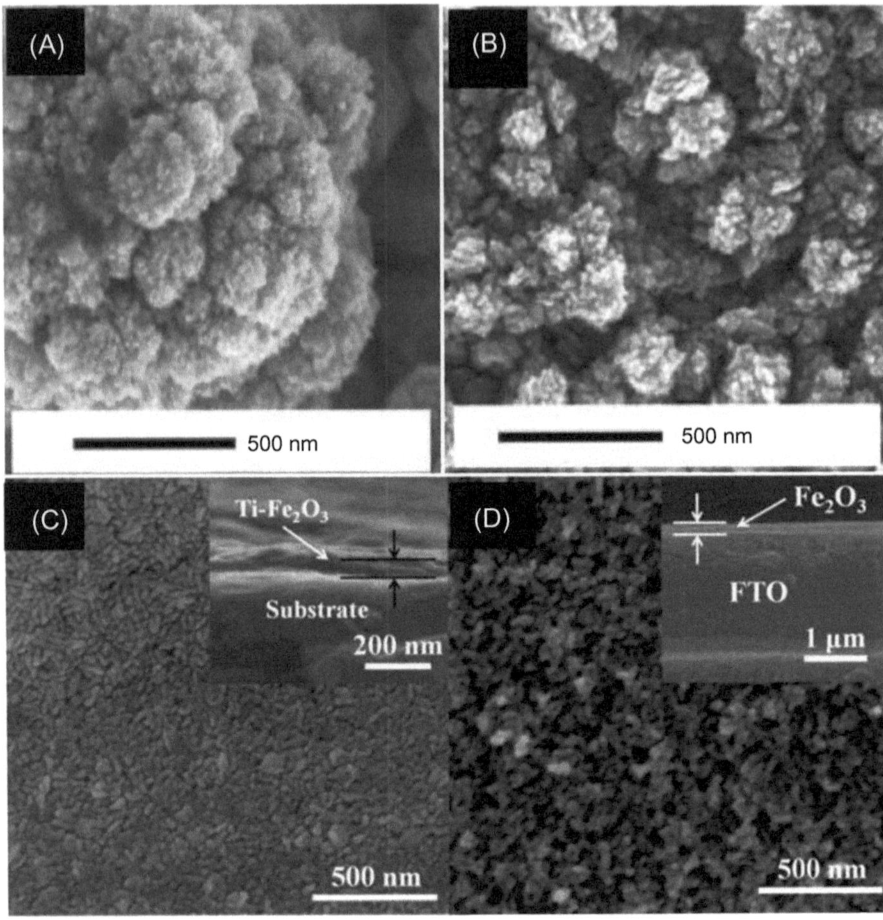

Fig. 8.23 SEM images of α-Fe_2O_3 films grown by APCVD on FTO with (A) no Ti; (B) 0.8% Ti; SEM images of α-Fe_2O_3 films grown by APCVD on FTO; (C) α-Fe_2O_3/FTO; and (D) α-Fe_2O_3/Ti-450. The insets show the cross-section images of the corresponding samples.
A and B reproduced with permission from P. Zhang, A. Kleiman-Shwarsctein, Y.-S. Hu, J. Lefton, S. Sharma, A.J. Forman, E. McFarland, Energy Environ. Sci. 4 (2011) 1020–1028; C and D reproduced with permission from S. Li, P. Zhang, X. Song, L. Gao, Int. J. Hydrog. Energy 39 (2014) 14596–14603.

oxidation, which is attributed to facile electron transfer from TiO_2 [83]. A Ti^{4+}-doped Fe_2O_3 thin film corroded in an acid solution exhibited a 100 mV cathodic shift of the photocurrent onset potential and double the photocurrent of untreated samples [84]. Favoring the forward reaction was shown to account for the cathodic shift of the onset potential.

A simple p-n homojunction of Mg-doped α-Fe_2O_3 thin film formed by ALD shows a nominal 200 mV turn-on voltage shift toward the cathodic direction [85]. A Si/α-Fe_2O_3

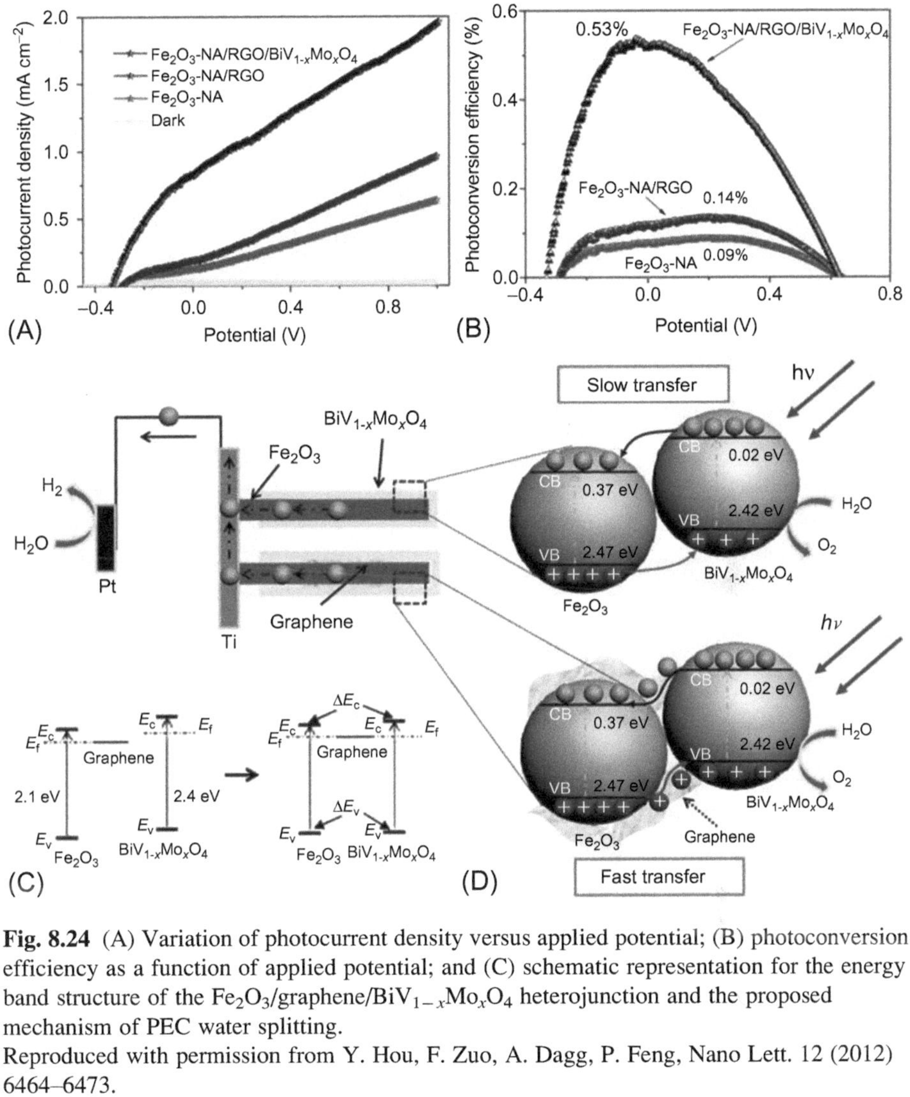

Fig. 8.24 (A) Variation of photocurrent density versus applied potential; (B) photoconversion efficiency as a function of applied potential; and (C) schematic representation for the energy band structure of the Fe_2O_3/graphene/$BiV_{1-x}Mo_xO_4$ heterojunction and the proposed mechanism of PEC water splitting.
Reproduced with permission from Y. Hou, F. Zuo, A. Dagg, P. Feng, Nano Lett. 12 (2012) 6464–6473.

dual absorbing heterojunction has been fabricated by α-Fe_2O_3 deposition (using ALD) on vertically aligned Si nanowires, and exhibits a cathodic shift of about 400 mV [86]. A heterojunction array consisting of Fe_2O_3/graphene/$BiV_{1-x}Mo_xO_4$ core/shell NRs used for the PEC water-splitting reaction delivers a photocurrent density of $1.97\,mA\,cm^{-2}$ at 1.0 V versus Ag/AgCl and a maximum photoconversion efficiency of 0.53% at −0.04 V versus Ag/AgCl under Xe irradiation (Fig. 8.24A and B); this is nearly four times that of bare α-Fe_2O_3 [87]. The improved PEC benefits from enhanced light absorption and separation of the photogenerated carriers at the

Fe_2O_3/graphene/$BiV_{1-x}Mo_xO_4$ interfaces. The possible charge transport mechanism across the heterojunction is schematically shown in Fig. 8.24C.

Co-Pi cocatalyst has been electrodeposited on a high-surface-area α-Fe_2O_3 photoanode, causing a cathodic shift of >350 mV in the onset potential for PEC water oxidation and an enhanced IPCE at 450 nm (Fig. 8.25A–C) [88]. The catalytic effect of $Ni(OH)_2$ on α-Fe_2O_3 for water splitting has been studied also (photocurrent shown in Fig. 8.25D) [89]. A two-step catalytic mechanism was proposed, entailing the fast initial oxidation of Ni^{2+} to Ni^{3+}, followed by the slow oxidation of Ni^{3+} to Ni^{4+} (Fig. 8.25E). However, the catalytic effect of the Ni(II) catalyst is limited by the slow formation of Ni^{4+}. When α-Fe_2O_3 was modified with IrO_2 nanoparticles (c.2 nm diameter), a 200 mV shift was observed in the photocurrent onset and the photocurrent improved from 3.45 to 3.75 mA cm^{-2} [90]. Co_3O_4 particles have been grown on Fe_2O_3 NR arrays using a hydrothermal process in situ [91]. A maximum photocurrent of 1.20 mA cm^{-2} and a shift in the onset potential of c.40 mV were observed at a Co^{2+} concentration of 5% (Fig. 8.25F and G), due to the high surface irregularity, larger Co_3O_4/hematite interfacial area, and smaller Co_3O_4 particle size. When Ni-Bi OECs were photodeposited on a photoanode consisting of Fe_2O_3 NRs, a cathodic shift of >200 mV was observed in the onset potential for water oxidation, and a ninefold increase in photocurrent was observed at 0.86 V versus RHE, compared with the bare Fe_2O_3 photoanode (Fig. 8.25H and I) [92].

The surface of a Fe_2O_3 photoanode can also be decorated with FeOOH by photoelectrodeposition [93]. A cathodic shift of 140 mV in the photocurrent onset potential is observed, and photocurrent is nearly four times that of bare hematite photoanodes. This can be ascribed to the high reaction surface area for the nanostructured morphology and high electrocatalytic activity of FeOOH, which increase the number of photogenerated holes involved in the water oxidation reaction and accelerate the kinetics of water splitting.

8.2.2.1.5 Copper(I) oxide, Cu_2O

Unlike the above-mentioned metal oxides, Cu_2O is a typical p-type semiconductor and is potentially the best candidate for photocathode material, due to its low cost, nontoxicity, narrow bandgap of 2.0 eV, and high theoretical maximum STH of 18% under AM 1.5 light. However, the instability of Cu_2O in water under illumination is a major limiting factor over its application and efficiency. Much work has already been undertaken to take advantage of its merits and avoid the shortcomings. The best results are obtained with film systems under an applied bias, rather than a suspension system.

A highly active oxide photocathode with the composition $Cu_2O/ZnO/Al_2O_3/TiO_2$/Pt for PEC water reduction was constructed of electrodeposited cuprous oxide, nanolayers of Al-doped zinc oxide and titanium oxide deposited by ALD, and electrodeposited Pt nanoparticles (Fig. 8.26A and B) [94]. It displays excellent photoresponse and stability. The maximum photocurrent of −7.6 mA cm^{-2} is obtained at 0 V versus RHE at a mild pH, lasting for >1 h during the water reduction reaction, which is mainly ascribed to the high conductivity of ZnO/Al_2O_3 (ZnO:Al) and

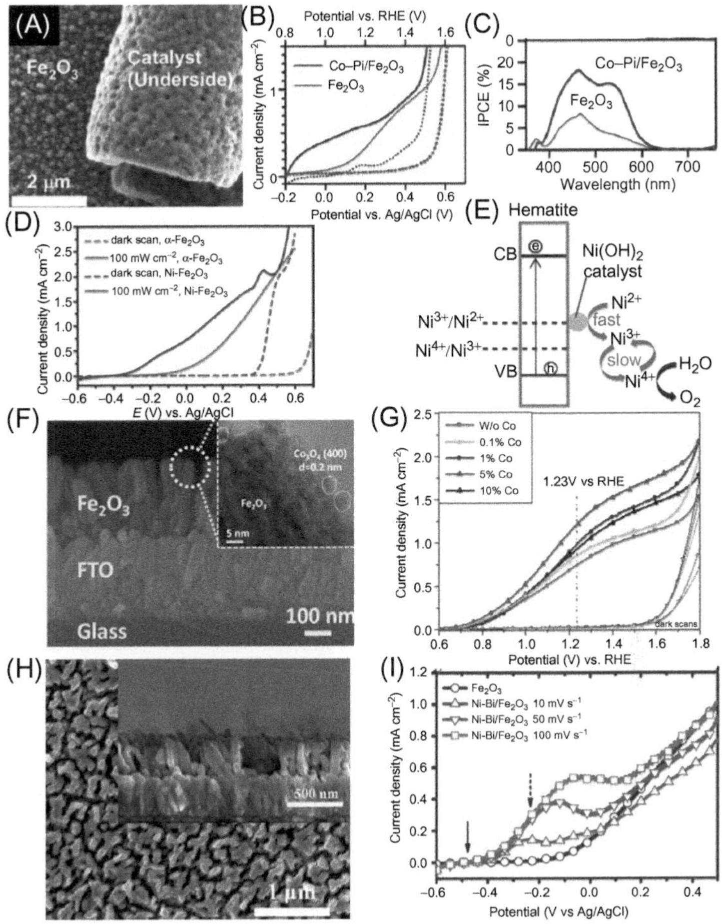

Fig. 8.25 (A) The Co-Pi underside topology conforms to the R-Fe_2O_3 mesostructure; (B) $I-V$ curves of α-Fe_2O_3 (red) and Co-Pi/α-Fe_2O_3 (blue) photoanodes; (C) IPCE spectra for α-Fe_2O_3 and Co-Pi/α-Fe_2O_3; (D) linear sweep voltammograms collected for α-Fe_2O_3 and Ni-α-Fe_2O_3 in the dark *(dashed line)* and under light illumination *(solid line)*; (E) schematic diagram illustrating the proposed catalytic mechanism of $Ni(OH)_2$ on hematite for PEC water oxidation; (F) image of Co_3O_4 decorated Fe_2O_3. Inset is the HRTEM image; (G) $I-V$ curves of hematite photoanode with different amounts of Co^{2+} added in situ; (H) SEM images of hematite NRs; and (I) linear sweep voltammograms of Ni-Bi/Fe_2O_3 photoanodes under illumination measured in 1 M NaOH electrolyte (pH 13.6).

A, B, and C reproduced with permission from D.K. Zhong, J. Sun, H. Inumaru, D.R. Gamelin, J. Am. Chem. Soc. 131 (2009) 6086–6087; D and E reproduced with permission from G. Wang, Y. Ling, X. Lu, T. Zhai, F. Qian, Y. Tong, Y. Li, Nanoscale 5 (2013) 4129–4133; F and G reproduced with permission from L. Xi, P.D. Tran, S.Y. Chiam, P.S. Bassi, W.F. Mak, H.K. Mulmudi, S.K. Batabyal, J. Barber, J.S.C. Loo, L.H. Wong, J. Phys. Chem. C 116 (2012) 13884–13889; H and I reproduced with permission from Y.-R. Hong, Z. Liu, S.F.B.S.A. Al-Bukhari, C.J.J. Lee, D.L. Yung, D. Chi, T.S.A. Hor, Chem. Commun. 47 (2011) 10653–10655.

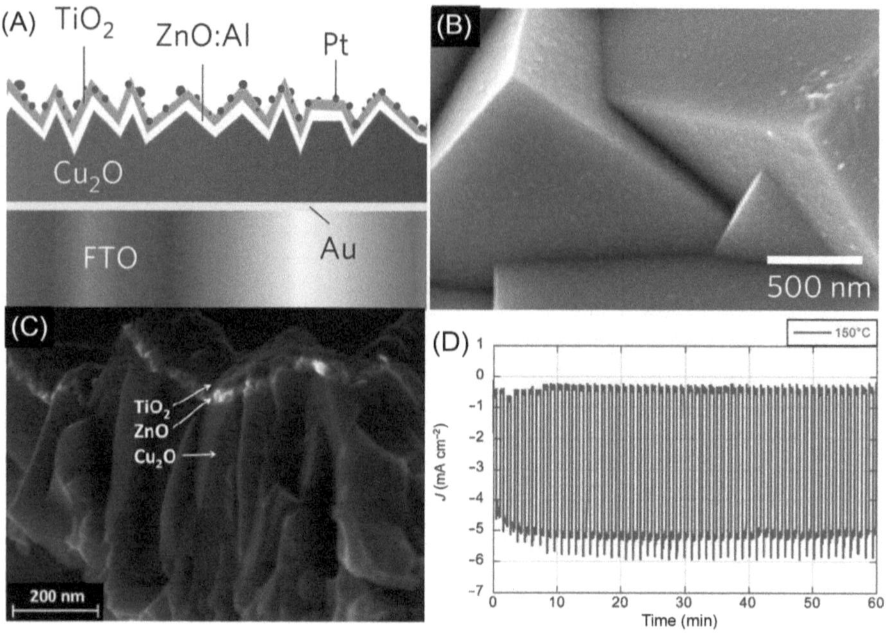

Fig. 8.26 The surface-protected Cu_2O electrode: (A) schematic representation of the electrode structure and (B) SEM image showing a top view of the electrode after atomic layer deposition (ALD) followed by electrodeposition of Pt nanoparticles. (C) Cross-sectional SEM image of a sample with TiO_2 deposited at 150°C. The protective layers of ZnO and TiO_2 conformally coat the sample surface; (D) stability test at 0 V versus RHE in the same electrolyte.
A and B reproduced with permission from A. Paracchino, V. Laporte, K. Sivula, M. Grätzel, E. Thimsen, Nat. Mater. 10 (2011) 456–461; C and D reproduced with permission from A. Paracchino, N. Mathews, T. Hisatomi, M. Stefik, S.D. Tilley, M. Grätzel, Energy Environ. Sci. 5 (2012) 8673–8681.

protection afforded by the TiO_2. The protective overlayers of Al:ZnO and TiO_2 were utilized to fulfill the requirements of favorable band alignment and chemical stability. After a long-time PEC reaction, the 62% stability obtained is due to the deposition of a semicrystalline TiO_2 overlayer (Fig. 8.26C) [95]. >10 h of testing was possible without re-platination (Fig. 8.26D).

In further investigation of junctions, MoS_{2+x} has been loaded on the surface of Cu_2O/*n*-AZO/TiO_2 by PEC deposition as a hydrogen evolution catalyst [96]. The photocurrent obtained with the Cu_2O/*n*-AZO/TiO_2/MoS_{2+x} electrode was $-5.7\,mA\,cm^{-2}$ at 0 V versus RHE (pH 1.0) under simulated AM 1.5 solar illumination. Furthermore, in comparison with the unadorned Cu_2O/*n*-AZO/TiO_2/Pt photocathode, the Cu_2O/n-AZO/TiO_2/MoS_{2+x} electrode exhibited enhanced stability in acidic environments. MoS_2 and Ni-Mo have also been used as hydrogen evolution reaction (HER) catalysts on a protected Cu_2O electrode [97]. The highest photocurrent obtained with a Cu_2O-based photocathode is $-6.3\,mA\,cm^{-2}$ at 0 V versus RHE in 1 M KOH electrolyte,

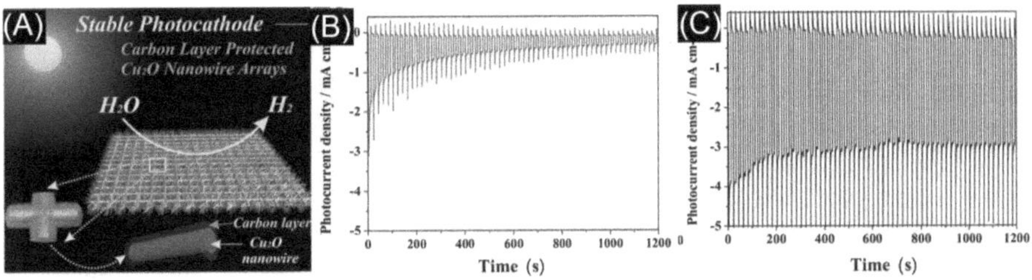

Fig. 8.27 (A) Schematic image of carbon-protected Cu_2O NW arrays on Cu mesh; density decay curves at 0 V versus RHE in chopped light illumination of (B) bare Cu_2O sample and (C) carbon-protected Cu_2O sample.
Reproduced with permission from Z. Zhang, R. Dua, L. Zhang, H. Zhu, H. Zhang, P. Wang, ACS Nano 7 (2013) 1709–1717.

representing a major advance in attempts to replace Pt as a cost-effective HER catalyst. A PEC cell consisting of Cu_2O nanowire photocathode modified with a thin film of NiO_x coupled to a WO_3 nanosheet photoanode has been prepared for overall water splitting [98]. The optimized NiO_x/Cu_2O electrode delivered a photocurrent of $-4.98\,mA\,cm^{-2}$ at -0.33 V, and $-0.56\,mA\,cm^{-2}$ at 0.1 V versus NHE under $26\,mW\,cm^{-2}$ illumination at pH 6; it had a threefold increased photostability.

Copper(II) oxide (CuO) can be used as a protective coating on a thin layer of Cu_2O [99]. A highly stable photocurrent of $-1.54\,mA\,cm^{-2}$ at 0 V versus RHE at a mild pH under illumination of AM 1.5 G is produced, which is more than two times that obtained with a bare Cu_2O electrode ($-0.65\,mA\,cm^{-2}$). The stability of the coated photocathode increased from 30.1% to 74.4%. In the Cu_2O/CuO structure, CuO not only minimizes Cu_2O photocorrosion but also inhibits the recombination of photogenerated electrons and holes at Cu_2O, which accounts for the enhanced stability and PEC activity of the Cu_2O/CuO composite. A Cu_2O/TiO_2 p-n heterojunction photoelectrode has been fabricated by depositing different Cu_2O nanoparticles on TiO_2 NTs via an ultrasonication-assisted sequential CBD. It exhibited a maximum photocurrent of $-4\,mA\,cm^{-2}$ at 1 V versus SCE in Na_2SO_4 electrolyte, which is ascribed to electron injection from Cu_2O into TiO_2 [100]. A protective carbon coating on Cu_2O nanowire arrays synthesized from copper mesh has been formed using glucose as the carbon precursor [101]. It exhibited remarkably improved photostability as well as considerably enhanced photocurrent density (Fig. 8.27). A maximum photocurrent of $-3.95\,mA\,cm^{-2}$ under illumination of AM 1.5 G was detected. The photostability increases more than sixfold, compared with a bare Cu_2O electrode.

8.2.2.2 Nonmetal oxides

Nonmetal oxides for solar water splitting have been described for many decades. Among the nonmetals, Si is a good semiconductor if it is protected properly. A Si electrode protected with TiO_2 deposited by ALD produces an excellent photocurrent of over $10\,mA\,cm^{-2}$ under AM 1.5 light in acidic conditions, and slightly weaker photocurrents in basic and neutral conditions [102]. The nitride Ta_3N_5 also forms a good photoanode. Utilizing a nanostructure design and modification with water oxidation catalysts, its activity and stability have been improved, delivering a photocurrent of $1\,mA\,cm^{-2}$ [103, 104]. Among the metal oxynitrides, TaON is a representative semiconductor with a narrower bandgap than the corresponding oxide. A photocurrent of about $1\,mA\,cm^{-2}$ was achieved [105]. CdS is another typical n-type semiconductor with a bandgap of 2.4 eV and theoretical maximum STH of 9%. It has been shown to be an almost ideal semiconductor for solar water splitting. However, the instability and toxicity of elemental Cd remain significant obstacles to its commercial application [44].

Most nonmetal oxides are photocathode semiconductors. Among the p-type III–V semiconductors, InP and GaP have been the most widely investigated, with narrow bandgaps of 1.34 and 2.2 eV, respectively. However, as they contain the scarce element In and toxic Ga, their large-scale commercial application is limited [31]. P-type CdTe and $CuIn_{1-x}Ga_xSe_2$ (CIGS) have been investigated as photocathodes for many years. The CdTe photocathode displays improved stability and hydrogen production efficiency [106]. CIGS has a bandgap of 1.0–1.68 eV, and an excellent photocurrent of $12\,mA\,cm^{-2}$ at 0 V versus RHE is obtained after modifying it with CdS/Pt [107]. Considering the high cost of In and toxicity of Ga, Cu_2ZnSnS_4 (CZTS) has been investigated intensively as an alternative candidate semiconductor due to its high absorption coefficient of over $10^4\,cm^{-1}$, an optimal bandgap of approximately 1.5 eV, natural abundance, and environmental friendliness. The reported solar-to-H_2 conversion efficiency is 1.2% on a co-sputtered CZTS photocathode with the composition $Pt/TiO_2/CdS/CZTS/Mo$ [108], and the instability issue can be overcome by depositing protective layers using ALD [109, 110].

8.3 Improvements in efficiency for solar water splitting

8.3.1 Modification and nanostructure to improve light harvesting

8.3.1.1 Modification of band structure for enhanced light harvesting by doping with elements

Light harvesting can be enhanced by tuning the band structure of a semiconductor by doping with elements. Taking TiO_2 as a typical example, doping TiO_2 with the elements N, S, C, and transition metals narrows the bandgap and cause a redshift of the light absorption edge, which improves light harvesting of TiO_2 and results in enhanced efficiency for solar water splitting. However, doping with elements may cause crystal defects and form deep-level defects, such as recombination centers of photogenerated electron-hole pairs. Therefore, the increase of IPCE and STH is limited [44]. For some semiconductors with narrow bandgaps, for example, Fe_2O_3 and $BiVO_4$, nonisovalent elements are often doped, which can increase the carrier concentration and the minority carrier diffusion length [111]. Decreasing the effective mass of minority carriers helps to increase the diffusion length of minority carriers. The photocurrent can be improved.

8.3.1.2 Nanostructure for enhanced light harvesting

Many nanostructures of photoelectrodes have been reported recently for water splitting. They consist mostly of planar structures [94], nanotubes [112], NRs [87, 113], nanowires [86], and mesoporous structures [114]. However, an integrated 3D structure called a "hierarchical structure" has attracted serious attention; this consists of a higher dimension configuration assembled from lower dimensional nanomaterials including nanoparticles (0D), nanowires/rods/tubes [one-dimensional (1D)], and nanosheets [two-dimensional (2D)] to invoke a 3D image [115]. These special structures provide a longer effective path for the photons to be absorbed in the photoelectrodes by stronger scattering and trapping enhancement, which can result in better light harvesting and further improve the efficiency for solar water splitting.

8.3.1.3 Surface modification for enhanced light harvesting

Depositing plasmonic metallic nanoparticles on the surface of a semiconductor is also an effective surface modification for improving light harvesting. Plasmonic-enhanced water splitting was first reported in 2004. TiO_2 thin films were immersed into electrolyte to deposit gold and silver nanoparticles, and the plasmonic effect is attributed to the transfer of photo-induced electrons from the CB of the metal to the CB of the semiconductor [116]. Gold and silver particles have also been deposited on the surfaces of Fe_2O_3 and N-doped TiO_2 electrodes [117, 118], resulting in PEC enhancement that is attributed to plasmon resonance energy transfer from the metal nanoparticles to the semiconductor, which is due to the spectral overlap between the plasmonic metal and the semiconductor [119]. Silver and gold are considered the best plasmonic metals so far, and have been deposited on the surfaces of CeO_2, CdS, and WO_3 [120–122].

8.3.2 Carrier separation and transport improvement

8.3.2.1 p-n junction

Forming a p-n junction is a highly effective approach to improve carrier separation and transport in semiconductors. A p-n junction usually consists of two types of semiconductor, p-type and n-type. When a p-n junction is established, a local electric field will be formed across the space charge region. Photogenerated electrons and holes migrate to the p-type semiconductor side and to the n-type semiconductor side, respectively, under the electric force. The charge carriers diffuse to the space charge region. Photogenerated electrons and holes are separated effectively, resulting in an enhanced efficiency for solar water splitting [123].

8.3.2.2 Heterojunction

The recombination of photo-induced electron-hole pairs usually competes with the separation of photogenerated carriers, which is unfavorable for increasing the efficiency for water splitting. For example, Si nanoarrays serve as effective charge collection frameworks, and therefore some types of semiconductor (TiO_2,α-Fe_2O_3) have been coated on the surface of Si nanoarrays [27, 90, 102]. These types of heterojunction structure show improved efficiency for water splitting due to the rapid transfer of photogenerated electrons through the Si arrays to the coated semiconductors.

8.3.2.3 Cocatalysts

In the solar water-splitting reaction, a low overpotential is beneficial for overcoming the kinetic limitation and driving the desired chemical reaction. A suitable cocatalyst should have a low overpotential for water reduction/oxidation and long-term stability. The most common noble metals (Pt, Ru, and Pd) have been loaded on the surface of semiconductors to improve hydrogen evolution. MoS_x, W-Ni-P, W-Cu, and Ni-Fe can serve as water reduction cocatalysts [111]. In addition, IrO_2, Co_3O_4, and amorphous cobalt-phosphate (Co-Pi), are loaded on the semiconductor surface to promote water oxidation and hence increase the reaction kinetics and efficiency for water splitting.

8.3.2.4 Nanostructure control

The main nanostructures can be classified into four groups: 0D nanocrystals, 1D NRs/tubes/wires, 2D nanosheets and films, and 3D hierarchical nanostructures, possessing different surface areas, particle sizes, crystallinity, and film thicknesses. The planar (2D) structure with a limited surface area and a long charge diffusion length is often applied in photocathodes. A relatively long distance on the micrometer scale is required to accomplish the separate migration of photogenerated electrons and holes to the surface and substrate. In the 1D structure, the carriers migrate over a short distance along the axes of the NRs/tubes/wires. 2D structures with large surface areas can provide physical support to obtain composites with other catalysts. The 3D structure provides a large surface area, short transport distance, and a curved transport path for carriers [124].

8.3.2.5 Graphene-based composite

The water-splitting efficiency can also be improved by forming a graphene/semiconductor composite system. In such a system, electrons in the CB of the semiconductor will be injected into graphene oxide while leaving holes in the VB of the semiconductor when the semiconductor absorbs light, effectively accomplishing the separation of photogenerated electron-hole pairs. Graphene can be a support substrate for catalysts due to its 2D platform structure. Graphene has been investigated so far as an effective electron collector and/or transporter for separating photogenerated electron-hole pairs by combining with $BiVO_4$, TiO_2, CdS, and ZnO for water splitting, achieving enhanced efficiency with small amounts of graphene [125].

8.3.3 Charge utilization improvement

8.3.3.1 Increase of the photocatalytic active surface

The solar water-splitting reaction takes place on the surface of the semiconductor, and ample surface area for reaction sites is very important. There are two effective approaches to increase the surface area. One is designing a nanostructure. The other is modifying the surface of the semiconductor with nanoparticles. Both will improve the efficiency for solar water splitting. The resulting semiconductor prepared using these approaches will possess qualities such as:

(1) the surface area is increased using different 1D, 2D, and 3D nanostructures;
(2) the surface area can be increased significantly due to the large specific surface area of nanoparticles on the surface of the semiconductor;
(3) the nanoparticles on the semiconductor surface have a high surface energy which favors the absorption of catalytic species; and
(4) a surface heterostructure can be formed between the nanoparticles and semiconductor, enhancing the separation of photogenerated electron-hole pairs [123].

8.3.3.2 Increase of active facet exposure

It is well known that different facets of crystals possess different chemical activities, which can determine the corresponding properties. The ranking of the surface energy of TiO_2 is reported as γ (001) (0.90 J m^{-2}) > γ (100) (0.53 J m^{-2}) > γ (101) (0.44 J m^{-2}) [123, 126], which means that the most energetic (001) facet possesses the best photocatalytic activity. However, it is difficult to access exposed (001) facets in TiO_2, and many approaches have been employed, such as controlling the growth process by adding surfactants, and tuning the synthesis procedure and the crystallization process.

Semiconductor crystal facet engineering has been reported recently, establishing that photoexcited electrons and holes may be driven to different crystal facets. For example, the rutile {011} and anatase {001} faces of TiO_2 provide reaction sites for oxidation, while the rutile {110} and anatase {101} facets offer sites for reduction [127]. A similar case has been reported on the {010} and {110} crystal facets of monoclinic $BiVO_4$ (mentioned in Section 8.2.2.1). Spatial separation of photogenerated electrons and holes takes place separately on the {010} and {110} facets of $BiVO_4$ under photo-irradiation. The reduction and oxidation cocatalysts are selectively deposited on the {010} and {110} facets to achieve reduction and oxidation reactions, respectively, resulting in highly improved activity in the photocatalytic and photoelectrocatalytic water oxidation reactions [70, 71]. Therefore, the efficiency of the solar water-splitting process can be improved by exposing the specific active surfaces.

8.4 Outlook

Solar water splitting has been investigated for several decades and great achievements and good efficiencies have been obtained. The current state of research on semiconductors for solar water splitting, including factors that influence the efficiency have been summarized and generally classified into three types: light-harvesting efficiency, carrier separation and transport efficiency, and charge utilization, which can all be improved by various methods. For instance, light-harvesting efficiency can be increased by doping with elements, designing nanostructures, modifying the surface with plasmonic metallic nanoparticles; carrier separation and transportation efficiency can be improved by forming p-n junctions and heterojunctions, loading cocatalysts, controlling the nanostructure, and combining with graphene; charge utilization efficiency can be improved by increasing the photocatalytic active surface and active facet exposure.

However, the state of the art still is a distance away from achieving practical application, and several breakthroughs are needed:

1. Developing novel single semiconductors with better light harvesting and carrier transport and other desired properties. Future semiconductors for solar water splitting must possess an optimal bandgap of c.2.0 eV, a proper band edge position, high absorption coefficient, low cost, good stability, environment friendliness, and long minority carrier diffusion length. Materials genome engineering is a new methodology used to discover novel PEC materials. A high-throughput procedure is used to synthesize a library of materials with certain predicted compositions of novel semiconductor materials. The library of materials can be screened by high-throughput characterization techniques to obtain the best compound.
2. Preparing semiconductors on a large scale. To date, most photocatalysts and photoelectrodes have been fabricated on an experimental scale. Recently, a printed photocatalyst sheet with a STH energy conversion efficiency exceeding 1% was prepared by screen printing [128]. In the future, scalable semiconductors and materials for solar water splitting will be fabricated using commercial inkjet printers and screen printing.
3. Developing photocathode-photoanode tandem devices. The efficiency of photocathode-photoanode tandem devices is lower than that of hybrid tandem cells and photovoltaic-based devices, because of the instability of the photocathodes. However, protective layers deposited by ALD have been shown to overcome the poor stability of photocathodes, and impressive advances have been achieved in recent years. Therefore, the use of photocathode-photoanode tandem devices may increase again in the near future.
4. Developing plasmonic nanostructures. The metal/oxide composite structure utilizing the metal effect should also be addressed. This structure is different from the conventional optical absorption through semiconductors. The Au/TiO_2 composite has shown reasonable performance in photocatalysis [116]. The efficiency of the metal plasmonic effect is still not comparable with traditional semiconductor materials, but with combination of the metal surface plasmon resonance effect the semiconductor might create a novel pathway for finding an ideal photocatalyst.

In summary, the materials genome process can be an efficient way for the discovery of novel materials. By combination of surface passivation layers, the performance and chemical stability of a single PEC material can be improved significantly. The combination of two or more semiconductors through a *Z*-scheme process will utilize the merits of different photocatalysts and improve the performance. The metal surface plasmon resonance effect is a new beneficial factor to be considered to further increase the PEC efficiency of semiconductors by the direct provision of photogenerated charges; local thermal effects probably also play a role because a higher temperature also promotes the PEC reaction on the surface. Therefore, further innovation and development are required to achieve practical and commercial application for solar water splitting.

References

[1] British Petroleum (BP), BP Statistical Review of World Energy, http://www.bp.com/content/dam/bp/pdf/energy-economics/statistical-review-2015/bp-statistical-review-of-world-energy-2015-full-report.pdf, 2015. Accessed 6 May 2017.
[2] T. Hisatomi, J. Kubota, K. Domen, Chem. Soc. Rev. 43 (2014) 7520–7535.
[3] A. Fujishima, K. Honda, Nature 238 (5358) (1972) 37–38.
[4] A.L. Linsebigler, G. Lu, J.T. Yates, Chem. Rev. 95 (1995) 735–758.
[5] J.A. Glasscock, Nanostructured Materials for Photoelectrochemical Hydrogen Production Using Sunlight (Ph.D. Thesis), University of New South Wales, 2008.
[6] M. Grätzel, Nature 414 (6861) (2001) 338–344.
[7] J. Li, N. Wu, Cat. Sci. Technol. 4 (2014) 4440.
[8] S. Liu, Photocatalytic Hydrogen Production with Iron Oxide Under Solar Irradiation (M.Sc. Thesis), Queensland University of Technology, 2010.
[9] A.J. Nozik, Annu. Rev. Phys. Chem. 29 (1978) 189–222.
[10] R. Asahi, T. Morikawa, T. Ohwaki, K. Aoki, Y. Taga, Science 293 (5528) (2001) 269–271.
[11] G. Liu, L.-C. Yin, J. Wang, P. Niu, C. Zhen, Y. Xie, H.-M. Cheng, Energy Environ. Sci. 5 (2012) 9603–9610.
[12] X. Chen, C. Li, M. Grätzel, R. Kostecki, S.S. Mao, Chem. Soc. Rev. 41 (2012) 7909–7937.
[13] W. Choi, A. Termin, M.R. Hoffmann, Angew. Chem. Int. Ed. Eng. 33 (1994) 1091–1092.
[14] Z. Luo, Q.-H. Gao, J. Photochem. Photobiol. A Chem. 63 (1992) 367–375.
[15] Y. Wang, H. Cheng, Y. Hao, J. Ma, W. Li, S. Cai, J. Mater. Sci. 34 (1999) 3721–3729.
[16] H. Luo, T. Takata, Y. Lee, J. Zhao, K. Domen, Yushan, Chem. Mater. 16 (2004) 846–849.
[17] S.U.M. Khan, M. Al-Shahry, W.B. Ingler, Science 297 (5590) (2002) 2243–2245.
[18] Z. Zhang, X. Yang, M.N. Hedhili, E. Ahmed, L. Shi, P. Wang, ACS Appl. Mater. Interfaces 6 (2014) 691–696.
[19] F. Zuo, L. Wang, T. Wu, Z. Zhang, D. Borchardt, P. Feng, J. Am. Chem. Soc. 132 (2010) 11856–11857.
[20] S. Hoang, S.P. Berglund, N.T. Hahn, A.J. Bard, C.B. Mullins, J. Am. Chem. Soc. 134 (2012) 3659–3662.
[21] X. Chen, L. Liu, P.Y. Yu, S.S. Mao, Science 331 (6018) (2011) 746–750.
[22] G.K. Mor, K. Shankar, M. Paulose, O.K. Varghese, C.A. Grimes, Nano Lett. 5 (2005) 191–195.
[23] M. Liu, N. de Leon Snapp, H. Park, Chem. Sci. 2 (2011) 80–87.
[24] M. Paulose, K. Shankar, S. Yoriya, H.E. Prakasam, O.K. Varghese, G.K. Mor, T. A. Latempa, A. Fitzgerald, C.A. Grimes, J. Phys. Chem. B 110 (2006) 16179–16184.
[25] P. Hartmann, D.-K. Lee, B.M. Smarsly, J. Janek, ACS Nano 4 (2010) 3147–3154.
[26] Z. Li, C. Yao, Y. Yu, Z. Cai, X.D. Wang, Adv. Mater. 26 (2014) 2262–2267.
[27] Z. Zhang, P. Wang, Energy Environ. Sci. 5 (2012) 6506–6512.
[28] I.S. Cho, Z. Chen, A.J. Forman, D.R. Kim, P.M. Rao, T.F. Jaramillo, X. Zheng, Nano Lett. 11 (2011) 4978–4984.
[29] G. Liu, C. Sun, H.G. Yang, S.C. Smith, L. Wang, G.Q. Lu, H.-M. Cheng, Chem. Commun. 46 (2010) 755–757.
[30] C. Liu, J. Tang, H.M. Chen, B. Liu, P. Yang, Nano Lett. 13 (2013) 2989–2992.
[31] L. Rovelli, K.R. Thampi, Solar water splitting using semiconductor systems, in: J. Sá (Ed.), Fuel Production with Heterogeneous Catalysis, Taylor & Francis CRC Press, Boca Raton, 2014, , pp. 2–62.

[32] M.G. Walter, E.L. Warren, J.R. McKone, S.W. Boettcher, Q. Mi, E.A. Santori, N. S. Lewis, Chem. Rev. 110 (2010) 6446–6473.
[33] H. Wang, T. Lindgren, J. He, A. Hagfeldt, S.E. Lindquist, J. Phys. Chem. B 104 (2000) 5686–5696.
[34] C. Santato, M. Ulmann, J. Augustynski, J. Phys. Chem. B 105 (2001) 936–940.
[35] C. Santato, M. Odziemkowski, M. Ulmann, J. Augustynski, J. Am. Chem. Soc. 123 (2001) 10639–10649.
[36] S.T. Berger, A. Ghicov, P. Schmuki, Appl. Phys. Lett. 88 (2006) 203119.
[37] B. Yang, P.R.F. Barnes, W. Bertram, V. Luca, J. Mater. Chem. 17 (2007) 2722–2729.
[38] B. Yang, Y. Zhang, E. Drabarek, P.R.F. Barnes, V. Luca, Chem. Mater. 19 (2007) 5664–5672.
[39] B. Marsen, E.L. Miller, D. Paluselli, R.E. Rocheleau, Int. J. Hydrog. Energy 32 (2007) 3110–3115.
[40] V. Cristino, S. Caramori, R. Argazzi, L. Meda, G.L. Marra, C.A. Bignozzi, Langmuir 27 (2011) 7276–7284.
[41] J. Su, X. Feng, J.D. Sloppy, L. Guo, C.A. Grimes, Nano Lett. 11 (2011) 203–208.
[42] J.K. Kim, K. Shin, S.M. Cho, T.-W. Lee, J.H. Park, Energy Environ. Sci. 4 (2011) 1465–1470.
[43] X. Zhang, X. Lu, Y. Shen, J. Han, L. Yuan, L. Gong, Z. Xu, X. Bai, M. Wei, Y. Tong, Y. Gao, J. Chen, J. Zhou, Z.L. Wang, Chem. Commun. 47 (2011) 5804–5806.
[44] S.J.A. Moniz, S.A. Shevlin, D.J. Martin, Z.-X. Guo, J. Tang, Energy Environ. Sci. 8 (2015) 731–759.
[45] J. Lin, P. Hu, Y. Zhang, M. Fan, Z. He, C.K. Ngaw, J.S.C. Loo, D. Liao, T.T.Y. Tan, RSC Adv. 3 (2013) 9330–9336.
[46] W. Kim, T. Tachikawa, D. Monllor-Satoca, H. Kim, T. Majima, W. Choi, Energy Environ. Sci. 6 (2013) 3732–3739.
[47] S.S.K. Ma, K. Maeda, R. Abe, K. Domen, Energy Environ. Sci. 5 (2012) 8390–8397.
[48] N. Naseri, M. Qorbani, H. Kim, W. Choi, A.Z. Moshfegh, J. Phys. Chem. C 119 (2015) 1271–1279.
[49] Y. Hou, F. Zuo, A.P. Dagg, J. Liu, P. Feng, Adv. Mater. 26 (2014) 5043–5049.
[50] Z.G. Zhao, Z.F. Liu, M. Miyauchi, Chem. Commun. 46 (2010) 3321–3323.
[51] G. Xi, J. Ye, Q. Ma, N. Su, H. Bai, C. Wang, J. Am. Chem. Soc. 134 (2012) 6508–6511.
[52] C. Feng, S. Wang, B. Geng, Nanoscale 3 (2011) 3695–3699.
[53] M.D. Bhatt, J.S. Lee, J. Mater. Chem. A 3 (2015) 10632–10659.
[54] A. Kudo, K. Ueda, H. Kato, I. Mikami, Catal. Lett. 53 (1998) 229–230.
[55] Y. Park, K.J. McDonald, K.-S. Choi, Chem. Soc. Rev. 42 (2013) 2321–2337.
[56] K. Sayama, A. Nomura, Z. Zou, R. Abe, Y. Abe, H. Arakawa, Chem. Commun. (2003) 2908–2909.
[57] W. Luo, Z. Yang, Z. Li, J. Zhang, J. Liu, Z. Zhao, Z. Wang, S. Yan, T. Yu, Z. Zou, Energy Environ. Sci. 4 (2011) 4046–4051.
[58] D.K. Zhong, S. Choi, D.R. Gamelin, J. Am. Chem. Soc. 133 (2011) 18370–18377.
[59] S.K. Pilli, T.E. Furtak, L.D. Brown, T.G. Deutsch, J.A. Turner, A.M. Herring, Energy Environ. Sci. 4 (2011) 5028–5034.
[60] F.F. Abdi, R. van de Krol, J. Phys. Chem. C 116 (2012) 9398–9404.
[61] W. Luo, Z. Wang, L. Wan, Z. Li, T. Yu, Z. Zou, J. Phys. D Appl. Phys. 43 (2010) 405402.
[62] H. Ye, J. Lee, J.S. Jang, A.J. Bard, J. Phys. Chem. C 114 (2010) 13322–13328.
[63] H. Ye, H.S. Park, A.J. Bard, J. Phys. Chem. C 115 (2011) 12464–12470.
[64] H.S. Park, K.E. Kweon, H. Ye, E. Paek, G.S. Hwang, A.J. Bard, J. Phys. Chem. C 115 (2011) 17870–17879.

[65] K.P. Parmar, H.J. Kang, A. Bist, P. Dua, J.S. Jang, J.S. Lee, ChemSusChem 5 (2012) 1926–1934.
[66] G. Wang, Y. Ling, X. Lu, F. Qian, Y. Tong, J.Z. Zhang, V. Lordi, C.R. Leao, Y. Li, J. Phys. Chem. C 117 (2013) 10957–10964.
[67] W.J. Jo, J.-W. Jang, K. Kong, H.J. Kang, J.Y. Kim, H. Jun, K.P.S. Parmar, J.S. Lee, Angew. Chem. Int. Ed. 51 (2012) 3147–3151.
[68] S. Obregón, G. Colón, Appl. Catal. B Environ. 152–153 (2014) 328–334.
[69] S. Huang, N. Zhu, Z. Lou, L. Gu, C. Miao, H. Yuan, A. Shan, Nanoscale 6 (2014) 1362–1368.
[70] R. Li, F. Zhang, D. Wang, J. Yang, M. Li, J. Zhu, X. Zhou, H. Han, C. Li, Nat. Commun. 4 (2013) 1432.
[71] R. Li, H. Han, F. Zhang, D. Wang, C. Li, Energy Environ. Sci. 7 (2014) 1369–1376.
[72] J. Su, L. Guo, N. Bao, C.A. Grimes, Nano Lett. 11 (2011) 1928–1933.
[73] P.M. Rao, L. Cai, C. Liu, I.S. Cho, C.H. Lee, J.M. Weisse, P. Yang, X. Zheng, Nano Lett. 14 (2014) 1099–1105.
[74] Y. Pihosh, I. Turkevych, K. Mawatari, T. Asai, T. Hisatomi, J. Uemura, M. Tosa, K. Shimamura, J. Kubota, K. Domen, T. Kitamori, Small 10 (2014) 3692–3699.
[75] I. Cesar, A. Kay, J.A. Gonzalez Martinez, M. Grätzel, J. Am. Chem. Soc. 128 (2006) 4582–4583.
[76] J.A. Glasscock, P.R.F. Barnes, I.C. Plumb, N. Savvides, J. Phys. Chem. C 111 (2007) 16477–16488.
[77] P. Zhang, A. Kleiman-Shwarsctein, Y.-S. Hu, J. Lefton, S. Sharma, A.J. Forman, E. McFarland, Energy Environ. Sci. 4 (2011) 1020–1028.
[78] S. Li, P. Zhang, X. Song, L. Gao, Int. J. Hydrog. Energy 39 (2014) 14596–14603.
[79] K. Sivula, F.L. Formal, M. Grätzel, Chem. Mater. 21 (2009) 2862–2867.
[80] W. Luo, T. Yu, Y. Wang, Z. Li, J. Ye, Z. Zou, J. Phys. D Appl. Phys. 40 (2007) 1091.
[81] D.K. Bora, A. Braun, R. Erni, U. Müller, M. Döbeli, E.C. Constable, Phys. Chem. Chem. Phys. 15 (2013) 12648–12659.
[82] J. Li, F. Meng, S. Suri, W. Ding, F. Huang, N. Wu, Chem. Commun. (2012) 8213–8215.
[83] S.J.A. Moniz, S.A. Shevlin, X. An, Z.-X. Guo, J. Tang, Chem. Eur. J. 20 (2014) 15571–15579.
[84] D. Cao, W. Luo, J. Feng, X. Zhao, Z. Li, Z. Zou, Energy Environ. Sci. 7 (2014) 752–759.
[85] Y. Lin, Y. Xu, M.T. Mayer, Z.I. Simpson, G. McMahon, S. Zhou, D. Wang, J. Am. Chem. Soc. 134 (2012) 5508–5511.
[86] M.T. Mayer, C. Du, D. Wang, J. Am. Chem. Soc. 134 (2012) 12406–12409.
[87] Y. Hou, F. Zuo, A. Dagg, P. Feng, Nano Lett. 12 (2012) 6464–6473.
[88] D.K. Zhong, J. Sun, H. Inumaru, D.R. Gamelin, J. Am. Chem. Soc. 131 (2009) 6086–6087.
[89] G. Wang, Y. Ling, X. Lu, T. Zhai, F. Qian, Y. Tong, Y. Li, Nanoscale 5 (2013) 4129–4133.
[90] S.D. Tilley, M. Cornuz, K. Sivula, M. Grätzel, Angew. Chem. 122 (2010) 6549–6552.
[91] L. Xi, P.D. Tran, S.Y. Chiam, P.S. Bassi, W.F. Mak, H.K. Mulmudi, S.K. Batabyal, J. Barber, J.S.C. Loo, L.H. Wong, J. Phys. Chem. C 116 (2012) 13884–13889.
[92] Y.-R. Hong, Z. Liu, S.F.B.S.A. Al-Bukhari, C.J.J. Lee, D.L. Yung, D. Chi, T.S.A. Hor, Chem. Commun. 47 (2011) 10653–10655.
[93] Q. Yu, X. Meng, T. Wang, P. Li, J. Ye, Adv. Funct. Mater. 25 (2015) 2686–2692.
[94] A. Paracchino, V. Laporte, K. Sivula, M. Grätzel, E. Thimsen, Nat. Mater. 10 (2011) 456–461.

[95] A. Paracchino, N. Mathews, T. Hisatomi, M. Stefik, S.D. Tilley, M. Grätzel, Energy Environ. Sci. 5 (2012) 8673–8681.
[96] C.G. Morales-Guio, S.D. Tilley, H. Vrubel, M. Grätzel, X. Hu, Nat. Commun. 5 (2014) 3059.
[97] C.G. Morales-Guio, L. Liardet, M.T. Mayer, S.D. Tilley, M. Grätzel, X. Hu, Angew. Chem. Int. Ed. Eng. 54 (2015) 664–667.
[98] C.-Y. Lin, Y.-H. Lai, D. Mersch, E. Reisner, Chem. Sci. 3 (2012) 3482–3487.
[99] Z. Zhang, P. Wang, J. Mater. Chem. 22 (2012) 2456–2464.
[100] M. Wang, L. Sun, Z. Lin, J. Cai, K. Xie, C. Lin, Energy Environ. Sci. 6 (2013) 1211–1220.
[101] Z. Zhang, R. Dua, L. Zhang, H. Zhu, H. Zhang, P. Wang, ACS Nano 7 (2013) 1709–1717.
[102] Y.W. Chen, J.D. Prange, S. Dühnen, Y. Park, M. Gunji, C.E.D. Chidsey, P.C. McIntyre, Nat. Mater. 10 (2011) 539–544.
[103] Y. Cong, H.S. Park, S. Wang, H.X. Dang, F.-R.F. Fan, C.B. Mullins, A.J. Bard, J. Phys. Chem. C 116 (2012) 14541–14550.
[104] H.X. Dang, N.T. Hahn, H.S. Park, A.J. Bard, C.B. Mullins, J. Phys. Chem. C 116 (2012) 19225–19232.
[105] M. Higashi, K. Domen, R. Abe, J. Am. Chem. Soc. 134 (2012) 6968–6971.
[106] K. Ohashi, K. Uosaki, J.O.M. Bockris, Int. J. Energy Res. 1 (1977) 25–30.
[107] Q. Huang, Z. Ye, X. Xiao, J. Mater. Chem. A 3 (2015) 15824–15837.
[108] D. Yokoyama, T. Minegishi, K. Jimbo, T. Hisatomi, G. Ma, M. Katayama, J. Kubota, H. Katagiri, K. Domen, Appl. Phys. Express 3 (2010) 101202.
[109] L. Rovelli, S.D. Tilley, K. Sivula, ACS Appl. Mater. Interfaces 5 (2013) 8018–8024.
[110] J. Wang, P. Zhang, X. Song, L. Gao, RSC Adv. 4 (2014) 21318–21324.
[111] Z. Li, W. Luo, M. Zhang, J. Feng, Z. Zou, Energy Environ. Sci. 6 (2013) 347–370.
[112] O.K. Varghese, M. Paulose, C.A. Grimes, Nat. Nanotechnol. 4 (2009) 592–597.
[113] Y. Hou, F. Zuo, A. Dagg, P. Feng, Angew. Chem. 125 (2013) 1286–1290.
[114] W. Hamd, S. Cobo, J. Fize, G. Baldinozzi, W. Schwartz, M. Reymermier, A. Pereira, M. Fontecave, V. Artero, C. Laberty-Robert, C. Sanchez, Phys. Chem. Chem. Phys. 14 (2012) 13224–13232.
[115] D. Lee, C.P. Grigoropoulos, Hierarchical nanostructures for photo-electro-chemical cells, in: S.H. Ko, C.P. Grigoropoulos (Eds.), Hierarchical Nanostructures for Energy Devices, 2015, RSC Nanostructures & Nanotechnology, Vol. 60, RSC, Cambridge, UK, 2015, pp. 174–203.
[116] Y. Tian, T. Tatsuma, Chem. Commun. 16 (2004) 1810–1811.
[117] E. Thimsen, F. Le Formal, M. Grätzel, S.C. Warren, Nano Lett. 11 (2011) 35–43.
[118] D.B. Ingram, P. Christopher, J.L. Bauer, S. Linic, ACS Catal. 1 (2011) 1441–1447.
[119] S.C. Warren, E. Thimsen, Energy Environ. Sci. 5 (2012) 5133–5146.
[120] A. Primo, T. Marino, A. Corma, R. Molinari, H. García, J. Am. Chem. Soc. 133 (2011) 6930–6933.
[121] T. Torimoto, H. Horibe, T. Kameyama, K. Okazaki, S. Ikeda, M. Matsumura, A. Ishikawa, H. Ishihara, J. Phys. Chem. Lett. 2 (2011) 2057–2062.
[122] R. Solarska, A. Królikowska, J. Augustyński, Angew. Chem. Int. Ed. Eng. 49 (2010) 7980–7983.
[123] J. Tian, Z. Zhao, A. Kumar, R.I. Boughton, H. Liu, Chem. Soc. Rev. 43 (2014) 6920–6937.
[124] P. Zhang, L. Gao, X. Song, J. Sun, Adv. Mater. 27 (2015) 562–568.
[125] H.M. Chen, C.K. Chen, R.S. Liu, L. Zhang, J. Zhang, D.P. Wilkinson, Chem. Soc. Rev. 41 (2012) 5654–5671.

[126] E. Grabowska, M. Diak, M. Marchelek, A. Zaleska, Appl. Catal. B Environ. 156–157 (2014) 213–235.
[127] T. Ohno, K. Sarukawa, M. Matsumura, New J. Chem. 26 (2002) 1167–1170.
[128] Q. Wang, T. Hisatomi, Q. Jia, H. Tokudome, M. Zhong, C. Wang, Z. Pan, T. Takata, M. Nakabayashi, N. Shibata, Y. Li, I.D. Sharp, A. Kudo, T. Yamada, K. Domen, Nat. Mater. 15 (2016) 611–615.

Metal oxides for hydrogen storage

9

Bingwei Chen, Zaichun Liu*, Chunyang Li*, Yusong Zhu*, Lijun Fu*, Yuping Wu*,‡, Teunis van Ree†*
*State Key Laboratory of Materials-Oriented Chemical Engineering, School of Energy Science and Engineering, Institute for Electrochemical Energy Storage, Nanjing Tech University, Nanjing, People's Republic of China, †Department of Chemistry, University of Venda, Thohoyandou, South Africa, ‡College of Science, Hunan Agricultural University, Changsha, China

9.1 Introduction

Energy is a vital resource in everyone's daily experience. The rapid population growth and increasing individual wealth are driving the increasing demand for energy. By 2035 the world's population will reach 8.7 billion, with 1.6 billion more people than today demanding energy [1]. The increasing energy demand and the decline of existing fossil fuel supplies are on conflicting routes, and the consumption of conventional fossil fuels leads to concern about greenhouse gas emissions that contribute to climate change. In addition, there is the negative impact of combustion-associated pollutants on human health. Therefore, clean and sustainable alternative energy sources must be found to replace the existing nonrenewable fossil fuels. The global generation of renewable power has grown rapidly, from 895 GW at the beginning of 2004 to 1560 GW at the end of 2013 [1]; however, many renewable power projects suffer from the disadvantage of the distance from the demand site, so that transmission of renewable energy presents a difficulty. Furthermore, renewable power sources are inherently intermittent and fluctuating, so that power availability is unpredictable, with an unavoidable disparity between renewable power generation and load demand. Therefore, the growth in decentralized power plants consisting of PV arrays and wind farms will influence significantly the stability of the distribution grid. Load shedding is often an undesirable side-effect preventing further growth of renewable power supply. To encourage growth in renewables and avoid many of the associated challenges, energy storage on a large scale must be considered.

Many energy storage mechanisms have been proposed and studied, such as batteries, capacitors, compressed air, flywheels, pumped hydro, and others. Hydrogen is another likely candidate for energy storage and as a clean-burning fuel. Hydrogen is an important industrial raw material used in ammonia production, petroleum refining, metal beneficiation, and in the food industry, but hydrogen can also provide energy to all sectors of the economy, including transportation and construction. Potentially, the use of hydrogen will lead to a low carbon industry, sometimes called the hydrogen economy [2].

Hydrogen storage forms an integral part of the hydrogen economy, and it is therefore necessary to develop safe, reliable, efficient, and effective hydrogen storage

Metal Oxides in Energy Technologies. https://doi.org/10.1016/B978-0-12-811167-3.00009-2

systems. Hydrogen has a high gravimetric energy density but a low volumetric energy density. This means that at normal temperature and pressure, 5-kg H_2 gas, for example, would fill a balloon 5 m in diameter, similar to the volume of an inflated hot-air balloon [3]. Consequently, the emphasis in hydrogen storage applications differs depending on the application. For example, for use in motor cars the gravimetric density is a crucial factor because the hydrogen storage system should fit inside the vehicle without adding too much weight. A reasonable driving distance should also be provided. Therefore, in transport applications it would be preferable to store H_2 at high density in a safe, light container, whereas in stationary applications a high volumetric density would be acceptable because the weight would not influence the system's efficiency.

Typically, three approaches to storing hydrogen can be distinguished:

1. storage as a compressed gas;
2. storage as cryogenic liquid hydrogen; and
3. materials-based or solid-state storage.

Of these, storing hydrogen as a compressed gas or in its liquid form are the best developed and most widely used methods. Materials-based and solid-state storage systems are still being developed and depend on the availability of new advanced materials. The US Department of Energy (US DoE) maintains technical and economic targets for hydrogen storage (Table 9.1 [4]). Ideally, 7.5 wt% gravimetric density and 7.0 wt% volumetric density are to be attained, at a cost of $266/kg.

Materials-based hydrogen storage has been considered as a long-term solution [5]. Regarding both the PVT filling and adsorption effects, improved volumetric hydrogen storage capacities in materials-packed gas cylinders compared with cylinders without

Table 9.1 Targets and current status of H_2 storage technologies

Storage targets	Gravimetric kWh/kg (kgH_2/kg system)	Volumetric kWh/L (kgH_2/L system)	Cost $ / kWh($/$kgH_2$)
2017	1.8 (0.055)	1.3 (0.040)	$ 12 ($ 400)
Ultimate	205 (0.075)	2.3 (0.070)	$ 8 ($ 266)
Projected H_2 storage system performance	Gravimetric (kWh/kg)	Volumetric (kWh/L)	Cost ($/kWh)
700 bar compressed (Type IV)	1.5	0.8	17
350 bar compressed (Type IV)	1.8	0.6	13
Metal hydride ($NaAlH_4$)	0.4	0.4	TBD
Sorbent (MOF-5, 100 bar)	1.1	0.7	16

Reproduced with permission from N.T. Stetson, Hydrogen storage program area, in: 2014 Annual Merit Review and Peer Evaluation, DOE, 2014.

any materials can be expected. Depending on the type of materials, the storage system is based on either physical sorption (physisorption) as in the case of porous materials or chemical sorption (chemisorption) as in the case of metal hydrides. Physisorption involves weaker binding energies of 4–10 kJ mol^{-1} and weak van der Waals forces, while chemisorption involves stronger binding energies of 50–100 kJ mol^{-1} in complex hydrides and metal hydrides, where the hydrogen molecule is first dissociated on the material surface, followed by diffusion of the hydrogen atoms into the metal host lattice [6]. In comparison with chemisorption, physisorption processes are easy to handle, are reversible due to the low interaction energy, and possess fast adsorption-desorption kinetics [7]. In porous materials such as zeolites, porous carbon structures, and metal-organic frameworks (MOFs), hydrogen is physisorbed on the surface of the pores, and the capacity depends mainly on the surface area, pore volume, working pressure, and working temperature. Typically, many of the physisorption-based materials have acceptable hydrogen storage capacities at cryogenic temperatures (77 K) and high pressures, but their capacities drop to below 1 wt% at ambient temperature and pressure in the range 50–100 bar [8]. Then the economic feasibility of storing hydrogen based on physisorption at cryogenic temperatures becomes an issue that should also be addressed. Given the binding energy of hydrogen on the surface of a certain material is 10 kJ mol^{-1} H_2, 6 kg H_2 will generate 30 MJ heat. If the 77 K temperature is maintained by liquid nitrogen, which has a heat of vaporization of 5.6 kJ mol^{-1}, then 5400 mol (150 kg) liquid nitrogen would be needed to circulate around the cylinder. Such a large amount of liquid nitrogen for cooling purposes would lead to difficulties for the engineering work at a later stage. Usually for chemisorption with stronger binding energies, formation of hydrogen compounds occurs under elevated hydrogen pressure, generating heat as a result of the exothermic reaction. To release the hydrogen, the pressure is reduced and heat is applied to the material. Although some metal hydrides can operate reversibly near ambient temperature, their gravimetric hydrogen storage capacities are lower than 3 wt% [9]. Some light metal hydrides and complex hydrides possess high gravimetric storage capacities (e.g., 7.6 wt% for MgH_2 and 18 wt% for $LiBH_4$), but releasing hydrogen from these hydrides requires high temperatures.

Magnesium hydride may appear as one of the most promising metal hydrides because of its high gravimetric hydrogen storage capacity (7.6 wt%), its reversibility, and low cost [10]. However, it has a number of shortcomings, such as its high decomposition temperature and unfavorable desorption/absorption kinetics [11]. To improve the performance of magnesium hydride in the field of solid-state hydrogen storage, several new techniques have been pioneered to improve the surface and kinetics by ball milling, adding a catalyst [12,13], and using the destabilization principle [14,15]. Adding a catalyst to magnesium hydride during ball milling significantly improved its hydrogen sorption characteristics, as was shown by doping metals [16], alloys [17], metal oxides [18,19], metal halides [20,21], carbon materials [22,23], and metal hydrides [24] into magnesium hydride. The addition of transition metal oxides (TMOs) has markedly improved the absorption-desorption kinetics of magnesium hydride, which can be ascribed to the effective catalysis shown by transition metal ions and compounds that are known to have a high affinity for hydrogen

[25–27]. Many TMOs are active catalysts because they enable the dissociation of hydrogen molecules and the recombination of hydrogen atoms to the molecular state. The additive oxide phase probably aids in overcoming the resistance of the MgO layer formed on the surface of the magnesium hydride particles, thus improving the reaction kinetics [28].

Recently, TMOs have come to the fore because of their good electrical and ionic conductivity, reversible capacity, and mechanical stability [29–31]. The improved performance is reached by use of double metal oxide species, which take part in several redox reactions during operations when drawing current. New TMO phases and crystal structures are being investigated for use as hydrogen storage materials [32]. Metal hydrides can be mixed with metal oxides to enhance the hydrogen storage ability, while the metal oxides themselves can also be applied as hydrogen storage materials. Particularly, the use of nano-additives as catalysts has been shown to improve the hydrogen cycling ability of light metal hydrides. In this chapter, we summarize the hydrogenation properties of MgH_2 doped with nano-size ZrO_2, γ-$Mg(BH_4)_2$ modified by MoO_3 and TiO_2, a MgH_2 and $4MgH_2$-Li_3AlH_6 system doped with Co_2NiO, and a MgH_2-$LiAlH_4$ composite doped with $MnFe_2O$. The addition of these TMOs has been shown to improve the absorption-desorption kinetics of magnesium hydride, which is the result of the highly effective catalysis shown by the transition metal compounds, due to the high affinity of transition metal cations for hydrogen. Nanostructured TMOs such as TiO_2, ZnO, SnO_2, $Zn_2V_2O_7$ hierarchical nanostructures (HNs), and hollow microflowers (HMFs) of Nb_3O_7F, also can act as good catalysts by enabling the dissociation of hydrogen molecules and the recombination of hydrogen atoms to molecules. Despite the many recent advances, much of the potential of such materials as yet remains untapped.

9.2 Surface modifiers for metal hydrides

9.2.1 MgH_2 doping with nano-size ZrO_2 catalyst

In recent research by Chen et al. MgH_2-ZrO_2 composites were prepared in molar ratios of 99:1 and 95:5, respectively, by high energy ball milling [33]. SEM images (Fig. 9.1) show that ZrO_2 facilitates the alloying of MgH_2-ZrO_2 composites and the Zr catalyst is uniformly distributed. As can be seen, the Zr catalyst enables the synthesis of MgH_2-ZrO_2 composites with a narrow and uniform size distribution. Moreover, in both samples the particle size is less than 2 μm. In the TEM images (Fig. 9.2) the final grain sizes of the MgH_2 samples after ball milling with the two different molar ratios of ZrO_2 can be compared. The sample milled with 5 mol% ZrO_2 (Fig. 9.2B) has a smaller grain size than the sample milled with 1 mol% ZrO_2 (Fig. 9.2A). The TEM results also show that the MgH_2-ZrO_2 sample with 5 mol% ZrO_2 has an average grain size of 71.2 nm, while that with 1 mol% ZrO_2 has an average grain size of 87.6 nm. These results confirm that ZrO_2 is broken down into smaller fragments during the milling process, as it is also seen when transition metals and oxides such as Nb_2O_5, V_2O_5, TiO_2, Sc_2O_3, Cr_2O_3, Mn_2O_3, and Fe_3O_4 are milled [26]. The zirconia

(A) SEM image of 99:1 (B) SEM image of 95:5

Fig. 9.1 SEM images of MgH_2-ZrO_2 composites with molar ratios of (A) 99:1 and (B) 95:5. Reproduced by permission from B.H. Chen, Y.S. Chuang, C.K. Chen, Improving the hydrogenation properties of MgH_2 at room temperature by doping with nano-size ZrO_2 catalyst, J. Alloys Compd. 655 (2016) 21–27.

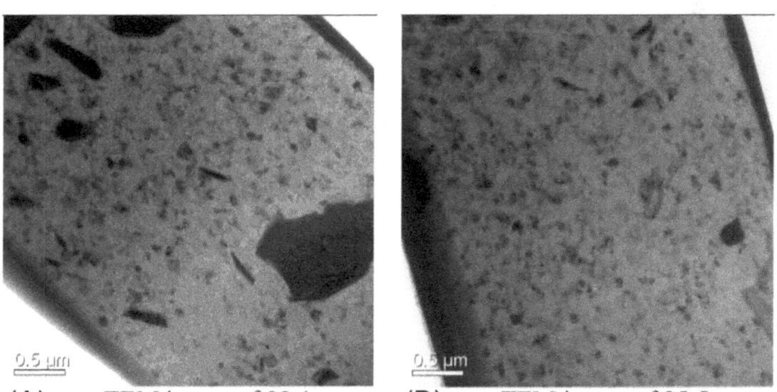

(A) TEM image of 99:1 (B) TEM image of 95:5

Fig. 9.2 TEM images of MgH_2-ZrO_2 composites with molar ratios of (A) 99:1 and (B) 95:5. Reproduced by permission from B.H. Chen, Y.S. Chuang, C.K. Chen, Improving the hydrogenation properties of MgH_2 at room temperature by doping with nano-size ZrO_2 catalyst, J. Alloys Compd. 655 (2016) 21–27.

is also distributed evenly throughout the Mg matrix, resulting in a nearly ideal hydrogenation catalyst.

The hydrogenation isotherms of the two MgH_2-ZrO_2 composites at a temperature of 298 K (Fig. 9.3) are very similar during the initial stages of the absorption process. This indicates that ZrO_2 does shorten the hydrogen absorption time efficiently, and the initial absorption rate of the MgH_2 system is insensitive to the amount of ZrO_2 added in the milling process. As shown by theoretical calculations, the use of nanoparticles rather than bulk or coarse hydride particles can alter the hydrogen uptake thermodynamics [34]. Moreover, the thermodynamics are dominated by the energy differences

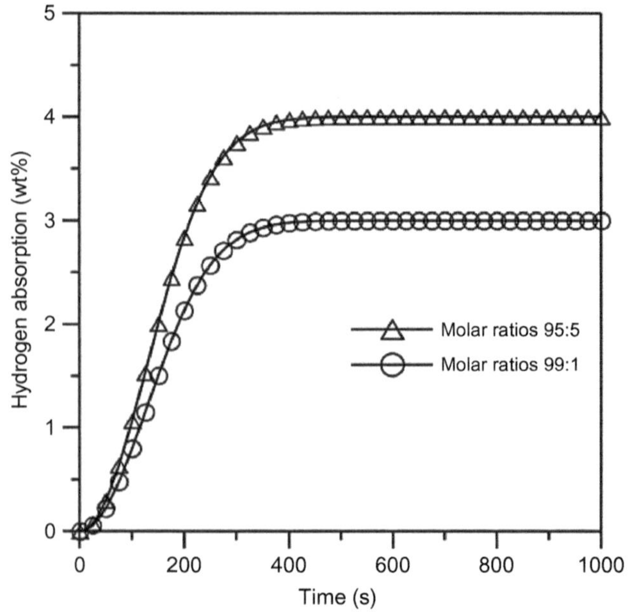

Fig. 9.3 Isothermal hydrogenation profiles of two MgH_2-ZrO_2 composites with molar ratios of 99:1 and 95:5, respectively, at a temperature of 298 K.
Reproduced by permission from B.H. Chen, Y.S. Chuang, C.K. Chen, Improving the hydrogenation properties of MgH_2 at room temperature by doping with nano-size ZrO_2 catalyst, J. Alloys Compd. 655 (2016) 21–27.

between the metal and its hydride. The hydrogenation reaction is also affected by the grain size. Interestingly, although the two materials had the same initial absorption rate, the composite with a 95:5 M ratio eventually absorbed a greater amount of hydrogen (i.e., 4.0 wt%) than that with a 99:1 (i.e., 3.0 wt%) molar ratio. These findings are envisaged to contribute substantially to the development of new hydrogen storage systems in applications such as hydrogen fuel cells, heat pumps, cooling systems, and hydrogen thermal compressors.

9.2.2 γ-$Mg(BH_4)_2$ modified by MoO_3 and TiO_2

Magnesium borohydride has great potential as a hydrogen storage material, but two vexing problems hampering its practical use are still its kinetics and reversibility [35,36]. $Mg(BH_4)_2$ is decomposed over several stages [37], and depending on the conditions, different magnesium polyboranes play a role. Magnesium dodecaborane ($MgB_{12}H_{12}$) is a stable intermediate ultimately oxidized to MgB_2. The icosahedral framework of $MgB_{12}H_{12}$ is geometrically similar to the boron elemental structure, which may be the reason why the rehydrogenation of $MgB_{12}H_{12}$ to $Mg(BH_4)_2$ is hindered [38]. However, it may be possible to synthesize $Mg(BH_4)_2$ from lower polyboranes at practical hydrogen pressures and temperatures [39]. Another way to improve the reversibility of $Mg(BH_4)_2$ dehydrogenation may be to design a stabilizer that could prevent the formation of dodecaborane.

Therefore, before $Mg(BH_4)_2$ can be used practically, a crucial problem is to discover effective materials for enhanced decomposition and formation of $Mg(BH_4)_2$. To promote reversible hydrogen storage, a number of TMOs have been studied as additives. For example, MoO_3 reacts with water to form H^+ ions [40], and TiO_2 has shown promise for use in energy production as a photocatalyst. Accordingly, Hauback proposed the use of both MoO_3 and TiO_2 to enhance reversible hydrogen sorption in γ-$Mg(BH_4)_2$ [41].

Commercial TiO_2 powder consists of a mixture of flakes with sizes varying from 10 to 50 μm, as shown by SEM (Fig. 9.4A). The titania flakes were reduced by high-energy ball milling to small aggregates composed of single 50-nm sized particles (Fig. 9.4B). The ball milling therefore reduced the particle size about 1000-fold. Concerning the hydrogen desorption-absorption properties of the γ-$Mg(BH_4)_2$-TiO_2 mixture prepared with premilled TiO_2 (Fig. 9.5B), the first vacuum desorption in vacuum removed almost 4 wt% H_2 after 40 h. After rehydrogenation under 140 bar H_2 followed by dehydrogenation as before, about 2.4 wt% H_2 was released over 70 h, which is considered to be the reversible hydrogen storage capacity.

In the case of the MoO_3 additive, the hydrogen desorption kinetics of a γ-Mg $(BH_4)_2$-MoO_3 mixture at a molar ratio of 1:0.02 made with premilled MoO_3 was compared with those of commercial γ-$Mg(BH_4)_2$ (Fig. 9.5A). The initial hydrogen desorption cycle of the γ-$Mg(BH_4)_2$-MoO_3 releases about 5.1 and 5.6 wt% H_2 over 10 and 15 h, respectively, at 267°C. The commercial γ-$Mg(BH_4)_2$ releases somewhat less, about 4.6 and 5.2 wt% H_2 during the same times at 263°C. Both samples release about 3 wt% hydrogen over the first 2.8 h and about 5.8 wt% H_2 over 25 h. Moreover, three approximate regions can be discerned in the desorption curves (from 0 to 2.8, 2.8–10,

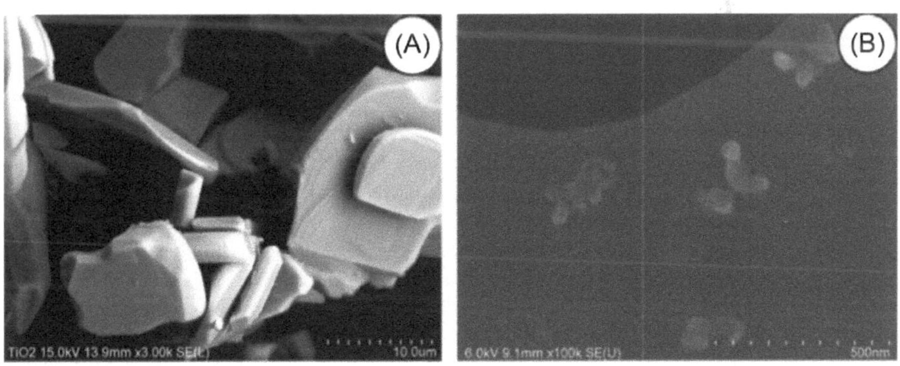

Fig. 9.4 SEM images of (A) as-received and (B) high-energy milled TiO_2 powder. Reproduced by permission from I. Saldan, C. Frommen, I. Llamas-Jansa, G.N. Kalantzopoulos, S. Hino, B. Arstad, R.H. Heyn, O. Zavoratynska, S. Deledda, M.H. Sørby, H. Fjellvåg, B.C. Hauback, Hydrogen storage properties of γ-$Mg(BH_4)_2$ modified by MoO_3 and TiO_2, Int. J. Hydrogen Energy 40 (2015) 12286–12293.

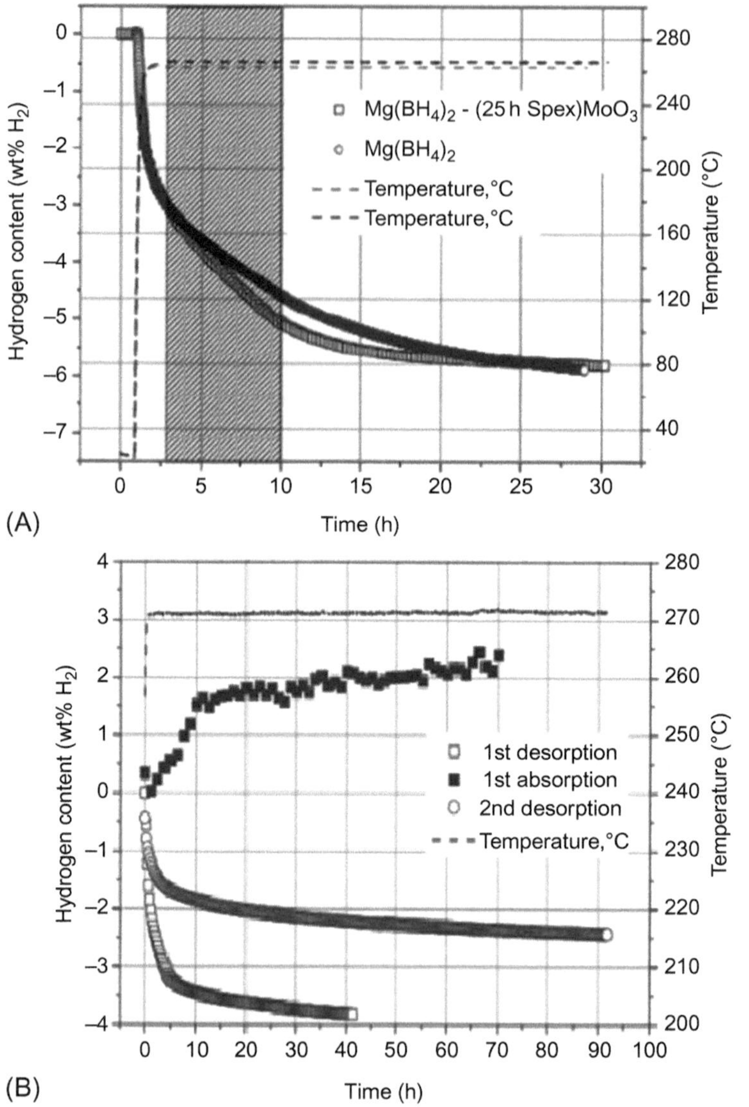

Fig. 9.5 (A) The first hydrogen desorption of as-received γ-$Mg(BH_4)_2$ and γ-$Mg(BH_4)_2$-MoO_3 mixture in a molar ratio of 1:0.02 prepared in a planetary micro-mill with previously high-energy milled (25 h Spex) MoO_3; (B) hydrogen desorption-absorption cycling of the γ-$Mg(BH_4)_2$-TiO_2 mixture in a molar ratio of 1:0.02 prepared by planetary micro-mill with previously high-energy milled TiO_2.
Reproduced by permission from I. Saldan, C. Frommen, I. Llamas-Jansa, G.N. Kalantzopoulos, S. Hino, B. Arstad, R.H. Heyn, O. Zavoratynska, S. Deledda, M.H. Sørby, H. Fjellvåg, B.C. Hauback, Hydrogen storage properties of γ-$Mg(BH_4)_2$ modified by MoO_3 and TiO_2, Int. J. Hydrogen Energy 40 (2015) 12286–12293.

and 10–25 h), with the middle phase apparently proceeding at a constant rate. This suggests that hydrogen desorption occurs over several steps, with the formation of relatively stable intermediates during the different stages. The quantity of hydrogen desorbed from the γ-$Mg(BH_4)_2$-(25 h Spex)-MoO_3 mixture is linearly related to the desorption time between 2.8 and 10 h (shaded area in Fig. 9.5A).

To summarize, for the two TMOs investigated as improvement additives, the most distinct effect of TiO_2 and MoO_3 on the decomposition of γ-$Mg(BH_4)_2$ was found to be the result of high-energy milling of the additives. This observation may be the result of the higher entropy of the smaller particles leading to a higher equilibrium pressure and destabilization of the hydrogenated sites of the mixtures. A solvent that disaggregates the oxide during milling may assist by more uniform positioning of the oxide nanoparticles around the γ-$Mg(BH_4)_2$ particles.

9.2.3 Doping with Co_2NiO catalyst

9.2.3.1 MgH_2 and 10 wt% Co_2NiO

The additive effect of Co_2NiO nanoparticles on the hydrogen storage properties of MgH_2 was investigated by Ismail [42]. MgH_2 mixed with 10 wt% Co_2NiO was prepared by ball milling, and the hydrogen storage properties and reaction mechanism of the product were examined. The SEM image of commercial MgH_2 (Fig. 9.6A) shows that the particles were angular shaped with sizes larger than 100 μm. The MgH_2 sample subjected to 1 h ball milling shows an extreme decrease in MgH_2 particle size, with a nonhomogeneous particle size (Fig. 9.6B). Some agglomeration was seen in the sample, but after adding Co_2NiO by ball milling, the particle sizes decreased compared to the milled MgH_2 sample (Fig. 9.6C). Apparently the hardness of the Co_2NiO helped in breaking the MgH_2 particles into smaller sizes, increasing the specific surface area and shortening the diffusion path length of the hydrogen, thus achieving a minimum desorption onset temperature [43].

The isothermal desorption curve for the milled MgH_2 and MgH_2 + 10 wt% Co_2NiO composite measured at 320°C under 1.0 atm pressure (Fig. 9.7A) shows that doping with 10 wt% Co_2NiO released about 2.5 wt% hydrogen within 6 min. In contrast, very little hydrogen was released over the same period by the milled MgH_2, demonstrating the important effect that Co_2NiO had on improving the dehydrogenation kinetics of MgH_2. The isothermal rehydrogenation kinetics for milled MgH_2 and MgH_2 + 10 wt% Co_2NiO samples are shown in Fig. 9.7B. The samples were treated at a constant temperature of 320°C under 30.0 atm hydrogen pressure. MgH_2 doped with 10 wt% Co_2NiO absorbed about 2.5 wt% hydrogen during the first 1.7 min, with a final hydrogen absorption of 3.75 wt%. In contrast, milled MgH_2 took about 3.4 min to absorb the same amount of hydrogen, with the final hydrogen absorption of 3.78 wt%. The doped sample therefore showed better rehydrogenation kinetics than as-milled MgH_2. All these results show that the performance of MgH_2 can also be improved by doping it with Co_2NiO.

Fig. 9.6 The SEM images of (A) as-received MgH_2, (B) milled MgH_2, and (C) $MgH_2 + 10\,wt\%$ Co_2NiO.
Reproduced by permission from N. Juahir, N.S. Mustafa, A.M. Sinin, M. Ismail, Improved hydrogen storage properties of MgH_2 by addition of Co_2NiO nanoparticles, RSC Adv. 5 (2015) 60983–60989.

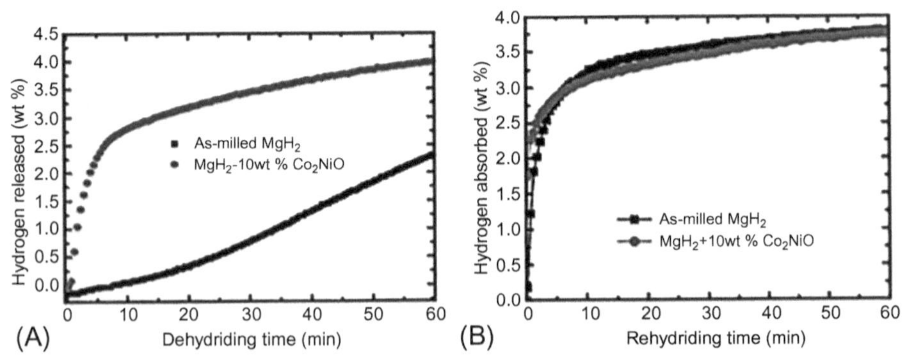

Fig. 9.7 (A) Isothermal dehydrogenation kinetics of the milled MgH_2 and $MgH_2 + 10\,wt\%$ Co_2NiO composite at 320°C; (B) isothermal rehydrogenation kinetics of the milled MgH_2 and doped $MgH_2 + 10\,wt\%$ Co_2NiO composite at 320°C under 30.0 atm hydrogen pressure.
Reproduced by permission from N. Juahir, N.S. Mustafa, A.M. Sinin, M. Ismail, Improved hydrogen storage properties of MgH_2 by addition of Co_2NiO nanoparticles, RSC Adv. 5 (2015) 60983–60989.

9.2.3.2 $4MgH_2$-Li_3AlH_6 system doped with Co_2NiO

Ismail et al. [44] initiated the study of MgH_2-Li_3AlH_6 systems doped with 10wt% Co_2NiO. Adding Co_2NiO to the $4MgH_2$-Li_3AlH_6 composite (Fig. 9.8B) produced particles that were smaller than the un-doped $4MgH_2$-Li_3AlH_6 particles (Fig. 9.8A). This may have been due to the hardness of the Co_2NiO particles, which may have helped to break the MgH_2-Li_3AlH_6 composite particles into smaller pieces. The smaller the particle size, the better the hydrogen absorption or desorption, because the diffusion path lengths of the hydrogen are shortened on average, and the particles' total reactive surface area is increased [45].

As shown in Fig. 9.9A, the hydrogen absorption properties of the doped composite were better than those of the un-doped composite. The $4MgH_2$-Li_3AlH_6-10wt% Co_2NiO composite attained 4.35wt% hydrogen absorption capacity after 60min, whereas the $4MgH_2$-Li_3AlH_6 composite absorbed only 2.7wt% H_2 over the same period. These results show that the addition of Co_2NiO also improves the rehydrogenation kinetics of the $4MgH_2$-Li_3AlH_6 system.

The dehydrogenation kinetics of the $4MgH_2$-Li_3AlH_6-10wt% Co_2NiO composite were also studied; isothermal dehydrogenation of the doped and un-doped MgH_2-Li_3AlH_6 at 320°C for 60min (Fig. 9.9B) shows that the $4MgH_2$-Li_3AlH_6-10wt% Co_2NiO composite released about 3.1wt% of hydrogen over 60min, while the $4MgH_2$-Li_3AlH_6 composite only desorbed about 2.35wt% hydrogen during the same period. For the rate of the dehydrogenation process, the doped composite was desorbed faster than the un-doped composite. Comparing these results, it can be seen that the Co_2NiO additive makes an important contribution to improvement of the dehydrogenation kinetics of the MgH_2-Li_3AlH_6 system.

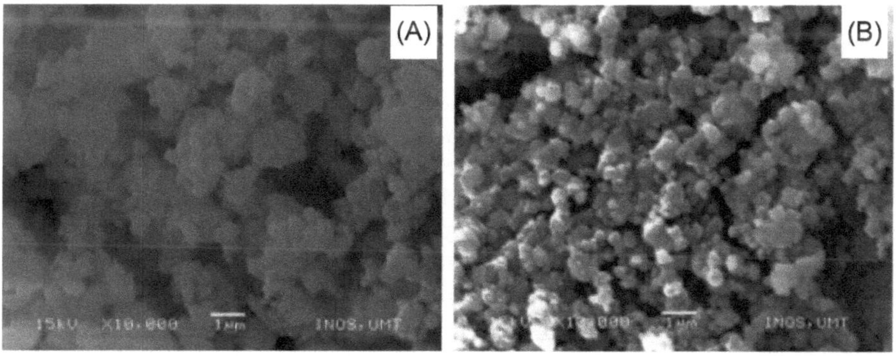

Fig. 9.8 SEM images of (A) milled $4MgH_2$-Li_3AlH_6, and (B) milled $4MgH_2$-Li_3AlH_6-10wt% Co_2NiO.
Reproduced by permission from F.A. Halim Yap, M.H.I. Ishak, M. Ismail, Enhanced hydrogen storage performance of destabilized $4MgH_2$-Li_3AlH_6 system doped with Co_2NiO nanopowder, Int. J. Hydrogen Energy 40 (2015) 10131–10138.

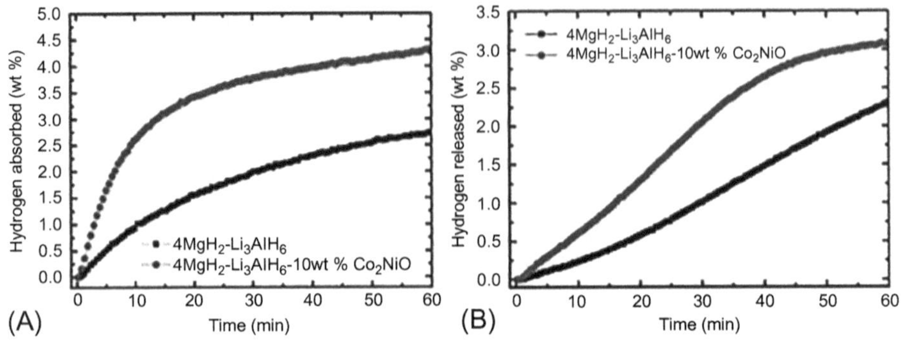

Fig. 9.9 (A) Isothermal rehydrogenation of $4MgH_2$-Li_3AlH_6 and $4MgH_2$-Li_3AlH_6-10 wt% Co_2NiO composites at 320°C under 33 atm H_2; (B) Isothermal dehydrogenation of $4MgH_2$-Li_3AlH_6 and $4MgH_2$-Li_3AlH_6-10 wt% Co_2NiO composites at 320°C. Reproduced by permission from F.A. Halim Yap, M.H.I. Ishak, M. Ismail, Enhanced hydrogen storage performance of destabilized $4MgH_2$-Li_3AlH_6 system doped with Co_2NiO nanopowder, Int. J. Hydrogen Energy 40 (2015) 10131–10138.

9.2.4 MgH_2-$LiAlH_4$ composite doped with $MnFe_2O_4$

In a study by Volinsky et al., $MnFe_2O_4$ catalyst nanoparticles were used to study their effect on the hydrogen storage properties of the MgH_2-$LiAlH_4$ system produced by high-energy ball milling [46].

The excellent acceleration of the dehydrogenation kinetics of the ball-milled MgH_2-$LiAlH_4$ sample by the $MnFe_2O_4$ catalyst nanoparticles was demonstrated by comparison of the isothermal desorption behavior of the composite with and without $MnFe_2O_4$ catalyst, under identical reaction conditions. Fig. 9.10A and B show the dehydrogenation curves of ball-milled MgH_2-$LiAlH_4$ and MgH_2-$LiAlH_4$+5 mol% $MnFe_2O_4$ samples at 200°C and 300°C under 0.1 MPa, respectively. The ball-milled MgH_2-$LiAlH_4$ system and the MgH_2-$LiAlH_4$+5 mol% $MnFe_2O_4$ sample released 0.94 and 2.91 wt% hydrogen, respectively, at 200°C in 1800 s under 0.1 MPa pressure (Fig. 9.10A). On further increase of the temperature to 300°C, the MgH_2-$LiAlH_4$+5 mol% $MnFe_2O_4$ sample desorbed 4.09 wt% hydrogen within 15 s, which is much more than that of the MgH_2-$LiAlH_4$ sample (0.31 wt%). In contrast, about 1800 s are required for the ball-milled MgH_2-$LiAlH_4$ sample to release 4.08 wt% hydrogen under the same conditions, demonstrating the superiority of $MnFe_2O_4$ in improving the MgH_2-$LiAlH_4$ dehydrogenation kinetics. Therefore, significant improvement of MgH_2-$LiAlH_4$ desorption kinetics can be obtained by adding $MnFe_2O_4$ nanopowders.

9.3 Oxides as host for hydrogen storage

9.3.1 TiO_2 for hydrogen storage

Titanium dioxide, TiO_2, has been widely used in pigments and paints, in optical coatings for laser mirrors, interference filters, etc. Its photocatalytic properties (such as decomposing water into hydrogen and oxygen on TiO_2 electrodes) were discovered

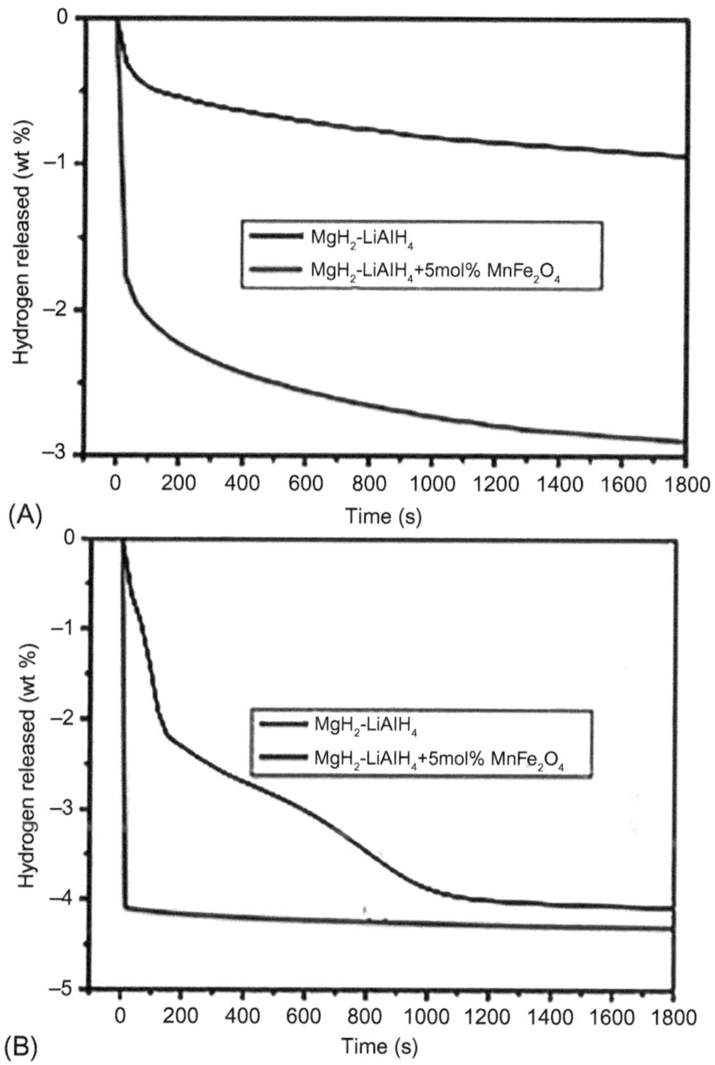

Fig. 9.10 Isothermal desorption curves of MgH_2-$LiAlH_4$ and MgH_2-$LiAlH_4$ + 5 mol% $MnFe_2O_4$ samples at (A) 200°C and (B) 300°C.

Reproduced by permission from Q. Wan, P. Li, Z.L. Li, F.Q. Zhai, X.H. Qu, A.A. Volinsky, Improved hydrogen storage performance of MgH_2-$LiAlH_4$ composite by addition of $MnFe_2O_4$, J. Phys. Chem. C 117 (2013) 26940–26947.

by Fujishima and Honda in 1972 [47]. TiO_2 nowadays finds various novel applications in photo-electrochemistry, photocatalysis, solar cells, and gas sensors [48]. An increased interest developed recently in TiO_2 for the development of nanostructured forms (nanotubes, nanorods, etc.) for renewable energy sources and the hydrogen economy. For example, TiO_2 nanotubes grown on Ti substrates by anodization could be used for hydrogen storage [49]. Metal-insulator-metal (MIM) structures such as Ti/TiO_2/Pt were proposed for the resistance random-access memories (ReRAM). Light-induced reversible metal-semiconductor transition at room temperature has been demonstrated for TiO_x nanoparticles, making it a promising material for the high-density storage media.

Ding [50] improved the catalytic activity of titania using a one-step electroless plating method to produce a tungsten-modified CoB/Ag-TiO_2 catalyst. The influence of the catalyst on the hydrolysis of 1 wt% $NaBH_4$ solution by 5 wt% NaOH was investigated and the amount of tungsten added was optimized. The morphology of the as-prepared 3W-3CoB/Ag-TiO_2 catalyst can be seen in Fig. 9.11A. The catalyst particle size is in the 35–45 nm range, which is close to that of the TiO_2 and Ag-TiO_2 supports. The hydrolysis reaction generates a steady H_2 stream and leaves the morphology of the 3W-3CoB/Ag-TiO_2 catalyst nearly unchanged (Fig. 9.11B and inset), demonstrating the improved stability of 3W-3CoB/Ag-TiO_2 over unsupported CoB catalyst.

The kinetics study of the 3W-3CoB/Ag-TiO_2 catalyst resulted in better insight into the $NaBH_4$ hydrolysis process. In addition, the catalyst's influence on the reaction mechanism was clarified by means of the obtained activation energy and kinetic equation. The H_2 generation $V-t$ curves at different temperatures for the 3W-3CoB/Ag-TiO_2 and unsupported CoB catalysts (Fig. 9.12A and B) show clearly that the H_2 generation rate of both catalysts increases with increasing temperature. The activation

Fig. 9.11 SEM images of (A) as-prepared 3W-3CoB/Ag-TiO_2, and (B) used 3W-3CoB/Ag-TiO_2. The inset SEM images of (A) and (B) are the high magnification images of as-prepared and used 3W-3CoB/Ag-TiO_2.
Reproduced with permission from X. Shen, Q. Wang, S. Guo, B. Liu, Z. Sun, Z. Zhang, Z. Wang, B. Zhao, W. Ding, W-modified CoB supported on Ag-activated TiO_2 for hydrogen generation from alkaline $NaBH_4$ solution, Int. J. Hydrog. Energy 40 (2015) 6346–6357.

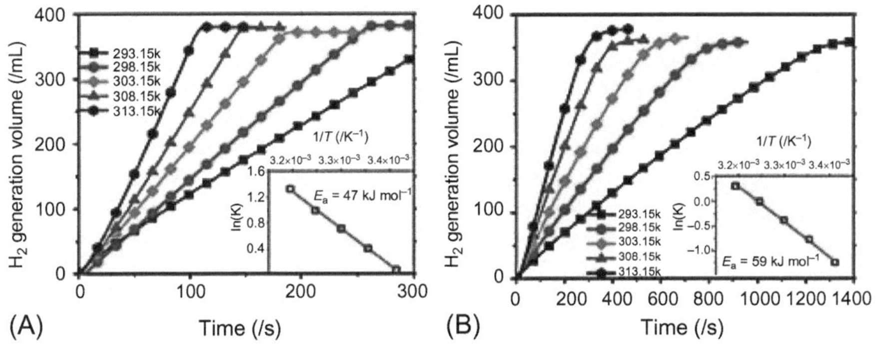

Fig. 9.12 Kinetics curves of (A) 3W-3CoB/Ag-TiO_2 and (B) unsupported CoB for hydrolysis of 1 wt% $NaBH_4$ + 5 wt% NaOH solution at different temperatures.
Reproduced by permission from X. Shen, Q. Wang, S. Guo, B. Liu, Z. Sun, Z. Zhang, Z. Wang, B. Zhao, W. Ding, W-modified CoB supported on Ag-activated TiO_2 for hydrogen generation from alkaline $NaBH_4$ solution, Int. J. Hydrogen Energy 40 (2015) 6346–6357.

energies of the catalysts were derived from the Arrhenius plots (shown in the insets of Fig. 9.12A and B) and found to be 47 kJ mol^{-1} for the 3W-3CoB/Ag-TiO_2 catalyst, which is lower than the 59 kJ mol^{-1} obtained for the unsupported CoB catalyst. The 3W-3CoB/Ag-TiO_2 catalyst possesses a distinct textural structure, with apparent electron transferring effects among Co^0, B^0, and W^0, highly concentrated surface active sites, and surface WO_x Lewis acid. These factors may all contribute to lowering the activation energy.

9.3.2 ZnO for hydrogen storage

Zinc oxide is an II–VI semiconductor compound with a hexagonal wurtzite crystal structure, due to its wide bandgap of 3.37 eV and large exciton binding energy of 60 meV; it is a highly likely photonic material in the ultraviolet spectrum. Since the discovery of carbon nanotubes, ZnO nanobelts and other one-dimensional (1D) nanostructures such as Si nanowires [51] and GaN nanowires have attracted the attention of researchers because of their unique physical and chemical properties. Many of the 1D nanostructures have been synthesized by the metal-catalysis-assisted growth approach, where metal nanocrystals act as catalysts and play a crucial part in forming liquid alloy droplets. Metal impurities may be generated unintentionally during crystal growth.

The concept of hydrogen storage in 1D nanostructures such as carbon nanotubes has appealed to many investigators. Yu et al. reported a simple and rapid process for the mass production of 99.999% pure ZnO nanowires by means of evaporation of zinc pellets at 900°C [52]. The SEM of the synthesized ZnO sample (Fig. 9.13A) shows randomly distributed nanowires with a uniform diameter of about 20 nm.

The hydrogen storage capacity of the nanowires is 0.83 wt% at about 3.03 MPa (Fig. 9.13B). The hydrogen storage capacity of the ZnO nanowires increases linearly with the pressure, suggesting that even more hydrogen may be stored at pressures above 3.03 MPa. Subsequently the pressure of the reaction chamber was lowered

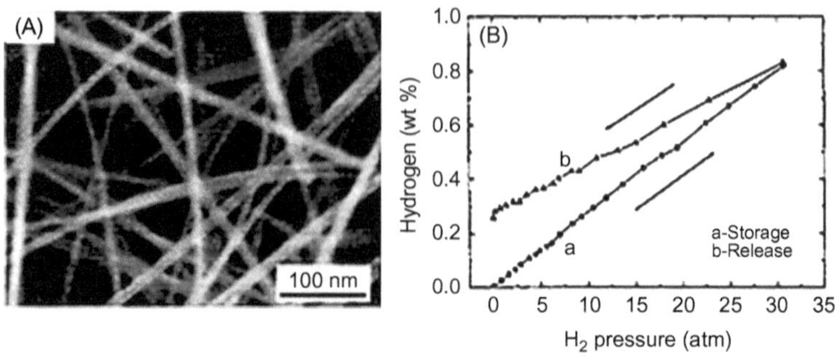

Fig. 9.13 (A) SEM images of the synthesized ZnO nanowires; (B) hydrogen storage and release of the ZnO nanowires as a function of pressure.
Reproduced by permission from Q. Wan, C.L. Lin, X.B. Yu, T.H. Wang, Room-temperature hydrogen storage characteristics of ZnO nanowires, Appl. Phys. Lett. 84 (2004) 124–126.

gradually to 1 atm (1.013105 Pa) to release the hydrogen. Curve b (Fig. 9.13B) shows that about 71% of the stored hydrogen is released at standard pressure. The pressure change over time was also monitored. The hydrogen uptake by both samples was very quick, with nearly 95% of the total capacity achieved in 300 s.

9.3.3 SnO_2 porous microspheres for hydrogen storage

SnO_2, an n-type semiconductor with a wide bandgap of 3.6 eV, has become of great interest for hydrogen storage. Various morphologies of SnO_2 nanostructures have been synthesized recently, including nanotubes, nanorods, multilayer structures, nanoparticles, and hollow structures [53, 54]. The hollow SnO_2 structures are especially interesting because of the wide range of applications in the fields of gas sensing, solar cells, photocatalysis, controlled drug delivery and release, and lithium-ion batteries (LIBs) [55]. Li et al. reported the first solvothermal synthesis of flower-like SnO_2 porous microspheres self-assembled as well-defined hollow nanospheres [56].

As shown in Fig. 9.14A, the magnified SEM image of the flower-like SnO_2 microspheres shows that the flower-like SnO_2 microspheres are self-assembled by well-defined hollow nanospheres with a rough surface and 300–500 nm in diameter. Moreover, it can be observed from the inset of Fig. 9.14A that the wall of the hollow nanospheres is composed of uniform nanoparticles, leading to an overall porous structure. The TEM image of a single microsphere shown in Fig. 9.14B reveals that the flower-like SnO_2 microsphere is completely self-assembled as high-density well-defined hollow nanospheres. Moreover, Fig. 9.14B shows that the wall of the SnO_2 hollow nanospheres is highly transparent to an electron beam, which shows that the wall is very thin.

The hydrogen storage capability of the synthesized product was measured at 373 K at 4 MPa pressure in one cycle; the hydrogen absorption curve (Fig. 9.15) shows that hydrogen storage increases rapidly during the first 40 min and then increases more

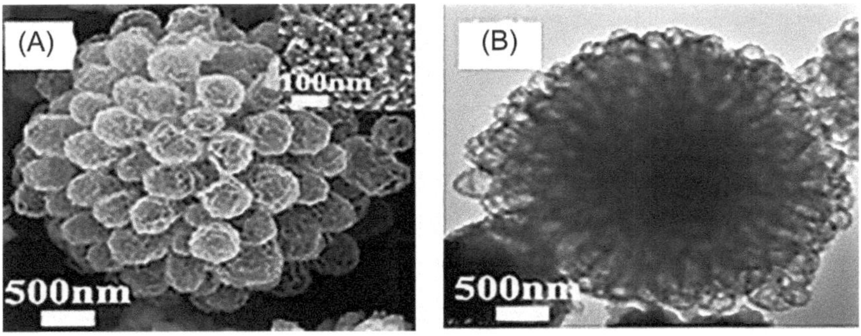

Fig. 9.14 (A) SEM and (B) TEM images of the as-prepared flower-like microspheres. Reproduced by permission from Z.Y. Wang, F.P. Wang, M.Y. Li, M.Z. Iqbal, Q.U.A. Javed, Y.Z. Lu, Synthesis, characterization and hydrogen storage characteristics of flower-like SnO_2 porous microspheres, Mater. Lett. 119 (2014) 36–38.

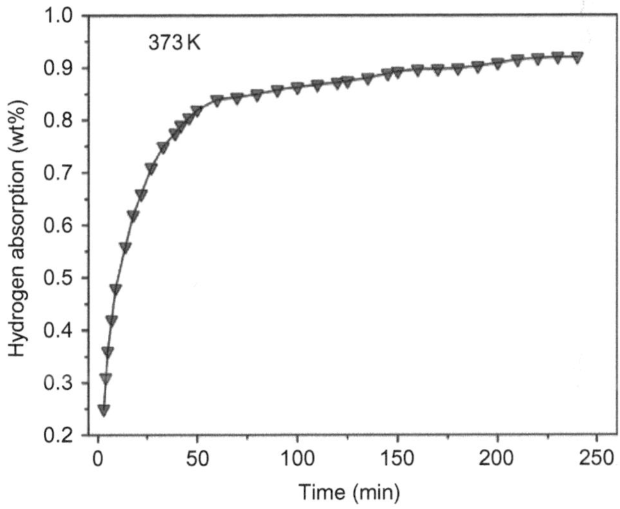

Fig. 9.15 Hydrogen absorption curve for flower-like SnO_2 hollow microspheres. Reproduced by permission from Z.Y. Wang, F.P. Wang, M.Y. Li, M.Z. Iqbal, Q.U.A. Javed, Y.Z. Lu, Synthesis, characterization and hydrogen storage characteristics of flower-like SnO_2 porous microspheres, Mater. Lett. 119 (2014) 36–38.

slowly with time. The maximum hydrogen storage capacity of the as-prepared flower-like SnO_2 microspheres was found to be approximately 0.92 wt%.

The porous and hollow structure of the flower-like SnO_2 microspheres is expected to result in a high concentration of defects. In addition, the high hydrogen absorption is probably associated with the porous and hollow structure of the flower-like SnO_2

microspheres, which may enhance hydrogen diffusion and improve the hydrogen storage capability. These distinctive properties show that the flower-like SnO_2 microspheres could be a leading contender for green energy storage and gas sensing applications.

9.3.4 Binary metal oxides for hydrogen storage

9.3.4.1 Novel $Zn_2V_2O_7$ HNs for hydrogen storage

The mixed oxides of vanadium are showing promise as electrochemical supercapacitor and hydrogen storage materials [57]. Because of their layered structures and excellent kinetics, the mixed metal vanadates are also being investigated for use in optical and magnetic instruments, catalysis, LIBs, and in supercapacitors [58–60]. Engineering novel vanadium-based TMO crystal structures may open new horizons for a variety of energy storage and conversion devices. Khan et al. were the first to synthesize $Zn_2V_2O_7$ HNs applying a facile, template free and inexpensive method [61]. $Zn_2V_2O_7$ is a self-activated yellow light-emitting phosphor used to make white LEDs and is also used as an anode material for LIBs [62]. First results indicate that it has superior hydrogen storage properties compared to other materials [61].

The FESEM images of the $Zn_2V_2O_7$ HNs (Fig. 9.16A and B) show the typical nanostructures consisting of a combination of petal-like structures. The petals are dense and about 50–65 nm in diameter (Fig. 9.16C), with lengths around tens of microns. The HRTEM image of the HNs (Fig. 9.16D) confirms the highly crystalline nature of the material. The nano-crystalline nature of the HNs is confirmed by the occasional concentric ring-like shapes in the SAED pattern (Fig. 9.16D).

Storage capacity measurements were carried out at 5 MPa at 473 and 573 K, respectively (Fig. 9.17). The maximum values obtained for hydrogen absorption by the $Zn_2V_2O_7$ HNs were calculated as 1.21 wt% at 473 K and 1.81 wt% at 573 K, respectively. The material absorbed 85% hydrogen during the first 12 min at 473 K, whereas 84.1% hydrogen was stored during the first 15 min at 573 K.

9.3.4.2 Novel hollow microflowers of Nb_3O_7F for hydrogen storage

Tanveer et al. used a two-step, facile surfactant-free hydrothermal route prepared novel Nb_3O_7F HMFs. Time-dependent experiments were performed which suggested that Nb_3O_7F HMFs were formed due to an Ostwald-ripening process [63].

SEM images of the Nb_3O_7F HMFs are shown in Fig. 9.18A–D at different resolutions. These novel microflowers with hollow interior were synthesized at 160°C for 24 h by a hydrothermal approach. The microflowers consisted of 1D hollow microrods that were assembled to form three-dimensional microflowers. The average length of the HMF is 2.72 μm and the average internal diameter is 509.5 nm, as obtained from the SEM. Hydrogen storage values of 0.79 wt% at 373 K and 1.08 wt% at 473 K were obtained. The hydrogen storage curves are shown in Fig. 9.19. The maximum

Fig. 9.16 (A and B) FESEM images of HNs; (C and D) TEM and HRTEM pattern of the HNs. The inset in (D) shows the SAED pattern.
Reproduced by permission from K.F. Butt, C.B. Cao, F. Idrees, M. Tahir, R. Hussain, W.S. Khan, Novel $Zn_2V_2O_7$ hierarchical nanostructures: optical and hydrogen storage properties, Int. J. Hydrogen Energy 40 (2015) 9359–9364.

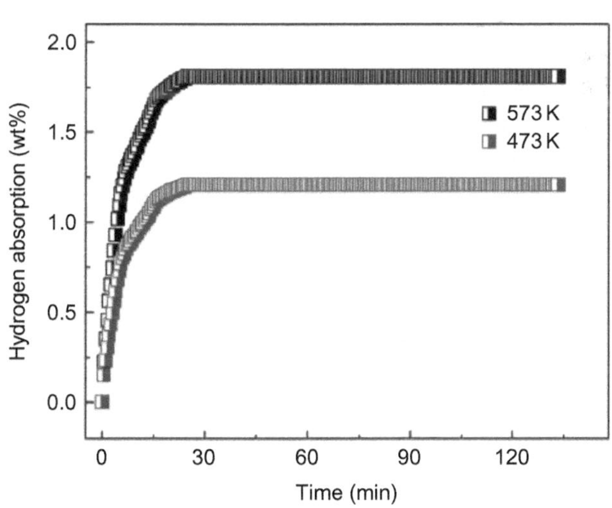

Fig. 9.17 Hydrogen storage measurements of $Zn_2V_2O_7$ HNs.
Reproduced by permission from K.F. Butt, C.B. Cao, F. Idrees, M. Tahir, R. Hussain, W.S. Khan, Novel $Zn_2V_2O_7$ hierarchical nanostructures: optical and hydrogen storage properties, Int. J. Hydrogen Energy 40 (2015) 9359–9364.

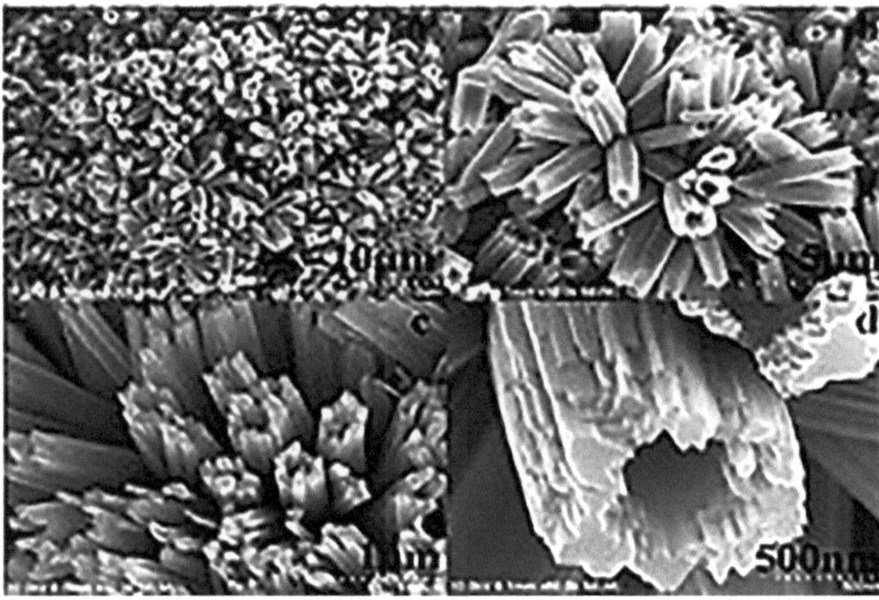

Fig. 9.18 (A–D) FESEM images of Nb_3O_7F HMF at different resolutions.
Reproduced by permission from F. Idrees, C.B. Cao, F.K. Butt, M. Tahir, I. Shakir, M. Rizwan, Synthesis of novel hollow microflowers (NHMF) of Nb_3O_7F, their optical and hydrogen storage properties, Int. J. Hydrogen Energy 39 (2014) 13174–13179.

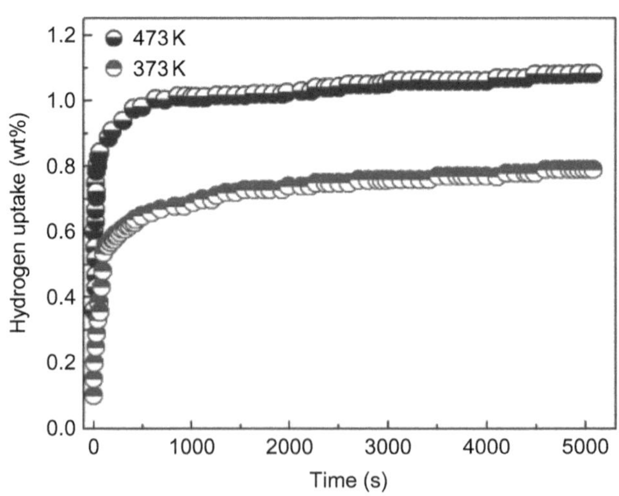

Fig. 9.19 Hydrogen storage measurements of Nb_3O_7F HMF at 373 and 473K.
Reproduced by permission from F. Idrees, C.B. Cao, F.K. Butt, M. Tahir, I. Shakir, M. Rizwan, Synthesis of novel hollow microflowers (NHMF) of Nb_3O_7F, their optical and hydrogen storage properties, Int. J. Hydrogen Energy 39 (2014) 13174–13179.

hydrogen storage was 1.08 wt%. The enhanced hydrogen storage characteristic is possibly due to the special hollow hierarchical Nb_3O_7F structures.

9.4 Outlook

There is a critical necessity to find clean alternative and sustainable energy sources to replace existing nonrenewable fossil fuels. There is worldwide a rapid growth in the development of renewable power generation, and the hydrogen economy is an increasingly attractive idea to investigate.

Magnesium and magnesium-based alloys are very encouraging prospective hydrogen storage materials for renewable and clean energy applications. They absorb about 7.6 wt% hydrogen. Also, MgH_2 is relatively light and inexpensive compared with other metal hydrides. However, the practical application of magnesium-based alloys has been hindered by their slow reaction kinetics and high dissociation temperatures. Therefore, many attempts have been made to improve the sorption properties, for example, by making nanocrystalline powders, synthesizing composite metal hydrides, and adding catalysts.

TMO and magnesium hydride-TMO materials have been identified to be attractive for hydrogen storage. Some studies have also highlighted the enhancement of hydrogen storage capacities, thermostability and hydrostability that can be achieved on TMO and magnesium hydride-TMO materials. To date, most research strategies have been focused on ways of achieving better performance of TMO-based physical sorption at ambient temperature.

Acknowledgments

Financial support from the National Materials Genome Project (2016YFB0700600), China National Fund for Distinguished Youth Scientists (51026004), and NSFC (U1601214 and 51502137) is gratefully acknowledged.

References

[1] REN21, Renewables 2014 Global Status Report, REN21 Secretariat, Paris, 2014.

[2] F. Zhang, P.C. Zhao, M. Niu, J. Maddy, The survey of key technologies in hydrogen energy storage, Int. J. Hydrog. Energy 41 (2016) 14535–14552.

[3] M. Ball, M. Wietschel, The Hydrogen Economy: Opportunities and Challenges, Cambridge University Press, Cambridge, UK, 2009.

[4] N.T. Stetson, Hydrogen storage program area, 2014 Annual Merit Review and Peer Evaluation, DOE, 2014.

[5] S. McWhorter, C. Read, G. Ordaz, N. Stetson, Materials-based hydrogen storage: attributes for near-term, early market PEM fuel cells, Curr. Opin. Solid State Mater. Sci. 15 (2011) 29–38.

[6] B. Sakintuna, F. Lamari-Darkrim, M. Hirscher, Metal hydride materials for solid hydrogen storage: a review, Int. J. Hydrog. Energy 32 (2007) 1121–1140.

[7] A.M. Martınew, G. Serfert, Adsorption of molecular hydrogen on nanostructured surfaces, Rev. Cub. Fis. 31 (2014) 32–34.

[8] D.P. Broom, C.J. Webb, K.E. Hurst, P.A. Parilla, T. Gennett, C.M. Brown, et al., Outlook and challenges for hydrogen storage in nanoporous materials, Appl. Phys. A Mater. Sci. Process. 122 (2016) 151.

[9] W. Oelerich, T. Klassen, R. Bormann, Metal oxides as catalysts for improved hydrogen sorption in nanocrystalline Mg based materials, J. Alloys Compd. 315 (2001) 237–242.

[10] G. Liang, J. Huot, S. Boily, A. van Neste, R. Schulz, Catalytic effect of transition metals on hydrogen sorption in nanocrystalline ball milled MgH_2-Tm (Tm = Ti, V, Mn, Fe and Ni) systems, J. Alloys Compd. 292 (1999) 247–252.

[11] I.P. Jain, C. Lal, A. Jain, Hydrogen storage in Mg: a most promising material, Int. J. Hydrog. Energy 35 (2010) 5133–5144.

[12] M. Ismail, Influence of different amounts of $FeCl_3$ on decomposition and hydrogen sorption kinetics of MgH_2, Int. J. Hydrog. Energy 39 (2014) 2567–2574.

[13] H. Gasan, O.N. Celik, N. Aydinbeyli, Y.M. Yaman, Effect of V, Nb, Ti and graphite additions on the hydrogen desorption temperature of magnesium hydride, Int. J. Hydrog. Energy 37 (2012) 1912–1918.

[14] M. Ismail, Y. Zhao, X.B. Yu, J.F. Mao, S.X. Dou, The hydrogen storage properties and reaction mechanism of the MgH_2-$NaAlH_4$ composite system, Int. J. Hydrog. Energy 36 (2011) 9045–9050.

[15] M. Ismail, Y. Zhao, S.X. Dou, An investigation on the hydrogen storage properties and reaction mechanism of the destabilized MgH_2-Na_3AlH_6 (4:1) system, Int. J. Hydrog. Energy 38 (2013) 1478–1483.

[16] A. Ranjbar, Z.P. Guo, X.B. Yu, D. Attard, A. Calka, H.K. Liu, Effects of SiC nanoparticles with and without Ni on the hydrogen storage properties of MgH_2, Int. J. Hydrog. Energy 34 (2009) 7263–7268.

[17] X.B. Yu, Z.X. Yang, H.K. Liu, D.M. Grant, G.S. Walker, The effect of a Ti-V-based BCC alloy as a catalyst on the hydrogen storage properties of MgH_2, Int. J. Hydrog. Energy 35 (2010) 6338–6344.

[18] M. Baricco, M.W. Rahman, S. Livraghi, A. Castellero, S. Enzo, E. Giamello, Effects of $BaRuO_3$ addition on hydrogen desorption in MgH_2, J. Alloys Compd. 536 (2012) S216–S221.

[19] K.S. Jung, E.Y. Lee, K.S. Lee, Catalytic effects of metal oxide on hydrogen absorption of magnesium metal hydride, J. Alloys Compd. 421 (2006) 179–184.

[20] N.S. Mustafa, M. Ismail, Influence of K_2TiF_6 additive on the hydrogen sorption properties of MgH_2, Int. J. Hydrog. Energy 39 (2014) 15563–15569.

[21] M. Ismail, Effect of $LaCl_3$ addition on the hydrogen storage properties of MgH_2, Energy 79 (2015) 177–182.

[22] A. Ranjbar, M. Ismail, Z.P. Guo, X.B. Yu, H.K. Liu, Effects of CNTs on the hydrogen storage properties of MgH_2 and MgH_2-BCC composite, Int. J. Hydrog. Energy 35 (2010) 7821–7826.

[23] C.Z. Wu, P. Wang, X. Yao, C. Liu, D.M. Chen, G.Q. Lu, H.M. Cheng, Hydrogen storage properties of MgH_2/SWNT composite prepared by ball milling, J. Alloys Compd. 420 (2006) 278–282.

[24] A.R. Yavari, J.F.R. de Castro, G. Vaughan, G. Heunen, Structural evolution and metastable phase detection in MgH_2-5%NbH nanocomposite during in-situ H-desorption in a synchrotron beam, J. Alloys Compd. 353 (2003) 246–251.

[25] D.L. Croston, D.M. Grant, G.S. Walker, The catalytic effect of titanium oxide based additives on the dehydrogenation and hydrogenation of milled MgH_2, J. Alloys Compd. 492 (2010) 251–258.

[26] W. Oelerich, T. Klassen, R. Bormann, Metal oxides as catalysts for improved hydrogen sorption in nanocrystalline Mg-based materials, J. Alloys Compd. 315 (2001) 237–242.

[27] H. Simchi, A. Kaflou, A. Simchi, Synergetic effect of Ni and Nb_2O_5 on dehydrogenation properties of nanostructured MgH_2 synthesized by high-energy mechanical alloying, Int. J. Hydrog. Energy 34 (2009) 7724–7730.
[28] M.W. Rahman, A. Castellero, S. Enzo, S. Livraghi, E. Giamello, M. Baricco, Effect of Mg-Nb oxides addition on hydrogen sorption in MgH_2, J. Alloys Compd. 509 (2011) S438–S443.
[29] F.K. Butt, C. Cao, R. Ahmed, W.S. Khan, T. Cao, N. Bidin, et al., Synthesis of novel ZnV_2O_4 spinel oxide nanosheets and their hydrogen storage properties, CrystEngComm 16 (2014) 894–899.
[30] F.K. Butt, C. Cao, Q. Wan, P. Li, F. Idrees, M. Tahir, et al., Synthesis, evolution and hydrogen storage properties of ZnV_2O_4 glomerulus nano/microspheres: a prospective material for energy storage, Int. J. Hydrog. Energy 39 (2014) 7842–7851.
[31] F.K. Butt, M. Tahir, C. Cao, F. Idrees, R. Ahmed, W.S. Khan, et al., Synthesis of novel ZnV_2O_4 hierarchical nanospheres and their applications as electrochemical supercapacitor and hydrogen storage material, ACS Appl. Mater. Interfaces 6 (2014) 13635–13641.
[32] L. Yu, L. Zhang, H.B. Wu, G. Zhang, X.W. Lou, Controlled synthesis of hierarchical $Co_xMn_{3}\text{-}xO_4$ array micro-nanostructures with tunable morphology and composition as integrated electrodes for lithium-ion batteries, Energy Environ. Sci. 6 (2013) 2664–2671.
[33] B.H. Chen, Y.S. Chuang, C.K. Chen, Improving the hydrogenation properties of MgH_2 at room temperature by doping with nano-size ZrO_2 catalyst, J. Alloys Compd. 655 (2016) 21–27.
[34] V. Berube, G. Chen, M.S. Dresselhaus, Impact of nanostructuring on the enthalpy of formation of metal hydrides, Int. J. Hydrog. Energy 33 (2008) 4122–4131.
[35] H.W. Li, Y. Yan, S.I. Orimo, A. Zuttel, C.M. Jensen, Recent progress in metal borohydrides for hydrogen storage, Energies 4 (2011) 185–214.
[36] J. Wang, H.W. Li, P. Chen, Amides and borohydrides for high capacity solid-state hydrogen storage-materials design and kinetic improvements, MRS Bull. 38 (2013) 480–487.
[37] G. Soloveychik, Y. Gao, J. Rijssenbeek, M. Andrus, S. Kniajanski, J.S. Bowman, et al., Magnesium borohydride as hydrogen storage material: properties and dehydrogenation pathway of unsolvated $Mg(BH)_4$, Int. J. Hydrog. Energy 34 (2009) 916–928.
[38] H.W. Li, K. Miwa, N. Ohba, T. Fujita, T. Sato, Y. Yan, et al., Formation of an intermediate compound with a $B_{12}H_{12}$ cluster: experimental and theoretical studies on magnesium borohydride $Mg(BH_4)_2$, Nano 20 (2009) 204013–204020.
[39] M. Chong, A. Karkamkar, T. Autrey, S.I. Orimo, S. Jalisatgi, C.M. Jensen, Reversible dehydrogenation of magnesium borohydride to magnesium triborane in the solid state under moderate conditions, Chem. Commun. 47 (2011) 1330–1332.
[40] C. Zollfrank, K. Gutbrod, P. Wechsler, J.P. Guggenbichler, Antimicrobial activity of transition metal acid MoO_3 prevents microbial growth on material surfaces, Mater. Sci. Eng. C 32 (2012) 47–49.
[41] I. Saldan, C. Frommen, I. Llamas-Jansa, G.N. Kalantzopoulos, S. Hino, B. Arstad, R. H. Heyn, O. Zavoratynska, S. Deledda, M.H. Sørby, H. Fjellvåg, B.C. Hauback, Hydrogen storage properties of $\gamma\text{-}Mg(BH_4)_2$ modified by MoO_3 and TiO_2, Int. J. Hydrog. Energy 40 (2015) 12286–12293.
[42] N. Juahir, N.S. Mustafa, A.M. Sinin, M. Ismail, Improved hydrogen storage properties of MgH_2 by addition of Co_2NiO nanoparticles, RSC Adv. 5 (2015) 60983–60989.
[43] N.S. Mustafa, M. Ismail, Influence of K_2TiF_6 additive on the hydrogen sorption properties of MgH_2, Int. J. Hydrog. Energy 39 (2014) 15563–15569.
[44] F.A. Halim Yap, M.H.I. Ishak, M. Ismail, Enhanced hydrogen storage performance of destabilized $4MgH_2\text{-}Li_3AlH_6$ system doped with Co_2NiO nanopowder, Int. J. Hydrog. Energy 40 (2015) 10131–10138.

[45] N.S. Mustafa, M. Ismail, Enhanced hydrogen storage properties of $4MgH_2$-$LiAlH_4$ composite system by doping with Fe_2O_3 nanopowder, Int. J. Hydrog. Energy 39 (2014) 7834–7841.
[46] Q. Wan, P. Li, Z.L. Li, F.Q. Zhai, X.H. Qu, A.A. Volinsky, Improved hydrogen storage performance of MgH_2-$LiAlH_4$ composite by addition of $MnFe_2O_4$, J. Phys. Chem. C 117 (2013) 26940–26947.
[47] A. Fujishima, K. Honda, Electrochemical photolysis of water at a semiconductor electrode, Nature 238 (1972) 37–38.
[48] B. O'Regan, M. Grätzel, A low-cost, high-efficiency solar cell based on dye-sensitized colloidal TiO_2 films, Nature 353 (1991) 737–740.
[49] P. Pillai, K.S. Raja, M. Misra, Electrochemical storage of hydrogen in nanotubular TiO_2 arrays, J. Power Sources 161 (2006) 524–530.
[50] X. Shen, Q. Wang, S. Guo, B. Liu, Z. Sun, Z. Zhang, Z. Wang, B. Zhao, W. Ding, W-modified CoB supported on Ag-activated TiO_2 for hydrogen generation from alkaline $NaBH_4$ solution, Int. J. Hydrog. Energy 40 (2015) 6346–6357.
[51] A.M. Morales, C.M. Lieber, A laser ablation method for the synthesis of crystalline semiconductor nanowires, Science 279 (1998) 208–211.
[52] Q. Wan, C.L. Lin, X.B. Yu, T.H. Wang, Room-temperature hydrogen storage characteristics of ZnO nanowires, Appl. Phys. Lett. 84 (2004) 124–126.
[53] S. Saipriya, R. Singh, Effect of spacer layer thickness on optical properties of $[SnO_2/Mn]_n$ and $[SnO_2/Co]_n$ discontinuous multilayers, Mater. Lett. 71 (2012) 157–159.
[54] S.Q. Liu, Y.X. Li, M.J. Xie, X.F. Guo, W.J. Ji, W.P. Ding, One-pot synthesis of intestine-like SnO_2/TiO_2 hollow nanostructures, Mater. Lett. 64 (2010) 402–404.
[55] C.L. Zheng, X.Z. Zheng, Z.S. Hong, X.K. Ding, M.D. Wei, Template-free synthesis of SnO_2 nanostructural hollow spheres covered by nanorods, Mater. Lett. 65 (2011) 1645–1647.
[56] Z.Y. Wang, F.P. Wang, M.Y. Li, M.Z. Iqbal, Q.U.A. Javed, Y.Z. Lu, Synthesis, characterization and hydrogen storage characteristics of flower-like SnO_2 porous microspheres, Mater. Lett. 119 (2014) 36–38.
[57] F.K. Butt, M. Tahir, C. Cao, F. Idrees, R. Ahmed, W.S. Khan, et al., Synthesis of novel ZnV_2O_4 hierarchical nanospheres and their applications as electrochemical supercapacitor and hydrogen storage material, ACS Appl. Mater. Interfaces 6 (2014) 13635–13641.
[58] R. Lu, M.L. Li, J. Lei, W.J. Bo, Q.M. Qiang, Y. Ying, Template free synthesis of $BiVO_4$ nanostructures. II. Relationship between various microstructures for monoclinic $BiVO_4$ and their photocatalytic activity for the degradation of rhodamine B under visible light, Nanotechnology 20 (2009) 405602.
[59] B. Peng, Z. Fan, X. Qiu, L. Jiang, G.H. Tang, H.D. Ford, et al., A novel transparent vanadate glass for use in fiber optics, Adv. Mater. 17 (2005) 857–859.
[60] Z. Chen, W. Huang, D. Lu, R. Zhao, H. Chen, Hydrothermal synthesis and electrochemical properties of crystalline $Zn_2V_2O_7$ nanorods, Mater. Lett. 107 (2013) 35–38.
[61] K.F. Butt, C.B. Cao, F. Idrees, M. Tahir, R. Hussain, W.S. Khan, Novel $Zn_2V_2O_7$ hierarchical nanostructures: optical and hydrogen storage properties, Int. J. Hydrog. Energy 40 (2015) 9359–9364.
[62] S.P. Kuang, Y. Meng, J. Liu, Z.C. Wu, L.S. Zhao, A new self-activated yellow-emitting phosphor $Zn_2V_2O_7$ for white LED, Optik—Int. J. Light Electron Opt. 124 (2013) 5517–5519.
[63] F. Idrees, C.B. Cao, F.K. Butt, M. Tahir, I. Shakir, M. Rizwan, Synthesis of novel hollow microflowers (NHMF) of Nb_3O_7F, their optical and hydrogen storage properties, Int. J. Hydrog. Energy 39 (2014) 13174–13179.

Requirements for efficient metal oxide photocatalysts for CO_2 reduction

10

Jennifer Strunk
Leibniz Institute for Catalysis at the University of Rostock, Rostock, Germany

10.1 Introduction

Within the current discussion concerning changing raw material and energy supply, there is no doubt that one molecule around us is present in large excess: carbon dioxide. Although still fairly diluted in air, the CO_2 concentration has not dropped below 400 ppm in the year 2016, which marks a record high [1]. As product of any combustion process of organic molecules, CO_2 is widely produced as by-product in the energy, transport, and production sectors. For example, for a regular passenger vehicle with a fuel consumption of $\sim$7 L per 100 km, a quick estimation of carbon emissions yields an amount of approximately 150 g CO_2 emitted for each single kilometer. It is now widely accepted that CO_2 is one of the main contributors to global warming. Although CO_2 is continuously removed from air by the natural carbon cycle, consisting of photosynthesis in plants as well as inorganic processes, the natural carbon cycle cannot keep up with humankind's enormous CO_2 emissions. Recycling carbon dioxide to fuel (additives) or raw materials for the chemical industry using just the energy of sunlight would be the ideal way to address both the reduction of carbon emissions and the supply of renewable resources for energy and production [2,3]. This is not a new idea: A few chemical processes have already used CO_2 as reactant for decades, for example for the synthesis of urea, salicylic acid, and cyclic organic carbonates [4]. It is also well known that in the industrial methanol synthesis over Cu/ZnO catalysts, CO_2 rather than CO is the primary carbon reactant in the synthesis gas feed [5–7], and modern research initiatives aim at making methanol solely from CO_2 and regeneratively produced H_2 [5]. However, industrial processes such as methanol synthesis currently rely on fossil fuels and traditional synthesis gas generation, so they do not contribute to CO_2 recycling. Thus, other ways have to be found to convert CO_2 without using fossil fuels, and without producing additional carbon emissions.

The first attempts to activate and react CO_2 photoelectrochemically or photocatalytically already date back to the late 1970s and the pioneering works of Halmann [8] and Inoue et al. [9]. Regardless, during the past almost 40 years, significant progress that would make such a process viable on the industrial scale has not yet been achieved [2]. Due to the high thermodynamic and kinetic stability of the CO_2 molecule [4], only extremely low yields have been obtained in previous studies.

Metal Oxides in Energy Technologies. https://doi.org/10.1016/B978-0-12-811167-3.00010-9

A compilation by Kondratenko et al. [5] revealed that in most studies reported between the 1980s and 2013s, the yields of the product methane by photocatalytic processes did not exceed $10\,\mu mol\,g^{-1}\,h^{-1}$. This has also been discussed in the review by Habisreutinger et al. [2]. This is far below a productivity that would attract the attention of the chemical industry. Apart from the main challenge to increase product yields, current photocatalysts that can convert CO_2 also face low selectivity, in which a variety of products such as carbon monoxide, methane, methanol, formic acid, and even longer hydrocarbon chains are often obtained in the same process.

In this chapter, the materials commonly studied in photocatalytic CO_2 reduction are discussed, revealing that the vast majority of them are based on oxides. In the following parts of the chapter, the principles of photocatalysis in general and of photocatalytic CO_2 reduction in particular are briefly explained. Possible adsorption modes of CO_2 on oxides and suggested mechanisms of photocatalytic CO_2 reduction are then discussed, with a particular focus on titanium dioxide-based materials. This is then followed by some remarks on photoreactors and reaction conditions that are important for reliable studies of oxide-based and other materials in photocatalytic CO_2 reduction. In the final sections of this chapter, the current state of the art of photocatalytic CO_2 reduction using some commonly used oxide materials is described.

10.2 Overview of photocatalysts for CO_2 reduction

To obtain an overview of photocatalyst materials studied for CO_2 reduction in the past, a structured literature search was performed, considering all original research papers on photocatalytic CO_2 reduction written in English that were published until the end of November 2014. Conference contributions and review papers were neglected. Only purely inorganic heterogeneous photocatalysts were considered. Anchored or tethered complexes with a ligand sphere on solid surfaces, as well as photocatalysts with major organic components (e.g., metal–organic frameworks, dye-sensitized systems) were neglected. Pure TiO_2 was not counted in a specific publication when it was only used as a reference material.

In total, 277 papers were considered, in which 443 photocatalysts were studied. The tested photocatalyst systems can be grouped into ten material classes:

- *Bulk TiO_2 based*, for example, pure crystalline phases, TiO_2 with (noble) metal cocatalysts, bulk- and surface-doped TiO_2 (total: 48%);
- *Pure* (simple) *oxides* other than TiO_2 (e.g., MgO, ZrO_2, ZnO, Cu_xO) (total: 6%);
- *Complex oxides*, such as mixed oxides, perovskite and pyrochlore structures, for example, titanates, niobates, or tungstates (total 14%);
- Systems based on *isolated titanate* species with tetrahedral TiO_4 geometry (total 7%; more than three quarters of all publications by Masakazu Anpo's group);
- *Nonisolated TiO_2/SiO_2* systems, that is, where TiO_x species were supported on or mixed with silica, but are not present as isolated photoactive TiO_4 species (total 5%);
- Pure *sulfides*, for example, CdS, ZnS, and Bi_2S_3 (total 4%);
- *Carbon-based materials*, for example, C_3N_4, graphene oxide, and carbon nanoparticles (total 3%);

- *Layered double hydroxides* (LDHs, total 2%);
- *Other semiconductors*, for example, GaN and SiC (total 2%);
- *Other isolated metal cations* or bimetallic purely inorganic complexes functioning as photoactive species, for example, Zn-Ti/SiO_2, Cr-Ti/SiO_2, and Zr-O-Co/SiO_2 (total 1%).

The remaining 9% of systems consists of mixed materials (e.g., mixtures of pure oxides, semiconductors supported on complex oxides, etc.; total: 5%) and miscellaneous materials, not belonging to any category (e.g., Ag@AgBr/CNT, NiO_x-Ta_2O_5/graphene, etc.; total: 4%). The percentage of systems studied in each category is visualized in Fig. 10.1.

It becomes evident that almost half of all studied systems were based on any of the forms of pure TiO_2 (48%). Non-TiO_2 oxides and complex oxides comprise another 20% of all studied systems. A further 13% are made up of isolated photoactive TiO_4 species or bulk TiO_2 in a silica matrix, so those systems are also purely oxide materials. The mixed systems, contributing 5% of all studied systems, also contain at least one oxide component. Thus, in summary, 86% of all photocatalyst systems studied until November 2014 for photocatalytic CO_2 reduction were oxide based.

Over recent 2 years (2015 and 2016), research has diverged into a variety of different directions, in which particularly mixed systems and complex composites of multiple materials have been studied, partially in the form of Z-scheme systems. It is thus almost impossible to define material classes. However, a few materials stand out and were studied more frequently, either by themselves or in composites, particularly carbon-based materials, such as graphitic carbon nitride [10–23], graphene, and others [11,12,24–34]. Returning the focus on oxides, materials such as In_2O_3 (as oxide or dopant) [35–43], copper oxides [27–30,44–47], sodium tantalates [48,49],

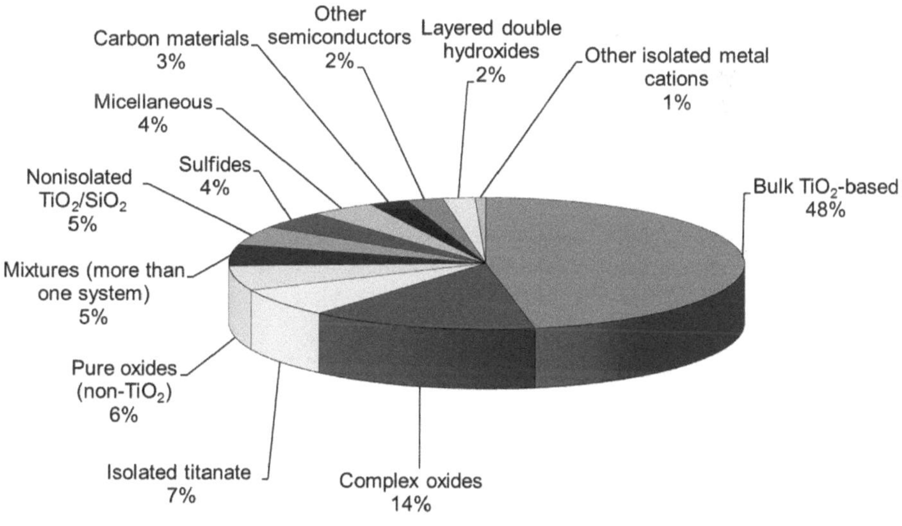

Fig. 10.1 Overview of heterogeneous photocatalyst systems studied for photocatalytic CO_2 reduction until November 2014; specific criteria of literature search: see text.

strontium titanates [47,50,51], and LDH-derived structures [52–58] were frequently applied, either pure or in composites. Furthermore, (nanostructure-engineered and doped) TiO_2 continues to be an interesting and frequently studied material [32–38,44,53,59–75]. Some insight into the different material classes listed above will be given near the end of this chapter. It needs to be noted early on, however, that only for TiO_2 and related materials (e.g., the isolated titanate), some progress toward understanding of the mode of action could be achieved so far [2]. Still, none of the studies on TiO_2-related materials has yet shown significant (i.e., order of magnitude) improvements in yields. Studies of all other systems are mostly standing on themselves, and the potential of these systems appears underexplored. Before the behavior of some of the materials listed above in photocatalytic CO_2 reduction is considered, it is useful first to take a look at the general principles of photocatalytic processes.

10.3 Basics of heterogeneous photocatalysis

A material that is to be used as heterogeneous photocatalyst needs to fulfill a number of functions. Light needs to be absorbed, ideally in the visible region, and charge carriers need to be generated, separated, and migrate to the surface of the material. Apart from those suitable photophysical properties the material needs to feature suitable catalytic properties. Active sites need to be present on the surface where reactants can bind and react with the photogenerated charges to the desired products, ideally with high reaction rates and low activation energy [2]. If a certain material does not feature efficient catalytic active sites on the surface, it is possible to add a cocatalyst [2], which facilitates either the oxidation reaction (hole transfer from semiconductor to reactant) or the reduction reaction (electron transfer from semiconductor to reactant). In the following, the requirements listed above will be briefly introduced. They are also visualized in Fig. 10.2. For a more detailed discussion, the reader is referred to previous excellent publications [2,3,76–81].

The processes occurring upon photoexcitation in a semiconductor can be considered from different perspectives (Fig. 10.2). The spatial view looks at the migration of the charge carriers and their further reactions, such as recombination or redox reactions (Fig. 10.2, left). The energetic view considers the processes occurring within the band structure of the semiconductor (Fig. 10.2, top right). Lastly, the processes occurring at the surface, such as the electron transfer onto an adsorbed acceptor molecule need to be considered (Fig. 10.2, bottom right). All those processes are discussed below.

For a semiconductor, the light energy that can be absorbed by the material is related to its bandgap. The band structure in semiconductors arises from the overlap of near infinite electronic states of the constituent atoms. Those electronic states fully occupied by electrons in the ground state at $T = 0\,K$ form the valence band, and the unoccupied states form the conduction band. In between, a gap exists in which no electronic states are available, which is referred to as the bandgap [2,76,82]. Photoexcitation with light of photon energy larger than the bandgap excites an electron from the valence band to an excited state in the conduction band (Fig. 10.2, top right).

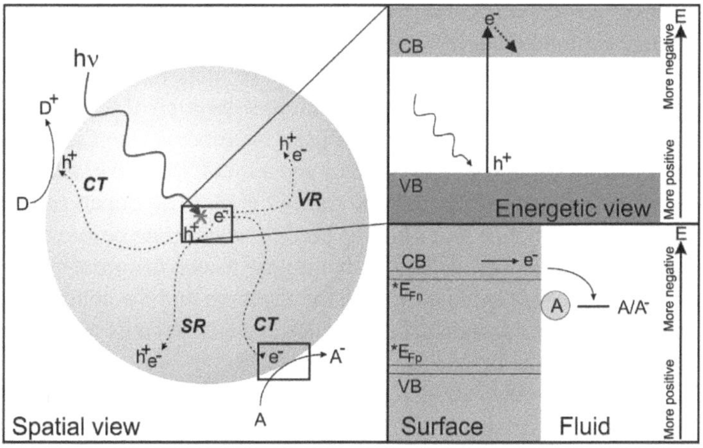

Fig. 10.2 Schematic representation of the processes occurring in heterogeneous photocatalysis. *Left*: Spatial view of charge carrier generation (e^-, h^+) upon irradiation with light of suitable energy, with *VR* = volume recombination, *SR* = surface recombination, *CT* = charge transfer processes. *Top right*: Energetic view of the excitation of an electron from the valence band (VB) to the conduction band (CB) and relaxation processes to the bottom of the conduction band. *Bottom right*: Processes occurring at the surface of a photocatalytic nanomaterial under irradiation, including the formation of quasi-Fermi levels and the transfer of an electron onto an adsorbed acceptor molecule A with a reduction potential less negative than the CB minimum. Based on A.L. Linsebigler, G. Lu, J.T. Yates, Chem. Rev. 95 (1995) 735 (left part) and R. Beránek, Adv. Phys. Chem. (2011) 786759 (bottom right).

Regardless of the exact energy state within the conduction band in which the electron will end up, it will relax practically immediately to the bottom of the conduction band [2,76,83]. The loss of an electron leaves behind an empty state at the top of the valence band, referred to as a hole (h^+). Consequently, the most relevant electronic states for photocatalysis are those at the top of the valence band and the bottom of the conduction band (Fig. 10.2, top right). In TiO_2, the top of the valence band is made up by the oxygen 2p states, while the lowest unoccupied states forming the bottom of the conduction band are Ti 3d states [78,84]. Semiconductors are often classified in two groups, those with a direct bandgap (e.g., TiO_2 rutile) and those with an indirect one (e.g., TiO_2 anatase). When optical transition (excitation and recombination) does not require a change in momentum, the transition is termed as being direct. For indirect optical transitions, both a photon and a phonon need to be involved. For more information, the reader is referred to Refs. [79,85,86].

In semiconductors, the charge carriers generated upon photoexcitation can migrate rather freely through the solid. It is desired that they reach the surface to participate in charge transfer reactions (Fig. 10.2, left, *CT* = charge transfer). However, charge carriers are lost due to recombination phenomena of a photoexcited electron and a hole, which can happen both on the surface (Fig. 10.2, left, *SR* = surface recombination) and in the bulk of the semiconductor (Fig. 10.2, left, *VR* = volume recombination).

This lowers the achievable efficiency of the photocatalytic reaction [2]; therefore, the recombination rate should be as low as possible. Recombination events are influenced by the mobility of charges and by charge trapping. Defects within the semiconductor lattice or an interface with another material that traps one type of the charges (electron or hole sink) can act as recombination sites [2]. Conversely, interfaces between different semiconductors [3] or a semiconductor and a metal [2,3] may also be favorable to separate charges. The term "diffusion length," a material-dependent quantity, is defined as the average distance that a charge carrier can migrate before recombination takes place [2]. In this respect, it may be beneficial to use nanomaterials for photocatalysis: if the particle size is smaller, then the distance that the charge carriers need to migrate before they reach the surface is shorter and the ratio with respect to the diffusion length is more favorable [87].

It is not only the size of the bandgap that determines whether a material can be used for a specific photocatalytic process. The energy levels also need to be properly positioned so that both the reduction and the oxidation reaction are thermodynamically allowed. The thermodynamically spontaneous process for electrons is to fall down to lower energy levels, so a reduction reaction on the surface of a semiconductor is only possible if the excited electron at the bottom of the conduction band is situated at a higher energy compared to the acceptor level of the molecule to be reduced. In other words, the conduction band minimum needs to be at a more negative potential on the electrochemical energy scale than the reduction potential of the acceptor molecule (Fig. 10.2, bottom right). Conversely, if an electron in a donor molecule is supposed to fill up a hole in the valence band, thereby oxidizing the donor, then the valence band maximum needs to be situated at a more positive potential than the oxidation potential of the donor [2,76,88]. This is often summed up in the statement that the band edges of the semiconductor need to straddle the redox potentials of the desired reactions [85]. It is often not sufficient that the potential of the conduction band minimum is just above (more negative than) the potential of the reduction reaction, and that the valence band maximum is just more positive than the oxidation reaction. Kinetic barriers of the reaction require providing an overpotential on each side so that the reaction can run with high rate [2]. The better the catalytic function of the photocatalyst, the lower will be the kinetic barrier and thus the required overpotential, but without any overpotential high rates are not achievable even with good catalysts [2]. It should be kept in mind that taking the conduction band minimum and valence band maximum as decisive factors for thermodynamic feasibility is a simplification, since precisely the quasi-Fermi levels of electrons ($^*E_{Fn}$) and holes ($^*E_{Fp}$) in the semiconductor need to be considered [2,84] (Fig. 10.2, bottom right). However, taking the position of the band edges is a reasonable approximation, particularly for TiO_2 and other n-type semiconductors [2,84].

Apart from existing sufficiently long and reaching the material's surface, the charges generated must be used for specific chemical reactions, ideally with a high rate. Consequently, the material needs to feature appropriate catalytic active sites on the surface that adsorb the reactants and stabilize relevant reaction intermediates. This condition is similar to classical (thermal) heterogeneous catalysis: by adsorption of at least one of the reactants on its surface, the heterogeneous catalyst offers an energetically more favorable reaction pathway and lowers the activation barrier of the process. Knowledge-guided design of the catalytic functionality of a photocatalyst for carbon dioxide reduction is currently still hindered by the fact that the preferred reaction pathway is not yet fully clear. Possible adsorbates and reaction intermediates are discussed in detail below, together with possible reaction mechanisms.

The principles outlined above refer to semiconductors as photocatalysts. However, another important class of oxide-based heterogeneous photocatalysts for CO_2 reduction consists of isolated photoactive (transition) metal cations on or in an inert oxide-based matrix, usually silica. For an isolated photoactive species, photoexcitation promotes an electron from the highest occupied molecular orbital (HOMO) to the excited state in the lowest unoccupied molecular orbital (LUMO). The necessary energy for this process is determined by the gap between HOMO and LUMO. For an isolated tetrahedral TiO_4 (titanate) species (Fig. 10.3, right), photoexcitation can be understood in terms of a ligand-to-metal charge transfer, in which an electron is transferred from one of the oxygen ligands to the central Ti cation [89] (Eq. 10.1):

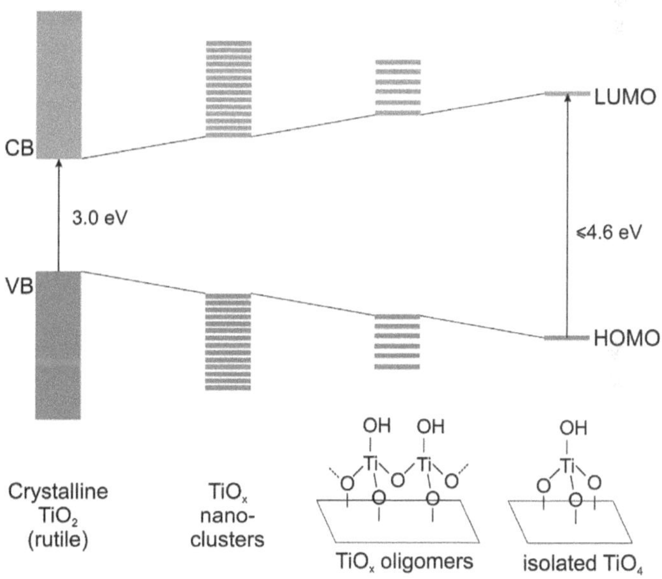

Fig. 10.3 Schematic representation of the quantum size effect; development of electronic states when going from the semiconductor TiO_2 (thermodynamically stable rutile phase with direct bandgap) over nanoclusters to TiO_2 oligomers and to the isolated titanate species on inert supports.
Modified from M. Matsuoka, M. Anpo, Local structures, excited states, and photocatalytic reactivities of "single-site" Ti-oxide photocatalysts constructed within zeolites or mesoporous materials, in: M. Anpo, P.V. Kamat (Eds.), Environmentally Benign Photocatalysts: Applications of Titanium Oxide Based Materials, Springer Science-Business Media, New York, 2010; D.W. Bahnemann, C. Kormann, M.R. Hoffmann, J. Phys. Chem. 91 (1987) 3789.

$$\left[Ti^{4+}-O^{2-}\right]^{h\nu}\ {}_{h\nu}\left[Ti^{3+}-O^{-}\right] \quad (10.1)$$

If the photoactive species is supported on an insulating support, neither type of charge carrier can be transported away. Instead, both electron and hole are localized at the photoactive species, requiring that the same species also needs to be the active catalyst. Only if conducting supports are used, or if noble metals are brought in close proximity, then electrons may be transferred to or from the photoactive species.

If materials based on the same transition metal cation are considered—most importantly bulk TiO_2 and the isolated tetrahedral titanate (TiO_4) species—then the bandgap excitation in the semiconductor always requires less energy than the corresponding HOMO-LUMO excitation in the isolated photoactive species (Fig. 10.3). This is a consequence of the interacting electronic states from neighboring atoms in a crystal lattice, leading to the formation of wide bands in extended semiconductors [82]. Since both the occupied states (the valence band) and the unoccupied states (the conduction band) develop into a broad and practically continuous band of energy states, the gap between them is narrower than the gap between a single HOMO and LUMO of an isolated species. Once the dimensions of a semiconductor become small enough, it becomes noticeable that less and less electronic states overlap, so that the band structure transforms more and more into an assortment of discrete electronic levels. This is then reflected in an increase of the energy required for photoexcitation [80,90], an observation commonly referred to as the *quantum size effect* [76,89] (Fig. 10.3). Considering the thermodynamic requirements that the excited electron should always be at a more negative potential than that of the reduction reaction to be carried out, and the hole at a more positive potential than the oxidation reaction, the benefit of using isolated tetrahedral titanate sites as compared to bulk TiO_2 can be easily rationalized with the more negative and positive potentials of the electron in the LUMO and the hole in the HOMO compared to photogenerated charge carriers in the conduction band and the valence band of crystalline TiO_2, respectively.

Many publications highlight nanosize and nanostructuring of oxides as possible means for the photocatalyst improvement [91–99]. Looking at this from a physical and chemical point of view, nanostructuring of the photocatalyst is beneficial for two reasons: firstly, the distance that the charges need to migrate before they reach the surface is shorter, which is favorable with respect to the carrier diffusion length. Secondly, nanostructured materials have a high surface-to-volume ratio, so that they may offer more catalytic active sites on the surface [2,80,87,91]. However, it should also be considered that catalytic active sites do not necessarily scale simply with the total surface area. An example is pure zinc oxide as (thermal) heterogeneous catalyst in methanol synthesis, where a simple increase in surface area does not increase catalytic activity further, since well-developed polar surfaces are required to expose catalytically highly active sites, such as oxygen vacancies [100–102]. It has also been suggested that the presence of polar surfaces on ZnO is beneficial for its activity in photocatalytic dye degradation reactions [103]. Furthermore, when nanostructured materials are used in photocatalysis, it must also be considered that a larger amount of defects is likely to be present in the material, and that interparticle charge transfer may be slowed down. For a detailed discussion of the effect of nanostructuring, the reader is referred to Ref. [97]. In any material, defects can be grouped into *point defects*, which are discussed below; *linear defects*, for example, dislocations; *planar defects*, in particular surfaces and grain boundaries; and *spatial defects*, such as small nanoparticles in a host phase. The large group of *point defects* in oxides comprises cation vacancies, oxygen vacancies, interstitial cations, electronic defects, and foreign ions in lattice or interstitial sites. Interstitial oxygen, on the other hand, is extremely rare. To satisfy the condition of charge neutrality of the crystal, charges added to or removed from the crystal must be internally compensated for [79]. While all defects will affect charge carrier lifetime, that is, charge migration and recombination rates, it is particularly relevant to consider oxygen vacancies as defects in oxides: TiO_2 is a native n-type semiconductor due to a natural occurrence of oxygen vacancies. This is related to the presence of Ti^{3+} cations [79,81], because the formation of oxygen vacancies by removal of oxygen from the lattice is accompanied by the transfer of two electrons to the neighboring titanium cations, reducing them from the +4 to the +3 oxidation state [104–106]. In ZnO, an "empty" V_O^{2+} vacancy may trap either one electron (V_O^{+}) or two electrons (V_O^{0}) [107]. While it was believed for a long time that oxygen vacancies are also the source of intrinsic n-type semiconducting behavior of ZnO, recent density functional theory calculations and electron paramagnetic resonance investigations came to the conclusion that this assignment cannot be correct. Instead, the n-type conductivity may be related to unintentional doping, for example with interstitial hydrogen [108]. While the topic of defects in oxides is certainly relevant for considerations of photocatalytic activity, a detailed discussion is outside the scope of this chapter. Instead, it will just be pointed out that for the case of TiO_2 (and Cu/TiO_2), the presence of oxygen vacancies, in particular on the brookite polymorph, has been suggested by the group of Ying Li to be beneficial for its activity in photocatalytic CO_2 reduction [109,110]. On the basis of high-level computational results, such as excited-state ab initio calculations of CO_2 adsorbed on clusters from the (010), (101), and (001) anatase surface planes, Indrakanti et al. [111] even came to the conclusion that on TiO_2 electron transfer to CO_2 is possible solely in the presence of oxygen vacancies. Only then, a strongly bound, bent adsorbate species can be formed from CO_2, and electron transfer becomes possible. While this is certainly an important piece of information, it should be kept in mind that oxygen vacancies or Ti^{3+} may also participate in stoichiometric reactions [111] in which CO_2 is reduced and/or oxygen atoms are removed from CO_2, so to put forward such a hypothesis requires careful evaluation of the catalytic nature of the overall process.

10.4 Involved (elementary) reactions and energetic requirements of CO_2 reduction

It has recently been suggested that photocatalysts for CO_2 reduction could only be improved substantially if an in-depth understanding of the reaction mechanism is achieved. However, such a level of understanding has not been reached [2]. Different reaction pathways may be in operation dependent on the reaction conditions and employed catalysts, and some fundamental questions concerning the elementary step mechanism have not yet been solved.

In the carbon dioxide molecule, the carbon atom is in the +4 oxidation state. For the formation of methane, it needs to be reduced by a total of eight electrons to reach the oxidation state −4. Not only is this an eight-electron reaction, but it also requires the removal of both oxygen atoms from the molecule. Formation of methane is not the only possible reaction. The C_1 reaction products observed in photocatalytic CO_2 reduction also comprise carbon monoxide and formic acid (two-electron processes), formaldehyde (four-electron process), and methanol (six-electron process). Products containing two or more carbon atoms have also been observed, such as ethane, ethylene, ethanol, oxalic acid, acetaldehyde, or higher hydrocarbons and alcohols [2]. The ideal reducing agent and proton source for the formation of hydrocarbons and oxygenates from CO_2 would be plain water [112], but then also the water splitting reaction needs to proceed concomitantly. In any case, reaction stoichiometry requires the formation of products from the surplus oxygen anions and holes, ideally gaseous O_2, but another potential product would be hydrogen peroxide [2].

Since CO_2 and H_2O are thermodynamically extremely stable, forming any C- and H-containing products from those two molecules is always an endergonic reaction ($\Delta G > 0$), that is, the reaction needs to proceed thermodynamically uphill [2,112]. It has become common practice to refer to this reaction as being "photocatalytic," but the correct term would be "photosynthetic" [2]. (Heterogeneous) Catalysts, by definition, can only accelerate reactions that are anyhow thermodynamically downhill ($\Delta G < 0$) [113], since they only lower the activation energy by offering an energetically more favorable reaction pathway on their surface. The possibility to carry out a photosynthetic (endergonic) reaction with a photocatalyst originates from the fact that the energy is provided by the incident photons [2] by the generation of an excited state, and the fact that the oxidation and reduction reaction are spatially separated [113]. This is the reason for the thermodynamic requirement outlined above, stating that the conduction band minimum of the semiconductor always needs to be at a more negative potential compared to the reduction potential of the acceptor molecule, while at the same time the valence band maximum must be at a more positive potential than the oxidation potential of the donor molecule. Only if this condition is fulfilled, the separate half reactions, reduction of the acceptor and oxidation of the donor, are thermodynamically allowed [113].

As mentioned above, reaction stoichiometry requires that both photogenerated electrons and holes are consumed, and that oxygen (or hydrogen peroxide) is formed as by-product from an oxidation reaction of one or both oxygen anions originating from the CO_2 molecule and/or the secondary reactant H_2O. The following Eqs. (10.2)–(10.4) exemplify the formation of methane in an eight-electron process. Overall reaction:

$$CO_2 + 2\,H_2O\,^{h\nu,} \qquad {cat}\,_{h\nu,\mathrm{cat}}\; CH_4 + 2\,O_2 \qquad (10.2)$$

Reduction:

$$CO_2 + 8\,H^+ + 8\,e^- \rightarrow CH_4 + 2\,H_2O \tag{10.3}$$

Oxidation:

$$2\,H_2O \rightarrow O_2 + 4\,H^+ + 4\,e^- \tag{10.4}$$

However, with few exceptions (e.g., Refs. [46,109,114,115]), the formation of oxygen is usually not reported in the literature, either because it was not studied at all, or because oxygen was not formed. This challenge of oxygen evolution is related to photocatalytic or photoelectrochemical water splitting [116], where this step, involving four holes per oxygen molecule, is thought to be the more demanding step in the overall process [2], associated with high kinetic barriers, or, in other words, with high overpotentials [117]. Apart from the difficulty to form the oxygen, it may also not desorb and remain on the surface of the photocatalyst. Alternatively, other products may be formed instead of gaseous dioxygen, such as hydroxyl or superoxide radicals. Due to their high reactivity, those species might oxidize products or intermediates of photocatalytic CO_2 reduction [2] and thereby strongly limit achievable yields. In this respect, the rapid degradation of hydrocarbons on photocatalysts for CO_2 reduction has been demonstrated previously [118]. Highly efficient catalysts for oxygen evolution, for example iridium or ruthenium oxide, are extremely rare and expensive, again strongly limiting the potential for large-scale application. All in all, it should be stated that the oxidation reaction is the much less studied process in photocatalytic CO_2 reduction [2,119], so that the fate of the holes and the surplus oxygen atoms is still virtually unknown.

The one-electron reduction of CO_2 to the anion radical $CO_2^{-}\cdot$ requires a potential of $-1.9\,eV$ and is therefore an extremely challenging reaction. This is due to the high chemical stability and extremely low electron affinity of the CO_2 molecule [2,3]. Electron addition requires breaking molecular symmetry. The formerly linear CO_2 molecule bends after taking up an electron due to repulsion of the additional electron located at the carbon atom and the free electron pairs of the oxygen atoms. Since such processes are energetically highly unfavorable, the LUMO of the CO_2 molecule is located at very high energy [4]. The conduction band of virtually no semiconductor is appropriately positioned to allow this reaction without any applied bias [2].

So, in a pure photocatalytic process in the presence of a gas phase above a semiconductor, this elementary step is not expected to occur. However, it should be considered that adsorption processes of CO_2 on the semiconductor surface may already bend and partially charge the CO_2 molecule, thus destabilizing it. This lowers the LUMO and makes electron transfer reactions slightly more favorable [2,4]. Alternatively, instead of a single electron transfer, proton-assisted transfers of multiple electrons may occur. With respect to the required potential, they are more easily achieved from a thermodynamic viewpoint [2,3]. However, it appears unreasonable to assume that such complicated processes involving multiple charge carriers and many reaction

Table 10.1 Some reactions potentially related to photocatalytic CO_2 reduction, hydrogen and oxygen evolution together with their redox potentials at pH = 7 in aqueous solution.

Reaction	E^0 (V vs NHE)
$2\,H_2O \longrightarrow O_2+4H^++4e^-$	+0.81
$2H^++2e^- \longrightarrow H_2$	−0.42
$CO_2+e^- \longrightarrow CO_2^{\bullet-}$	−1.90
$CO_2+2H^++2e^- \longrightarrow HCOOH$	−0.61
$CO_2+2H^++2e^- \longrightarrow CO+H_2O$	−0.53
$CO_2+4H^++4e^- \longrightarrow H_2CO+H_2O$	−0.48
$CO_2+6H^++6e^- \longrightarrow CH_3OH+H_2O$	−0.38
$CO_2+8H^++8e^- \longrightarrow CH_4+2H_2O$	−0.24

(Values of standard potentials taken from L. Guo, Y. Wang, T. He, Chem. Rec. 16 (2016) 1918.)

partners are elementary steps, especially because hardly any evidence for such concerted reactions has so far been obtained [2]. Table 10.1 presents an overview of reactions that have been suggested to be potentially related to photocatalytic reduction of CO_2 and evolution of H_2 and O_2, together with their redox potentials at pH = 7 [3].

Studies of reaction mechanisms in photocatalysis in general and in CO_2 reduction in particular are vastly complicated by the fact that reaction intermediates are often radical species with high reactivity and short lifetime. The high reactivity implies numerous possibilities in which they can reach further, leading to strongly branched and complex reaction networks [2]. The short lifetimes make it difficult to detect all relevant intermediates.

Already the first stages of any reaction mechanism of photocatalytic CO_2 reduction are a matter of debate. In general, additional to simply linearly physisorbed CO_2, (reactive) adsorption of CO_2 on oxide surfaces can generate a variety of adsorbed surface species, namely carbonates, bicarbonates, carboxylates, and formate species. A very thorough discussion of CO_2 adsorption on oxides has been provided by Busca and Lorenzelli [120]. The possible structures that can be formed on oxide surfaces are visualized in Fig. 10.4 together with the original enumeration from Ref. [120].

Another possible surface intermediate is the carboxyl species (not shown in Fig. 10.4), an actual organic acid functional group (*COOH; asterisk indicates an adsorbed species). Regarding the composition, carboxyl species are similar to formates (*HCOO), but in carboxyl species the hydrogen atom is attached to an oxygen atom instead of the central carbon atom. It should also be noted that on zinc oxide an unusual tridentate carbonate species has been detected previously [121].

While carbonate species are formed whenever a surface (lattice) oxygen atom is involved in CO_2 adsorption, bicarbonates can be formed from a reaction of CO_2 with a surface hydroxyl group. The formation of formates requires some activated form of hydrogen. The different adsorbates can be distinguished by vibrational spectroscopy, as it has been discussed in detail in Ref. [120]. For TiO_2 in photocatalytic CO_2 reduction, enhancing the formation of (bi)carbonates on the surface by addition of sodium cations has recently been shown to be detrimental for

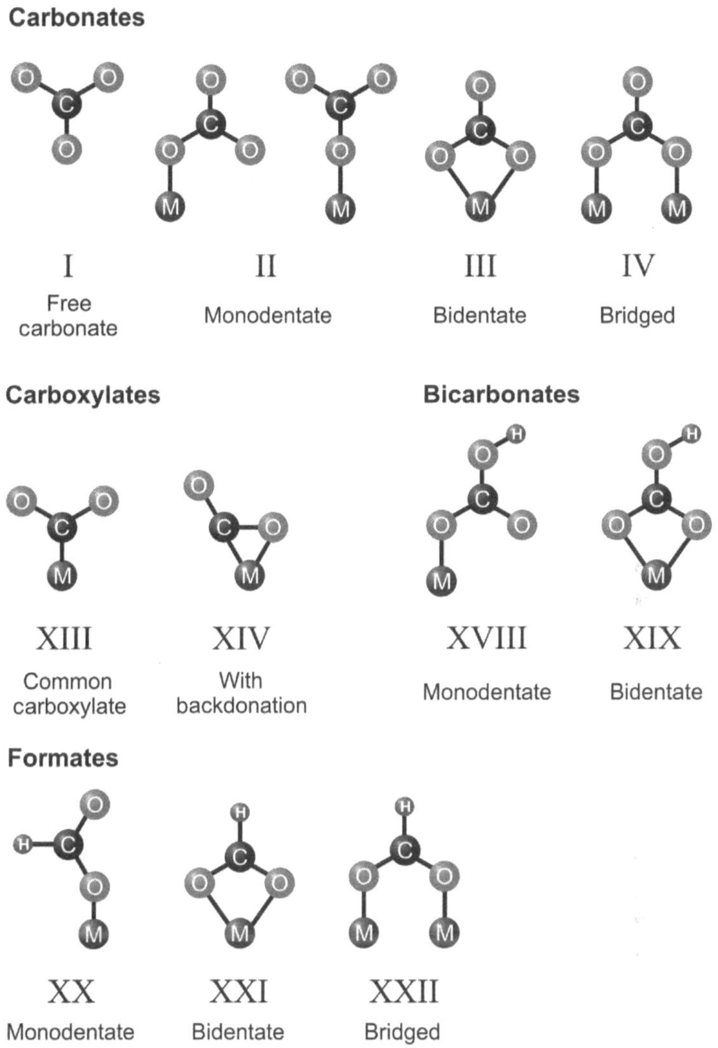

Fig. 10.4 Possible structures of adsorbates formed upon CO_2 adsorption on oxide surfaces together with the original classification and enumeration by Busca and Lorenzelli [120].

methane production [59]. Furthermore, for Cu/TiO_2, where the participation of carbonaceous impurities on the surface in photocatalytic CO_2 reduction was clearly demonstrated, it was still found that surface carbonates decompose mostly to CO_2 under illumination [122]. It is thus possible that the formation of (bi)carbonate on the surface is detrimental rather than beneficial, either because those species are too stable for any further reaction, or because they act as hole traps rather than attracting electrons needed for a reduction reaction [2,123]. On the other hand, for a doping of TiO_2 with bismuth it has been found that a stronger adsorption of CO_2 was beneficial to increase the yields of methane [62]. An enhanced adsorption of CO_2 was also discussed as one reason for the improved activity of TiO_2 coated with a thin silica shell, although the presence of Ti—O—Si bonds was also suggested to play a role in causing an improved charge carrier lifetime [73]. In contrast to this finding, acidification of the TiO_2 surface by treatment with sulfuric acid increased methane yields as well. In those nanosheet samples, with predominantly exposed {001} and {101} facets, the improvement caused by acid treatment was attributed to the formation of surface hydroxyl groups and oxygen vacancies/Ti^{3+} species. This then increased charge carrier lifetime [66].

So far, actual reaction mechanisms were only discussed for TiO_2, because for this material the most detailed studies have been performed. Currently (2016), the mechanistic steps are not unambiguously clear, but three main routes have been discussed in the literature [2,124–126], which are visualized in Fig. 10.5 and are discussed below.

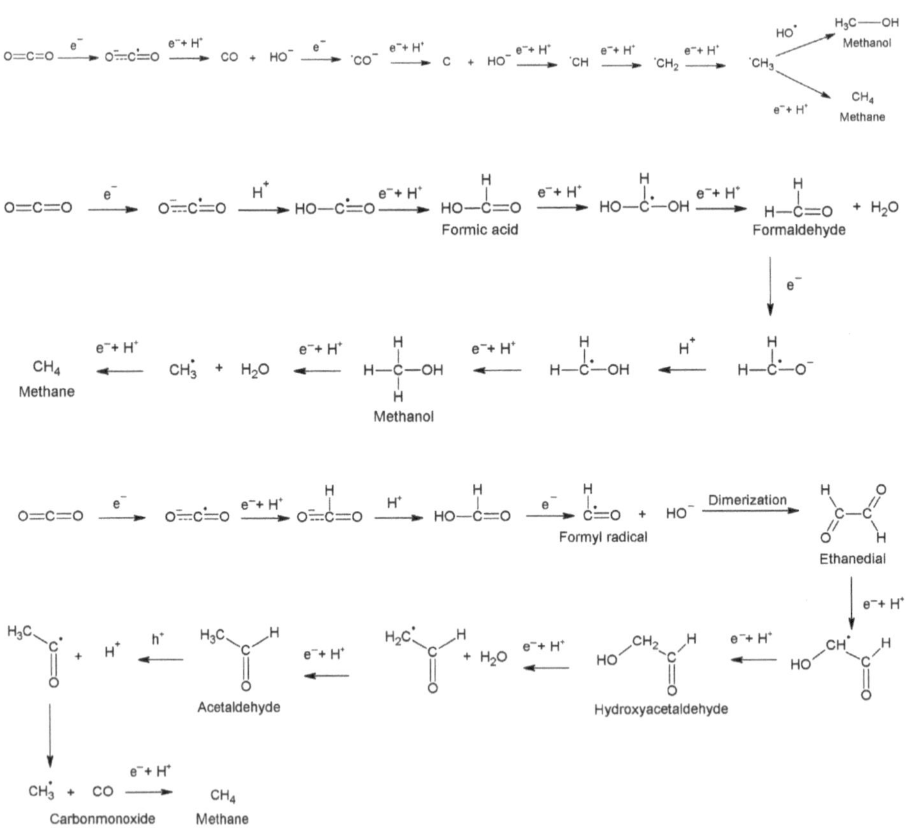

Fig. 10.5 Schematic representation of three proposed pathways of photocatalytic CO_2 reduction.
Modified from S. Cao, J. Low, J. Yu, M. Jaroniec, Adv. Mater. 27 (2015) 2150; K.A.S. Fernando, S. Sahu, Y. Liu, W.K. Lewis, E.A. Guliants, A. Jafariyan, P. Wang, C.E. Bunker, Y.-P. Sun, ACS Appl. Mater. Interfaces. 7 (2015) 8363.

Following the nomenclature in Ref. [2], all three mechanisms are named after a unique intermediate, being referred to as *formaldehyde pathway*, *carbene pathway*, and *glyoxal pathway*. It must be noted that the observation of different intermediates and thus reaction mechanisms may be related to the considerably varying reaction conditions applied in previous studies, in particular with respect to light intensity, liquid- or gas-phase reactor operation, or the concentration of CO_2, H_2O, and potential other additives. In the following discussion, the different mechanisms will be briefly described.

10.4.1 Carbene pathway

This mechanism has been supported already in very early studies by the group of M. Anpo for isolated titanate species on silica as photocatalysts, in particular based on EPR detection of C, H, and CH_3 radicals [127,128]. The anion radical $CO_2^{-}\cdot$ is initially split to yield CO and a hydroxyl species. With the assistance of another proton and electron transfer, the second oxygen atom is also removed as hydroxyl species, and carbon atoms are obtained. These are successively hydrogenated by hydrogen radicals until a methyl radical (CH_3) is obtained. This can be hydrogenated to yield methane, or it can undergo a reaction with a hydroxyl radical to yield methanol. So, different from the formaldehyde mechanism, methanol and methane are not formed consecutively, but in parallel reactions.

10.4.2 Formaldehyde pathway

This mechanism starts from the anion radical $CO_2^{-}\cdot$, from which a carboxylate (COOH) and then formic acid is formed. A dihydroxymethyl [$HC(OH)_2$] radical formed by another H radical addition then dehydrates to the key intermediate formaldehyde (H_2CO). Another electron and proton transfer leads to the formation of hydroxymethylene (H_2COH), from which methanol is consecutively formed. The further reduction of the methanol leads to the formation of methyl radicals (CH_3) with release of water, and finally the formation of methane [2]. This mechanism has been described as the "simplest" reaction scheme, because it can be considered a sequence of four consecutive two-electron proton-coupled reactions [125]. It should be noted that although the formaldehyde pathway has been discussed in detail in Ref. [125], the same article provides convincing evidence against this reaction pathway.

10.4.3 Glyoxal pathway

This reaction pathway, also named for a key reaction intermediate, is considerably different from all others since it involves the intermediate formation of a C—C bond. It is believed that formyl radicals (CHO·) are formed initially. The reaction sequence resembles the formaldehyde mechanism up to the formation of formic acid. However, since formic acid has a strong tendency to be oxidized rather than further reduced, Shkrob et al. [124,125] concluded that only a concerted process involving electron transfer from the semiconductor CB to the acid coupled to a transfer of an oxygen atom from a hydroxyl group to a surface titanium atom is a feasible process for its further reaction. Formyl radicals then undergo a dimerization to glyoxal [$(CHO)_2$], which is afterwards further hydrogenated to hydroxyacetaldehyde and acetaldehyde. The splitting of acetaldehyde then yields CO and a methyl radical. The latter is then hydrogenated to yield methane. This reaction pathway does not involve methanol as intermediate or product at all, but it implies that the formation of methane is coupled to the formation of CO as by-product. Another feature of this mechanism is that it involves both reduction and oxidation steps. It is thus possible only on semiconductor powders in a photocatalytic process, and not in photoelectrochemical processes, where reduction and oxidation reactions are separated on photocathode and photoanode, respectively [2].

All mechanisms discussed above have been proposed for titanium-oxide based photocatalysts, either for crystalline TiO_2, or for isolated tetrahedral titanate species supported on or in silica. It needs to be noted that although those two photocatalysts are discussed together here, the reaction mechanism on both of them is likely very different, because TiO_2 is a semiconductor, while the isolated titanate species behaves like a molecular photocatalyst. On the other hand, an important role of four-coordinate Ti atoms also in TiO_2 has been suggested on the basis of computational results [129], so similarities might exist.

The carbene pathway is often proposed with respect to product formation on isolated titanate species. This appears reasonable, considering that only one electron-hole pair resulting from HOMO-LUMO excitation is located at each photoactive site. Accumulation of several electrons or holes for multielectron steps is thus not possible. In this respect CO has already been observed as primary single-photon, two-electron product on isolated titanate species [130]. Although formic acid, formaldehyde and methanol have frequently been reported as intermediates or products of photocatalytic CO_2 reduction on the semiconductor TiO_2 and related materials [9,63,131–136], the formaldehyde mechanism, although often proposed, is not beyond doubt. It has not been verified experimentally [2], and in a kinetic study [137] the concentration profiles of methanol and methane production did not match a consecutive formation according to the formaldehyde mechanism. In this kinetic study, the parallel formation as suggested in the carbene pathway was thus more likely [137]. In a related manner, Dey et al. [138] and Koirala et al. [139] found that methanol is preferably oxidized on TiO_2, and it was concluded that CH_4 formation did not proceed through a methanol intermediate. Both methanol and formaldehyde were found by Shkrob et al. to be primarily oxidized instead of being reduced, so after their formation they may again act as sacrificial hole scavengers [125]. Glyoxal was instead suggested as potential intermediate that is more likely to be reduced than oxidized, leading to the suggestion of the glyoxal pathway [125]. Very recently, the reactions of formic acid, formaldehyde and methanol on TiO_2 were studied under similar reaction conditions as applied in photocatalytic CO_2 reduction, but from none of these molecules, the formation of methane was observed [59]. When instead, acetic acid or acetaldehyde was offered to the photocatalyst, the formation of methane took place, and product distributions resembled those of photocatalytic CO_2 reduction [59]. These results strongly support the glyoxal mechanism as suggested by Shkrob et al. [125]. Furthermore, acetaldehyde has also been found previously as a product of CO_2 reduction, for example using doped TiO_2 [63].

A conceptually very different mechanism for photocatalytic CO_2 reduction to CH_4 has been suggested by Look and Gafney, not for TiO_2, but for amorphous tungsten oxides on porous Vycor glass [140]. The role of the oxide in the observation of the formation of methane was explained by an *excited-state acid-base mechanism*. In this mechanism, the oxide is not a source of electrons. Instead, it shuttles electrons and holes between acidic and basic regions on its surface, where the reduction of chemisorbed CO_2 and the splitting of water can occur thermodynamically more favorably, possibly even exergonically. This hypothesis is derived based on the estimation of reaction potentials (free energies) for the reactions given in Eqs. (10.3), (10.4) as function of pH. While the former becomes more and more exergonic with higher acidity (lower pH), the latter becomes more favorable at increasingly basic pH. For reaction (10.4) to become exergonic, however, a hypothetical pH of 21 would be required. Look and Gafney suggest that the population of the WO_3 conduction band induces a charge polarization that leads to the formation of electron-deficient (acidic) and electron-rich (basic) regions in the metal oxide, where reactions (10.3), (10.4) can occur more readily. Chemisorbed CO_2, possibly in the form of formate species, is suggested to react preferably compared to physisorbed CO_2. This mechanism can also explain why amorphous WO_3 can drive CO_2 reduction, although the conduction band minimum ($+0.31 \pm 0.11$ V) is by far not negative enough to transfer a single electron onto CO_2 (-1.9 V) [140].

In summary, it cannot be conclusively decided which reaction mechanism may be in operation on the different photocatalysts, not even for the simple case of TiO_2. However, the discussion given above may be a guideline for further research on TiO_2 and other oxide materials.

10.5 Some remarks concerning photoreactors and reaction conditions

As discussed above, significant progress is still needed with respect to the development of photocatalysts and the insight into the reaction mechanism. A prerequisite to achieve these goals is the gathering of reliable data sets. This is a significant challenge, considering that the yields currently obtained in photocatalytic CO_2 reduction on virtually any photoactive material are very low. A key issue in this respect is appropriate and reliable reaction conditions. It has been shown impressively by Yang et al. [122,131] that reaction conditions of highest purity and extensive photocatalytic cleaning of the photocatalysts are required to obtain correct product yields. Similarly, Cybula et al. [60] have shown for the case of Ag-TiO_2, that hydrocarbon formation originated from contaminants in the catalyst and not from the offered $^{13}CO_2$ in the gas phase. Product formation over bare TiO_2 and other noble metal-loaded TiO_2 photocatalysts was also observed in inert

(CO_2-free) atmosphere, in spite of extensive calcination prior to the photoreaction, which is clear proof that carbon sources other than CO_2 served as reactant [60,122,131]. For Au/TiO_2/SBA-15 prepared by photodeposition in aqueous methanol solution it has been shown that the amounts of methane and ethane liberated in a cleaning experiment greatly exceeded those formed in the actual photoreaction [133]. It may be envisioned that residues from catalyst synthesis (e.g., metal-organic precursors), naturally adsorbed (bi)carbonates on oxide surfaces, or even grease from sealant materials can react photocatalytically to form "products" such as carbon monoxide, methane, methanol, or other hydrocarbons. If such processes occur, the potential of the studied material to photocatalytically convert CO_2 into useful chemicals is greatly overestimated.

There are different possibilities to assure that the detected products have actually been formed from CO_2. Ideally, the mass balance of the reaction should be solved, that is, stoichiometric amounts of methane and oxygen should be detected, ideally together with the corresponding conversion of CO_2. Since the quantification of extremely small amounts of oxygen and very small CO_2 conversions is very challenging, this ideal approach is often not feasible. Instead, a very elegant method is the use of $^{13}CO_2$ as reactant and the detection of ^{13}C-labeled products [2,60,122]. This is only possible, though, if product detection is sensitive to the molecular mass of the products, that is, by using a mass spectrometer, or if products are detected by their vibrational bands and the corresponding shift due to isotopic labeling. If other means of trace gas analysis are used, for example, a gas chromatograph with a flame-ionization detector for the detection of hydrocarbons, then the change of the carbon isotope will remain undetected. In that case, it is critical to conduct a blank experiment in which CO_2 is not fed to the reactor. It is common practice to perform blank experiments in photocatalysis, for example, a dark experiment to assure that the reaction is not thermally induced, or a reaction without catalyst to assure that it is not a pure photoreaction. However, for CO_2 reduction, it is critical to perform an experiment in which all reaction conditions are exactly the same as in the actual CO_2 reduction experiment (e.g., light intensity, amount and pretreatment of catalyst, flow rates, sealant materials), but in which the CO_2 has been exchanged by an inert gas such as helium or nitrogen. In such an experiment, products must not be detected, because they cannot be formed from CO_2 and instead must be formed from impurities. If product formation is detected, then the amounts of products detected in the actual CO_2 reduction must be corrected by the amounts detected in the blank experiment to prevent the overestimation of product formation.

Similar reaction conditions to such a blank experiment can be used for the cleaning of the photocatalyst: if a photoactive material is illuminated in the presence of water (but no CO_2) then the photogenerated charges will degrade any organic impurities present in the reacting system. Such a cleaning procedure can be carried out until no more carbon-containing products are detected or until a stable baseline is reached. When CO_2 is afterwards added to the reaction mixture with the clean photocatalyst, and an increase in product formation is detected, then this is strong evidence for product formation from CO_2 [131,133].

10.6 Examples of non-TiO_2 oxide materials for CO_2 reduction

Since the above discussion mainly concerned TiO_2, some characteristic of other classes of oxide materials in CO_2 reduction is discussed in the following. It should be noted that it is often not possible to distinguish clearly which role each of the components of a complex composite material fulfills, that is, that of a photoabsorber, that of a (co-) catalyst, or possibly that of a redox mediator. Thus, this section is ordered with respect to the materials, regardless of which function(s) they take over. Furthermore, focus is put on the product distribution and the nature of the employed material or composite. It makes little sense to compare the total yields, since the reaction conditions are usually not comparable. Such a description of material classes can never be complete, owing to already roughly 40 years of research on CO_2 reduction. Therefore, only a few examples are highlighted here that have been of particular interest in recent years.

10.6.1 *Indium oxide*

All studies on indium oxide for CO_2 reduction have been published very recently. In an initial report [141] of a composite of In_2O_3 and g-C_3N_4 activity tests were performed both for the hydrogen evolution in the presence of a sacrificial reagent and for the CO_2 reduction. Methane was the only hydrocarbon product reported. The three times higher yield compared to g-C_3N_4 and even four times higher yield compared to pure In_2O_3 was attributed to charge separation between the two materials, whereby the lower CB edge of In_2O_3 caused the excited electrons to end up in this material. The holes, on the other hand, ended up in the graphitic carbon nitride. The deposition of 0.5% Pt additionally doubled the methane yields [141]. On its own, In_2O_3 has mostly been studied in the hydroxylated, oxygen deficient form, $In_2O_{3-x}(OH)_y$ [142]. The reaction conditions, other than for most studies highlighted above, were those of the reverse water gas shift (RWGS) reaction, so the co-reactant was not water, but hydrogen, and expected products are carbon monoxide and water. Indeed, large amounts of CO are formed, at 150°C under light irradiation, even under simulated solar light and with just visible light (>420 nm). The material was found stable for a reaction time of 4 days. ^{13}C-labeling was employed to confirm that CO was formed from CO_2. Small amounts of CH_4 formed with and without light irradiation were instead found to originate from impurities. In this study, the activity was already traced back to the presence of surface oxygen vacancies and hydroxyl groups. In a further study [38], spectroscopic and kinetic evidence, together with computational results, provided further insight into the exact roles of those surface species. Analogously to molecular frustrated Lewis pair sites, the hydroxyls on the surface are Lewis bases, and their adjacent indium sites—next to oxygen vacancies—are Lewis acids. On those sites, hydrogen can be heterolytically split and CO_2 can be adsorbed. CO and water are then formed, whereby CO_2 dissociation is the rate-limiting step [38]. In the excited state, Lewis acidic and basic characters are enhanced, thus leading to a lowered activation barrier under irradiation [40]. From a more practical point of view, it was found that the choice of the precursor for the $In_2O_{3-x}(OH)_y$ influenced photocatalytic activity, with indium hydroxide producing the most active photocatalysts [41]. Very recently, the defective hydroxylated indium oxide has been used for photocatalytic RWGS in a hybrid device, in which it was coated on silicon nanowires. This led to a ~6-fold increase in CO yields [43]. Alternatively, the indium oxide itself can be used in the form of nanorods [42]. Tahir et al. [35–38] instead used indium oxide as an additive to TiO_2. Both the reactions of CO_2 with water and with hydrogen were studied. In the reaction with water, doping with indium markedly increased the yield of CH_4 and longer hydrocarbons (up to C_3), while CO was the main product without indium [38]. The increased activity was attributed to improved interfacial charge transfer. Product formation was explained with a mechanism strongly resembling the carbene mechanism explained above. Rates were fitted with a Langmuir-Hinshelwood kinetic model [38]. In the reaction with hydrogen at temperatures at or above 120°C [143], selectivity switched back to CO as main product. The yield of CO was further improved by adding copper (oxide) to the photocatalyst. In this study, an additional significant increase in yields was brought about by an improved reactor design [143]. When nickel was added to In-doped TiO_2 instead of Cu, CH_4 was the main product in a reaction with water, while in a reaction with hydrogen CO was predominantly formed [36]. Again, the improved reactor design significantly increased CO yields [37]. In a very recent study, methanol was also found as by-product on Cu- and In-codoped TiO_2 [35]. When gold was added to In-TiO_2, then the photocatalyst was 99% selective for the formation of CO in the photocatalytic RWSG [144].

10.6.2 Copper oxides

The use of copper and its oxides (particularly Cu_2O) in photocatalytic CO_2 reduction is straightforward: Concerning Cu_2O as photoabsorber, it is a p-type semiconductor with a bandgap situated at negative potentials, so that electron transfer should be favorable on these materials [30,145]. Concerning the functionality as catalytic active site, metallic copper is known as classical heterogeneous catalyst for the synthesis of methanol from CO_2 in synthesis gas since the 1960s [57]. Indeed, for either CuO or Cu_2O nanorods in a composite with reduced graphene oxide (rGO), methanol was reported as main product [27]. CuO showed better performance than Cu_2O. Surprisingly, in Ref. [27] it was concluded that the rGO is responsible for the reduction reaction, not the copper oxide, since the electronic levels favor charge carrier transport in this direction. In that case, however, copper (oxide) should not act as a catalyst for the formation of methanol from CO_2, but, rather, for the oxidation reaction. In another study of Cu_2O deposited on multilayer graphene, ethanol was found as main carbon product, aside from the observation of the formation of hydrogen [28]. Reactions were carried out either in a slurry or in a capillary reactor. In this study, Cu_2O was suggested as catalytic active site, and the formation of ethanol was suggested to involve the intermediate formation of formate species [28]. Methanol was instead observed as the main product using a more complex architecture, in which a coaxial Cu_2O nanowire array was covered with an ultrathin graphene shell onto which bimetallic Au—Cu particles were deposited [30]. It was concluded that the electrons from the Cu_2O, which functioned as photoabsorber, were transferred through the graphene layer to the Au—Cu nanoparticles which then functioned as catalytic active site for the formation of methanol. Oxygen evolution was suggested to proceed on the Cu_2O nanowires [30]. In a study with a related system of carbon-coated Cu_2O nanorods, but without additionally deposited metallic nanoparticles, CH_4 and C_2H_4 were detected as products, and carbon labeling was employed to prove the origin of those products from CO_2 [45]. In both Refs. [30,45] it was also discussed that one of the functions of the carbon coating is to prevent photocorrosion of Cu_2O, which is otherwise a severe issue. In a direct *Z*-scheme system of n-type α-Fe_2O_3 and p-type Cu_2O, on the other hand, CO was reported as predominant product [46]. In this very thorough study, the function of the two materials was clearly identified, in which the Cu_2O acted as reduction site and the Fe_2O_3 as oxidation site. Even the stoichiometric by-product oxygen was detected and quantified in Ref. [46], providing strong evidence for the formation of CO and O_2 from CO_2 in a catalytic process. Using a composite from Cu_2O and carbon quantum dots [29], methanol was the only significant product detected by GC and NMR. Here, too, Cu_2O is suggested to be the reduction site/catalyst, while the oxidation reaction is suggested to proceed on the carbon material [29]. In a composite with Cu_xO nanoparticles on strontium titanate nanotubes, it was clearly shown that the copper oxide took over the function of a cocatalyst [47]. Carbon monoxide was the main product in this study, in which the reactions were carried out in basic bicarbonate solution. Isotope labeling was used to verity that CO had been produced from CO_2, and labeling of oxygen was employed to trace the origin of gaseous oxygen back to water oxidation. It was found that the solution used for copper deposition needed to be extremely dilute, with a concentration around 0.0005 wt%, to obtain the best results [47].

10.6.3 Strontium titanate

As already outlined in the previous paragraph, a Cu_xO-$SrTiO_3$ composite led to the formation of CO as main product from the reduction of CO_2. At the same time it stoichiometrically oxidized water. In this system, $SrTiO_3$ acted as photoabsorber, and potentially as oxidation site [47]. A composite made from ZnTe and $SrTiO_3$ produced CH_4 as primary product under irradiation with visible light ($\geq$420 nm) [146]. Although ZnTe is a p-type material, which would be beneficial for reduction reactions, the authors concluded that band alignment would favor electron transfer from this material to the $SrTiO_3$. Only ZnTe, with a bandgap of ~2.24 eV, can be excited with visible light. In this study, $SrTiO_3$ did not act as photoabsorber itself, because it can absorb only UV light, and consequently it produced no products under irradiation with visible light. Although oxygen was not detected in Ref. [146], the authors suggested that O_2 should be produced from water oxidation at the ZnTe, because of the holes left behind in this material. To enhance visible light absorption, a study was conducted in which some of the Ti^{4+} cations were replaced by less electronegative elements, that is, Fe, Ni, and Co ions [51]. It should be noted that either Pt or RuO_2 was additionally deposited as cocatalyst. CH_4 was the only organic product. The sample doped with Co (+Pt cocatalyst) showed the best performance, and at the same time, in comparison with the other samples, the largest amounts of CO_2 could be adsorbed on the Co-doped sample. Hydrothermal synthesis yielded better samples than solid-state synthesis [51]. Apart from $SrTiO_3$, the layered perovskite material $Sr_3Ti_2O_7$ has also been studied previously [50]. The material was additionally doped with iron, nitrogen, and/or sulfur. Methanol was found as main product, but also ethanol, acetaldehyde, ethane, and ethylene were detected in trace amounts. The sample promoted with all elements listed above (Fe, N, S) showed the highest product yields. Unfortunately, no blank experiment under the same reaction conditions, but without CO_2, was mentioned in Ref. [50], and oxygen, although it was detected, was not quantified.

10.6.4 Alkali tantalates

In a study from 2010, Teramura et al. [147] studied different $ATaO_3$ tantalates, with A = Li, Na, K. The reaction studied was not CO_2 reduction with water as coreactant, but with hydrogen. CO was found as only product. $LiTaO_3$ showed the highest activity, and it also had the largest bandgap with 4.9 eV. Thus, photocatalytic activity was found to be correlated with the optical bandgap values, that is, the wider the bandgap, the higher the CO production. Furthermore, $LiTaO_3$ chemisorbed significantly larger amounts of CO_2 than the other two tantalates [147]. In a later study [49] specifically focused on $NaTaO_3$, hydrogen was added to the $CO_2 + H_2O$ reaction mixture. Different cocatalysts were deposited on the tantalate surface. The Ru-modified sample showed highest yields of methane, while the Pt-loaded sample produced predominantly CO. Without the addition of hydrogen, the Ru-modified sample was almost inactive. In this study, it was concluded from the results of isotope labeling studies with deuterium that water is still a major source for protons in the product methane, and that hydrogen was instead needed to convert peroxide intermediates into water. With hydrogen in the reaction mixture, the Ru-modified sample showed stable productivity for 24 h, while the formation of CO on the Pt-modified sample started to decline after several hours. This was attributed to the poisoning of Pt with CO molecules [49]. In another study on $NaTaO_3$, CuO was used as cocatalyst [48]. In this study, the reaction was carried out in isopropanol, which at the same time helped to dissolve CO_2 and functioned as sacrificial reagent to trap holes. Both the formation of methanol as reduction product from CO_2 and the formation of acetone as oxidation product from isopropanol were observed. The optimal loading was determined as 2 wt % Cu. Mechanistically, it was concluded that the CuO acted as reduction site, while isopropanol oxidation occurred on the tantalate [48]. Li et al. [148] synthesized $KTaO_3$ nanoflakes in a solvent mixture of hexane and water. In a reaction system containing only CO_2 and H_2O, the nanoflakes as well as cubic reference $KTaO_3$ samples were all nearly 100% selective for the formation of CO from CO_2, with only traces of CH_4 being formed as by-product. However, large amounts of hydrogen were formed as non-carbon-containing by-product. The nanoflakes showed about an order of magnitude larger yields than conventional $KTaO_3$. The formation of oxygen was also detected in this study. The loading of silver as cocatalyst changed reaction selectivity toward the formation of CO, while the formation of hydrogen was strongly suppressed. On the other hand, Pt as cocatalyst showed the opposite trend [148].

10.7 Concluding remarks

In this chapter, a general overview was provided of the requirements to fabricate efficient metal oxide photocatalysts for the reduction of CO_2. In this field, research can nowadays draw upon an already large pool of data, since >80% of all studies published until November 2014 have been carried out on oxide-based materials. In very recent times, carbon materials have gained increasing attention, but oxides continue to be in the focus, particularly the oxides of indium and copper, as well as titanates and tantalates. Looking at the general requirements for an active photocatalyst, the most challenging aspect clearly is the very negative LUMO energy and high thermodynamic stability of the CO_2 molecule. Other reaction pathways involving multiple electrons and protons are not situated that negative on the electrochemical energy scale, however, photoabsorbers are needed with rather negative band positions. Considering the catalytic functionality, it is questionable whether the formation of (bi)carbonates on the photocatalyst surface is beneficial, but undoubtedly the CO_2 molecule needs to be activated in some way by adsorption processes. While three major reaction pathways are discussed for TiO_2 and other photocatalyst classes, the exact pathway, in operation as function of the specific reaction conditions, is not yet clear. Reliable studies under clean conditions and with all possible blank experiments are needed, now and in the future, to shed more light on the elementary steps, if possible aided by in situ spectroscopy. Together with the knowledge on promising material classes that has already been obtained this can provide a basis to finally develop materials suitable for large-scale carbon dioxide recycling to useful chemicals.

References

[1] *https://scripps.ucsd.edu/programs/keelingcurve/*. Accessed 6 December 2016.
[2] S.N. Habisreutinger, L. Schmidt-Mende, J.K. Stolarczyk, Angew. Chem. Int. Ed. 52 (2013) 7372.
[3] L. Guo, Y. Wang, T. He, Chem. Rec. 16 (2016) 1918.
[4] H.J. Freund, M.W. Roberts, Surf. Sci. Rep. 25 (1996) 225.
[5] E.V. Kondratenko, G. Mul, J. Baltrusaitis, G.O. Larrazábal, J. Pérez-Ramirez, Energy Environ. Sci. 6 (2013) 3112.
[6] A.Y. Rozovskii, G.I. Lin, Top. Catal. 22 (2003) 137.
[7] G.C. Chinchen, P.J. Denny, D.G. Parker, M.S. Spencer, D.A. Whan, Appl. Catal. 30 (1987) 333.
[8] M. Halmann, Nature 275 (1978) 115.
[9] T. Inoue, A. Fujishima, S. Konishi, K. Honda, Nature 277 (1979) 637.
[10] Y. Zheng, L. Lin, X. Ye, F. Guo, X. Wang, Angew. Chem. Int. Ed. 53 (2014) 11926.
[11] W.-J. Ong, L.-L. Tan, S.-P. Chai, S.-T. Yong, Chem. Commun. 51 (2015) 858.
[12] W.-J. Ong, L.-L. Tan, S.-P. Chai, S.-T. Yong, A.R. Mohamed, Nano Energy 13 (2015) 757.
[13] W.-J. Ong, L.-L. Tan, S.-P. Chai, S.-T. Yong, Dalton Trans. 44 (2015) 1249.

[14] J. Qin, S. Wang, H. Ren, Y. Hou, X. Wang, Appl. Catal. B Environ. 179 (2015) 1.
[15] W. Yu, D. Xu, T. Peng, J. Mater. Chem. A 3 (2015) 19936.
[16] D. Zheng, C. Pang, X. Wang, Chem. Commun. 51 (2015) 17467.
[17] F. Raziq, Y. Qu, X. Zhang, M. Humayun, J. Wu, A. Zada, H. Yu, X. Sun, L. Jingo, J. Phys. Chem. C 120 (2016) 98.
[18] M. Li, L. Zhang, M. Wu, Y. Du, X. Fan, M. Wang, L. Zhang, Q. Kong, J. Shi, Nano Energy 19 (2016) 145.
[19] H. Shi, C. Zhang, C. Zhou, G. Chen, RSC Adv. 5 (2015) 93615.
[20] S. Yin, J. Han, T. Zhou, R. Xu, Cat. Sci. Technol. 5 (2015) 5048.
[21] S. Cao, J. Low, J. Yu, M. Jaroniec, Adv. Mater. 27 (2015) 2150.
[22] Y. He, L. Zhang, B. Teng, M. Fan, Environ. Sci. Technol. 49 (2015) 649.
[23] Y. Bai, T. Chen, P.Q. Wang, L. Wang, L. Ye, X. Shi, W. Bai, Sol. Energy Mater. Sol. Cells 157 (2016) 406.
[24] Q. Xiang, B. Cheng, J. Yu, Angew. Chem. Int. Ed. 54 (2015) 11350.
[25] K.A.S. Fernando, S. Sahu, Y. Liu, W.K. Lewis, E.A. Guliants, A. Jafariyan, P. Wang, C. E. Bunker, Y.-P. Sun, ACS Appl. Mater. Interfaces 7 (2015) 8363.
[26] J. Low, J. Yu, W. Ho, J. Phys. Chem. Lett. 6 (2015) 4244.
[27] R. Gusain, P. Kumar, O.P. Sharma, S.L. Jain, O.P. Khatri, Appl. Catal. B Environ. 181 (2016) 352.
[28] L. Hurtado, R. Natividad, H. Garcia, Catal. Commun. 84 (2016) 30.
[29] H. Li, X. Zhang, D.R. MacFarlane, Adv. Energy Mater. 5 (2015) 1401077.
[30] J. Hou, H. Cheng, O. Takeda, H. Zhu, Angew. Chem. Int. Ed. 54 (2015) 8480.
[31] L. Liu, Ceram. Int. 42 (2016) 12516.
[32] Z. Xiong, Y. Luo, Y. Zhao, J. Zhang, C. Zheng, J.C.S. Wu, Phys. Chem. Chem. Phys. 18 (2016) 13186.
[33] Y. Yang, M. Qiu, L. Liu, Ceram. Int. 42 (2016) 15081.
[34] Z. Xiong, Y. Luo, Y. Zhao, J. Zhang, C. Zheng, J.C.S. Wu, Phys. Chem. Chem. Phys. 18 (2016) 13186.
[35] M. Tahir, B. Tahir, N.A.S. Amin, H. Alias, Appl. Surf. Sci. 389 (2016) 46.
[36] M. Tahir, B. Tahir, N.A.S. Amin, A. Muhammad, Energy Convers. Manag. 119 (2016) 368.
[37] M. Tahir, N.A.S. Amin, Chem. Eng. J. 285 (2016) 635.
[38] M. Tahir, N.A.S. Amin, Appl. Catal. B Environ. 162 (2015) 98.
[39] K.K. Ghuman, T.E. Wood, L.B. Hoch, C.A. Mims, G.A. Ozin, C.V. Singh, Phys. Chem. Chem. Phys. 17 (2015) 14623.
[40] K.K. Ghuman, L.B. Hoch, P. Szymanski, J.Y.Y. Loh, N.P. Kherani, M.A. El-Sayed, G. A. Ozin, C.V. Singh, J. Am. Chem. Soc. 138 (2016) 1206.
[41] L.B. Hoch, L. He, Q. Qiao, K. Liao, L.W. Reyes, Y. Zhu, G.A. Ozin, Chem. Mater. 28 (2016) 4160.
[42] L. He, T.E. Wood, B. Wu, Y. Dong, L.B. Hoch, L.M. Reyes, D. Wang, C. Kübel, C. Qian, J. Jia, K. Liao, P.G. O'Brien, A. Sandhel, J.Y.Y. Loh, P. Szymanski, N.P. Kherani, T. C. Sum, C.A. Mims, G.A. Ozin, ACS Nano 10 (2016) 5578.
[43] L.B. Hoch, P.G. O'Brien, A. Jelle, A. Sandhel, D.D. Perovic, C.A. Mims, G.A. Ozin, ACS Nano 10 (2016) 9017.
[44] S. Zhu, S. Liang, Y. Tong, X. An, J. Long, X. Fu, X. Wang, Phys. Chem. Chem. Phys. 17 (2015) 9761.
[45] L. Yu, G. Li, X. Zhang, X. Ba, G. Shi, Y. Li, P.K. Wong, J.C. Yu, Y. Yu, ACS Catal. 6 (2016) 6444.

[46] J.-C. Wang, L. Zhang, W.-X. Fang, J. Ren, Y.-Y. Li, H.-C. Yao, J.-S. Wang, Z.-J. Li, ACS Appl. Mater. Interfaces 7 (2015) 8631.
[47] S. Shoji, G. Yin, M. Nishikawa, D. Atarashi, E. Sakai, M. Miyauchi, Chem. Phys. Lett. 658 (2016) 309.
[48] T. Xiang, F. Xin, J. Chen, Y. Wang, X. Yin, X. Shao, Beilstein J. Nanotechnol. 7 (2016) 776.
[49] M. Li, P. Li, K. Chang, T. Wang, L. Liu, Q. Kang, S. Ouyang, J. Ye, Chem. Commun. 51 (2015) 7645.
[50] V. Jeyalakshmi, R. Mahalakshmy, K. Ramesh, P.V.C. Rao, N.V. Choudary, G.S. Ganesh, K. Thirunavukkarasu, K.R. Krishnamurthy, B. Viswanathan, RSC Adv. 5 (2015) 5958.
[51] J. Kou, J. Gao, Z. Li, H. Yu, Y. Zhou, Z. Zou, Catal. Lett. 145 (2015) 640.
[52] Y. Zhao, X. Jia, G.I.N. Waterhouse, L.-Z. Wu, C.-H. Tung, D. O'Hare, T. Zhang, Adv. Energy Mater. 6 (2016) 1501974.
[53] H. Zhao, J. Xu, L. Liu, G. Rao, C. Zhao, Y. Li, J. CO_2 Util. 15 (2016) 15.
[54] S. Iguchi, K. Teramura, S. Hosokawa, T. Tanaka, Appl. Catal. A Gen. 521 (2016) 160.
[55] M. Lv, H. Liu, J. Solid State Chem. 227 (2015) 232.
[56] C. Zhao, L. Liu, G. Rao, H. Zhao, L. Wang, J. Xu, Y. Li, Cat. Sci. Technol. 5 (2015) 3288.
[57] D. Saliba, A. Ezzeddine, R. Sougrat, N.M. Khashab, M. Hmadeh, M. Al-Ghoul, ChemSusChem 9 (2016) 800.
[58] Q. Guo, Q. Zhang, H. Wang, Z. Liu, Z. Zhao, Catal. Commun. 77 (2016) 118.
[59] A. Pougin, M. Dilla, J. Strunk, Phys. Chem. Chem. Phys. 18 (2016) 10809.
[60] A. Cybula, M. Klein, A. Zaleska, Appl. Catal. B Environ. 164 (2015) 433.
[61] K.-Y. Lee, K. Sato, A.R. Mohamed, Mater. Lett. 163 (2016) 240.
[62] J.-H. Lee, H. Lee, M. Kang, Mater. Lett. 178 (2016) 316.
[63] O. Ola, M.M. Maroto-Valer, Appl. Catal. A Gen. 502 (2015) 114.
[64] H. Park, H.-H. Ou, U. Kang, J. Choi, M. Hoffmann, Catal. Today 266 (2016) 153.
[65] X. Xin, T. Xu, L. Wang, C. Wang, Sci. Rep. 6 (2016) 23684.
[66] Z. He, J. Tang, J. Shen, J. Chen, S. Song, Appl. Surf. Sci. 364 (2016) 416.
[67] W. Qingli, Z. Zhaoguo, C. Xudong, H. Zhengfeng, D. Peimei, C. Yi, Z. Xiwen, J. CO2 Util. 12 (2015) 7.
[68] Y. Wang, J. Jiao, Z. Zhao, J. Liu, J. Li, G. Jiang, Y. Wang, A. Duan, Appl. Catal. B Environ. 179 (2015) 422.
[69] Y. Cao, Q. Li, C. Li, J. Li, J. Yang, Appl. Catal. B Environ. 198 (2016) 378.
[70] M.S. Akple, J. Low, Z. Qin, S. Wageh, A.A. Al-Ghamdi, J. Yu, S. Liu, Chin. J. Catal. 36 (2015) 2127.
[71] S. Qamar, F. Lei, L. Liang, S. Gao, K. Liu, Y. Sun, W. Ni, Y. Xie, Nano Energy 26 (2016) 692.
[72] P. Reñones, A. Moya, F. Fresno, L. Collado, J.J. Vilatela, V.A. de la Peña O'Shea, J. CO_2 Util. 15 (2016) 24.
[73] L. Yuan, C. Han, M. Pagliaro, Y.-J. Xu, J. Phys. Chem. C 120 (2016) 265.
[74] E. Korovin, D. Selishchev, D. Kozlov, Top. Catal. 59 (2016) 1292.
[75] H. Park, H.-H. Ou, A.J. Colussi, M.R. Hoffmann, J. Phys. Chem. A 119 (2015) 4658.
[76] A.L. Linsebigler, G. Lu, J.T. Yates, Chem. Rev. 95 (1995) 735.
[77] J. Schneider, M. Matsuoka, M. Takeuchi, J. Zhang, Y. Horiuchi, M. Anpo, D. W. Bahnemann, Chem. Rev. 114 (2014) 9919.
[78] R. van de Krol, Principles of photoelectochemical cells, in: R. van de Krol, M. Grätzel (Eds.), Photoelectrochemical Hydrogen Production, Springer Series Electronic Materials: Science & Technology, Springer Science-Business Media, New York, 2012.

[79] J. Nowotny, Oxide Semiconductors for Solar Energy Conversion–Titanium Dioxide, CRC Press, Taylor and Francis, Boca Raton, FL, 2012.
[80] M.A. Henderson, Surf. Sci. Rep. 66 (2011) 185.
[81] A. Fujishima, X. Zhang, D.A. Tryk, Surf. Sci. Rep. 63 (2008) 515.
[82] P.W. Atkins, Physical Chemistry, 5th ed., Oxford University Press, Melbourne, Tokyo, 1994.
[83] D.P. Colombo, R.M. Bowman, J. Phys. Chem. 100 (1996) 18445.
[84] R. Beránek, Adv. Phys. Chem. (2011) 786759.
[85] J. Li, N. Wu, Cat. Sci. Technol. 5 (2015) 1360.
[86] J. Zhang, P. Zhou, J. Liu, J. Yu, Phys. Chem. Chem. Phys. 16 (2014) 20382.
[87] S. Buller, J. Strunk, J. Energy Chem. 25 (2016) 171.
[88] O. Carp, C.L. Huismann, A. Reller, Prog. Solid State Chem. 32 (2004) 33.
[89] M. Matsuoka, M. Anpo, Local structures, excited states, and photocatalytic reactivities of "single-site" Ti-oxide photocatalysts constructed within zeolites or mesoporous materials, in: M. Anpo, P.V. Kamat (Eds.), Environmentally Benign Photocatalysts: Applications of Titanium Oxide Based Materials, Springer Science-Business Media, New York, 2010.
[90] D.W. Bahnemann, C. Kormann, M.R. Hoffmann, J. Phys. Chem. 91 (1987) 3789.
[91] J.A. Anta, Curr. Opin. Colloid Interface Sci. 17 (2012) 124.
[92] F.-X. Xiao, J. Miao, H.B. Tao, S.-F. Hung, H.-Y. Wang, H.B. Yang, J. Chen, R. Chen, B. Liu, Small 11 (2015) 2115.
[93] H.-M. Xu, H.-C. Wang, Y. Shen, Y.-H. Lin, C.-W. Nan, J. Adv. Ceram. 4 (2015) 159.
[94] J.S. Lee, J. Jang, J. Ind. Eng. Chem. 20 (2014) 363.
[95] J. Tian, Z. Zhao, A. Kumar, R.I. Boughton, H. Liu, Chem. Soc. Rev. 43 (2014) 6920.
[96] B. Weng, S. Liu, Z.-R. Tang, Y.-J. Xu, RSC Adv. 4 (2014) 12685.
[97] F.E. Osterloh, Chem. Soc. Rev. 42 (2013) 2294.
[98] B. Levy, J. Electroceram. 1:3 (1997) 239.
[99] D. Chen, X. Zhang, A.F. Lee, J. Mater. Chem. A 3 (2015) 14487.
[100] H. Wilmer, M. Kurtz, K.V. Klementiev, O.P. Tkachenko, W. Grünert, O. Hinrichsen, A. Birkner, S. Rabe, K. Merz, M. Driess, C. Wöll, M. Muhler, Phys. Chem. Chem. Phys. 5 (2003) 4736.
[101] M. Kurtz, J. Strunk, O. Hinrichsen, M. Muhler, K. Fink, B. Meyer, C. Wöll, Angew. Chem. Int. Ed. 44 (2005) 2790.
[102] S. Polarz, J. Strunk, V. Ischenko, M.W.E. van den Berg, O. Hinrichsen, M. Muhler, M. Driess, Angew. Chem. Int. Ed. 45 (2006) 2965.
[103] T. Mclaren, G. Valdes-Solis, S.C. Li, J. Tsang, Am. Chem. Soc. 131 (2009) 12540.
[104] A.T. Goodrow, J. Bell, Phys. Chem. C 112 (2008) 13204.
[105] P.L.D. Lindan, N.M. Harrison, M.J. Gillan, J.A. White, Phys. Rev. B 55 (1997) 15919.
[106] A.T. Paxton, L. Thiên-Nga, Phys. Rev. B 57 (1998) 1579.
[107] V. Ischenko, S. Polarz, D. Grote, V. Stavarache, K. Fink, M. Driess, Adv. Funct. Mater. 15 (2005) 1945.
[108] A. Janotti, C.G. Van de Walle, Rep. Prog. Phys. 72 (2009) 126501.
[109] L. Liu, H. Zhao, J.M. Andino, Y. Li, ACS Catal. 2 (2012) 1817.
[110] L. Liu, F. Gao, H. Zhao, Y. Li, Appl. Catal. B Environ. 134–135 (2013) 349.
[111] V.P. Indrakanti, H.H. Schobert, J.D. Kubicki, Energy Fuel 23 (2009) 5247.
[112] W. Tu, Y. Zhou, Z. Zou, Adv. Mater. 26 (2014) 4607.
[113] B. Ohtani, J Photochem. Photobiol. C: Photochem. Rev. 11 (2010) 157.
[114] K. Iizuka, T. Wato, Y. Miseki, K. Saito, A. Kodu, J. Am. Chem. Soc. 133 (2011) 20863.

[115] K. Teramura, S. Iguchi, Y. Mizuno, T. Shishido, T. Tanaka, Angew. Chem. Int. Ed. 51 (2012) 8008.
[116] D. Uner, M.M. Oymak, Catal. Today 181 (2012) 82.
[117] S. Ristig, N. Cibura, J. Strunk, Green 5 (2015) 23.
[118] M.S. Hamdy, R. Amrollahi, I. Sinev, B. Mei, G. Mul, J. Am. Chem. Soc. 136 (2014) 594.
[119] M. Marszewski, S. Cao, J. Yu, M. Jaroniec, Mater. Horiz. 2 (2015) 261.
[120] G. Busca, V. Lorenzelli, Mater. Chem. 7 (1982) 89.
[121] Y. Wang, R. Kováčik, B. Meyer, K. Kotsis, D. Stodt, V. Staemmler, H. Qiu, F. Traeger, D. Langenberg, M. Muhler, C. Wöll, Angew. Chem. Int. Ed. 46 (2007) 5624.
[122] C.C. Yang, Y.H. Yu, B. van der Linden, J.C.S. Wu, G. Mul, J. Am. Chem. Soc. 132 (2010) 8398–8406.
[123] N.M. Dimitrijevic, B.K. Vijayan, O.G. Poluektov, T. Rajh, K.A. Gray, H.Y. He, P. Zapol, J. Am. Chem. Soc. 133 (2011) 3964.
[124] I.A. Shkrob, N.M. Dimitrijevic, T.W. Marin, H. He, P. Zapol, J. Phys. Chem. C 116 (2012) 9461.
[125] I.A. Shkrob, T.W. Marin, H. He, P. Zapol, J. Phys. Chem. C 116 (2012) 9450.
[126] C. Liu, H. He, P. Zapol, L.A. Curtiss, Phys. Chem. Chem. Phys. 16 (2014) 26584.
[127] M. Anpo, K. Chiba, J. Mol. Catal. 74 (1992) 207.
[128] M. Anpo, H. Yamashita, Y. Ichihashi, S. Ehara, J. Electroanal. Chem. 396 (1995) 21.
[129] D. Lee, Y. Kanai, J. Am. Chem. Soc. 134 (2012) 20266.
[130] W. Lin, H. Han, H. Frei, J. Phys. Chem. B 108 (2004) 18269.
[131] M. Halmann, M. Ulman, B. Aurian-Blajeni, Sol. Energy 31 (1983) 429.
[132] C.C. Yang, J. Vernimmen, V. Meynen, P. Cool, G. Mul, J. Catal. 284 (2011) 1.
[133] B. Mei, A. Pougin, J. Strunk, J. Catal. 306 (2013) 184.
[134] K. Kočí, K. Matějů, L. Obalová, S. Krejčíková, Z. Lacný, D. Plachá, L. Čapek, A. Hospodková, O. Šolcová, Appl. Catal. B Environ. 96 (2010) 239.
[135] J.C.S. Wu, H.-M. Lin, Int. J. Photoenergy 7 (2005) 115.
[136] G.R. Dey, J. Nat. Gas Chem. 16 (2007) 217.
[137] L. Kočí, O.Š. Obalová, Chem. Process. Eng. 31 (2010) 395.
[138] G.R. Dey, A.D. Belapurkar, K. Kishore, J. Photochem. Photobiol. A Chem. 163 (2004) 503.
[139] R. Koirala, S. Docao, S.B. Lee, K.B. Yoon, Catal. Today 243 (2015) 235.
[140] J.C.S. Wu, H.-M. Lin, Int. J. Photoenergy 7 (2005) 115.
[141] S.-W. Cao, X.-F. Liu, Y.-P. Yuan, Z.-Y. Zhang, Y.-S. Liao, J. Fang, S.C.J. Loo, T. C. Sum, C. Xue, Appl. Catal. B Environ. 147 (2014) 940.
[142] L.B. Hoch, T.E. Wood, P.G. O'Brien, K. Liao, L.M. Reyes, C.A. Mims, G.A. Ozin, Adv. Sci. 1 (2014) 1400013.
[143] M. Tahir, N.A.S. Amin, Appl. Catal. A Gen. 493 (2015) 90.
[144] B. Tahir, M. Tahir, N.A.S. Amin, Appl. Surf. Sci. 338 (2015) 1.
[145] A. Paracchino, V. Laporte, K. Sivula, M. Grätzel, E. Thimsen, Nat. Mater. 10 (2011) 456.
[146] M.F. Ehsan, M.N. Ashiq, F. Bi, Y. Bi, S. Palanisamy, T. He, RSC Adv. 4 (2014) 48411.
[147] K. Teramura, S. Okuoka, H. Tsuneoka, T. Shishido, T. Tanaka, Appl. Catal. B Environ. 96 (2010) 565.
[148] K. Li, A.D. Handoko, M. Khraisheh, J. Tang, Nanoscale 6 (2014) 9767.

Metal oxide catalysts for biodiesel production

11

Umer Rashid[*,†], *Soroush Soltani*[*], *Saud Ibrahim Al-Resayes*[†], *Imededdine Arbi Nehdi*[†]
[*]Institute of Advanced Technology, Universiti Putra Malaysia, Serdang, Malaysia,
[†]Chemistry Department, College of Science, King Saud University, Riyadh, Saudi Arabia

11.1 Introduction

Energy crises have been currently confronting the human life because of the quick depletion of oil reserves. On one hand, the human population and world's economy are increasing; while on the other hand, both of them are highly dependent on the transportation which is currently supplied by petroleum [1]. Despite the continuous increasing prices of petroleum-based fuels, more concerns are associated with the risk of public health and climate changes. These worrying issues have led to the exploration of eco-friendly alternatives [2].

Since Rudolf Diesel used seed oil as diesel fuel, intensive experiments have been carried out on producing fuels from a variety of feedstocks. Biodiesel is a mixture of mono alkyl esters, which has very close physicochemical characteristics to the current petroleum. It is a nontoxic environmentally friendly source of energy which can substantially decrease the carbon dioxide footprint by 78%.

Homogeneous acid catalysts have gained huge attention due to their high efficiency in a short reaction time. Although, these catalysts have lost their place in catalysis reaction due to some unfortunate drawbacks, such as difficult separation process, mass loss during washing treatment, and waste water. These drawbacks have led to the exploration of alternatives. Heterogeneous base catalysts have also received wide attention by most researches to be substituted for homogeneous catalysts. In comparison with homogeneous catalysts, heterogeneous base catalysts could highly reduce the cost of production due to regeneration and recycling properties. However, these catalysts require refined feedstocks which affect the final production cost. Thus, many researches focused on improvement of heterogeneous acid catalysts in order to gain high free fatty acid (FFA) conversion using waste and cheap feedstocks.

Heterogeneous mesostructured acid catalysts have been recently attracting attention for biodiesel production owing to their notable properties, such as large surface area, flexible pore size, and high thermal stability. The combination of a large surface area and large pore size simplify the diffusion of the reactants into mesopore channels which consequently increases the accessibility of the reactants to the majority of the active sites.

Metal Oxides in Energy Technologies. https://doi.org/10.1016/B978-0-12-811167-3.00011-0

Mesoporous solid acid catalysts can be synthesized differently via different synthetic methods. The five common synthesis techniques include: self-assembly process, sol–gel, incipient wetness impregnation, coprecipitation, and hydrothermal [3].

The aim of this chapter has been focused on various synthesis pathways to generate efficient heterogeneous solid acid catalysts. This chapter briefly reviews the importance of heterogeneous catalysis reaction. Besides that, some important synthetic processes to prepare heterogeneous acid catalysts are vastly reviewed. Moreover, the catalytic activity and recyclability of the different types of solid acid catalysts are discussed.

11.2 Biodiesel

Biodiesel is commonly generated via different synthesis methods, such as the direct blending of vegetable oils, thermal cracking, microemulsion, supercritical, transesterification, and esterification, of which each method has its pros and cons.

Transesterification is the most common method to convert triglyceride (TG) to fatty acid methyl ester (FAME) in the presence of an alcohol and a proper catalyst. In case of using feedstocks with high FFAs, such as waste cooking oils (WCOs), simultaneous transesterification and esterification of TGs and FFAs take place, respectively. This chapter is mostly focused on the esterification reaction. Esterification chemically occurs as 1 mol of FFA mixed with 3 mol of alcohol over a suitable catalyst and results in 1 mol FAME and 1 mol of water. The general equation of esterification reaction with methanol is depicted as follows [4].

$$\underset{\text{FFA}}{R{-}COOH} + \underset{\text{Methanol}}{CH_3OH} \xleftrightarrow{\text{Catalyst}} \underset{\text{Methyl Ester}}{R{-}COO{-}CH_3} + \underset{\text{Water}}{H_2O}$$

11.3 Problems with homogeneous acid and heterogeneous base catalysts

The conversion of FFA content to FAME over homogeneous acid catalysts needs the high-temperature equipment. The purification process is another challenge for these type of catalysts which requires a huge amount of water. Another challenge faced by homogeneous acid catalysts is being nonrecyclable which highly affects the cost of production [5]. In the case of using heterogeneous base catalysts, interaction between the alkali catalysts and the FFA results in the soap formation. Consequently, the formed soaps solidify into a semisolid mass which causes deactivation of the catalyst. Thereby, transesterification in the presence of heterogeneous base catalysts requires refined raw materials which possess quiet low FFA (around 0.5% *w*/w) [6, 7]. Prior to the transesterification, a preheating treatment can be applied in order to reduce the moisture content through the reaction. Leaching of the alkaline catalyst is another main concern which significantly reduces the catalytic activity [8].

11.4 Heterogeneous acid catalysts

The heterogeneous acid catalysts were established as alternatives for homogeneous acid catalysts and heterogeneous base catalysts in order to avoid the mentioned obstacles. These catalysts are highly active through the esterification of inexpensive and readily available feedstocks. Besides that, the solid acid catalysts are able to catalyze simultaneous transesterification and esterification of TGs and FFAs, respectively. Other advantages of the heterogeneous acid catalysts consist of: low corrosion issue, less impurity of FAME because of easy separation process, and simple washing process. These types of catalysts can be regenerated and further recycled for several reaction runs with negligible loss of activity. The hydrophobic surface of the heterogeneous acid catalysts make them highly stable against water which minimizes the deactivation of the catalysts [9, 10].

Based on the above discussion, it is understood why the solid acid catalysts are getting great attention. The heterogeneous acid catalysts are highly active due to the presence of high acid site strength and minimal leaching of the active site, and minimal undesirable side reaction. Besides, that, the faster reaction rate causes a high reduction of the cost of production. So far, a large number of the metal oxide acid catalysts have been synthesized, such as mixed metal oxides, mesoporous silica oxides [11], sulfated zirconia [12], and zeolites [13].

11.4.1 Biodiesel production via solid acid catalysis

Recently, a great deal of efforts have been made by many researches via improvements in solid acid catalysis [14]. The variety of acidic sites on the heterogeneous acid catalyst may lead to the high activity of these catalysts. However, the process of esterification or transesterification over solid acid catalysts is similar to homogeneous acid catalysts. The mechanism is shown in Fig. 11.1. On the other hand, the high interaction between the surface hydrophobicity catalyst and the active sites of the heterogeneous acid catalysts make it different from homogeneous catalysis [15].

11.5 Synthesis methods to produce solid acid catalysts

Some important synthetic routes to prepare heterogeneous catalysts are included, such as self-assembly, sol-gel, coprecipitation, incipient wetness impregnation, and hydrothermal.

11.5.1 Self-assembly processing

The self-assembly method is a convenient route to make catalysts by organizing particles as a result of multiple interactions of the components. It is the main challenge to control the size and shape of the particles in order to form a desired structure. The mechanism of self-assembly structures follows two steps [16, 17]: (i) placing dispersed particles into the substrate via controlling the interactions between the

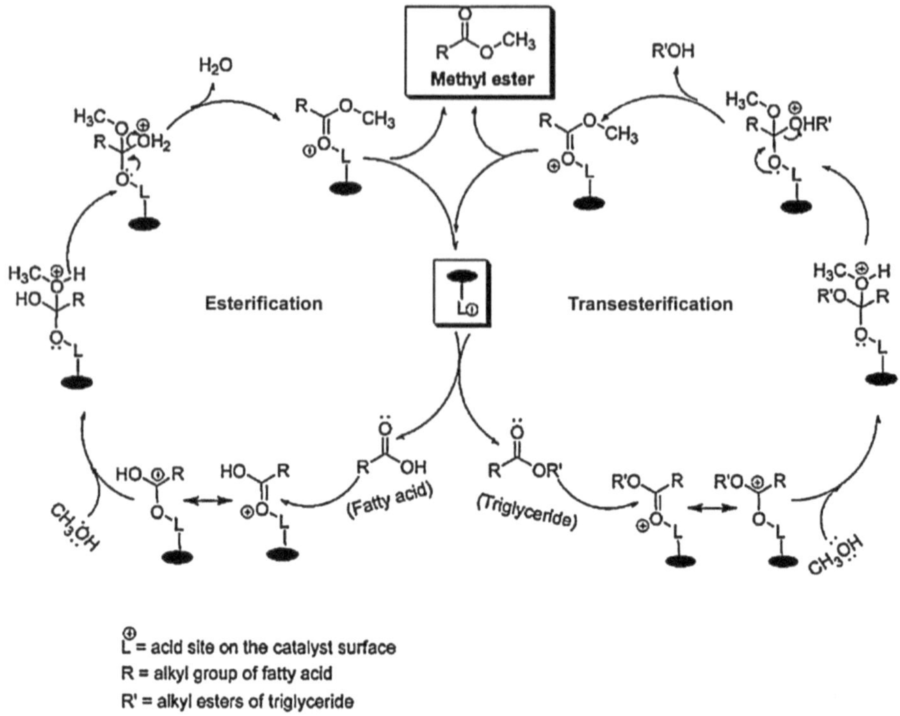

Fig. 11.1 Solid acid catalysis through simultaneous transesterification and esterification [5].

hydrophobic and hydrophilic particles on the surface and (ii) linking the particle and substrate in order to prevent the displacement of the nanoparticles. On the other hand, catalyst synthesis via the self-assembly method usually takes a long aging time (up to 7 days) to complete the reaction. It can be the best synthesis method when the low reaction temperature (room temperature) is considered [18].

Liu et al. [19] synthesized both mesoporous carbon and mesoporous silica nanocomposites using TEOS as a silica precursor, phenol formaldehyde (PF) resins as a polymer precursor, and F127 as triblock copolymers, using the co-assembly route (Fig. 11.2). The removal process was achieved by calcination and etching processes for the mesoporous carbon and mesoporous silica, respectively.

11.5.2 Hydrothermal

The hydrothermal autoclaving method is a heterogeneous efficient reaction to synthesize nano-size heterogeneous catalysts. Typically, the hydrothermal method is used at the reactions which require high temperature and high pressure [20]. It is known as one of the fastest and the most powerful synthesis pathways in chemical processing. Some of the important advantages of the hydrothermal method include fast speed, high surface contact, high mass transfer, and high yield. It is considered as the most eco-friendly synthesis method because of a sealed autoclave system condition.

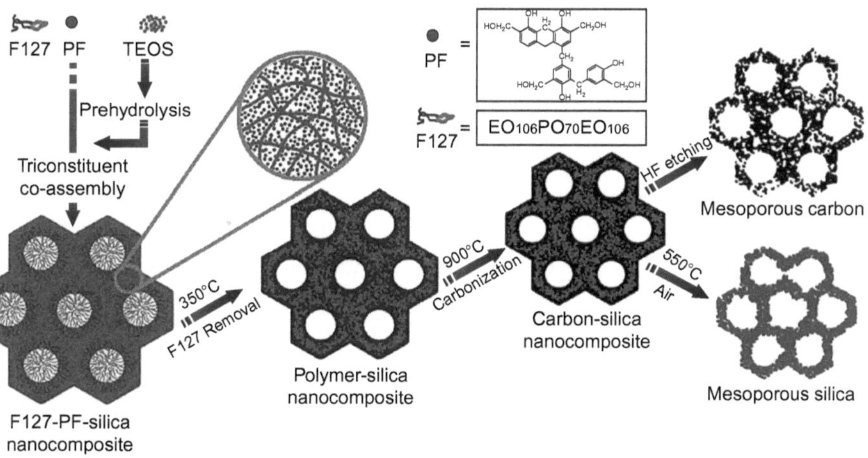

Fig. 11.2 Mesoporous carbon and mesoporous silica synthesis, using co-assembly route [19].

The high pressure and high temperature inside the autoclave cause the high interaction between the particles and induce the quick polymerization of them. Consequently, it leads to the periodic assembly between precursor particles.

At low autoclave temperatures, there is an inadequate energy of the intramolecular collision of the hydration which leads to the failure of the construction of the nanoparticles. Raising the temperature speeds up the dehydration of the intermolecular bands and improves the creation of nanoparticles [21]. Besides that, the higher reaction temperature improves the saturation which consequently effects the crystal growth rate [22].

Each synthetic route has its own drawbacks. However, the hydrothermal autoclaving method is a fast and efficient pathway to synthesize nano-size catalysts; on the other hand, higher energy is required for the high pressure and high temperature. Besides that, the acid treatment is not possible which may affect the equipment piping via corrosion.

Xie et al. [23] functionalized SBA-15 with $Ca(NO_3)_2$ and $(NH_4)_6Mo_7O_{24}$ to fabricate mesoporous CaO-MoO_3-SBA-15, using the hydrothermal method. Various calcination temperatures (623–823 K) were examined to optimize the catalytic activity of the prepared sample. It was shown that the oil conversion was increased by increasing the calcination temperature. At the calcination temperature of 623 K, the oil conversion was only 48.3%; while by increasing the temperature to 823 K, the oil conversion was maximized to 83.2%.

11.5.3 Sol-gel

Another method to fabricate heterogeneous catalysts is sol–gel. In this process, a gel is formed in the presence of initial materials and a proper solvent. The interaction between the solvent and the hydroxyl groups on the surface of the particles causes the hydrolysis reaction. The hydrolysis reaction is followed by the removal of the solvent.

The particle's surface is sheltered by ionic groups, which is expressed as the surface potential. From the hydroxides, the surface potential will be assessed by the reactions with the ions H^+ and OH^-. Hence, the surface potential is pH dependent.

$$M{-}OH + H^+ \rightarrow M{-}OH_2{}^+ \quad (11.I)$$

$$M{-}OH + OH^- \rightarrow M{-}O^- + H_2O \quad (11.II)$$

In the sol-gel process, by varying the amount of the solvent, the morphology and size of the particles can be controlled. Thereby, interaction between the particles can be enhanced in the presence of a higher amount of solvent which provides a hydrophilic surface area [24, 25].

Wu et al. [26] established a sol-gel method to synthesize the mesoporous Al-SBA-15 (Si/Al ratio of 5) as a support without a hydrothermal condition. The prepared sample was further impregnated with K_2SiO_3, K_2CO_3, and KAc as active components to prepare the heterogeneous catalyst. However, doping Al on SBA-15 had no significant impact on the morphology of the mesoporous silica. On the other hand, it acted as an excellent protection of the pore structure of the supported catalyst. It might be attributed to the creation of Al—O—Si units in the context of Al-SBA-15 which declined the concentration of the Si—OH groups in the calcination process. Consequently, the thermal stability of the mesoporous Al-SBA-15 was significantly increased. By gradually increasing the K_2SiO_3 concentration, the surface area of K_2SiO_3/Al-SBA-15 was reduced because the surface area and pores of the support were covered through the preparation process. In the case of Al—O > Si—O (bond length), the porosity of the mesoporous Al-SBA-15 increased. Additionally, the loading potassium compound had a huge impact on the basicity and catalytic activity of AlSBA-15 of which K_2SiO_3 had the superior activity in comparison with K_2CO_3 and KAc. The basicity of the Al-SBA-15 was also increased by loading a higher amount of potassium compounds. Transesterification was conducted using Jatropha oil under a batch condition. The highest conversion gained was 95% with K_2SiO_3 loading on the Al-SBA-15 support. To examine the catalyst steadiness, the catalysts were recycled for 5 runs which resulted in the superior catalytic activity of the K_2SiO_3 loading on the AlSBA-15 in comparison with the other catalysts.

In another research, Jeon et al. [27] hydrothermally functionalized mesoporous MgO using the sol-gel method. The experiment was performed in the presence of a PDMS-PEO copolymer as a template (t-MgO) and further compared with non-templated MgO(nt-MgO). The Brunauer-Emmett-Teller (BET) surface area was calculated as 32.9 and 79.6 $m^2\,g^{-1}$ for n-MgO and t-MgO, respectively. It revealed that the BET surface area of the templated sample was 2.4 times larger than the non-templated one. The crystallinity degree of the two prepared samples was investigated by XRD analysis. The crystal size was determined by using the Scherrer equation; they were 5.7 and 31.3 nm for t-MgO and nt-MgO, respectively. The smaller crystallinity of the t-MgO sample might be attributed to the presence of PDMS-PEO which significantly declined the particle accumulation through the hydrothermal and calcination processes. The basic site strength was determined with CO_2-TPD, which revealed that the basic strength of t-MgO was 2.5 times higher than the nt-MgO catalyst. The higher number of active sites had a huge impact on the biodiesel yield. The TGA plot shows that the polymer chains started to decompose within the range of 250–350°C and the catalyst was completely lost at 500°C. It revealed that the prepared catalyst was not stable at a high temperature. The transesterification reaction was performed to determine the catalytic activity of the t-MgO mesoporous catalyst. The advantage of the PDMS-PEO templated MgO was reported to be the rapid transportation structure, which resulted in the 96.5% conversion of canola oil to biodiesel via the transesterification process.

11.5.4 Incipient wetness impregnation

Incipient wetness impregnation is also called the dry impregnation method to fabricate heterogeneous catalysts. Typically, the metal precursor is dissolved in an organic solution. In this process, the metal is introduced to the catalyst support, then the capillary action causes absorption of the solution into the pores. The pure catalyst is obtained after the postcalcination process which eliminates the presence of volatile compounds inside the solution.

In another study, which was carried out by Li et al., Li and Rudolph [28], three types of mesoporous silica (SBA-15, MCM-41, and KIT-6) were used as the support which was further incorporated with MgO in two ways—impregnating and in-situ-coating methods. In comparison, the in-situ-coating method caused lower blocking of the mesostructure than the impregnating method, which was attributed to the smooth formation of the MgO_x particles on the mesoporous silica. Besides that, the porosity of the samples prepared with the in-situ-coating method was higher than the samples with the other method. It might have happened due to the lower deposition of the Mg particles on the surface area by the in-situ-coating method. Among the three types of silica sources, MCM-41-MgO had the highest surface area ($391\,m^2\,g^{-1}$), followed by SBA-15-MgO ($252\,m^2\,g^{-1}$) and then, KIT-6-MgO ($112\,m^2\,g^{-1}$). On the other hand, the catalyst with the highest surface area had the lowest pore diameter, which was in accordance with other reports [29, 30]. The catalytic activity was examined by the transesterification of blended vegetable oils. Surprisingly, SBA-15-MgO showed higher catalytic activities than the others, even MCM-41-MgO, which had a higher surface area. Thereby, the higher conversion (96%) was obtained as SBA-15-MgO catalyzed the reaction.

Barros et al. [31] functionalized a mesoporous SBA-15 catalyst with MgO and ZnO via the incipient wetness method. The results revealed that MgO-SBA-15 possessed higher surface area ($553\,m^2\,g^{-1}$) than ZnO-SBA-15 ($515\,m^2\,g^{-1}$). The catalytic activity was carried out via the esterification of lauric acid. The results showed that MgO-SBA-15 with a higher surface area gave a higher FFA conversion (81%) than the ZnO-SBA-15 (with the FFA conversion of 65%).

11.5.5 Coprecipitation

Coprecipitation is another pathway to synthesize heterogeneous catalysts. The coprecipitation process includes heating the precipitates and liquor after precipitation. The formation of the mixed metal oxide catalyst mainly depends on the presence of the impurity in the precipitate. An ionic lattice plays a prominent part in the adsorption of ions to form a compound with the opposite ion charge.

Dehkordi and Ghasemi [32] used the coprecipitation method in order to synthesize mixed metal oxides (CaO-ZrO_2) with different molar ratios. It was noted that by increasing the metal ratio (0.1, 0.25, 0.5, 0.75, and 1.0), the basicity was increased (0.13, 0.28, 0.54, 0.83, and 1.10 mmol g^{-1}), respectively. However, by raising the metal ratio, the surface area was decreased (35.2, 18.9, 7.3, 8.7, and 10.2 $m^2 g^{-1}$), respectively. The catalytic activity was evaluated through the transesterification of WCO. It was observed that by increasing the metal ratio, the FAME yield was increased. It was concluded that the effect of the basicity on the FAME was more important than the surface area.

In other research [33], SBA-15 was functionalized with ammonium chloride organosilane by using the coprecipitation method to prepare the tetraalkylammonium hydroxide-functionalized SBA-15 materials (SBA-15-pr-NR_3OH). The surface area and pore size of the prepared SBA-15 support were calculated from the N_2 adsorption-desorption isotherm, 810 $m^2 g^{-1}$ and 6.75 nm, respectively. The textural property of SBQ-15 was drastically affected via the functionalization and was reduced to 271 $m^2 g^{-1}$ and 5.38, respectively. The catalytic activity of the prepared product was tested through the transesterification of soybean under various reaction conditions. The very high yield (99.4%) was obtained under the suitable reflux condition, including: catalyst concentration 2.5 wt%, methanol:oil molar ratio of 12:1 and reaction time of 30 min. The prepared sample was recovered for several recycling times without even a negligible loss of activity. The same researchers [34], fabricated a SBA-15 silica-supported basic ionic liquid mesoporous catalyst via the functionalizing of 4-butyl-1,2,4-triazolium hydroxide onto SBA-15 silica, under hydrothermal conditions. The synthesized SBA-15-pr-ILOH was used for the catalytic reaction in order to convert soybean oil to biodiesel. The surface area and pore size of the prepared SBA-15 support were calculated at 590 $m^2 g^{-1}$ and 6.64 nm, respectively. After the functionalization, the textural property of SBQ-15 was drastically reduced to 341 $m^2 g^{-1}$ and 5.58 nm, respectively. Despite the dropping of the textural properties, the conversion rate (95.4 wt%) was still high, which was attributed to the high basicity of the synthesized catalyst (2.33 mmol g^{-1}).

11.6 Effect of synthesis conditions on characteristics

The textural characteristics play a decisive role on the performance of mesoporous catalysts through the catalysis reaction. The textural properties strongly depend on the synthetic conditions [35]. Therefore, textural adjustment of mesoporous materials makes them more applicable in many fields of research. Several factors, such as pH value and postcalcination temperature, have significant influence on the textural characteristics of mesoporous catalysts [36], which are discussed below.

11.6.1 Effect of pH value

It is reported that the pH value of the prepared precursor has a huge impact on the morphology of the final product [37]. Based on reports, the surface area deceases at higher pH concentrations, which is attributed to the high agglomeration of the particles [24]. At lower doses of pH, the adsorption system goes naturally through the hydrogen bonding [38]. The surface modification is strongly associated to the pH value of the reaction which is the main reason for the occurrence of Ostward repening. The dissociation constant (pK_a) and isoelectric point (iep) are two important parameters in the surface modification. Although, at a pH lower than the iep, the surface of the modifier will be surrounded by a minority of hydroxyl groups and majority of positive charges. In this situation, the modifier will only be attached to the surface particles through strong hydrogen bonding. On the other hand, at a pH lower than the pK_a, the modifier will not be dissociated; but at a pH higher than the pK_a, the modifier will be dissociated between the OH_2^+ surface particles and the dissociated particles. Thereby, the modifier particles will be linked to the particle surface under the dehydration conditions [39].

Ureas with weak alkaline properties, are mostly used as a reducing agent, which reduces the electron concentration in the host area and results in electron depletion in the host area [40]. A urea solution under a hydrothermal condition might produce gas (bubbles). Nanoparticles agglomerate around the formed bubbles which results in the surface energy reduction. This process goes on till the construction of solid spheres [37, 40]. By increasing the concentration of the reducing agent, the nanoparticle size will be decreased [41].

Yu et al. [42] immobilized the Candida rugose lipase (CRL) on mesoporous silica in order to catalyze the esterification of linoleic acid. In this study, the effect of the pH value (varying from 6.0 to 8.0) on the catalyst fabrication and FFA conversion was studied. Typically, the negative charge on the CRL adsorbed to the negative charge of silica particles. On the other hand, the repulsive interaction between silica and CRL particles may not influence the interaction. Thereby, the CRL particles may adsorb to the silica particles through hydrogen bonding. The rate of adsorption affects the activity of the catalyst through the esterification reaction. The result revealed that the activity was higher with the pH value of 7.0. It was attributed to the microenvironment of the immobilized CRL.

11.6.2 Effect of the calcination temperature

Calcination treatment is an integral part during fabrication and activation of the heterogeneous catalysts. Postannealing treatment affects the textural and structural characteristics of the preliminary catalysts [43].

Gao et al. [44] synthesized sulfonated ordered mesoporous carbons using the self-assembly surfactant F127 and phenolic, and later annealed it under different calcination temperatures (400°C, 500°C, 600°C, 700°C, and 800°C). It was observed that increasing the calcination temperature caused high enhancement of the textural properties. On the other hand, by increasing the calcination temperature, the acidity was decreased from 2.28 to 0.83 mmol g^{-1}, respectively. The impact of various annealing temperatures on the esterification of oleic acid was examined under the following conditions: 0.5 g of catalyst, oleic acid to methanol 1:10, 120°C for 8 h. According to the results, the highest conversion rate was achieved at 96.3% over the calcined catalysts at 500°C.

Similar results were achieved by Hasan et al. [45]. Hasan and his coworkers fabricated a sulfonated mesoporous silica used for the esterification of oleic acid. The impact of various annealing temperatures (300°C, 400°C, and 500°C) on FFA conversion was studied under optimized reaction conditions: methanol amount of 10, 2 mL of oleic acid, 0.15 g of the catalyst, at 90°C for 5 h. The results exposed that the conversion rate decreased by raising the annealing temperature from 300°C to 500°C. Therefore, the optimum calcination temperature was selected to be 300°C.

11.7 Effect of reaction parameters on the catalytic activity of heterogeneous catalysts

The effect of different parameters on the catalytic activity of heterogeneous catalysts, such as alcohol-to-oil molar ratio, catalyst concentration, reaction time, and reaction temperature, is detailed below:

Liu et al. [46] hydrothermally synthesized magnetically active ordered mesoporous resin (MOMR) and further functionalized it with the sulfonic group. The catalytic activity of MOMC-SO_3H was further examined through the esterification of acetic acid. The esterification was carried out at 90°C for 4 h, using only 0.01 g of the catalyst. The maximum FFA conversion was obtained at 96.8% over the synthesized catalyst. It was concluded that 0.01 g of the MOMC-SO_3H catalyst was potential to boost the conversion rate up to 96.8%.

Saravanan et al. [47] prepared a sulfonated zirconia (SZ) catalyst with the surface area of 111 m^2 g^{-1}, pore volume of 0.095 cm^3 g^{-1}, pore diameter of 4.7 nm, and acid density of 1.65 mmol g^{-1}. The catalytic activity was examined through the esterification of palmitic acid. The experiments were carried out under the following reaction conditions: 1 wt% amount of the catalyst, MeOH (60 mmol), oil (3 mmol), and reaction temperature of 60°C for 7 h. The highest FAME yield was achieved 80% over the mesoporous SZ catalyst.

Peng et al. [48] synthesized the CMK-3-873-SO_3H catalyst in order to catalyze the esterification of oleic acid. The catalyst possessed a surface area of 880 m^2 g^{-1}, pore volume of 0.47 cm^3 g^{-1}, and acid density of 0.91 mmol g^{-1}. The catalytic activity was carried out through the esterification of FFA (catalyst amount 0.1 g, methanol concentration 50 mmol, oleic acid amount 5 mmol, reaction temperature 80°C, and reaction time 6 h). The highest FAME yield was achieved at 60% under optimum reaction conditions, which was attributed to the combination of remarkable textural properties and a hydrophobic surface.

Carmo et al. [49] synthesized mesoporous Al-MCM-41 through the one-pot functionalization of silica sources with aluminum chloride hexahydrate ($AlCl_3$). In this research, the optimum Si/Al ratio was found to be 8 with a surface area of 713 $m^2 g^{-1}$ and pore volume of 0.67 $cm^3 g^{-1}$. The catalytic activity was tested through the esterification of palmitic acid. In each reaction, 0.6 wt% of the catalyst was exerted with a methanol-to-oil molar ratio of 60:1 at 120°C for 2 h. The highest FFA conversion was achieved at 79% over the mesoporous Al-MCM-41.

11.8 Catalyst stability

To commercialize the application of the catalyst for biodiesel production, the stability and, especially, recyclability of the catalyst should be enhanced. A stable catalyst should never be dissolved in methanol and oil, so it will remain in its entirety at the end of the reaction [50]. Table 11.1 summarized the recyclability of different heterogeneous catalysts with different characteristics.

Wu et al. [51] used the direct synthesis for encapsulating the lacunary polyoxometalate (LPOM) into HMS to synthesize LPOM-HMS hybrid catalysts. The polymeric mesoporous catalyst possessed surface areas ranging from 626 to 878 $m^2 g^{-1}$ and pore size of 3.8 nm. The esterification of acetic acid was carried out in order to test the catalytic activity of the synthesized catalyst using a reflux system (catalyst amount of 0.3 g, *n*-butanol-to-oil molar ratio of 3:1, reaction temperature of 115°C, and reaction time of 5 h). The highest conversion was achieved at 88.3% which dropped to 86.8% and 83.7% for the second and third recycle runs, and decreased by 11% after the third runs which was attributed to the leaching of LPOM. The high stability of the catalyst was attributed to the direct functionalization method.

Similar results were achieved by Zhang et al. [52] which used direct synthesis for encapsulating the 12-tungstophosphoric heteropolyacid (HPW) into a MIL-100(Fe)

Table 11.1 Comparison of recyclability of different heterogeneous catalysts with different characteristics

Catalyst	$^aS_{BET}$	^bV_p	^cD_p	^{d}Acid density	Cycle	Ref
LPOM-HMS	878	–	3.8	–	4	[51]
HPW@MIL-100(Fe)	1470	0.53	–	0.24	5	[52]
HSiW-SBA-15	818	–	5.5	–	5	[55]
H-PDVB-SSFBI	243	0.19	3.78	–	7	[56]
SA/MCM-41	324	0.412	2.7	–	6	[59]
Ph-SO_3H-MSM	1118	2.39	1.86	1.86	6	[57]
SO_3H-$ZnAl_2O_4$	352.39	0.13	3.10	1.95	8	[58]

[a]Specific surface areas were obtained from Brunauer, Emmet, and Teller method.
[b]Total pore volumes were calculated at $P/Po = 0.99$ on the N_2 adsorption isotherms.
[c]Average pores size were calculated from the N_2 desorption branch using the BJH model.
[d]Total acid density (mmol g^{-1}).

and mesoporous organic framework (MOF) in order to synthesize HPW@MIL-100 (Fe) hybrid catalysts. The synthesized catalyst possessed a high surface area of 1470 $m^2 g^{-1}$, pore volume of 0.53 $cm^3 g^{-1}$, and acid density of 0.24 $mmol g^{-1}$. The synthesized catalyst showed high catalytic activity and stability through the esterification of acetic acid via a reflux system. The highest yield was achieved at 96% under optimized reaction conditions (catalyst amount of 0.2 g, acetic acid amount of 0.2 mol, alcohol concentration of 0.1 mol, at 110°C for 12 h). The catalyst reusability was tested under optimum conditions, resulting in high catalytic activity after five recycle runs, without any loss of activity. The high stability of the catalyst was attributed to negligible leaching of the Fe or W from the catalyst which was attributed to the direct synthesis method.

In order to compare the effect of direct and postfunctionalizations, Kuwahara et al. [53] synthesized sulfonated mesoporous zirconosilicate using two methods: direct sulfonation and postsulfonation procedures. The catalytic activity was examined through the esterification of levulinic acid (ethanol-to-oil ratio of 1:10, 5.0 wt% of the catalyst, reaction temperature of 70°C, and reaction time of 24 h). The catalytic activity of both types of the catalysts had dropped after continual activities. The loss of activity was assigned to the elution of the sulfonate ions and de-structuration of the catalysts. According to the results, the samples functionalized by the postsulfonation treatment were more stable compared with the samples functionalized through the direct process. In another study Zhang et al. [54] successfully synthesized a mesoporous silica catalyst using a dual template (vinylpyrrolidone and triblock copolymer). The catalytic activity of the optimized mesoporous catalyst was examined through the esterification reaction. The results revealed that the highest FFA conversion was achieved at 90.15%. The catalytic activity dropped to 30% of the FFA conversion after the fifth reuse, which showed the poor stability of the catalyst.

Chen et al. [55] functionalized SBA-15 with 12-tungstosilicate (HSiW) in order to be used for the esterification of oleic acid in the presence of methanol. The synthesized mesoporous catalysts possessed high surface areas within the range of 402–818 $m^2 g^{-1}$ where the porosity varied from 5 to 5.5 nm. The catalytic activity was further tested, resulting in a maximum FAME yield of 89.7%. However, the yield dropped to 70% after the second cycle, the recycled catalyst possessed the surface area of 636 $m^2 g^{-1}$ with the pore diameter of 5.5 nm. The high stability of the HSiW-SBA-15 after the second runs was attributed to excellent textural properties. Then, the FAME yield dropped to 64.5% after the fifth cycles, which was attributed to the leaching of the HSiW species.

He et al. [56] fabricated a mesoporous flour-polymer mesoporous solid acid catalyst. The catalytic activity was tested through the esterification of acetic acid. The experiments were carried out under the following reaction condition: catalyst loading of 10 mol%, acetic acid amount of 1 mmol, and petanol amount of 4 mmol at 110°C for 4 h. The highest FFA conversion was found to be 95% under optimum conditions. The recyclability of the catalyst was tested which showed that the spent catalyst stayed active for seven cycles, giving a high yield of 85%. The recovered catalyst possessed the same textural properties with the fresh catalyst (surface area of 243 $m^2 g^{-1}$, pore volume of 0.19 $cm^3 g^{-1}$, and pore diameter of 3.78 nm). The high stability of the catalyst was attributed to excellent textural properties and less leakage of the active sites through the reaction.

Hasan et al. [45] synthesized a sulfonated mesoporous silica in order to catalyze the esterification of oleic acid. The catalyst stability was studied through the esterification reaction (0.15 g of the catalyst, 2 mL of oleic acid, and methanol amount of 10°C at 90°C for 5 h). According to the results, the catalytic activity did not drop even after the third run. The high stability of the catalyst was assigned to the covalent attaching of the SO_3H to Si—O— bond. Geng et al. [57] functionalized mesoporous carbon with Ph-SO_3H groups in order to catalyze the esterification of oleic acid. The optimum catalyst possessed a surface area of 1118 $m^2\,g^{-1}$, pore diameter of 1.86 nm, pore volume of 2.39 $cm^3\,g^{-1}$, and acid density of 1.86 mmol g^{-1}. The catalytic activity of the SC-Al-900 catalyst was tested through the esterification of FFA (catalyst concentration of 20 mL, oil amount of 1 g, methanol amount of 8 mL, reaction temperature of 65°C, and reaction time of 1 h). The highest conversion rate was achieved at 77.7% over the catalyst. The synthesized catalyst was able to stay active for six recycle reactions with insignificant loss of activity. The high FFA conversion was achieved at 72% after the sixth recycle time. The high stability of the catalyst was attributed to a strong attachment of the sulfonic groups to the active sites.

A polymeric mesoporous $ZnAl_2O_4$ material was hydrothermally synthesized, using the polyethylene glycol-assisted method [58]. The synthesized mesoporous material was further postfunctionalized by benzene-sulfonic acid. The stability of the SO_3H-$ZnAl_2O_4$ catalyst was examined via the esterification of palm fatty acid distillate (PFAD) under the following reaction conditions: methanol to PFAD molar ratio of 9:1, catalyst concentration of 1 wt%, reaction temperature of 100°C, and reaction time of 60 min. The synthesized catalyst was highly active for eight consecutive runs without substantial losses of activity. The excellent stability of the catalyst was associated to the strong attachment of the SO_3H groups to the active sites which prohibited the leaching of the SO_3H groups through the catalysis reaction (Fig. 11.3.).

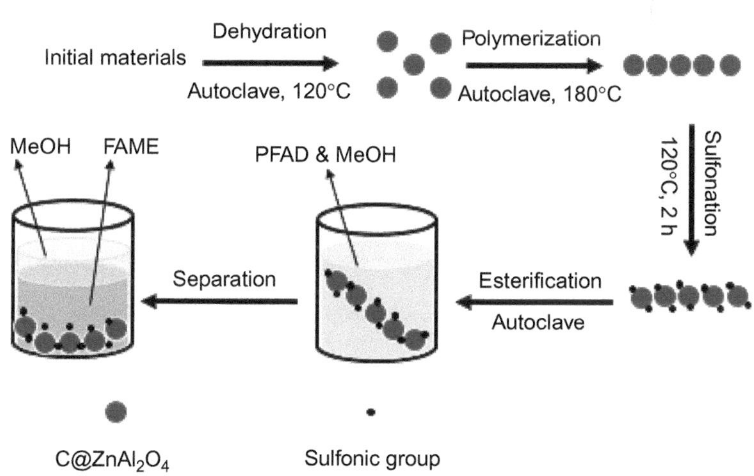

Fig. 11.3 Synthesis of the postsulfonation treatment of the polymeric mesoporous $ZnAl_2O_4$ catalyst [58].

11.9 Conclusion

This chapter reviewed the importance of various synthesis methods via heterogeneous catalyst preparation. Moreover, the effect of the catalytic activity of the heterogeneous catalysts through the production of biodiesel was vastly reviewed. The following conclusions were made:

- In comparison with heterogeneous base catalysts, heterogeneous acid catalysts are highly active due to the presence of high acid site strength and minimal leaching of the active site, and minimal undesirable side reaction.
- Among all the synthesis methods, the hydrothermal autoclaving method can provide fast processing speeds due to the high surface contact and high mass transfer.
- The textural properties strongly depend on the synthetic conditions, such as pH value and postcalcination temperature.
- The surface area deceases at higher pH concentrations, which is attributed to the high agglomeration of the particles.
- Postannealing treatment affects the textural and structural characteristics of the preliminary catalysts. Based on the results, the calcined catalysts at lower temperatures gave higher a conversion rate.
- To commercialize the application of the catalyst for biodiesel production, the recyclability of the catalyst is considered. A stable catalyst remains entirety intact at the end of the reaction.
- The stability of the catalyst was attributed to the synthesis method, functionalization method, and textural properties of the catalysts.

Acknowledgment

The authors extend their appreciation to the International Scientific Partnership Program (ISPP) at King Saud University for funding this research work through ISPP# 0035.

References

[1] S. Soltani, U. Rashid, R. Yunus, Y.H. Taufiq-Yap, Synthesis of biodiesel through catalytic transesterification of various feedstocks using fast solvothermal technology: a critical review, Catal. Rev. Sci. Eng. 57 (4) (2015) 1–29.

[2] G. Knothe, Biodiesel and renewable diesel: a comparison, Prog. Energy Combust. Sci. 36 (2010) 364–373.

[3] H. Wang, Z. Yao, X. Zhan, Y. Wu, M. Li, Preparation of highly dispersed W/ZrO_2–Al_2O_3 hydrodesulfurization catalysts at high WO_3 loading via a microwave hydrothermal method, Fuel 158 (2015) 918–926.

[4] S. Soltani, U. Rashid, S.I. Al-Resayes, I.A. Nehdi, Recent progress in synthesis and surface functionalization of mesoporous acidic heterogeneous catalysts for esterification of free fatty acid feedstocks: a review, Energy Convers. Manag. 141 (2016) 183–205.

[5] Y.M. Sani, W.M.A. Wan Daud, A.R. Abdul Aziz, Activity of solid acid catalysts for biodiesel production: a critical review, Appl. Catal. A Gen. 470 (2014) 140–161.

[6] G.N. Shao, R. Sheikh, A. Hilonga, J.E. Lee, Y.-H. Park, H.T. Kim, Biodiesel production by sulfated mesoporous titania–silica catalysts synthesized by the sol–gel process from less expensive precursors, Chem. Eng. J. 215-216 (2013) 600–607.

[7] J. Van Gerpen, Biodiesel processing and production, Fuel Process. Technol. 86 (2005) 1097–1107.
[8] Y.M. Sani, W.M.A. Wan Daud, A.R. Abdul Aziz, Solid acid-catalyzed biodiesel production from microalgal oil—the dual advantage, J. Environ. Chem. Eng. 1 (2013) 113–121.
[9] F. Guo, Z. Fang, C.C. Xu, R.L. Smith, Solid acid mediated hydrolysis of biomass for producing biofuels, Prog. Energy Combust. Sci. 38 (2012) 672–690.
[10] A.P.S. Chouhan, A.K. Sarma, Modern heterogeneous catalysts for biodiesel production: a comprehensive review, Renew. Sust. Energ. Rev. 15 (2011) 4378–4399.
[11] M.H. Valkenberg, W.F. Hölderich, Preparation and use of hybrid organic–inorganic catalysts, Catal. Rev. Sci. Eng. 44 (2002) 321–374.
[12] K. Saravanan, B. Tyagi, H.C. Bajaj, Sulfated zirconia: an efficient solid acid catalyst for esterification of myristic acid with short chain alcohols, Cat. Sci. Technol. 2 (2012) 2512–2520.
[13] A. Corma, State of the art and future challenges of zeolites as catalysts, J. Catal. 216 (2003) 298–312.
[14] A. Zięba, A. Drelinkiewicz, E.N. Konyushenko, J. Stejskal, Activity and stability of polyaniline-sulfate-based solid acid catalysts for the transesterification of triglycerides and esterification of fatty acids with methanol, Appl. Catal. A Gen. 383 (2010) 169–181.
[15] G. Morales, L.F. Bautista, J.A. Melero, J. Iglesias, R. Sánchez-Vázquez, Low-grade oils and fats: effect of several impurities on biodiesel production over sulfonic acid heterogeneous catalysts, Bioresour. Technol. 102 (2011) 9571–9578.
[16] M. Yoshimura, K. Byrappa, Hydrothermal processing of materials: past, present and future, J. Mater. Sci. 43 (2008) 2085–2103.
[17] P.A. Kralchevsky, N.D. Denkov, Capillary forces and structuring in layers of colloid particles, Curr. Opin. Colloid Interface Sci. 6 (2001) 383–401.
[18] L. Liu, F.Y. Wang, G.S. Shao, Z.Y. Yuan, A low-temperature autoclaving route to synthesize monolithic carbon materials with an ordered mesostructure, Carbon 48 (2010) 2089–2099.
[19] R. Liu, Y. Shi, Y. Wan, Y. Meng, F. Zhang, D. Gu, Triconstituent co-assembly to ordered mesostructured polymer–silica and carbon–silica nanocomposites and large-pore mesoporous carbons with high surface areas, J. Am. Chem. Soc. 128 (2006) 11652–11662.
[20] K. Byrappa, Hydrothermal Method Handbook, William Andrew, Oxford, 2013.
[21] M. Li, Q. Wu, M. Wen, J. Shi, A novel route for preparation of hollow carbon nanospheres without introducing template, Nanoscale Res. Lett. 4 (2009) 1365–1370.
[22] F.A. Harraz, Polyethylene glycol-assisted hydrothermal growth of magnetite nanowires: synthesis and magnetic properties, Physica E 40 (2008) 3131–3136.
[23] W. Xie, L. Zhao, Heterogeneous CaO–MoO_3–SBA-15 catalysts for biodiesel production from soybean oil, Energy Convers. Manag. 79 (2014) 34–42.
[24] S. Bagheri, Z.A.M. Hir, A.T. Yousefi, S.B.A. Hamid, Progress on mesoporous titanium dioxide: synthesis, modification and applications, Microporous Mesoporous Mater. 218 (2015) 206–222.
[25] S. Bagheri, F. Chekin, S.B.A. Hamid, Cobalt doped titanium dioxide nanoparticles: synthesis, characterization and electrocatalytic study, J. Chin. Chem. Soc. 61 (2014) 702–706.
[26] H. Wu, J. Zhang, Y. Liu, J. Zheng, Q. Wei, Biodiesel production from Jatropha oil using mesoporous molecular sieves supporting K_2SiO_3 as catalysts for transesterification, Fuel Process. Technol. 119 (2014) 114–120.
[27] H. Jeon, D.J. Kim, S.J. Kim, J.H. Kim, Synthesis of mesoporous MgO catalyst templated by a PDMS–PEO comb-like copolymer for biodiesel production, Fuel Process. Technol. 116 (2013) 325–331.

[28] E. Li, V. Rudolph, Transesterification of vegetable oil to biodiesel over MgO-functionalized mesoporous catalysts, Energy Fuel 22 (2008) 145–149.
[29] J. Yu, G. Wang, B. Cheng, M. Zhou, Effects ofhydrothermal temperature and time on the photocatalytic activity and microstructures ofbimodal mesoporous TiO_2 powders, Appl. Catal. B 69 (2007) 171–180.
[30] L. Zhang, J. Yan, M. Zhou, Y. Yang, Y.-N. Liu, Fabrication and photocatalytic properties of spheres-in-spheres $ZnO/ZnAl_2O_4$ composite hollow microspheres, Appl. Surf. Sci. 268 (2013) 237–245.
[31] S.D.T. Barros, A.V. Coelho, E.R. Lachter, R.A.S. San Gil, K. Dahmouche, M.I. da Silva, Esterification of lauric acid with butanol over mesoporous materials, Renew. Energy 50 (2013) 585–589.
[32] A. Molaei Dehkordi, M. Ghasemi, Transesterification of waste cooking oil to biodiesel using Ca and Zr mixed oxides as heterogeneous base catalysts, Fuel Process. Technol. 97 (2012) 45–51.
[33] W. Xie, M. Fan, Biodiesel production by transesterification using tetraalkylammonium hydroxides immobilized onto SBA-15 as a solid catalyst, Chem. Eng. J. 239 (2014) 60–67.
[34] W. Xie, L. Hu, X. Yang, Basic ionic liquid supported on mesoporous SBA-15 silica as an efficient heterogeneous catalyst for biodiesel production, Ind. Eng. Chem. Res. 54 (2015) 1505–1512.
[35] A. Galarneau, H. Cambon, F. Di Renzo, R. Ryoo, M. Choi, F. Fajula, Microporosity and connections between pores in SBA-15 mesostructured silicas as a function of the temperature of synthesis, New J. Chem. 27 (2003) 73–79.
[36] S.A. Al-Muhtaseb, J.A. Ritter, Preparation and properties of resorcinol-formaldehyde organic and carbon gels, Adv. Mater. 15 (2003) 101–114.
[37] X. Sun, Y. Li, Hollow carbonaceous capsules from glucose solution, J. Colloid Interface Sci. 291 (2005) 7–12.
[38] D. Santhiya, S. Subramanian, K.A. Natarajan, S.G. Malghan, Surface chemical studies on the competitive adsorption of poly (acrylic acid) and poly (vinyl alcohol) onto alumina, 153 (1999) 143–153.
[39] K. Byrappa, M. Yoshimura, Handbook of Hydrothermal Technology, William Andrew Publishing, Oxford, 2013.
[40] M. Yin, S. Liu, Preparation of ZnO hollow spheres with different surface roughness and their enhanced gas sensing property, Sensors Actuators B Chem. 197 (2014) 58–65.
[41] S. Chandra, A. Kumar, Modulation of synthetic parameters of novel zinc nanoparticles and reducing agent: powder X-ray diffraction, transmission electron microscopy and spectral studies, Spectrochim. Acta A Mol. Biomol. Spectrosc. 97 (2012) 935–941.
[42] W.H. Yu, D.S. Tong, M. Fang, P. Shao, C.H. Zhou, Immobilization of candida rugosa lipase on MSU-H type mesoporous silica for selective esterification of conjugated linoleic acid isomers with ethanol, J. Mol. Catal. B Enzym. 111 (2015) 43–50.
[43] M. Feyzi, A. Hassankhani, H.R. Rafiee, Preparation and characterization of $Cs/Al/Fe_3O_4$ nanocatalysts for biodiesel production, Energy Convers. Manag. 71 (2013) 62–68.
[44] Z. Gao, S. Tang, X. Cui, S. Tian, M. Zhang, Efficient mesoporous carbon-based solid catalyst for the esterification of oleic acid, Fuel 140 (2015) 669–676.
[45] Z. Hasan, J.W. Yoon, S.H. Jhung, Esterification and acetylation reactions over in situ synthesized mesoporous sulfonated silica, Chem. Eng. J. 278 (2015) 105–112.
[46] F. Liu, J. Sun, Q. Sun, L. Zhu, L. Wang, X. Meng, et al., High-temperature synthesis of magnetically active and SO_3H-functionalized ordered mesoporous carbon with good catalytic performance, Catal. Today 186 (2012) 115–120.

[47] K. Saravanan, B. Tyagi, R.S. Shukla, H.C. Bajaj, Esterification of palmitic acid with methanol over template-assisted mesoporous sulfated zirconia solid acid catalyst, Appl. Catal. B Environ. 172-173 (2015) 108–115.
[48] L. Peng, A. Philippaerts, X. Ke, J. Van Noyen, F. De Clippel, G. Van Tendeloo, Preparation of sulfonated ordered mesoporous carbon and its use for the esterification of fatty acids, Catal. Today 150 (2010) 140–146.
[49] A.C. Carmo, L.K.C. de Souza, C.E.F. da Costa, E. Longo, J.R. Zamian, G.N. da Rocha Filho, Production of biodiesel by esterification of palmitic acid over mesoporous aluminosilicate Al-MCM-41, Fuel 88 (2009) 461–468.
[50] Q. Liu, L. Wang, C. Wang, W. Qu, Z. Tian, H. Ma, et al., The effect of lanthanum doping on activity of Zn-Al spinel for transesterification, Appl. Catal. B Environ. 136-137 (2013) 210–217.
[51] N. Wu, B. Li, W. Ma, C. Han, Synthesis of lacunary polyoxometalate encapsulated into hexagonal mesoporous silica and their catalytic performance in esterification, Microporous Mesoporous Mater. 186 (2014) 155–162.
[52] F. Zhang, Y. Jin, J. Shi, Y. Zhong, W. Zhu, M.S. El-Shall, Polyoxometalates confined in the mesoporous cages of metal–organic framework MIL-100(Fe): efficient heterogeneous catalysts for esterification and acetalization reactions, Chem. Eng. J. 269 (2015) 236–244.
[53] Y. Kuwahara, T. Fujitani, H. Yamashita, Esterification of levulinic acid with ethanol over sulfated mesoporous zirconosilicates: influences of the preparation conditions on the structural properties and catalytic performances, Catal. Today 237 (2014) 18–28.
[54] J. Zhang, G. Zhou, B. Jiang, M. Zhao, Y. Zhang, Effect of polyvinylpyrrolidone on mesoporous silica morphology and esterification of lauric acid with 1-butanol catalyzed by immobilized enzyme, J. Solid State Chem. 213 (2014) 210–217.
[55] Y. Chen, Y. Cao, Y. Suo, G.-P. Zheng, X.-X. Guan, X.-C. Zheng, Mesoporous solid acid catalysts of 12-tungstosilicic acid anchored to SBA-15: characterization and catalytic properties for esterification of oleic acid with methanol, J. Taiwan Inst. Chem. Eng. 51 (2015) 186–192.
[56] M. He, J. Xu, Z.-H. Ma, H. Yuan, J. Ma, Facile synthesis of acid mesoporous fluoropolymer as water-tolerant catalyst for esterification, Microporous Mesoporous Mater. 211 (2015) 30–37.
[57] L. Geng, G. Yu, Y. Wang, Y. Zhu, Ph-SO_3H-modified mesoporous carbon as an efficient catalyst for the esterification of oleic acid, Appl. Catal. A Gen. 427-428 (2012) 137–144.
[58] S. Soltani, U. Rashid, R. Yunus, Y.H. Taufiq-Yap, Biodiesel production in the presence of sulfonated mesoporous $ZnAl_2O_4$ catalyst via esterification of palm fatty acid distillate (PFAD), Fuel 178 (2016) 253–262.
[59] H. Pan, J. Wang, L. Chen, G. Su, J. Cui, D. Meng, et al., Preparation of sulfated alumina supported on mesoporous MCM-41 silica and its application in esterification, Catal. Commun. 35 (2013) 27–31.

Solar-driven fuel production by metal-oxide thermochemical cycles

12

Xinhai Yuan[*,†], *Xiongwei Wu*[†], *Jing Wang*[‡], *Yuping Wu*[*,†], *Teunis van Ree*[§]
*State Key Laboratory of Materials-Oriented Chemical Engineering, School of Energy Science and Engineering, Institute for Electrochemical Energy Storage, Nanjing Tech University, Nanjing, People's Republic of China, †College of Science, Hunan Agricultural University, Changsha, China, ‡State Key Laboratory of Materials-Oriented Chemical Engineering, College of Chemistry and Chemical Engineering, Nanjing Tech University, Nanjing, China, §Department of Chemistry, University of Venda, Thohoyandou, South Africa

12.1 Introduction

Nowadays, the need to reduce anthropogenic emissions of greenhouse gases due to combustion of fossil fuels causing the increase of global average surface air temperatures and the need to meet an expanding global energy demand are the most pressing challenges faced by mankind [1, 2]. Solar energy, which is clean, renewable, plentiful, and distributed, offers an intelligent solution to both challenges [1]. One promising approach to harvest and store solar energy is the production of synthetic fuels, that is, solar-driven fuel, via high-temperature thermochemical processes. Solar-driven fuel has significant advantages for long-time storage and/or long-distance transportation of energy [3, 4]. It not only is an appealing supplement or alternative to fossil fuels, but also has the potential to alleviate concerns environment, security, and price volatility, which are associated with the production, transportation, and consumption of fossil fuels, especially imported petroleum [5].

The main pathways for conversion of sunlight, H_2O, and CO_2 to fuels are photosynthesis, electrolysis, photoelectrocatalysis, and solar thermochemical. Biophotosynthesis, although promising, is limited by its resource constraints as well as the inherently low efficiency [6]. Artificial photosynthesis, namely, biomimetic or bioinspired molecular approaches that create the desired fuel-producing functionality in isolation from a living organism, is an alternative that could resolve these problems, but is still faced with serious challenges in the realm of its basic research [7]. Electrolysis of water is an important tool for energy storage in wind and solar applications; however, there still exist some technical barriers, such as system efficiency and electricity cost, that wait to be resolved [8]. Photoelectrocatalysis, whose efficiency is impressive, represents a pathway by which solar fuel might be realized, but is still confronted with numerous scientific challenges, for instance, the overall energy conversion efficiency is still low, and far from practical application [9–11]. Solar-thermochemical

Metal Oxides in Energy Technologies. https://doi.org/10.1016/B978-0-12-811167-3.00012-2

cycles, using the solar resource directly and more fully as a source of energy to drive an endothermic chemical reaction and eliminating electricity generation, have the potential to be simpler and more efficient than photoelectrocatalysis processes [10–12].

Thermochemical cycles, using concentrated solar radiation to generate H_2 by splitting H_2O and generate CO by splitting CO_2, were first developed for the coupling with a nuclear heat source available below 1000°C in the 1960s, whereafter numerous theoretical candidates were studied [13]. Among the various thermochemical cycles, metal-oxide thermochemical cycles, which present the simplest possible system while introducing minimal engineering complexity and limiting fundamental and practical inefficiencies, are presently among the most advanced paths for solar-driven fuels production. Such cycles involve only two steps: a high-temperature thermal reduction [TR; producing oxygen gas] and fuel production at a lower temperature [14].

As a matter of fact, the exploitation of solar-aided ferrites-based thermochemical cycles for the splitting of CO_2 was proposed as early as 1980 but aimed at that time to producing carbon rather than CO as a source for hydrocarbon fuels [15]. Thermodynamic analysis showed that CO may be thermally decomposed to C and O_2 in a two-step reaction scheme like the above that involves the use of FeO and Fe_3O_4 [16].

In the meantime, numerous investigations were performed initially for WS and subsequently for carbon dioxide splitting (CDS) on other redox pairs such as Mn_3O_4/MnO, Co_3O_4/CoO, Nb_2O_5/NbO_2, In_2O_3/In, WO_3/W, and CdO/Cd [17–19]. Some newly proposed redox pairs, such as SnO_2/SnO, CeO_2/Ce_2O_3, GeO_2/GeO, MgO/Mg, etc., have been demonstrated in the most recent time [14].

In this chapter, we briefly review the items involved in the two-step reaction principle of metal-oxide thermochemical cycles. In addition to this, the factors that must be considered in the design of metal-oxide optimized for thermochemical cycles are discussed. Additionally, we will summarize the major developments in the area of metal-oxide thermochemical cycles for solar-driven fuels production, discourse pathways for developing improved materials. At the end of this chapter, an outlook is provided to point out latest research trends in designing nanostructured positive electrode materials. It is anticipated that this chapter can shed some light on the development of advanced metal-oxide thermochemical cycles for solar-driven fuel production.

12.2 Reaction principles of the two-step thermochemical metal-oxide cycles

The two-step reaction cycles are naturally the simplest multistep thermochemical water splitting (WS)/CDS methods, splitting the overall chemical reaction into two steps (Fig. 12.1) [14]:

Solar; endothermic step:

$$MO \rightarrow MO_{1-\delta} + \tfrac{1}{2}O_2\,(g) \tag{12.1}$$

Nonsolar; exothermic step:

$$MO_{1-\delta} + (\delta-\gamma)\,H_2O\,(g) + \gamma\,CO_2\,(g) \rightarrow MO + (\delta-\gamma)\,H_2\,(g) + \gamma\,CO\,(g) \tag{12.2}$$

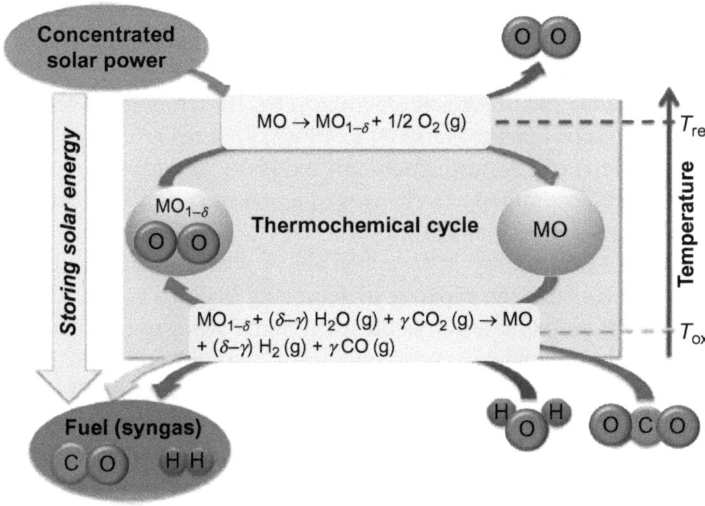

Fig. 12.1 The two-step metal-oxide thermochemical cycles for H_2/CO production from solar power and H_2O/CO_2.
Reproduced with permission from C. Agrafiotis, M. Roeb, C. Sattler, A review on solar thermal syngas production via redox pair-based water/carbon dioxide splitting thermochemical cycles, Renew. Sustain. Energy Rev. 42 (2015) 254–285.

Net reaction:

$$(\delta-\gamma)\,H_2O\,(g)+\gamma\,CO_2\,(g)\rightarrow(\delta-\gamma)\,H_2\,(g)+\gamma\,CO\,(g)+\tfrac{1}{2}O_2\,(g) \tag{12.3}$$

Eqs. (12.1), (12.2) illustrate a generic two-step metal oxide H_2O and CO_2 splitting cycle, it replaces the single thermal dissociation reaction by two more favorable reactions at different temperatures. The net result is the splitting of water or carbon dioxide (12.3). The solar step (12.1) is a highly endothermic TR of the metal oxide during which the metal oxide is reduced into a low valence oxide or dissociated to pure metal at high temperature using concentrated solar power to drive the reaction. The non-solar, exothermic step (12.2) is reoxidation of the reduced species in presence of H_2O and/or CO_2 to produce H_2 and/or CO and to regenerate the metal oxide.

In general, two-step cycles can be divided into two categories; volatile and nonvolatile. Nonvolatile cycles utilize metal oxides which remain in the solid state during reduction (i.e., Fe_3O_4(s) → FeO–Fe_3O_4(s) [20]), while volatile redox cycles consist of metal oxides that undergo gas-solid-phaser transitions (i.e., ZnO(s) → Zn (g) [21]). Generally speaking, volatile reactions have a greater oxygen exchange capability than nonvolatile reactions (more O_2 release/uptake, and thus fuel production, per mass of oxide) and reduction is thermodynamically more favorable due to the increased entropy generated resulting from gas-phase products. Nevertheless, the volatile products must be quenched rapidly to avoid recombination, and to date this issue has not been solved in an energetically efficient fashion [22].

Nonvolatile metal oxides cycles can be further categorized into two subcategories, namely, stoichiometric and nonstoichiometric cycles. Stoichiometric reactions generally form solid solutions upon reduction (i.e., $Fe_3O_4(s) \rightarrow FeO{-}Fe_3O_4(s)$ [8]) and Fe^{3+} is partially reduced to Fe^{2+}. Nonstoichiometric oxides remain crystallographically stable while the lattice accommodates changes in anion or cation vacancies concentrations (i.e., $CeO_2(s) \rightarrow CeO_{2-\delta}(s)$ [23]).

Each category of material/cycle described above possesses its own pros and cons. Stoichiometric cycles have the capacity of achieving high hydrogen production efficiencies per mass of redox material, as the maximum hydrogen production is proportionally related to the amount of oxygen subtracted from the oxide. On the other hand, in the case of nonstoichiometric cycles, only a rather small portion of oxygen participates in the redox reactions, thereby limiting hydrogen production potential. Based on the phase diagrams provided in the literature, the vacancy concentration (δ) that ceria can achieve at temperatures higher than 1000°C ~ 0.25, [24] a value that corresponds to a theoretical hydrogen yield of 2.9 mmol H_2/g. This value seems rather low in comparison with the theoretical yield that can be achieved by the Fe_3O_4/FeO couple (4.3 mmol H_2/g) or by the ZnO/Zn pair (12.3 mmol H_2/g). In practice, sufficient reduction of ceria is achieved only at significantly higher temperatures, while doping of ceria with other materials has also been reported in the literature as a potentially efficient means of achieving reduction at lower temperature [25] and therefore increases the oxygen storage capacity [26]. Main advantages of the nonstoichiometric mechanism are the faster kinetic rates achieved and the chemical and morphological stability [27] that leads to good cyclability. Common challenges with stoichiometric cycles are sintering and consequently deactivation of the materials since temperatures needed for reduction can often be higher than the melting point of at least one of the two oxides. This is the case for the reduction of Fe_3O_4 to FeO where the reduction temperature is >2200°C [28], while Fe_3O_4 melting point is 1538°C [29] and the melting point of FeO lies even lower. Such a reaction scheme is highly dependent on the available surface of the redox active material so that operation at these temperatures inevitably would lead to deactivation of the material due to loss of surface area caused by sintering.

Problems that are associated with the higher-temperature TR step are common to both WS and CDS cycles. Metal-oxide redox pairs, such as Fe_2O_3/Fe_3O_4, Co_3O_4/CoO, whose TR temperature is lower than their melting point, that could be easily thermally reduced under air atmosphere (intermediate partial O_2 pressure) at moderate temperatures the hydrogen yields during the hydrolysis step were too low to be of economic interest [30, 31].

Systems based on metal-oxide redox pairs, such as ZnO/Zn, Fe_3O_4/FeO, exhibiting high hydrogen yields during hydrolysis, require temperatures for TR above their melting points. One option to lower the TR temperatures of these oxides is by reducing the oxygen's partial pressure [14].

Thermodynamic calculations have identified the magnitude of oxygen partial pressure effects on the TR step equilibrium and have shown the prohibitive thermodynamics for performing this step under air atmosphere. In practice, most experimental set-ups and reactors use an inert gas to sweep the oxygen out of the reaction chamber and keep the atmosphere under a low-oxygen partial pressure. In this case though, recycling inert gas imposes additional energy penalties to the overall process. An alternative option is to reduce the metal oxide under vacuum total pressures [14]. Metal oxide reduction under low pressure was shown to be of special interest because the reaction rate is greatly enhanced and the required temperature to achieve a given reduction rate is significantly lowered. In contrast to inert gas dilution, a total pressure decrease also reduces the need for a diluent gas, thus simplifying the process and avoiding energy penalties associated with inert gas recycling [32].

12.3 Considerations in the design of metal-oxide materials

A number of physical and chemical phenomena influence material selection to a significant degree, including reaction thermodynamics, vaporization losses, microstructural stability, reaction kinetics, and transport phenomena [12].

12.3.1 Reaction thermodynamics

When designing a solar-driven gas-splitting process it is essential to first consider the thermodynamics of reactions (12.1)–(12.3). The upper limit for η_{max} is defined by the thermodynamics of the net reaction [Reaction (12.3)]. η_{max} can be defined in terms of T_{TR}, T_{OX}, and $T_{thermolysis}$, which are defined as the temperatures at which $\Delta G° = 0$ for Reactions (12.1), (12.2), and (12.3), respectively, subject to the constraints that $T_{OX} \geq 298\,K$ and $T_{TR} \leq T$ thermolysis.

$$\eta_{max} = \frac{[1 - 298/T_{thermolysis}]}{[1 - T_{OX}/T_{thermolysis}]}[1 - T_{OX}/T_{TR}] \quad (12.4)$$

It can be concluded from Fig. 12.2 that Reaction (12.1) is thermodynamically favorable when T_{TR} is high while Reaction (12.2) is favorable when T_{OX} is low. Then Eq. (12.4) reduces to a modified Carnot relationship when $T_{OX} > 298\,K$. With regard to any heat engine, the highest efficiencies could be obtained when T_{OR} is small and T_{TR} is large. Moreover, as T_{high} approaches T_{low}, the incremental effect of temperature difference on efficiency increases.

12.3.2 Vaporization losses [12]

The high temperature involved will cause the loss of active material by vaporization, which is and can be significant, even for refractory oxides such as ceria. Vaporization of the metal oxide will reduce the lifetime of the active material and lead to increased capital, maintenance, and operating costs. Additionally, condensation of the vaporized material, for example, on reactor windows or other components can lead to equipment failure.

Fig. 12.2 Simplified plot of the thermodynamics of thermolysis (3) and a hypothetical two-step thermochemical cycle (1,2). Reproduced with permission from J.E. Miller, A.H. McDaniel, M.D. Allendorf, Considerations in the design of materials for solar-driven fuel production using metal-oxide thermochemical cycles, Adv. Energy Mater. 4 (2) (2014) https://doi.org/10.1002/aenm.201300469.

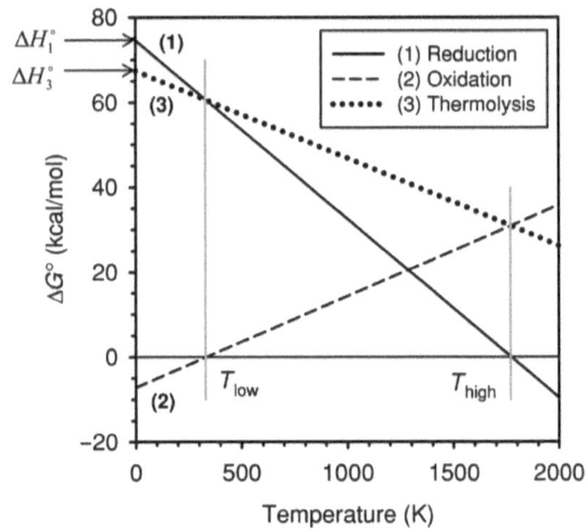

Such losses from both ferrites and ceria during TR can be significant at typical operating temperatures. An upper limit to the rate of mass loss (Me) can be estimated using the Langmuir sublimation model if the vapor pressures (P, in Torr) and molecular weights (MW) of the evaporating species are known at a given temperature:

$$\mathrm{Me} = 5.85 \times 10^2 \mathrm{P}\left(\frac{\mathrm{MW}}{T}\right)\frac{\mathrm{g}}{\mathrm{cm}^2\mathrm{S}} \tag{12.5}$$

This model assumes evaporation into a vacuum, that is, species that leave the surface do not build up in the gas phase to an equilibrium composition, but rather vanish from consideration. The model is a worst-case scenario; it is most reasonable in situations wherein relatively large gas flows or surface exposed to vacuum conditions are employed.

Mass loss rates can be minimized, but at the expense of the extent of reduction and the gas-splitting rate, both of which will decrease if T_{TR} and $T_{CDS/WS}$ are lowered. Clearly, the material loss rate that can be tolerated depends on the cost of material and associated maintenance.

12.3.3 Microstructural stability [12]

Using reactive oxides in morphologies other than the bulk, fully dense form may be desirable for practical implementation of thermochemical cycles, either for structural or chemical reasons. For example, if the gas-splitting reaction kinetics are limited by surface or interfacial area, as is the case for ferrites [33, 34], meso- or even microporous materials could be a way to achieve adequate production rates.

Similarly, materials with kinetics rate limited by bulk diffusion through a product layer may require a microstructure in which the grain size is sufficiently small that, on the timescale of reaction, this diffusion is fast [33, 35]. Such materials may need to be supported on other materials to achieve sufficient mechanical strength. However, reactions between materials at interfaces can result in material degradation and altered reaction rates. In addition, high reaction temperatures will promote grain growth by the various mechanisms that can disrupt or destroy the desired microstructure.

Material design with respect to microstructure can be guided by general principles derived from the field of high-temperature materials science, where it is found that the melting temperature (T_{mp}) of a material is a qualitative indication of its microstructural stability. It suggests that long-term microstructural stability for materials used in solar-driven thermochemical cycles will require melting points in excess of 3100 K, which is found in very few materials. Consequently, it is a virtual certainty that microstructural changes will occur that may lead to the mechanical failure or the loss of gas-splitting activity, thus limiting component lifetime. Coupled with the vaporization processes discussed above these microstructural phenomena may be the ultimate limits to the longevity of any active layer material used in a thermochemical cycle.

12.3.4 Reaction kinetics [12]

The discussion above shows that thermodynamic information alone is not sufficient to make informed choices about redox active materials because the extreme thermal conditions can induce undesirable physical changes in the solid. While the thermodynamics define the limits of a process, the reaction kinetics largely determine the reaction extents and efficiencies that may be realized in practice and are therefore an important consideration in the process and reactor design.

12.3.5 Transport phenomena [12]

The characteristic dimensions, mass and thermal diffusion lengths (l), in the active solid can be defined as

$$l_{TM} = \sqrt{D_{TM} t} \tag{12.6}$$

in which D is the diffusivity, t is time, and T and M refer to thermal or mass transport, respectively [12, 36]. The magnitude of these properties varies widely among different materials. In general, thermal diffusion rates are relatively uniform for metal oxides; the characteristic diffusion length l_T for these materials is 1–3 cm in 60 s. In contrast to thermal transport, the bulk transport of ions necessary for the redox chemistry is highly dependent on the material. Both the rate and identity of the diffusing species can vary from material to material.

12.3.6 Pathways to improvement

Because the very high reduction temperature required for the $MO/MO_{1-\delta}$ redox pairs (e.g., Fe_3O_4/FeO and ZnO/Zn) creates an engineering challenge, any possibility of lowering the reduction temperature is desirable. Typically, there are three ways that can lower the reduction temperature significantly: (a) mixed solid solutions between the good H_2 yields the redox system with the low reduction temperature redox system (e.g., ferrites); (b) stabilized on, or dissolved in, inert materials such as YSZ (yttria-stabilized zirconia) and ZrO_2; (c) producing a particle of a radius or film, of which the thickness is smaller than the reaction penetration depth, on porous zirconia substrates [12, 37].

12.4 Metal oxides for two-step thermochemical metal-oxide cycles

The most widely investigated metal oxides for two-step thermochemical cycles to produce solar fuel are volatile metal oxides, such as ZnO and SnO_2, and nonvolatile metal oxides, such as ferrites, hercynites, ceria, and perovskites.

12.4.1 Synthesis and characterization of metal-oxide materials [14]

Since nonvolatile metal oxides remain in the solid state throughout the process, their physicochemical characteristics such as specific surface area, particle size, intraparticle porosity, etc., become important for the realization of the reaction with the gaseous steam/CO_2. Therefore, many studies explored the various synthesis techniques to enhance such characteristics. It should be kept in mind, however, that the redox materials' issue in thermochemical cycles is not really the production of materials with high surface area, high porosity, or enhanced redox properties in their as-synthesized state, but on whether these properties are maintained during the actual operating conditions encountered in a "real" solar-aided operation involving very high temperatures and repeated thermal cycling. Further issues have to do with the "materials' screening" test procedures. A common problem for several studies on synthesis/characterization of such redox pairs is that the properties of the synthesized materials are often reported after synthesis or postsynthesis annealing conditions much milder than the actual ones during the subsequently performed TR-WS/CDS tests. Therefore it is often not ensured that either any organic additives employed during synthesis have been completely removed or any structural or phase transformations have been completed. Such phenomena are much more pronounced when particles of nanoscale features are thermally reduced and the original as-synthesized morphology eventually vanishes [33, 38]. Consequently, it becomes unclear what are really the phases and mechanisms that take place during either TR (especially) or in the reaction of the oxide with steam or CO_2. In fact, TR temperatures reported in some studies for particular compounds, significantly lower than the bulk of the published results with the same material families, should rather be attributed to in situ reduction by organic species present during the preparation step and not completely removed before the TR/splitting cycle. One way to avoid such ambiguities is before any TR-WS/CDS testing, to precalcine the synthesized materials at temperatures higher than those to be encountered in such tests.

12.4.2 Materials

The most widely investigated metal oxides for the two-step thermochemical cycles to produce solar fuel are volatile metal oxides, such as ZnO, and nonvolatile metal oxides, such as ferrites, hercynites, ceria, and perovskites.

12.4.2.1 ZnO

Among the many metal oxide/metal systems considered for the two-step WS/CDS process, the ZnO/Zn system is characterized by a combination of favorable thermodynamic properties, such as sufficiently nonprecious, high energy content per mass [39–41]. The two-step water-splitting cycle by a ZnO/Zn redox pair (the "ZnO process") was first proposed in 1977 and has been developed as described below [42].

First step: solar-driven endothermic reduction of ZnO:

$$ZnO_{(s)} \rightarrow Zn_{(g)} + \tfrac{1}{2}O_2, (T > 2000\,K, \Delta H = 456\,kJ/mol) \tag{12.7}$$

Second step: exothermic oxidation of Zn into ZnO and product gases:

$$Zn_{(g)} + H_2O \rightarrow ZnO_{(s)} + H_2, (T < 1300\,K, \Delta H = -104\,kJ/mol) \tag{12.8a}$$

$$Zn_{(g)} + CO_2 \rightarrow ZnO_{(s)} + CO, (T < 1300\,K, \Delta H = -67\,kJ/mol) \tag{12.8b}$$

The first step, during which ZnO reduce into Zn and ½ O_2 via solar decomposition, and the temperature for which the $\Delta G° = 0$ is 2235 K, is the most challenging part of this process. The product mixture needs to be quenched to avoid recombination, with the quenching efficiency being sensitive to the dilution ratio of Zn(g) in an inert gas flow and to the temperature of the surface on which the products are quenched. As a result, the final product leaving the solar thermolysis reactor/quencher, that is, the feed for the oxidation step, generally contains a substantial amount of ZnO observed to be as low as 6 mol% and as high as 85 mol% depending on reaction conditions and inert gas/Zn(g) dilution ratio.

The second step is the exothermic oxidation of Zn into ZnO and product gases (H_2/CO), which thermodynamically proceeds at temperatures below 1300 K under 1 bar. Laboratory studies were performed with steam bubbling through molten Zn in the range 723–773 K, but the continuous removal of ZnO(s) was thought to pose process complications [43].

It was shown that the presence of inert ZnO affects both the oxidation kinetics and the final asymptotic conversion of Zn [44]. The mixture of Zn and ZnO particles offered an effective inert support for preventing sintering and enabled simple and complete recovery of ZnO particles for recycling to the first solar step [41]. At 673 K about 95% of Zn was converted to the product after being diluted with 50 wt % ZnO, resembling the composition of a typical product mixture generated in a solar reactor [44].

12.4.2.2 Ferrites

The first two-step cycle based on the Fe_3O_4/FeO redox pair was proposed by Nakamura et al. [20]. The two-step cycle proceeds as follows:

$$Fe_3O_4 \rightarrow 3\,FeO + \frac{1}{2}O_2 \tag{12.9}$$

$$H_2O + 3\,FeO \rightarrow Fe_3O_4 + H_2 \tag{12.10}$$

The TR of magnetite (Fe_3O_4) to wustite (FeO) proceeds at temperatures above 2500 K under 1 bar, which is higher than the melting point of magnetite (1808 K) and wustite (1643 K), resulting in a liquid Fe_3O_4 and FeO phase and causing a rapid decrease of the iron oxide surface area and thus a deactivation of the material. Practically, this reaction can be driven at lower temperatures by reducing the oxygen partial pressure (p_{O2}). Milling of the redox metal oxide into a fine powder is required in the iron oxide process, especially after the high-temperature TR step. This is because the subsequent reactions are mass transfer limited reduced active surface area of the redox metal oxide results in a slower reaction rate and lower conversion. Many nonvolatile metal oxide processes must have the same problem. Therefore, an important challenge in the iron oxide process (and also in the ferrite process) is to develop the redox materials, the reaction devices, or a reactor setup in which the large active surface area of the metal oxide can be retained throughout high-temperature cyclic reactions [40, 45–47].

The iron-oxygen-phase diagram shows that a TR at 1623 K is possible if the oxygen partial pressure in the atmosphere is reduced to 10^{-7} bar [27]. But the production of such an atmosphere is rather energy-intensive and causes high process costs. Mixed solid solutions of Fe_3O_4/FeO and M_3O_4/MO (M = transition metal or alkaline earth metal) can be reduced at lower temperatures than the pure iron oxide system combining the high H_2 yields of the Fe_3O_4/FeO system with the low reduction temperatures of a M_3O_4/MO system. In these systems, iron is partially substituted in Fe_3O_4 by a transition metal, for example, manganese (Mn), cobalt (Co), nickel (Ni), or zinc (Zn), forming the so-called mixed iron oxides or ferrites of the form $(Fe_{1-x}M_x)_3O_4$ while the reduced phase $(Fe_{1-x}M_x)_{1-y}O$ is still capable of performing the hydrolysis reaction. Numerous ferrites have been investigated in the past years with respect to their water-splitting capability [40].

Moreover, ferrites faced the issue of melting and sintering where the particle size increases after a few cycles of operation and reduces overall hydrogen yield. In order to suppress this problem, ferrites are often stabilized on, or dissolved in, inert materials such as YSZ and ZrO_2. When supported on an inert material, the reduced phase Fe^{2+} dissolves in stabilized YSZ or ZrO_2 lattice and thus melting is avoided [35, 48].

12.4.2.3 Hercynites

In addition to ferrites, another spinel material family, the (doped) aluminum spinels of the formula $(A_x,B_{1-x})^{+2}Al_2^{+3}O_4$ with cations A and B being Fe and/or Cu have been proposed as thermochemical water splitters. The rationale is to exploit the redox effect of iron oxides in conjunction with the thermal stability of aluminum oxides. For instance, Al—Cu ferrites with the various Cu/Al/Fe cation ratios have been synthesized and tested [49]; it was reported that such systems exhibited lower TR temperatures than $NiFe_2O_4$ and $ZnFe_2O_4$. A mixed cobalt ferrite-hercynite system ($CoFe_2O_4$/$FeAl_2O_4$) has been proposed as an effective WS-TR system [50].

The hercynite occurred due to the reaction between $CoFe_2O_4$ and Al_2O_3 (the ferrite was deposited on an Al_2O_3 support) at the elevated temperatures employed during testing of the ferrite in cyclic experiments. In this case the redox pair reactions are as follows.

Very low TR temperatures (1213–200 K lower than that of $CoFe_2O_4$) have been reported. These were attributed to a reaction between the ferrite and Al_2O_3 resulting in the formation of the stable aluminates $FeAl_2O_4$ and $CoAl_2O_4$, that is thermodynamically more favorable than the formation of solid solutions or vacancies. Additionally, $CoFe_2O_4/Al_2O_3$ was capable of being cycled between 1473 K (TR) and 1273 K (WS) producing significant amounts of H_2 with no obvious changes in H_2 conversion. The studies were extended to CDS where under either 75 or 600 Torr total pressure, a $CoFe_2O_4$-coated Al_2O_3 material was capable of producing appreciable amounts of CO after TR at a temperature as low as 1633 K with consistent oxidation behavior up to 23 TRs [51]. These TR temperature values are approximately 100–150 K lower than values reported for ferrites or CeO_2.

12.4.2.4 Ceria

In 2006, a redox cycle based on stoichiometric reduction of CeO_2 to Ce_2O_3 was demonstrated via a solar reactor at pressures of 100–200 mbar and at 2273 K where CeO_2 was already in the molten state, and extensive sublimation was observed (>50%) [52].

Subsequently, a similar cycle based on the nonstoichiometric reduction of ceria ($CeO_2 \rightarrow CeO_{2-\delta}$), which does not require the extreme temperatures necessary for the stoichiometric reduction, was proposed [53].

TR of ceria is quite different from ferrites. The fluorite phase maintains its structure across a wide range of oxygen stoichiometry, spanning a composition range of $CeO_{2-\delta}$ for $\delta = 0$–0.24 at 1773 K. This leads to formation of oxygen vacancies that promote the transport of oxygen atoms through the material [54]. Sintering is less problematic for ceria because the melting point is considerably higher. Thus, it is not necessary to support or dissolve them within more stable structures [54]. While ceria is attractive from a kinetic and stability perspective, its oxygen storage capacity suffers compared to most other redox materials. For example, its reduction yields are roughly 1/8 of that of cobalt ferrite ($x = 0.15$), at $T = 1473$, $pO_2 = 10^{-10}$ bar [23]. As a result, for this cycle to achieve a high $\eta_{\text{solar-to-fuel}}$, and thus become economically feasible, efficient heat recuperation from either the solid or gas phases are absolutely necessary.

The thermodynamic and kinetic properties of ceria can be altered by doping its fluorite structure with alkaline earth metal, transition metal, and lanthanide oxides. Dopants considered for ceria-based cycles include +2 (Ca, Sr, Li), +3 (Sm, Gd, Y, Cr, Pr, La, Sc), and +4 (Zr, Hf) cations [27]. The introduction of +2 and +3 dopants result in stable intrinsic oxygen vacancies, and thus an increase in electronic mobilities within the ceria lattice [55–57].

Because diffusion of oxygen in ceria occurs via ambipolar diffusion of both electrons and ions, the introduction of intrinsic vacancies results in higher oxygen diffusion coefficients compared to pure ceria [58]. Le Gal and Abanades introduced La^{3+}, Y^{3+}, Pr^{3+}, and Gd^{3+} into Zr-doped ceria and report that oxidation rates are faster because of decreased diffusion limitations. Moreover, the addition of La^{3+} was shown to improve the thermal stability upon repetitive cycling [59]. However, +2 and +3 dopants have not been shown to increase reduction extents compared to pure ceria (at conditions relevant to thermochemical cycles, high T_{high} and pO_2), and thus their introduction is not expected to have a thermodynamic benefit [60, 61].

Tetravalent dopants, such as Zr^{4+} and Hf^{4+}, have been shown to have a positive effect on increasing reduction extents under conditions relevant to thermochemical cycles. Studies show that the reducibility of Ce^{4+} is increased with the introduction of tetravalent dopants (e.g., Zr^{4+}, Hf^{4+}) due to a decrease in the partial oxygen molar enthalpy. Several reports have indicated that oxidation of Zr-doped systems is more difficult than pure ceria, but it is not clear if this is purely a thermodynamic phenomenon and or also related to kinetic limitations [62].

Considering the entropy change, ΔS_{TR} shows that doping shifts the entropy to a more negative value at a given composition for all dopants, which is expected for lanthanide and alkaline earth dopants because the number of reducible oxygen atoms decreases as the result of vacancy creation. The large shift to more negative ΔS_{TR} due to 19% Zr doping is not expected, however, because this ion has the same +4 charge as Ce and therefore does not create vacancies [12].

Shah et al. concluded that at this dopant concentration local ordering to form pyrochlore-like structures leads to a dramatic reduction in entropy [62], which may explain the poor cyclability of Zr-doped ceria seen in some thermochemical WS experiments [63].

First-principles analysis shows the limit the utility of these doped cerias due to thermodynamic limitations, particularly for TR. Regarding efficiency, the doped and undoped ceria data points cluster near the same maximum efficiencies, approximate $65\% \pm 3\%$, further illustrating that the effects of doping are relatively minor [64].

Thermodynamic properties extracted from literature data using a defect interaction model of various dopants (such as Gd, Y, Sm, Ca, Sr oxide dopants) demonstrates that none of these dopants would improve the overall solar-to-fuel energy conversion efficiency [60].

12.4.2.5 Perovskites

Perovskites of the form $ABO_{3-\delta}$ remain a class of nonstoichiometric oxides that are widely unexplored for thermochemical cycles [27]. Dopants can be substituted on both A and B cation sites, and thus the number of potential material configurations is substantially greater than ceria-based systems. A thermodynamic analysis based on evaluation of oxygen nonstoichiometry data and extraction of partial enthalpies and entropies by Scheffe et al. has shown that LSM perovskites ($La_{1+x}Sr_xMnO_{3-\delta}$)

can increase oxygen exchange capacity compared to pure ceria [65]. Experiments indicated that reduction extents are considerably greater ($\approx$2 times by mass at T_{high} = 1773 K), but oxidation thermodynamics are less favorable, resulting in incomplete oxidation. Nevertheless total CO yields during CO_2 splitting cycles are still substantially greater than ceria. An even more promising class of perovskites was explored by McDaniel et al. [66]. By doping $LaAlO_{3-\delta}$ perovskites with $Mn^{2+/3+/4+}$ on B sites and Sr^{2+} on A sites, so called SLMA perovskites, they were able to increase H_2 yields by 9 × and CO yields by 6 × compared to ceria when reduced at 1523 K and oxidized at 1273 K. The temporal oxidation behavior of three different SLMA perovskites was compared to pure ceria. Remarkably, reaction kinetics are shown to be similar to ceria and stability has been demonstrated for up to 80 redox cycles, in contrast to ferrite-based systems.

12.5 New concepts to further improve the two-step thermochemical metal-oxide cycles

12.5.1 *Isothermal thermochemical cycles*

In the two-step nonisothermal thermochemical cycles, large temperature swings between the reduction and oxidation steps cause thermodynamic inefficiencies from the irreversible heat losses incurred upon cooling of the active material to the oxidation temperature and the heat required solely to reheat the active material to its reduction temperature. Additionally, thermal stresses arising from the rapid thermal cycling of the system over large temperature differences present engineering and materials challenges in the design of high-temperature redox systems. To overcome the problems associated with cyclic WS-TR temperature-swing operation mentioned above, a isothermal pressure-swing cycling was proposed [67, 68]. Isothermal pressure-swing cycling alleviates the thermomechanical challenges associated with rapid thermal swings and consequent reduced efficiency can be avoided, but penalizes fuel production per cycle [68–70].

However, the concept of isothermal cycles in a real-scale solar reactor will face the problem of the spatial competition between oxidation and reduction. It seems unavoidable that at some regions of the reactor at least local oxygen evolution will coexist with hydrogen evolution. As a matter of fact this phenomenon has already been experimentally observed but was attributed to other possible causes [71]. This at the best case can reduce the efficiency/product yield of the process due to recombination, and at the worst case could lead to explosive mixtures. Actually, this concept has been recently challenged also on the ground of thermodynamic calculations for the case of WS by the SNL group. In their study it is argued that isothermal H_2 production is impractical and inefficient, irrespective of reactor design or reactive oxide properties. Instead, an optimal temperature difference between cycle steps, for which efficiency is the highest, can be determined for a wide range of other operating parameters. The group advocates for a combination of well-targeted pressure and temperature swing, rather than either individually, as the most efficient mode of operation of a two-step thermochemical cycle for solar fuel production [72].

12.5.2 Chemical-looping thermochemical metal-oxide cycles

One option to overcome the design challenges associated with the two-temperature operation of the cycle and the thermodynamic limitations of isothermal operation with sweep gas is carbothermal reduction. Introducing a carbon source, for example, coke or methane, alters the system thermodynamics to enable reduction at lower temperature [14, 73].

Metal-oxide cycle with the partial oxidation of methane coupled to the MO_x reduction reaction (12.3) and subsequent oxidation according to Eq. (12.2) (Fig. 12.3):

$$MO_x + y\,CH_4 \rightarrow MO_{x-y} + y\,(CO + 2\,H_2) \tag{12.11}$$

The utilization of fuel ensures an enhanced catalytic reaction at lower temperature with significantly improved extent of metal-oxide reduction and, hence, larger oxygen-carrying capacity. The coupling of partial oxidation of CH_4 with MO_x redox cycle enables one to perform the entire reactions under isothermal or near isothermal conditions because the temperature of the reduction step is greatly decreased and both reactions can thus proceed at the same temperature [74]. Also, carbothermal reduction eliminates the need for an inert sweep gas or vacuum operation [75]. Furthermore, additional H_2 and CO are produced during the MO_x reduction step in the presence of CH_4 with a 2:1 ratio desired for conversion to liquid fuels via Fischer-Tropsch reactions or methanol synthesis [75].

12.5.3 Photo-thermochemical cycles

A novel photo-thermochemical cycle (PTC), in which the photochemical method is introduced into the two-step thermochemical cycle to eliminate the temperature limitation of the thermochemical cycle, for the dissociation of H_2O or CO_2 has been reported by Zhang et al. Inspired by the formation mechanism of photo-induced oxygen vacancies (V_O), the cycle can be divided into the following two reactions [76–78]:

$$MO \rightarrow MO_{1-\delta} + \tfrac{1}{2}O_2\,(g)\ \text{(photochemistry; } > 300\text{K)} \tag{12.12}$$

$$\begin{aligned} &MO_{1-\delta} + (\delta-\gamma)\,H_2O\,(g) + \gamma\,CO_2\,(g) \\ &\quad \rightarrow MO + (\delta-\gamma)\,H_2\,(g) + \gamma\,CO\,(g)\ \text{(thermochemistry; } > 573\text{K)} \end{aligned} \tag{12.13}$$

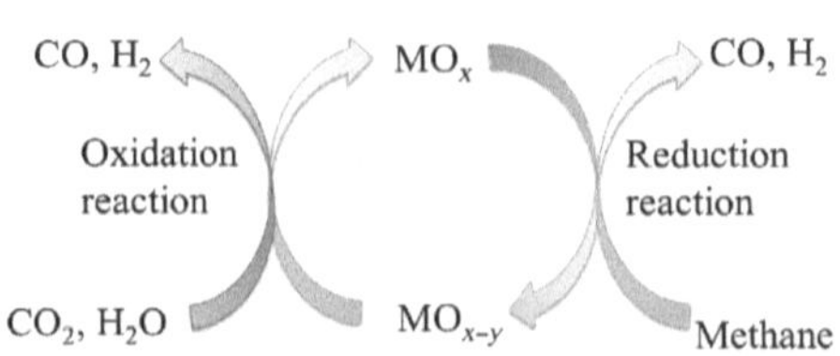

Fig. 12.3 General scheme of metal-oxide cycle with the partial oxidation of methane.

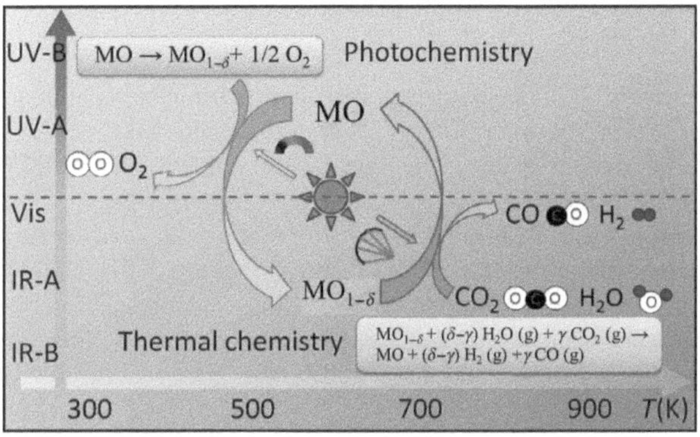

Fig. 12.4 Schematic illustration of the photo-thermochemical cycle of WS/CDS based on metal oxide. Reproduced with permission from Y. Zhang, C. Xu, J. Chen, et al., A novel photo-thermochemical cycle for the dissociation of CO_2 using solar energy, Appl. Energy 156 (2015) 223–229.

Photo-induced V_O are produced on the surface of the metal oxide (such as TiO_2) under UV in the first step [Reaction (12.12)]. The photo-induced V_O are then used to dissociate H_2O/CO_2 below 873 K in the second thermochemical step [Reaction (12.13)], which is similar to the second step in the thermochemical cycle. As a result, the overall reaction is the dissociation of H_2O/CO_2 and the production of H_2/CO and O_2. Therefore, the PTC has characteristics of both thermochemistry and photochemistry, thereby utilizing both solar luminous energy and thermal energy, as shown in Fig. 12.4 [77].

12.6 Outlook

This chapter reviewed the issues, such as reaction principles, design principles, synthesis and characterization, and material research progress, of the two-step WS/CDS processes involving redox metal oxides. Moreover, new concepts have been proposed and implemented to overcome the drawbacks of nonisothermal thermochemical cycles. Isothermal thermochemical cycles were supposed to replace nonisothermal thermochemical cycle to overcome the problems associated with the cyclic WS-TR temperature-swing operation. In order to lower the TR temperature, chemical-looping thermochemical metal-oxide cycles and PTCs were proposed.

It should be borne in mind that "technically simpler" concepts are in principle more attractive for the large-scale implementation and demonstration of any technology and in most cases a "trade-off" has to be made between theoretical targets and operation simplicity. Within this framework, perhaps "open-loop thermochemical cycles" or "photo-thermochemical cycles," which can be performed under much milder conditions, can be a viable process option for at least a transition period before syngas produced only by renewable resources.

Acknowledgment

Financial support from National Materials Genome Project (2016YFB0700600), and Distinguished Young Scientists Program of the National Natural Science Foundation of China (NSFC51425301, 21374021, and 51673096) is gratefully appreciated.

References

[1] J. Lapp, J.H. Davidson, W. Lipiński, Efficiency of two-step solar thermochemical non-stoichiometric redox cycles with heat recovery, Energy 37 (1) (2012) 591–600.
[2] B. Obama, The irreversible momentum of clean energy, Science 355 (6321) (2017) 126–129.
[3] R. Michalsky, V. Botu, C.M. Hargus, et al., Design principles for metal oxide redox materials for solar-driven isothermal fuel production, Adv. Energy Mater. 5 (7) (2015) 1401082.
[4] H. Miyaoka, Thermochemical water splitting by concentrated solar power, in: Solar to Chemical Energy Conversion, Springer International Publishing, Cham, Switzerland, 2016, , pp. 137–151.
[5] I. Ermanoski, N.P. Siegel, E.B. Stechel, A new reactor concept for efficient solar-thermochemical fuel production, J. Sol. Energy Eng. 135 (3) (2013) 031002.
[6] N.S. Lewis, D.G. Nocera, Powering the planet: chemical challenges in solar energy utilization, Proc. Natl. Acad. Sci. 103 (43) (2006) 15729–15735.
[7] D. Gust, T.A. Moore, A.L. Moore, Solar fuels via artificial photosynthesis, Acc. Chem. Res. 42 (12) (2009) 1890–1898.
[8] Cortney Mittelsteadt, Giner Inc, High-Temperature, High-Pressure Electrolysis, Doe Hydrogen and Fuel Cells Program: 2016 Annual Progress, II.B4, 1-4, https://www.hydrogen.energy.gov/pdfs/progress16/ii_b_4_mittelsteadt_2016.pdf, 2016. http://www.hydrogen.energy.
[9] R. Abe, Recent progress on photocatalytic and photoelectrochemical water splitting under visible light irradiation, J. Photochem. Photobiol C: Photochem. Rev. 11 (4) (2010) 179–209.
[10] J. Yang, D. Wang, H. Han, et al., Roles of cocatalysts in photocatalysis and photoelectrocatalysis, Acc. Chem. Res. 46 (8) (2013) 1900–1909.
[11] N.P. Siegel, J.E. Miller, I. Ermanoski, R.B. Diver, E.B. Stechel, Factors affecting the efficiency of solar driven metal oxide thermochemical cycles, Ind. Eng. Chem. Res. 52 (9) (2013) 3276–3286.
[12] J.E. Miller, A.H. McDaniel, M.D. Allendorf, Considerations in the design of materials for solar-driven fuel production using metal-oxide thermochemical cycles, Adv. Energy Mater. 4 (2) (2014), https://doi.org/10.1002/aenm.201300469.
[13] J.E. Funk, R.M. Reinstrom, Energy requirements in production of hydrogen from water, Ind. Eng. Chem. Process. Des. Dev. 5 (1966) 336–342.
[14] C. Agrafiotis, M. Roeb, C. Sattler, A review on solar thermal syngas production via redox pair-based water/carbon dioxide splitting thermochemical cycles, Renew. Sust. Energ. Rev. 42 (2015) 254–285.
[15] L.R. Martin, Use of solar energy to reduce carbon dioxide, Sol. Energy 24 (1980) 271–277.
[16] R.D. Palumbo, C.L. Larson, Production of C from CO_2 in a two-step solar process utilizing FeO and Fe_3O_4, Energy 15 (1990) 479–487.
[17] V.I. Anikeev, A.S. Bobrin, J. Ortner, et al., Catalytic thermochemical reactor/receiver for solar reforming of natural gas: design and performance, Sol. Energy 63 (2) (1998) 97–104.
[18] J.R. Rostrup-Nielsen, J. Sehested, J.K. Nørskov, Hydrogen and synthesis gas by steam-and CO_2 reforming, ChemInform 34 (17) (2002), https://doi.org/10.1002/chin.200317288.
[19] S.K. Ngoh, D. Njomo, An overview of hydrogen gas production from solar energy, Renew. Sust. Energ. Rev. 16 (9) (2012) 6782–6792.

[20] T. Nakamura, Hydrogen production from water utilizing solar heat at high temperatures, Sol. Energy 19 (5) (1977) 467–475.
[21] A. Steinfeld, Solar hydrogen production via a two-step water-splitting thermochemical cycle based on Zn/ZnO redox reactions, Int. J. Hydrog. Energy 27 (6) (2002) 611–619.
[22] I. Alxneit, Assessing the feasibility of separating a stoichiometric mixture of zinc vapor and oxygen by a fast quench—model calculations, Sol. Energy 82 (11) (2008) 959–964.
[23] R.J. Panlener, R.N. Blumenthal, J.E. Garnier, A thermodynamic study of non-stoichiometric cerium dioxide, J. Phys. Chem. Solids 36 (11) (1975) 1213–1222.
[24] L. Eyring, in: G. Meyer, L.R. Morss (Eds.), The Binary Lanthanide Oxides: Synthesis and Identification, Synthesis of Lanthanide and Actinide Compounds, Kluwer Academic Publishers, Dordrecht, 1991, , pp. 187–224.
[25] S. Abanades, A. Legal, A. Cordier, G. Peraudeau, G. Flamant, A. Julbe, Investigation of reactive cerium-based oxides for H_2 production by thermochemical two-step water-splitting, J. Mater. Sci. 45 (15) (2010) 4163–4173.
[26] M. Kuhn, S.R. Bishop, J.L.M. Rupp, H.L. Tuller, Structural characterization and oxygen nonstoichiometry of ceria-zirconia ($Ce_{1-x}Zr_xO_{2-\delta}$) solid solutions, Acta Mater. 61 (2013) 4277–4288.
[27] J.R. Scheffe, A. Steinfeld, Oxygen exchange materials for solar thermochemical splitting of H2O and CO_2: a review, Mater. Today 17 (7) (2014) 341–348.
[28] F. Sibieude, M. Ducarroir, A. Tofighiv, J. Ambriz, High temperature experiments with a solar furnace: the decomposition of Fe_3O_4, Mn_3O_4, CdO, Int. J. Hydrog. Energy 7 (1) (1982) 79–88.
[29] D.W. Green, R.H. Perry (Eds.), Perry's Chemical Engineers Handbook, eigth ed., McGraw-Hill, New York, 2008. ISBN: 9780071422949.
[30] M. Lundberg, Model calculations on some feasible two-step water splitting processes, Int. J. Hydrog. Energy 18 (5) (1993) 369–376.
[31] P. Charvin, et al., Hydrogen production by three-step solar thermochemical cycles using hydroxides and metal oxide systems, Energy Fuel 21 (2007) 2919–2928.
[32] G. Levêque, S. Abanades, Investigation of thermal and carbothermal reduction of volatile oxides (ZnO, SnO_2, GeO_2, and MgO) via solar-driven vacuum thermogravimetry for thermochemical production of solar fuels, Thermochim. Acta 605 (2015) 86–94.
[33] J.R. Scheffe, A.H. McDaniel, M.D. Allendorf, et al., Kinetics and mechanism of solar-thermochemical H_2 production by oxidation of a cobalt ferrite–zirconia composite, Energy Environ. Sci. 6 (3) (2013) 963–973.
[34] P.G. Loutzenhiser, M.E. Gálvez, I. Hischier, et al., CO_2 splitting via two-step solar thermochemical cycles with Zn/ZnO and FeO/Fe_3O_4 redox reactions II: kinetic analysis, Energy Fuel 23 (5) (2009) 2832–2839.
[35] E.N. Coker, J.A. Ohlhausen, A. Ambrosini, et al., Oxygen transport and isotopic exchange in iron oxide/YSZ thermochemically-active materials via splitting of $C(^{18}O)_2$ at high temperature studied by thermogravimetric analysis and secondary ion mass spectrometry, J. Mater. Chem. 22 (14) (2012) 6726–6732.
[36] R.B. Bird, W.E. Stewart, E.N. Lightfoot, Lightfoot Transport Phenomena, John Wiley and Sons, New York, 1976.
[37] J.E. Miller, A. Ambrosini, E.N. Coker, et al., Advancing oxide materials for thermochemical production of solar fuels, Energy Procedia 49 (2014) 2019–2026.
[38] J.E. Miller, A.H. McDaniel, M.D. Allendorf, Considerations in the design of materials for solar-driven fuel production using metal-oxide thermochemical cycles, Adv. Energy Mater. 4 (2014) 1300469, https://doi.org/10.1002/aenm.201300469.

[39] T. Rager, J. Golczewski, Solar-thermal zinc oxide reduction assisted by a second redox pair, Z. Phys. Chem. 219 (2) (2005) 235–246.

[40] T. Kodama, N. Gokon, Thermochemical cycles for high-temperature solar hydrogen production, Chem. Rev. 107 (10) (2007) 4048–4077.

[41] P.G. Loutzenhiser, A. Meier, A. Steinfeld, Review of the two-step H_2O/CO_2-splitting solar thermochemical cycle based on Zn/ZnO redox reactions, Materials 3 (11) (2010) 4922–4938.

[42] E. Koepf, W. Villasmil, A. Meier, Pilot-scale solar reactor operation and characterization for fuel production via the Zn/ZnO thermochemical cycle, Appl. Energy 165 (2016) 1004–1023.

[43] A. Berman, M. Epstein, The kinetics of hydrogen production in the oxidation of liquid zinc with water vapor, Int. J. Hydrog. Energy 25 (10) (2000) 957–967.

[44] A. Stamatiou, A. Steinfeld, Z.R. Jovanovic, On the effect of the presence of solid diluents during Zn oxidation by CO_2, Ind. Eng. Chem. Res. 52 (5) (2013) 1859–1869.

[45] L.S. Darken, R.W. Gurry, The system iron-oxygen. I. The wüstite field and related equilibria, J. Am. Chem. Soc. 67 (8) (1945) 1398–1412.

[46] O.N. Salmon, High temperature thermodynamics of the iron oxide system, J. Phys. Chem. 65 (3) (1961) 550–556.

[47] I. Bransky, A.Z. Hed, Thermogravimetric determination of the composition-oxygen partial pressure diagram of wustite ($Fe_{1\text{-}y}O$), J. Am. Ceram. Soc. 51 (4) (1968) 231.

[48] T. Kodama, N. Gokon, R. Yamamoto, Thermochemical two-step water splitting by ZrO_2-supported $Ni_xFe_{3-x}O_4$ for solar hydrogen production, Sol. Energy 82 (1) (2008) 73–79.

[49] H. Kaneko, T. Yokoyama, A. Fuse, et al., Synthesis of new ferrite, Al–Cu ferrite, and its oxygen deficiency for solar H_2 generation from H_2O, Int. J. Hydrog. Energy 31 (15) (2006) 2256–2265.

[50] J.R. Scheffe, J. Li, A.W. Weimer, A spinel ferrite/hercynite water-splitting redox cycle, Int. J. Hydrog. Energy 35 (2010) 3333–3340.

[51] D. Arifin, V.J. Aston, X. Liang, A.H. McDaniel, A.W. Weimer, $CoFe_2O_4$ on a porous Al_2O_3 nanostructure for solar thermochemical CO_2 splitting, Energy Environ. Sci. 5 (11) (2012) 9348–9443.

[52] S. Abanades, G. Flamant, Thermochemical hydrogen production from a two-step solar-driven water-splitting cycle based on cerium oxides, Sol. Energy 80 (12) (2006) 1611–1623.

[53] W.C. Chueh, S.M. Haile, Ceria as a thermochemical reaction medium for selectively generating syngas or methane from H_2O and CO_2, ChemSusChem 2 (8) (2009) 735–739.

[54] M. Mogensen, N.M. Sammes, G.A. Tompsett, Physical, chemical and electrochemical properties of pure and doped ceria, Solid State Ionics 129 (1) (2000) 63–94.

[55] L.J. Gauckler, M. Gödickemeier, D. Schneider, Nonstoichiometry and defect chemistry of ceria solid solutions, J. Electroceram. 1 (2) (1997) 165–172.

[56] T. Kobayashi, S. Wang, M. Dokiya, et al., Oxygen nonstoichiometry of $Ce_{1-y}Sm_yO_{2-0.5y-x}$ ($y=0.1, 0.2$), Solid State Ionics 126 (3) (1999) 349–357.

[57] S. Wang, T. Kobayashi, M. Dokiya, et al., Electrical and ionic conductivity of Gd-doped ceria, J. Electrochem. Soc. 147 (10) (2000) 3606–3609.

[58] W.C. Chueh, S.M. Haile, A thermochemical study of ceria: exploiting an old material for new modes of energy conversion and CO_2 mitigation, Philos. Trans. A Math. Phys. Eng. Sci. 368 (1923) (2010) 3269–3294.

[59] Q.L. Meng, C. Lee, T. Ishihara, et al., Reactivity of CeO_2-based ceramics for solar hydrogen production via a two-step water-splitting cycle with concentrated solar energy, Int. J. Hydrog. Energy 36 (21) (2011) 13435–13441.

[60] J.R. Scheffe, A. Steinfeld, Thermodynamic analysis of cerium-based oxides for solar thermochemical fuel production, Energy Fuel 26 (3) (2012) 1928–1936.
[61] A. Le Gal, S. Abanades, Dopant incorporation in ceria for enhanced water-splitting activity during solar thermochemical hydrogen generation, J. Phys. Chem. C 116 (25) (2012) 13516–13523.
[62] P.R. Shah, T. Kim, G. Zhou, et al., Evidence for entropy effects in the reduction of ceria-zirconia solutions, Chem. Mater. 18 (22) (2006) 5363–5369.
[63] A. Le Gal, S. Abanades, G. Flamant, CO_2 and H_2O splitting for thermochemical production of solar fuels using nonstoichiometric ceria and ceria/zirconia solid solutions, Energy Fuel 25 (10) (2011) 4836–4845.
[64] B. Meredig, C. Wolverton, First-principles thermodynamic framework for the evaluation of thermochemical H_2O- or CO_2-splitting materials, Phys. Rev. B 80 (24) (2009) 245119.
[65] J.R. Scheffe, D. Weibel, A. Steinfeld, Lanthanum–strontium–manganese perovskites as redox materials for solar thermochemical splitting of H_2O and CO_2, Energy Fuel 27 (8) (2013) 4250–4257.
[66] A.H. McDaniel, E.C. Miller, D. Arifin, et al., Sr- and Mn-doped $LaAlO_{3-\delta}$ for solar thermochemical H_2 and CO production, Energy Environ. Sci. 6 (8) (2013) 2424–2428.
[67] M. Roeb, C. Sattler, Isothermal water splitting, Science 341 (6145) (2013) 470–471.
[68] C.L. Muhich, B.W. Evanko, K.C. Weston, et al., Efficient generation of H_2 by splitting water with an isothermal redox cycle, Science 341 (6145) (2013) 540–542.
[69] R.B. Diver, J.E. Miller, M.D. Allendorf, et al., Solar thermochemical water-splitting ferrite-cycle heat engines, J. Sol. Energy Eng. 130 (4) (2008) 041001, https://doi.org/10.1115/1.2969781.
[70] R. Bader, L.J. Venstrom, J.H. Davidson, et al., Thermodynamic analysis of isothermal redox cycling of ceria for solar fuel production, Energy Fuel 27 (9) (2013) 5533–5544.
[71] Y. Hao, C.K. Yang, S.M. Haile, High-temperature isothermal chemical cycling for solar-driven fuel production, Phys. Chem. Chem. Phys. 15 (40) (2013) 17084–17092.
[72] I. Ermanoski, J. Miller, M. Allendorf, Efficiency maximization in solar-thermochemical fuel production: challenging the concept of isothermal water splitting, Phys. Chem. Chem. Phys. 16 (2014) 8418–8427.
[73] D. Yadav, R. Banerjee, A review of solar thermochemical processes, Renew. Sust. Energ. Rev. 54 (2016) 497–532.
[74] M.M. Nair, S. Abanades, Tailoring hybrid nonstoichiometric ceria redox cycle for combined solar methane reforming and thermochemical conversion of H_2O/CO_2, Energy Fuel 30 (7) (2016) 6050–6058.
[75] P.T. Krenzke, J.H. Davidson, Thermodynamic analysis of syngas production via the solar thermochemical cerium oxide redox cycle with methane-driven reduction, Energy Fuel 28 (6) (2014) 4088–4095.
[76] C. Xu, Y. Zhang, J. Chen, et al., Enhanced mechanism of the photo-thermochemical cycle based on effective Fe-doping TiO_2 films and DFT calculations, Appl. Catal. B Environ. 204 (2017) 324–334.
[77] Y. Zhang, C. Xu, J. Chen, et al., A novel photo-thermochemical cycle for the dissociation of CO_2 using solar energy, Appl. Energy 156 (2015) 223–229.
[78] Y. Zhang, J. Chen, C. Xu, et al., A novel photo-thermochemical cycle of water-splitting for hydrogen production based on TiO_{2-x}/TiO_2, Int. J. Hydrog. Energy 41 (4) (2016) 2215–2221.

Further reading

[79] K. Kalyanasundaram, M. Graetzel, Artificial photosynthesis: biomimetic approaches to solar energy conversion and storage, Curr. Opin. Biotechnol. 21 (3) (2010) 298–310.

Metal oxides in energy-saving smart windows

13

*Yanfang Wang**, Bingwei Chen*, Yusong Zhu*, Lijun Fu*, Yuping Wu*,‡, Teunis van Ree†*

*State Key Laboratory of Materials-Oriented Chemical Engineering, School of Energy Science and Engineering, Institute for Electrochemical Energy Storage, Nanjing Tech University, Nanjing, People's Republic of China, †Department of Chemistry, University of Venda, Thohoyandou, South Africa, ‡College of Science, Hunan Agricultural University, Changsha, China

Some 30%–40% of the energy generated in the world is used currently in lighting, ventilating, heating, cooling, and operations of buildings and other structures [1]. Especially air conditioners account for a large amount of energy consumed. Therefore, if the energy exchange between the interior and exterior of buildings through windowpanes could be more efficiently controlled, substantial energy savings would result. For example, a highly effective strategy would be to coat thin films on windows to minimize the effects of solar radiation and black-body radiation. For effective use of energy in such buildings, window glass is needed with a high solar transmittance but a low thermal emittance on cold days to allow visible light to enter and to prevent loss of heat. On the other hand, windows should at the same time be transparent to visible radiation and reflect infrared (IR) radiation on hot days, to keep the inside of the building cool. Generally, window glass with a solar-control coating, or the so-called "smart window," is designed for intelligent control of the amount of light and heat [mainly in the near-IR (NIR) region] passing through the window, in response to an external stimulus such as light (photochromic), heat (thermochromic), and electricity (electrochromic).

13.1 Thermochromic smart windows

Thermochromic materials usually refer to materials having the property to change from one color to another with changing temperature. Some thermochromic materials can restore their colors as the temperature recovers to the initial state and thus their change could be regarded as reversible. Those materials undergoing reversible color change are thought to be useful coatings for smart windows. Clearly, to be effective in real applications, these color changes must occur in a temperature region near room temperature (RT). Vanadium dioxide (VO_2), being able to undergo reversible structural changes from an IR-transparent semiconductive state to an IR-blocking metallic state at 68°C, has received particular attention because this characteristic could

Metal Oxides in Energy Technologies. https://doi.org/10.1016/B978-0-12-811167-3.00013-4

automatically control the entry of solar radiation into buildings or vehicles in response to environmental temperatures without extra external driving devices or complex installation [2]. Although a VO_2-coated smart window can block IR radiation when in the metallic state, its visible transparency remains almost unchanged [3]. In addition to their transparency to ultraviolet, IR, and visible light, the phase transition temperatures (T_c) of thermochromic coatings are also of great importance in smart windows.

13.1.1 VO_2

VO_2 is an important material for thermochromic smart windows, undergoing a reversible structural transition at a critical temperature (T_c) of 68°C. At temperatures below T_c, it exists in the semiconductive state with a monoclinic structure (M phase, space group $P2_1/c$) allowing IR transmission. At temperatures above T_c, the VO_2 undergoes a transition to the metallic state assuming a rutile structure (R phase, space group $P4_2/mnm$) resulting in high reflectivity or opacity in the IR region. Moreover, UV radiation is absorbed during the phase transition, whereas visible-light transmission is still allowed [1]. In addition, it is possible to adjust the critical temperature needed for the VO_2-phase transition close to RT using appropriate dopants or by reducing the crystal size (making use of the nanosize effect) [4, 5]. It is also possible to increase the visible-light transmittance with antireflective coatings or a combination of multilayered films. These characteristics collectively make VO_2-based materials some of the most suitable choices for use in energy-saving windows.

13.1.1.1 Preparation methods for VO_2

The usual synthesis methods for VO_2 films make use of gas-phase reactions such as sputtering deposition, chemical vapor deposition (CVD), pulsed laser deposition, and ion implantation. These methods offer exact management of parameters such as the oxygen partial pressure (which should be between 0.06 and 0.13 Pa) and film characteristics like thickness and microstructure. However, the application of these methods is constrained by the need for expensive equipment. Recent developments have shown that solution-based processing is a useful technique for the preparation of high-performance VO_2 films with improved optical properties at lower cost.

13.1.1.1.1 Sputtering deposition

VO_2 films can be made by reactive DC magnetron sputtering in a versatile thin-film deposition system [6, 7]. Ion-beam sputter deposited films show a strong dependence of growth characteristics and film properties on the substrate temperature as well as the oxygen present during deposition. Chain et al. deposited well-defined VO_2 films with grain sizes of 0.2 and 1 μm at 620°C. The optical hysteresis loops at 3.4 μm are nearly straight with complete switching taking place typically near 68°C and over a total range of 14–20°C (Fig. 13.1A). Larger grain size appears to be the dominant factor in achieving high contrast optical and electronic switching characteristics in VO_2. Increasing the beam oxygen fraction leads to films with increasing oxygen content and increasing IR transmittance (Fig. 13.1B).

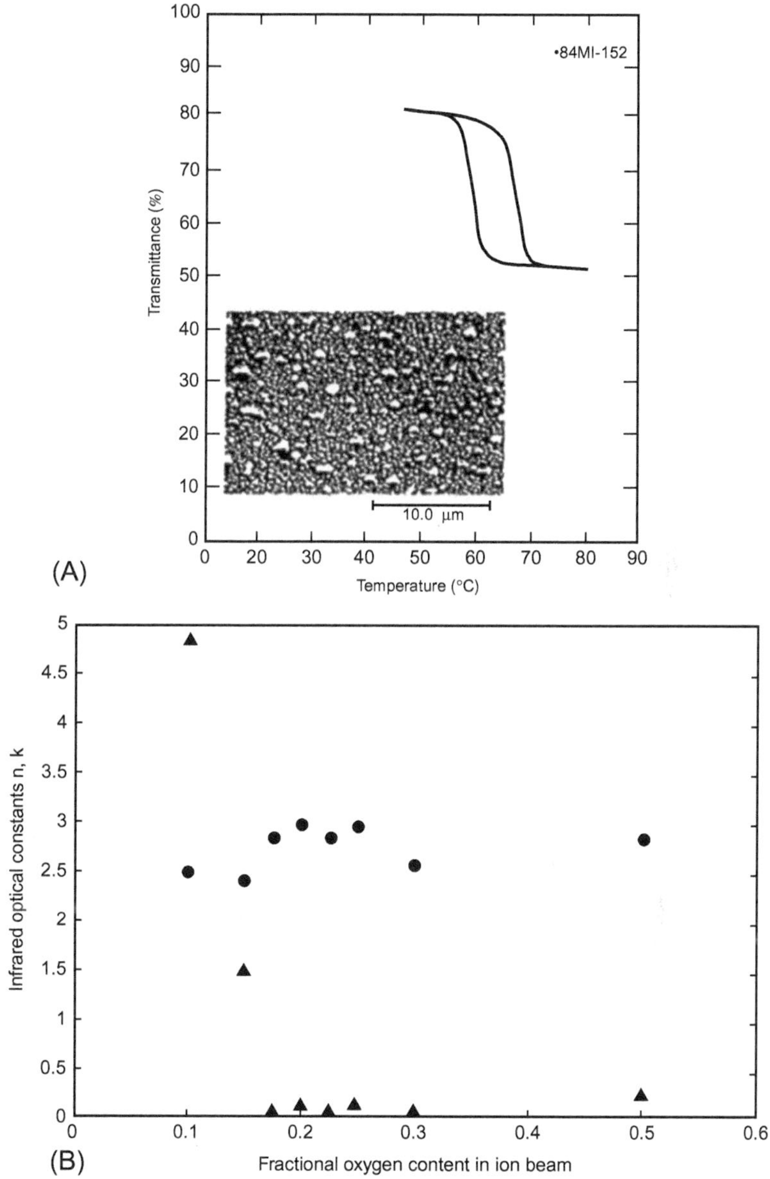

Fig. 13.1 (A) Transmittance spectrum and scanning electron micrograph of VO_2 film deposited at 620°C. (B) Variation of the RT optical constants at 3.4 μm for vanadium oxide films as a function of deposited oxygen content: ● refractive index n, and ▲ extinction coefficient k. Reproduced with permission from E.E. Chain, Optical properties of vanadium dioxide and vanadium pentoxide thin films, Appl. Opt. 30 (1991) 2782–2787.

13.1.1.1.2 Ion implantation

Lopez et al. [5] have demonstrated that VO_2 precipitates can form in amorphous SiO_2 by ion implantation. A high-purity-fused silica substrate was implanted at RT with vanadium and oxygen ions at various energies ranging between 100 and 300 keV, and 36 and 120 keV, respectively. The implantation energies were calculated and then selected to obtain superposition of both the vanadium and oxygen-implant distributions at the same depth in the near-surface region of the fused SiO_2 host. The development of the VO_2 precipitates succeeded using subsequent thermal annealing of the implanted specimens in an argon flow at atmospheric pressure and temperatures ranging from 700°C to 1000°C. The volume density of implanted atoms tended to decrease with increasing implantation energy because of the effects of straggling. Therefore, the optimal thermal annealing parameters to achieve controlled VO_2 precipitation tended to vary with temperature.

13.1.1.1.3 Solution-based method

Kang et al. [8] described a simple route to prepare VO_2 thermochromic films using a $VOCl_2$ solution with poly(vinylpyrrolidone) (PVP). Typically, vanadium pentoxide (V_2O_5) and hydrazine hydrochloride ($N_2H_4{\cdot}HCl$) were employed as starting materials to prepare a $VOCl_2$ solution. PVP, as a film-forming promoter, was added to the above solution to form a precursor. Films were coated on fused-silica substrates by spinning. Adding PVP to these solutions clearly improved the wettability of the solution, so that crack-free, uniform films were formed. Smooth PVP gel films covering the whole substrate were formed by spin-coating them with the aqueous PVP solution. In the PVP molecules, the amide nitrogen atoms partially donate their lone-pair electrons to the neighboring electrophilic carbonyl carbon atoms by orbital conjugation, resulting in carbonyl groups with a partial negative charge on carbon, and amide groups with a partial positive charge on nitrogen. The electrostatic interactions between oppositely charged groups conserve the integrity of gel films consisting of two polymerizing monomers, creating a cross-linked polymer gel film. PVP also improves the stability of the precursor solution, possibly due to the partially charged carbonyl groups bonding to VO^{2+} to form a relatively stable precursor solution (Fig. 13.2). These interactions between the polymer molecules through the oppositely charged functional groups, accompanied by the interactions between the carbonyl groups and the vanadyl ions, ensure the formation of high-quality cross-linked films after removal of the solvent.

13.1.1.2 Thermochromic performance of VO_2

The optical properties of VO_2 films are largely dependent on film morphology, grain size, and thickness. Minch et al. [9] studied the optical transmittance properties of VO_2 films to clarify the effects of grain size. As shown in Fig. 13.3A, the temperature dependence of the transmittance in forward and backward measurements at 1100 nm describes a hysteresis that is characteristic of the semiconductor-metal transition temperature (SMT) and NIR switching property (thermochromism) in VO_2. The resulting film morphologies remarkably demonstrate the effects of grain size on the

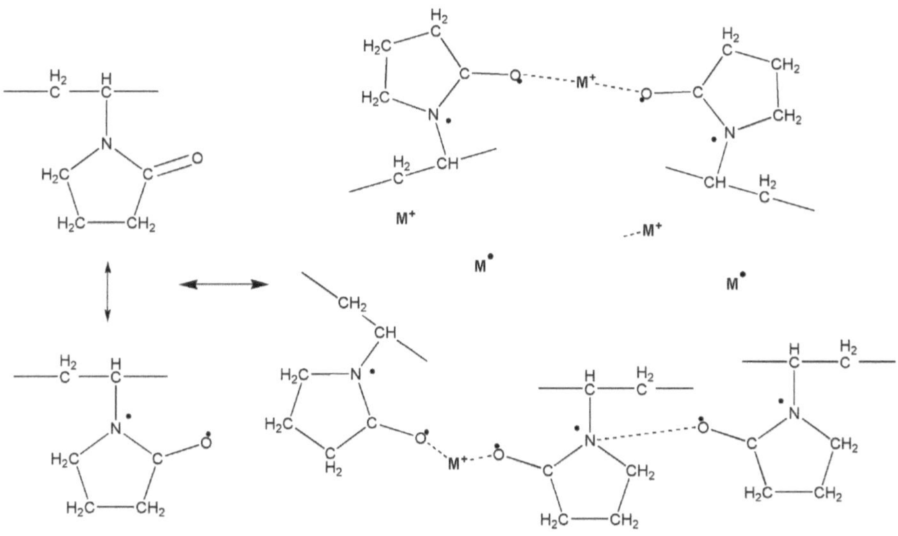

Fig. 13.2 Illustration of interactions in the cross-linked polymer gel film.
Redrawn with permission from L. Kang, Y. Gao, H. Luo, A novel solution process for the synthesis of VO_2 thin films with excellent thermochromic properties, ACS Appl. Mater. Interfaces 10 (2009) 2211–2218.

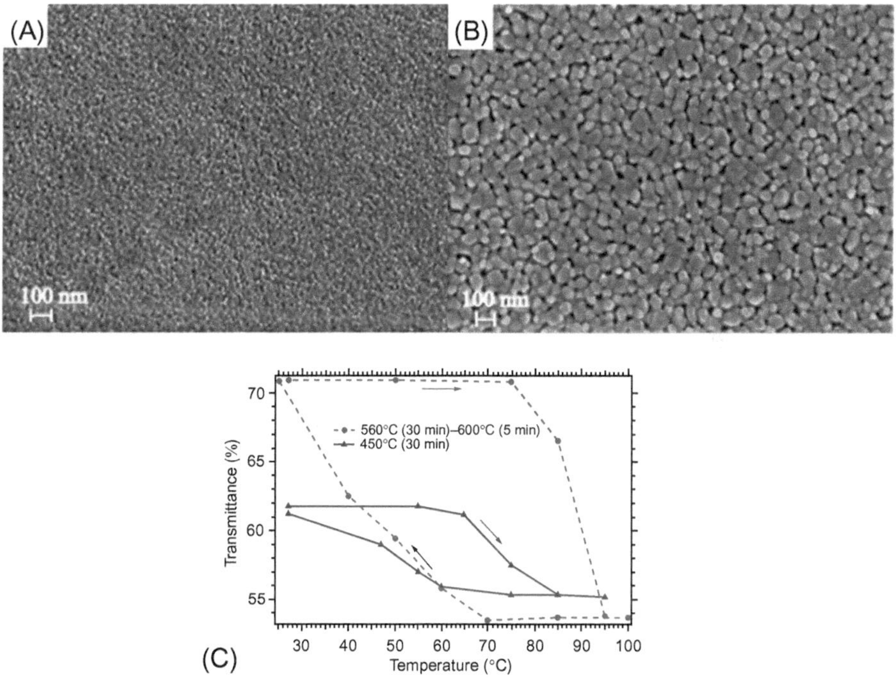

Fig. 13.3 Scanning electron microscope (SEM) images of VO_2 films annealed at (A) 450°C and (B) 560°C. (C) Hysteresis loops of the VO_2 films. *Arrows* show the reversible character of the transmittance change due to temperature-induced SMT. Hysteresis loops were obtained by measuring the transmittance at 1100 nm as a function of temperature. Reproduced with permission from R. Minch, K.R. Moonoosawmy, C.H. Solterbeck, M. Es-Souni, The influence of processing conditions on the morphology and thermochromic properties of vanadium oxide films, Thin Solid Films 556 (2014) 277–284.

Table 13.1 Comparison of particle sizes, annealing temperatures and parameters of hysteresis loops: transition temperature (T_{SMT}), width (ΔH), normalized amplitude (ΔA), and sharpness of transition (ΔT)

Sample	Average grain size (nm)	Annealing T (°C)	T_{SMT} (°C)	ΔH (°C)	ΔA (%)	ΔT (°C)
A	18	450	61	26	11	30
B	29	560	75	45	24	21

T_{SMT} is the estimated start temperature of the solar transmittance modulation (SMT).
(Reproduced with permission from R. Minch, K.R. Moonoosawmy, C.H. Solterbeck, M. Es-Souni, The influence of processing conditions on the morphology and thermochromic properties of vanadium oxide films, Thin Solid Films 556 (2014) 277–284.)

hysteresis curve (Fig. 13.3B). The different characteristics of the hysteresis loop such as transition temperature (T_{SMT}), width (ΔH), amplitude (ΔA), and sharpness of transition (ΔT) are summarized in Table 13.1 together with the grain size. The results demonstrate the grain size effects: the T_{SMT} decreases with decreasing grain size pointing to the heterogeneous nucleation character of the transformation that is aided by the abundance of grain boundaries.

Zhang et al. [1] showed that the optical properties of VO_2/fluorine-doped tin oxide (FTO) double-layered films depend on the film thickness. As evident in Fig. 13.4A, the visible transmittance (T_{vis}) of these films decreased clearly with increasing film thickness, a phenomenon mainly caused by visible light (380–780 nm) absorption. A compromise must be made between the solar transmittance modulation ability

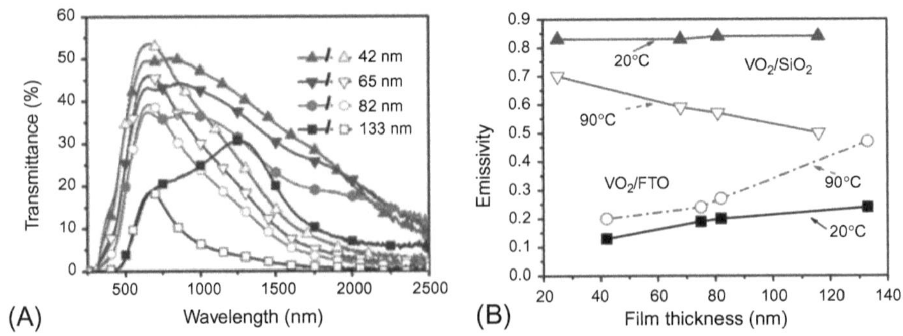

Fig. 13.4 (A) Transmittance spectra and (B) emissivity for the VO_2/FTO/substrate double-layered films with different thicknesses. Spectra with *solid* and *open* symbols were recorded at 20°C and 90°C, respectively.
Reproduced with permission from Z. Zhang, Y. Gao, H. Luo, L. Kang, Z. Chen, J. Du, M. Kanehira, Y. Zhang, Z.L. Wang, Solution-based fabrication of vanadium dioxide on F:SnO_2 substrates with largely enhanced thermochromism and low-emissivity for energy-saving applications, Energy Environ. Sci. 4 (2011) 4290–4297.

Table 13.2 Summary of optical properties of VO_2/FTO double-layer films with different thicknesses. The solar transmittance modulation (STM) was calculated from the transmittance spectra shown in Fig. 13.4A

Thickness (nm)	$T_{vis\text{-}20°C}$ (%)	$T_{vis\text{-}90°C}$ (%)	STM
42	39.6	35.9	3.5
65	34	28.6	4.9
82	28	23.9	4.8
133	14.9	8.2	7.1

(Reproduced with permission from Z. Zhang, Y. Gao, H. Luo, L. Kang, Z. Chen, J. Du, M. Kanehira, Y. Zhang, Z.L. Wang, Solution-based fabrication of vanadium dioxide on F:SnO_2 substrates with largely enhanced thermochromism and low-emissivity for energy-saving applications, Energy Environ. Sci. 4 (2011) 4290–4297)

and the visible transmittance (Table 13.2): increasing the film thickness produces the higher solar transmittance modulation ability but at the cost of lower visible light transmittance. The dependence of the emissivity on the film thickness is shown in Fig. 13.4B. This strong absorption, especially by the VO_2 R-phase, was confirmed by the rising tendency of the emissivity or absorption of VO_2/FTO double-layered films caused by increasing the VO_2 thickness.

As shown in Fig. 13.4A, the VO_2/FTO double-layered films displayed insufficient visible-light transparency and strong reflection in the visible region. A careful layer design is needed to boost the visible transmittance and at the same time the solar transmittance modulation. Commonly, an antireflective coating (ARC) is a feasible choice. Titanium oxide (TiO_2) could be used as an effective ARC to boost the visible light transmittance of VO_2 films due to its appropriate optical constants in the visible region ($2.2 < n < 2.5$ at 550 nm for TiO_2 with n optimized between 2.0 and 2.4). Furthermore, TiO_2 is transparent over a wide IR region, so that no significant absorption of IR radiation would occur. In fact, TiO_2 is one of the materials most extensively used in thermal insulating or heat-mirroring approaches. In practical smart-window applications, the TiO_2 layer performs several functions, such as photocatalysis and self-cleaning. In addition, a substantial improvement in the oxidation durability can be accomplished by the addition of TiO_2 capping layers. Fig. 13.5A shows the transmittance and reflectance spectra of a 55 nm VO_2 film with and without TiO_2 (60 nm thick). Clearly, the reflectance decreased significantly in the visible region with the TiO_2 coating, that is, from 32.6% to 23.1% at 550 nm RT; the total visible transmittance improved from 34.0% and 35.1% at 20°C to 44.0% and 38.2% at 90°C, respectively. Moreover, a great increase in the solar transmittance modulation ability (in the range 240–2600 nm) was also obtained: from 4.35% to 8.81% without and with the ARC, respectively (Fig. 13.5B). This is because of the large decrease in transmittance in the shorter wavelength NIR region, which carries more solar energy than longer-wavelength radiation, based on the solar irradiation spectrum for air mass 1.5. The addition of a TiO_2 ARC could work as an efficient functional approach to boost both the visible transmittance and the solar modulation ability.

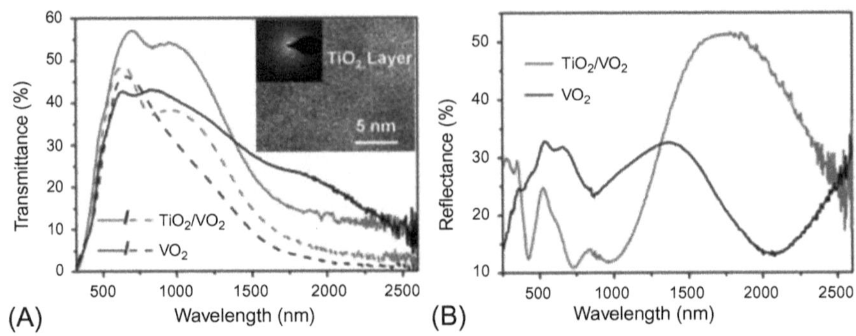

Fig. 13.5 (A) Transmittance and (B) reflectance spectra for the VO_2/FTO/substrate double-layered films without and with a TiO_2 layer. The solid and dotted lines are the spectra recorded at 20°C and 90°C, respectively. The inset in (A) shows the TEM image of the polycrystalline TiO_2 layer.
Reproduced with permission from Z. Zhang, Y. Gao, H. Luo, L. Kang, Z. Chen, J. Du, M. Kanehira, Y. Zhang, Z.L. Wang, Solution-based fabrication of vanadium dioxide on F:SnO_2 substrates with largely enhanced thermochromism and low-emissivity for energy-saving applications, Energy Environ. Sci. 4 (2011) 4290–4297.

13.1.2 Modifications of VO_2

Despite the abovementioned appealing properties, a low visible transparency and an unaesthetic yellow color limit the application of VO_2 smart windows in certain architectures, due to the intense absorption in the visible range (380–760 nm) for both the semiconducting and the metallic phases of VO_2. Up to now, introducing pores [10], design of multilayer films [6], composite films [11], and nanopowder-based solution coatings [12] have been effective approaches to enhance the visible transparency. However, substituting part of the vanadium cations by other elements may be a suitable way to enhance the visible transparency and diminish the yellow color of VO_2 thin films.

13.1.2.1 Zn doping

Jiang et al. [3] reported a Zn-doped VO_2 film with enhancing luminous transmittance (T_{lum}) and diminishing yellow color, benefiting its application in energy-efficient smart windows. Typically, $V_{1-x}Zn_xO_2$ thin films with a thickness of 80 nm were deposited on quartz glasses by reactive magnetron sputtering of V and Zn metal targets (99.9% purity).

Fig. 13.6 shows the AFM images of the prepared $V_{1-x}Zn_xO_2$ films. The undoped film is composed of relatively large crystal grains (a few hundreds of nanometers) and the crystal grains are faceted, indicating the high crystallinity of the film. The film for $x = 0.077$ seems to be very fine grained and the particle size is relatively homogenous. The surface appears smoother and the average grain size is much smaller (a few tens of nanometers) compared with the undoped film. The RMS roughness for Zn doping

Fig. 13.6 AFM images of $V_{1-x}Zn_xO_2$ films with different Zn dopant concentrations: (A) undoped, (B) $x=0.038$, and (C) $x=0.077$.
Reproduced with permission from M. Jiang, S. Bao, X. Cao, Y. Li, S. Li, H. Zhou, H. Luo, P. Jin, Improved luminous transmittance and diminished yellow color in VO_2 energy efficient smart thin films by Zn doping, Ceram. Int. 40 (2014) 6331–6334.

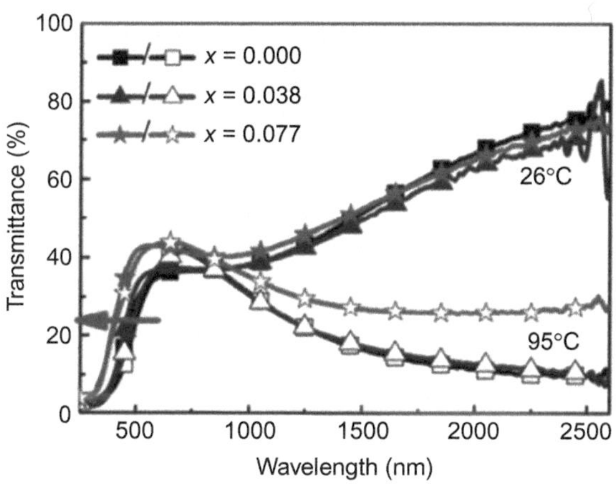

Fig. 13.7 Transmission spectra of $V_{1-x}Zn_xO_2$ films.
Reproduced with permission from M. Jiang, S. Bao, X. Cao, Y. Li, S. Li, H. Zhou, H. Luo, P. Jin, Improved luminous transmittance and diminished yellow color in VO_2 energy efficient smart thin films by Zn doping, Ceram. Int. 40 (2014) 6331–6334.

levels $x=0$, 0.038, and 0.077 was 6.8, 4.7, and 4.1 nm, respectively. Obviously, Zn doping had a strong effect on the morphology of the films. The addition of Zn depressed the growth of VO_2 particles, so that the average crystal grain size of the $V_{1-x}Zn_xO_2$ thin films decreases with increasing Zn content.

As evident in Fig. 13.7 and Table 13.3, the T_{lum} for these films increased with increasing x, which can be attributed mainly to the blue shift of the absorption edge. For $T < T_c$, the blue shift of the absorption edge was 75 nm as x increased from 0 to 0.077, with T_{lum} increasing from 32.7% to 42.5%, a 30% enhancement. For $T > T_c$, the corresponding blue shift and T_{lum} enhancement were 95 nm and 28.3%. The absorption-edge blue shift was attributed to the size effect, that is, diminishing particle size commonly induced a blue shift of the absorption edge.

Table 13.3 **Optical properties of prepared $V_{1-x}Zn_xO_2$ films**

X	$T_{lum\text{-}l}$ (%)	$T_{lum\text{-}h}$ (%)	Absorption edge (nm)	
			26°C	95°C
0.000	32.7	32.1	455	475
0.038	35.4	33.9	445	470
0.077	42.5	41.2	380	380

$T_{lum\text{-}l}$ and $T_{lum\text{-}h}$ denote the luminous transmittance at low (26°C) and high (95°C) temperatures. The absorption edge was obtained from the position of the maximum slope $(dT/d\lambda)_{max}$. T and λ denote the transmittance and wavelength in Fig. 13.6.
(Reproduced with permission from M. Jiang, S. Bao, X. Cao, Y. Li, S. Li, H. Zhou, H. Luo, P. Jin, Improved luminous transmittance and diminished yellow color in VO_2 energy efficient smart thin films by Zn doping, Ceram. Int. 40 (2014) 6331–6334.)

13.1.2.2 W doping

The T_c of VO_2 is slightly higher for practical applications although it is close to RT. Therefore, it is necessary to adjust T_c to ambient temperature (around 25°C). W-doping has been demonstrated to be an effective way to lower T_c by 23°C per 1 at.%. Huang et al. [4] synthesized W-VO_2 thin films on borosilicate glass substrates by reactive ion-beam sputtering; they had low phase transition temperatures at RT. Fig. 13.8A displays the temperature dependence of the resistivity of the W-VO_2 films. These films exhibit a phase transition temperature of 34°C, which is close to RT. The changes in IR transmittance at 2500 nm accompanied by the detecting temperatures of samples S1–S4 presented in Fig. 13.8B show transition characteristics similar to those of the resistances. The changes in transmission at T_c are significant for samples S2, S3, and S4, and the samples thus become opaque in the IR spectral range when heated to the metal phase. Inspired by this amazing progress, it is not hard to believe that there is still enough room for further improvement of thermochromic smart windows.

13.2 Electrochromic smart windows

Electrochromic (EC) materials, a term that usually refers to materials that are able to change colors as the electric field changes, have also been regarded as suitable coating materials for smart windows. In fact, many metal oxides undergo reversible redox reactions and color changes as electric fields change, allowing them to be employed in electrochromic devices (ECDs). Typical EC window devices consist of a multilayer coating (Fig. 13.9) comprising an active EC electrode, a counter electrode, an electrolyte layer between the two electrodes, two transparent conducting layers acting as electrical leads, and the supporting substrates [13]. The electrode layers are usually made of transition metal oxides. This is called a "battery-type" device, which is the standard geometry for most ECDs. In the off (bleached) state, cations such as Li^+ or H^+ are located in the electrolyte and in the counter electrode. When the device

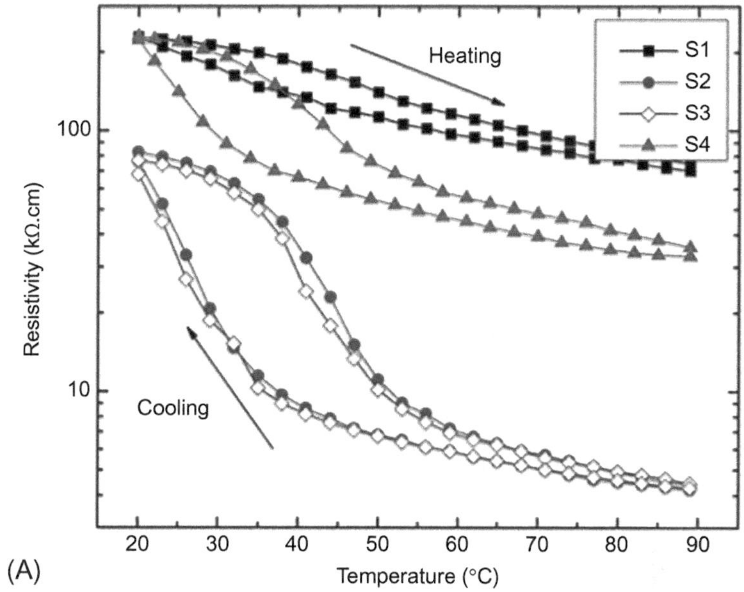

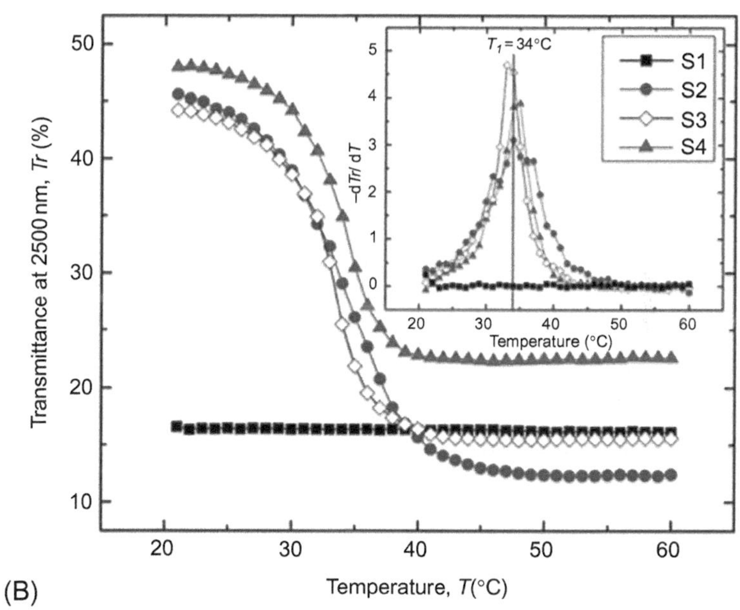

Fig. 13.8 (A) Resistance-temperature hysteresis loops and (B) transmission-temperature curves at 2500nm of W-VO_2 film samples (S1–S4). *Curves* S1–S4 denote the different annealing temperatures between 490°C and 610°C at 40°C increments. The inset in (B) shows the quantitative analysis using the derivatives of the heating curves.
Reproduced with permission from Z. Huang, C. Chen, C. Lv, S. Chen, Tungsten-doped vanadium dioxide thin films on borosilicate glass for smart window application, J. Alloys Comp. 564 (2013) 158–161.

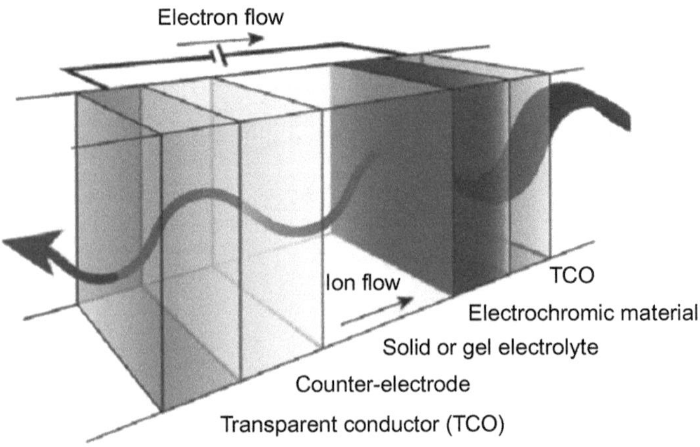

Fig. 13.9 Electrochromic device stack schematic, depicted in a *darkened* state. Electrons flow through an external circuit into the EC material, while ions flow through the electrolyte to compensate the electronic charge.
Reproduced with permission from E.L. Runnerstrom, A. Llordes, S.D. Lounis, D.J. Milliron, Nanostructured electrochromic smart windows: traditional materials and NIR-selective plasmonic nanocrystals, Chem. Commun. 50 (2014) 10555–10572.

is switched on, a voltage applied across the opposing conducting layers causes the cations to migrate by a Faradaic process from the counter electrode through the electrolyte to the EC electrode, changing its oxidation state and optical properties.

13.2.1 Cathode materials

13.2.1.1 WO_3

Tungsten trioxide is a widely researched EC material. Injection of a cathodic Faradaic charge reduces the tungsten, changing its color from a clear, transparent state to a dark blue, translucent state. The charges of the inserted electrons are balanced by protons or alkali cations in the electrolyte by insertion into the WO_3 octahedral sites, as shown by the following chemical Eq. [14]:

$$\underset{\text{(colorless)}}{WO_3 + xM^+ + xe^-} \rightleftharpoons \underset{\text{(blue)}}{M_xWO_3} \tag{13.1}$$

where M^+ is usually H^+, Li^+, or Na^+. Although the theoretical basis for the coloration is not yet fully understood, the current idea is that the injected electrons migrate to the available d band, causing new electronic transitions. In crystalline WO_3 the added electrons are delocalized, whereas the electrons in amorphous WO_3 are localized at metal sites, causing polaron optical absorption; this is best described as an electronic charge-transfer transition between neighboring metal sites with different valences. On the whole, amorphous WO_3 shows the best overall inorganic EC performance, because of its high coloration efficiency of over $50\,cm^2C^{-1}$, its short switching time, and excellent cycle life.

13.2.1.2 TiO_2

TiO_2 nanocrystals with tetragonal symmetry, which consists of TiO_6 octahedra, are regarded currently as the least expensive and most practical material for use in studies of electrochromism [15]. Also, nanostructured films of TiO_2 are widely used in ECDs, because the vacant sites formed by the shared edges in the TiO_2 structure are large enough to accommodate H^+ and Li^+ ions. The weak-ion accommodation ability, unfavorable band structures, and problematic charge carrier mobility of the TiO_2 films have led to mediocre EC responses, but coating or decorating with complementary EC TMOs seems to improve their EC properties significantly.

13.2.1.3 MoO_3

MoO_3 is generally found in two different crystal phases: α-MoO_3 and β-MoO_3 [16]. An important aspect of α-MoO_3 is its inherent layered nature that can easily accommodate a large number of positive ions, which can result in prospective bandgap manipulations. The α-MoO_3 has a bandgap in the range of 3.0–3.1 eV; therefore, if MoO_3 is used as a coating material, opportunities exist to modify the band structure of TiO_2, thereby enhancing ion intercalation, which can improve its EC properties (Fig. 13.10). Furthermore, using electron scavenging materials with appropriate electronic structures (complementing the MoO_3 electronic structure), can enhance charge transfer and then boost the EC performance. Furthermore, MoO_3 thin film has a gray feature in the colored state, which consists of a smooth absorption spectrum in the

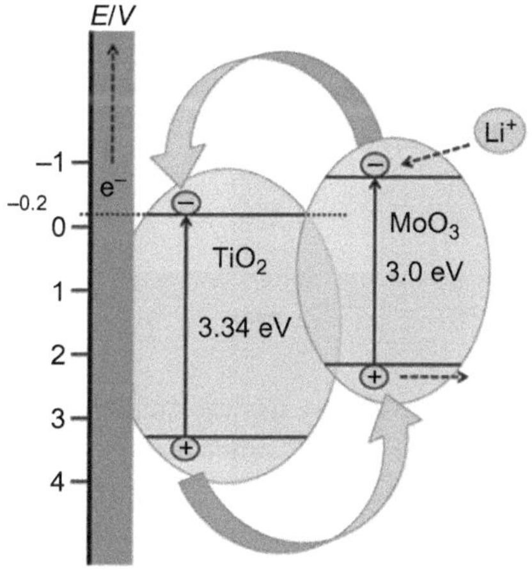

Fig. 13.10 Band structure coupling of TiO_2 and coated MoO_3.
Reproduced with permission from D.D Yao, M.R. Field, A.P. O'Mullane, K. Kalantar-zadeh, J.Z. Ou, Electrochromic properties of TiO_2 nanotubes coated with electrodeposited MoO_3, Nanoscale 5 (2013) 10353–10359.

visible spectrum. Its absorption band is much closer to the human eye's sensitivity, which means it is highly suitable for use in building lighting systems.

13.2.1.4 Other materials

Two other examples of transition metal oxides displaying cathodic coloration characteristics are Nb_2O_5 and Ta_2O_5 [13]. As is the case with WO_3, the crystal structures of these oxides also consist of edge- and corner-sharing MO_6 octahedra, enabling migration of ions through the chains of interstitial sites (or "tunnels") that exist due to the periodic structure of the MO_6 octahedra. The oxides also have similar electronic band structures, with empty d bands that become populated during cathodic charge injection, causing color changes by new intraband transitions.

13.2.2 Anode materials

Anodically coloring oxides are normally employed to serve in counter electrodes, coloring in a way complementary to the primary EC electrode. NiO is one of the most popular of these oxides, being transparent in its lithiated or reduced state and gray in its oxidized state. The coloration efficiencies of nickel oxide (NiO)-based thin films vary from around $30\,cm^2\,C^{-1}$ up to $100\,cm^2\,C^{-1}$ for mixed Ni-V-oxide films. Other widely used anodically coloring oxides are V_2O_5, IrO_2, and Co_3O_4.

13.2.2.1 NiO

NiO is an inexpensive and widely available EC material and its attractiveness for the mass production of ECDs is expected to grow [17]. It possesses EC properties that change the optical characteristics when a potential is applied and an electrical charge is inserted or removed. Ni(II) containing species are transparent, which is the case for NiO and $Ni(OH)_2$, whereas the Ni(III) containing oxides, such as Ni_2O_3 and NiOOH, are absorbing. The EC effect results from electron transitions between different atomic states, as shown by the following Eq. [18]:

$$NiO + OH^- \rightleftharpoons NiOOH + e^- \qquad (13.2)$$

13.2.2.2 V_2O_5

Layered V_2O_5 materials, with the ability of the vanadium ion to change its oxidation state, are highly likely candidates for EC applications [19, 20]. As a widely studied oxide that displays cathodic coloration as well as anodic coloration, V_2O_5 materials not only exhibit a good transmittance modulation but also show multicolor changes during the reversible lithium-ion insertion and removal processes [21]. However, the diffusion coefficient of Li^+ ions in V_2O_5 (10^{-13}–$10^{-12}\,cm^2\,s^{-1}$) and electrical conductivity (10^{-3}–$10^{-2}\,S\,cm^{-1}$) are both low [22], so that the intercalation process is quite slow. Therefore, a large surface area and a short diffusion distance are needed to achieve faster intercalation and extraction kinetics [23].

13.2.3 Preparation methods for EC films

The manufacture of EC films is based mainly on gas-phase reactions such as sputtering deposition and chemical or physical vapor deposition [13, 18]. Unfortunately, for the production of commercially viable consumer electronics, low cost, and versatile processing are requirements that cannot be ignored. Furthermore, the requisite patterning and multilayered structures impose additional requirements on the development of such devices. Inkjet printing is a very attractive and inexpensive technology for patterning EC materials on flexible plastic substrates [24]. EC materials can be deposited on a specific location, and thickness control is possible by varying the number of printed layers. It is a one-step process, inexpensive, and can be applied to a variety of substrates. These methods will be introduced by demonstrating how they are applied in the preparation of WO_3 films.

13.2.3.1 Sputtering deposition

Baloukas et al. [25] deposited porous WO_3 films in a specially made vacuum chamber by radio-frequency magnetron sputtering from a 50mm diameter WO_3 target at a power of 140W, using a mixture of Ar and O_2 (4:1 Ar to O_2 ratio). The base pressure was approximately 2.5×10^{-6} Torr. The WO_3 films were deposited on 2.5cm × 5cm tin-doped indium oxide (ITO)-coated glass substrates that had been cleaned with soap and deionized (DI) water. The obtained films showed good durability in aqueous H_2SO_4, but little transmission variation, slow electro-optical response times, and a diminishing bleached state. The porous films had short response times and large transmission variations, but degraded rather quickly.

13.2.3.2 Chemical vapor deposition

Gesheva et al. [26] prepared mixed molybdenum and tungsten oxide films using atmospheric pressure CVD in a horizontal quartz reactor with cold walls. The substrate temperature was maintained at 200°C by a high-frequency generator. The precursor was a mixture of the Mo and W carbonyls. After placing the precursor powders in a sublimator, immersed in a silicon oil bath, the mixture was heated by a magnetic stirrer-heater to the sublimation temperature (90°C for the $Mo(CO)_6$: $W(CO)_6$ mixture). The vaporized carbonyl decomposition products were transferred from the sublimator to the CVD reactor by an argon stream. The Ar to oxygen (the reactive gas) flow ratio was chosen as 1:16 and 1:32. After deposition (40min), the samples were annealed further for 1h in air at temperatures of 300–500°C. The obtained films were uniform and transparent, with their color efficiency attaining the maximum values of $120\,cm^2\,C^{-1}$, indicating that atmospheric pressure CVD is a useful method to produce EC films.

13.2.3.3 Physical vapor deposition

Yang et al. [27] fabricated ultrathin WO_3 films by evaporating WO_3 powder on FTO. The physical vapor deposition system was employed for WO_3 film evaporation. When the chamber pressure reached 4×10^{-4} Pa, WO_3 powder (50nm, 99.9%) loading on a tungsten boat was evaporated by applying a fixed current of 100 A for about 30 min, after which the FTO substrate was covered by a 300 nm transparent WO_3 film. This demonstrated the feasibility of fabricating WO_3 films on a large scale. In spite of the large-scale fabrication, the films exhibited good performance, with their optical transmittance abruptly decreasing from 91.3% to 15.1% at a wavelength of 633 nm when a voltage of −0.6 V was applied.

13.2.3.4 Inkjet printing

Layani et al. [24] demonstrated a versatile ink formulation with good printability and prepared EC WO_3 films by inkjet printing. The WO_3 nanoparticles were produced using a modified sol-gel route. The WO_3 ink consisted of a formulation with diethylene glycol *n*-butyl ether (DB) and DI water as the solvents (DB:DI water ratio 1:1). The WO_3 ink had a surface tension and viscosity suitable for inkjet printing and was stable under normal conditions for several weeks, without sedimentation or aggregation of ink particles. Precise, simple and reproducible patterning of the active layers is possible using inkjet printing. It is also possible to tune the contrast of the ECD by controlling the WO_3 nanoparticle concentration in the ink and the number of printed layers.

13.2.4 Electrochromic performance of ECDs

High-performance ECDs need high optical modulation, good optical memory, and chemical stability. NiO is a well-known and popular anodic EC material and tungsten trioxide (WO_3) is one of the most widely accepted cathodic EC materials; both materials have been exhaustively investigated. They are complementary electrode materials when integrated in an ECD using lithium-based electrolytes. They can be colored and bleached by opposite polarity, thereby increasing the optical modulation of the complementary device. The EC performance of different WO_3-NiO cells will be introduced as examples to illustrate how ECDs function.

Katase et al. [28] demonstrated a leakage-free water-based device in a nanogap parallel plate structure, which can be considered a nanosized electrochemical cell (Fig. 13.11A), enabling high electric-field applications for ion migration and effective protonation/deprotonation of the a-WO_3 layer. As shown in Fig. 13.11B, a gate voltage (V_g) application induces water electrolysis in the nanoporous calcium aluminate (CAN) film, and produces H^+ and OH^- ions, which move to protonate the a-WO_3 layer (Eq. 13.1) and hydroxylate the NiO layer (Eq. 13.2), respectively. The NiO counter layer is expected to work as the OH^- absorber, which maintains a better electrochemical balance and should improve the reversibility and reproducibility of device operation. Alternate positive and negative V_g applications induce the reversible protonation/deprotonation of the a-WO_3 layer, switching it from a transparent insulator to a dark blue conductor. Fig. 13.12A shows the optical transmission spectra of the device in the initial state (black line) and protonated states at +3 V (red line), +5 V (green line), and +10 V (blue line). The transmission (T) of the device was largely changed by the applied positive V_g. The initial device was, to some extent, transparent

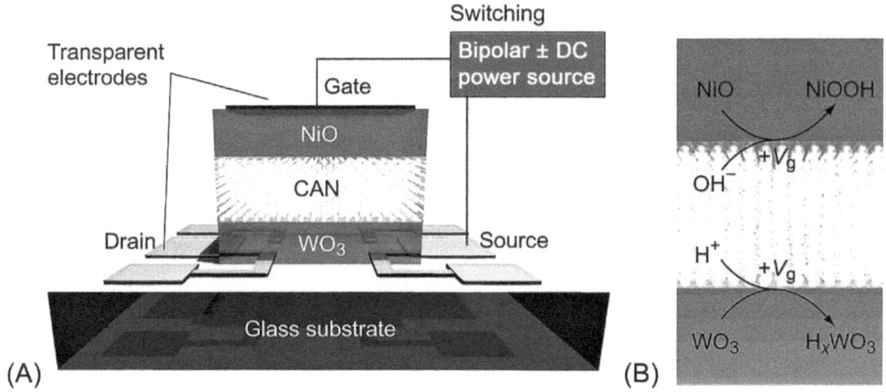

Fig. 13.11 (A) Schematic structure of three-terminal TFT-type device, composed of a-WO_3 (80 nm), CAN (300 nm), and NiO (20 nm)/ITO (20 nm) layers. Transparent ITO thin films are used for all the electrodes. (B) Device operation mechanism. During the positive V_g application, protonation of the a-WO_3 layer and hydroxylation of the NiO layer occur simultaneously. Conversely, a-H_xWO_3 and NiOOH return to a-WO_3 and NiO, respectively, during the negative V_g application.
Reproduced with permission from T. Katase, T. Onozato, M. Hirono, T. Mizuno, H. Ohta, A transparent electrochromic metal-insulator switching device with three-terminal transistor geometry, Sci. Rep. 6 (2016) 25819.

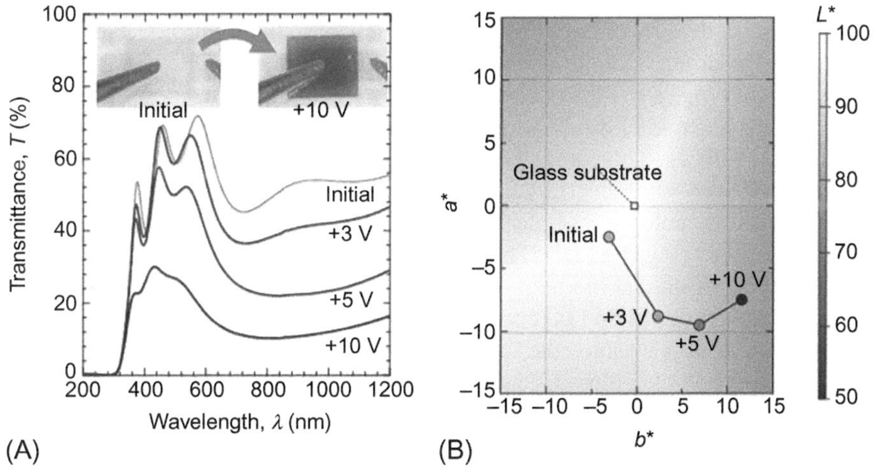

Fig. 13.12 (A) Optical transmission spectra of the a-WO_3-CAN-NiO device. A significant decrease in the transmission is seen with increasing V_g. The inset shows photographs of the device; after the application of $V_g = +10$ V, the device becomes opaque/*dark blue (right)*, whereas the device is colorless transparent at the initial state *(left)*. (B) $L^*a^*b^*$ color space plot of the device at each state in (A).
Reproduced with permission from T. Katase, T. Onozato, M. Hirono, T. Mizuno, H. Ohta, A transparent electrochromic metal-insulator switching device with three-terminal transistor geometry, Sci. Rep. 6 (2016) 25819.

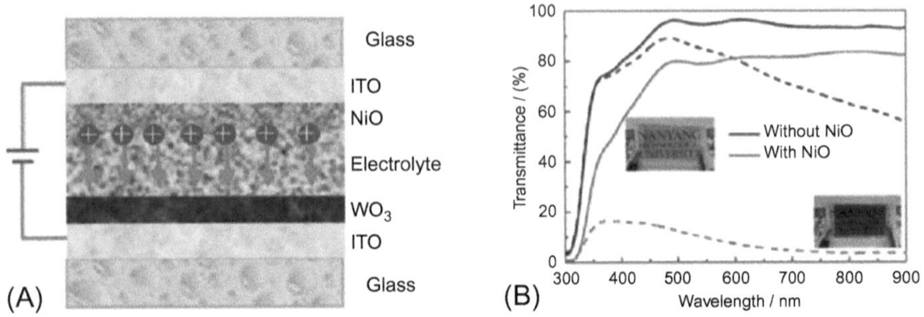

Fig. 13.13 (A) Setup of a solid ECD with NiO as ion-storage layer, and (B) transmission spectra of the solid with and without NiO as the ion-storage layer in the bleached (2.5 V) and colored (−2.5 V) states over the 300–900 nm wavelength range (photos of the solid ECD with NiO as the ion-storage layer in the bleached state and the colored state are shown in the insets). Reproduced with permission from G. Cai, P. Darmawan, M. Cui, J. Chen, X. Wang, A.L. Eh, S. Magdassi, P.S. Lee, Inkjet-printed all solid-state electrochromic devices based on NiO/WO_3 nanoparticle complementary electrodes, Nanoscale 8 (2016) 348–357.

in the visible light region. After protonation, T at $\lambda = 700$ nm reduced to 24% that of the initial state (transmission modulation of 35%). The inset shows the device at the initial and protonated states (+10 V); the color clearly changes from transparent colorless to opaque dark blue, as can be seen in $L^*a^*b^*$ color space Fig. 13.12B), reflecting the optical modulation of the a-WO_3 channel by RT protonation.

To solve the problem of electrolyte leakage, a solid-state ECD may be assembled, improving handling, providing greater reliability and reducing the cost of packaging. Cai et al. [18] assembled a full solid-state ECD using a NiO film as the ion-storage layer, a WO_3 film as EC layer, and poly(methyl methacrylate) (PMMA)-based polymer electrolytes as solid electrolyte (Fig. 13.13A). The transmittance spectra of the solid ECDs in the colored and bleached states were measured for 30 s at −2.5 and 2.5 V, respectively (Fig. 13.13B). The solid device using a NiO film as the ion-storage layer displayed an optical modulation of 75% at 633 nm, which is much higher than that of a device using a plain ITO complementary electrode without NiO film (18.8%). The insets in Fig. 13.13B show the solid device using NiO film as the ion-storage layer in the bleached and the colored states. The enhancement of EC performance may be ascribed to the synergistic action of the NiO and WO_3 films, because the NiO film acts not only as an ion-storage layer, but also as a complementary EC layer.

13.3 Outlook

Over the last couple of decades, metal oxides have fascinated academic and industrial researchers with their potential use in energy-saving smart windows. Unfortunately, there has been no sustained progress, and a dramatic break-through has not yet been achieved. The reason for the slower development of smart windows than, for example, information and communication technology, lies in the inherent difficulties of fabricating devices that rely on several high-cost technologies. However, as nanotechnology matures, developments may gain the necessary momentum needed to fulfill their promise. Nanomaterials offer these devices improved charge capacity, higher coloration efficiency, and more resilience. Industry is also presented an opportunity to use nanotechnology to migrate from energy intensive and expensive vacuum deposition methods to solution processing, which could reduce significantly the cost of smart windows.

Acknowledgment

Financial support from the National Materials Genome Project (2016YFB0700600), China National Funds for Distinguished Youth Scientists (51026004) and NSFC (U1601214 and 51502137) is gratefully acknowledged.

References

[1] Z. Zhang, Y. Gao, H. Luo, L. Kang, Z. Chen, J. Du, M. Kanehira, Y. Zhang, Z.L. Wang, Solution-based fabrication of vanadium dioxide on F:SnO_2 substrates with largely enhanced thermochromism and low-emissivity for energy-saving applications, Energy Environ. Sci. 4 (2011) 4290–4297 (and references therein).

[2] Z. Chen, Y. Gao, L. Kang, J. Du, Z. Zhang, H. Luo, H. Miao, G. Tang, VO_2-based double-layered films for smart windows: optical design, all-solution preparation and improved properties, Sol. Energy Mater. Sol. Cells 95 (2011) 2677–2684.

[3] M. Jiang, S. Bao, X. Cao, Y. Li, S. Li, H. Zhou, H. Luo, P. Jin, Improved luminous transmittance and diminished yellow color in VO_2 energy efficient smart thin films by Zn doping, Ceram. Int. 40 (2014) 6331–6334.

[4] Z. Huang, C. Chen, C. Lv, S. Chen, Tungsten-doped vanadium dioxide thin films on borosilicate glass for smart window application, J. Alloys Comp. 564 (2013) 158–161.

[5] R. Lopez, L.A. Boatner, T.E. Haynes, L.C. Feldman, R.F. Haglund, Synthesis and characterization of size-controlled vanadium dioxide nanocrystals in a fused silica matrix, J. Appl. Phys. 92 (2002) 4031–4036.

[6] N.R. Mlyuka, G.A. Niklasson, C.G. Granqvist, Thermochromic multilayer films of VO_2 and TiO_2 with enhanced transmittance, Sol. Energy Mat. Sol. Cells 93 (2009) 1685–1687.

[7] E.E. Chain, Optical properties of vanadium dioxide and vanadium pentoxide thin films, Appl. Opt. 30 (1991) 2782–2787.

[8] L. Kang, Y. Gao, H. Luo, A novel solution process for the synthesis of VO_2 thin films with excellent thermochromic properties, ACS Appl. Mater. Interfaces (10) (2009) 2211–2218.

[9] R. Minch, K.R. Moonoosawmy, C.H. Solterbeck, M. Es-Souni, The influence of processing conditions on the morphology and thermochromic properties of vanadium oxide films, Thin Solid Films 556 (2014) 277–284.

[10] L.T. Kang, Y.F. Gao, H.J. Luo, Z. Chen, J. Du, Z.T. Zhang, Nanoporous thermochromic VO_2 films with low optical constants, enhanced luminous transmittance and thermochromic properties, ACS Appl. Mater. Interfaces 3 (2011) 135–138.

[11] J. Du, Y.F. Gao, H.J. Luo, Z.T. Zhang, L.T. Kang, Z. Chen, Formation and metal-to-insulator transition properties of VO_2–ZrV_2O_7 composite films by polymer-assisted deposition, Sol. Energy Mater. Sol. Cells 95 (2011) 1604–1609.

[12] Y.F. Gao, S.B. Wang, H.J. Luo, L. Dai, C.X. Cao, Y.L. Liu, Z. Chen, M. Kanehira, Enhanced chemical stability of VO_2 nanoparticles by the formation of SiO_2/VO_2 core/shell structures and the application to transparent and flexible VO_2-based composite foils with excellent thermochromic properties for solar heat control, Energy Environ. Sci. 5 (2012) 6104–6110.

[13] E.L. Runnerstrom, A. Llordes, S.D. Lounis, D.J. Milliron, Nanostructured electrochromic smart windows: traditional materials and NIR-selective plasmonic nanocrystals, Chem. Commun. 50 (2014) 10555–10572.

[14] J. Neto, R.M. Torresi, S. Torresi, Electrochromic behavior of WO_3 nanoplate thin films in acid aqueous solution and a protic ionic liquid, J. Electroanal. Chem. 765 (2016) 111–117.

[15] H. Yu, Y. Li, L. Zhao, G. Li, J. Li, H. Rong, Z. Liu, Novel MoO_3-TiO_2 composite nanorods films with improved electrochromic performance, Mater. Lett. 169 (2016) 65–68.

[16] D.D. Yao, M.R. Field, A.P. O'Mullane, K. Kalantar-zadeh, J.Z. Ou, Electrochromic properties of TiO_2 nanotubes coated with electrodeposited MoO_3, Nanoscale 5 (2013) 10353–10359.

[17] D. Dong, W. Wang, G. Dong, F. Zhang, Y. He, H. Yu, F. Liu, M. Wang, X. Diao, Electrochromic properties and performance of NiO_x films and their corresponding all-thin-film flexible devices prepared by reactive DC magnetron sputtering, Appl. Surf. Sci. 383 (2016) 49–56.

[18] G. Cai, P. Darmawan, M. Cui, J. Chen, X. Wang, A.L. Eh, S. Magdassi, P.S. Lee, Inkjet-printed all solid-state electrochromic devices based on NiO/WO_3 nanoparticle complementary electrodes, Nanoscale 8 (2016) 348–357.

[19] J. Chu, Z. Kong, D. Lu, W. Zhang, X. Wang, Y. Yu, S. Li, X. Wang, S. Xiong, J. Ma, Hydrothermal synthesis of vanadium oxide nanorods and their electrochromic performance, Mater. Lett. 166 (2016) 179–182.

[20] Z. Tong, N. Li, H. Lv, Y. Tian, H. Qu, X. Zhang, J. Zhao, Y. Li, Annealing synthesis of coralline V_2O_5 nanorod architecture for multicolor energy-efficient electrochromic device, Sol. Energy Mater. Sol. Cells 146 (2015) 135–143.

[21] M. Scherer, L. Li, P. Cunha, O. Scherman, U. Steiner, Enhanced electrochromism in gyroid-structured vanadium pentoxide, Adv. Mater. 24 (2012) 1217–1221.

[22] E. Portiron, A. Salle, A. Varbaere, Y. Piffard, F. Guyomard, Electrochemically synthesized vanadium oxides as lithium insertion hosts, Electrochim. Acta 45 (1999) 197–214.

[23] Z. Tong, J. Hao, K. Zhang, J. Zhao, B. Su, Y. Li, Improved electrochromic performance and lithium diffusion coefficient in three-dimensionally ordered macroporous V_2O_5 films, J. Mater. Chem. C 2 (2015) 3651–3658.

[24] M. Layani, P. Darmawan, W.L. Foo, L. Liu, A. Kamyshny, D. Mandler, S. Magdassi, P. S. Lee, Nanostructured electrochromic films by inkjet printing on large area and flexible transparent silver electrodes, Nanoscale 6 (2014) 4572–4576.

[25] B. Baloukas, J.M. Lamarre, L. Martinu, Electrochromic interference filters fabricated from dense and porous tungsten oxide films, Sol. Energy Mater. Sol. Cells 95 (2010) 807–815.

[26] K.A. Gesheva, T.M. Ivanova, G. Bodurov, Transition metal oxide films: technology and "Smart Windows" electrochromic device performance, Prog. Org. Coat. 74 (2011) 635–639.

[27] P. Yang, P. Sun, Z. Chai, L. Huang, X. Cai, S. Tan, J. Song, W. Mai, Large-scale fabrication of pseudocapacitive glass windows that combine electrochromism and energy storage, Angew. Chem. Int. Ed. 53 (2014) 11935–11939.

[28] T. Katase, T. Onozato, M. Hirono, T. Mizuno, H. Ohta, A transparent electrochromic metal-insulator switching device with three-terminal transistor geometry, Sci. Rep. 6 (2016). 25819.

Metal oxide-based superconductors in AC power transportation and transformation

14

*Yanfang Wang**, *Chunyang Li**, *Weibin Zhou**, *Yusong Zhu**, *Lijun Fu**, *Yuping Wu**,‡, *Teunis van Ree*†
*State Key Laboratory of Materials-Oriented Chemical Engineering, School of Energy Science and Engineering, Institute for Electrochemical Energy Storage, Nanjing Tech University, Nanjing, People's Republic of China, †Department of Chemistry, University of Venda, Thohoyandou, South Africa, ‡College of Science, Hunan Agricultural University, Changsha, China

Superconductors, materials that lose their electrical resistance under a specific temperature, were first discovered by Kamerlingh Onnes in 1911. At the time, metals like Hg, Nb, and Pb were the dominant materials in this field. However, the superconductivity disappeared at fields below 50mT, which makes their practical applications unreachable. Significant advances in understanding of the magnetic properties of superconductors were made in the 1930s, when London and Shubnikov were able to unravel the crucial differences between type I and type II superconductors. In type I superconductors, the current flows at the surface of the material, and superconductivity is negated by weak fields, whereas type II superconductors are capable of carrying bulk supercurrent at high fields [1]. At the end of 1986, Bednorz and Muller reported their discovery of superconductivity at 30K in the layered cuprate, $LaBa_2CuO_{4-x}$ [2]. Subsequently, more and more metal oxides have been discovered for use as high-T_c superconductors (HTSs), such as other cuprates [3], and most recently iron-based oxides [4].

14.1 Some parameters for superconductors

Materials known as superconductors usually have three fundamental properties: superconductivity, perfect diamagnetism, and tunneling effect. Superconductivity, also known as the "zero-resistance" effect, is the phenomenon that a material loses resistance below a specific temperature. This temperature is called the critical transition temperature (T_c). Materials that show superconductivity at temperatures higher than liquid–nitrogen (LN) temperature (77K) are usually called high-temperature superconductors. Perfect diamagnetism, also called the Meissner effect, means that magnetic lines do not pass through superconductors in a magnetic field. The strongest magnetic field strength at which superconductors exclude all magnetic lines is

Metal Oxides in Energy Technologies. https://doi.org/10.1016/B978-0-12-811167-3.00014-6

regarded as the critical magnetic field strength (H_c). The tunneling effect, also called Josephson effect, refers to the phenomenon that electrons can pass through the insulated layer between two superconductive layers if the interlayer is thin enough.

14.2 Early stage metal oxide-based superconductors

14.2.1 Li-Ti-O ternary system

Prior to the discovery of cuprate-based superconductors, only a small number of oxides were known to exhibit superconductivity. In 1973, the high-temperature superconductivity of a Li-Ti-O system was first reported by Johnston et al., with onset temperatures as high as 13.7 K [5]. They employed arc melting (Li_2CO_3, TiO_2, and Ti metal) as well as a sintering mixture ($Li_2Ti_2O_5$ and Ti_2O_3) to produce multiphase samples and define the various ranges of T_c in the Li-Ti-O system. They found that the Li-Ti-O system existed as a spinel structure with a homogeneity ranging from $Li_{1.33}Ti_{1.67}O_4$ ($a = 8.36$ Å) to approximately $Li_{0.8}Ti_{2.2}O_4$ ($a \sim 8.40$ Å). There also appeared to be a small homogeneity range for the metal–oxygen ratio. The lithium and/or oxygen-rich compositions had a white to light-blue color, and did not become superconducting down to 1.5 K. However, as the Ti/Li and/or Ti/O ratio increased toward the ideal spinel composition $LiTi_2O_4$, the samples became dark blue to blue-black in color, and became superconducting in the temperature range 7–13.7 K (Fig. 14.1).

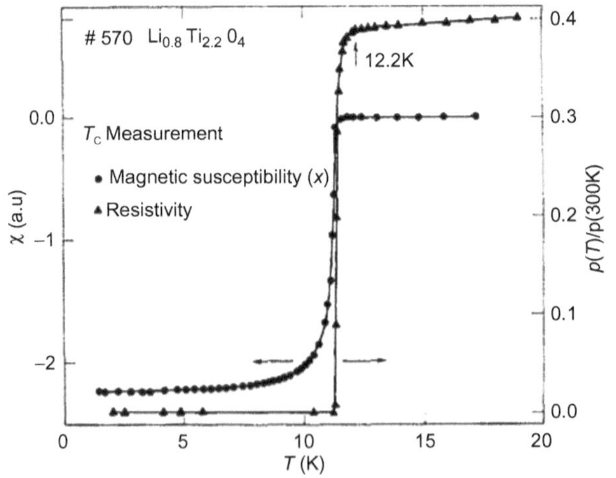

Fig. 14.1 Low-temperature magnetic susceptibility and resistivity data for $Li_{0.8}Ti_{2.2}O_4$ prepared by sintering $Li_2Ti_2O_5$ and Ti_2O_3 at 800°C for 1 day.
Reproduced by permission from D.C. Johnston, H. Prakash, W.H. Zachariasen, R. Viswanathan, High temperature superconductivity in the Li-Ti-O ternary system, Mater. Res. Bull. 8 (1973) 777–784.

14.2.2 $SrTiO_3$

$SrTiO_3$ was regarded as an "ideal" cubic perovskite-type compound, consisting of warped bands at the center zone rather than many valleys at or near the Brillouin-zone boundaries along the (100) direction. Matteiss [6] reported the occurrence of a cubic-to-tetragonal phase transition of $SrTiO_3$ at about 110 K, as shown in Fig. 14.2. Koonce and Cohen [7] reported that $SrTiO_3$ samples exhibited superconductivity despite their small carrier concentrations (Fig. 14.3).

Inspired by the unique properties of superconducting materials, Binnig et al. produced free carriers by direct doping of $SrTiO_3$ crystals with Nb and proved its two-

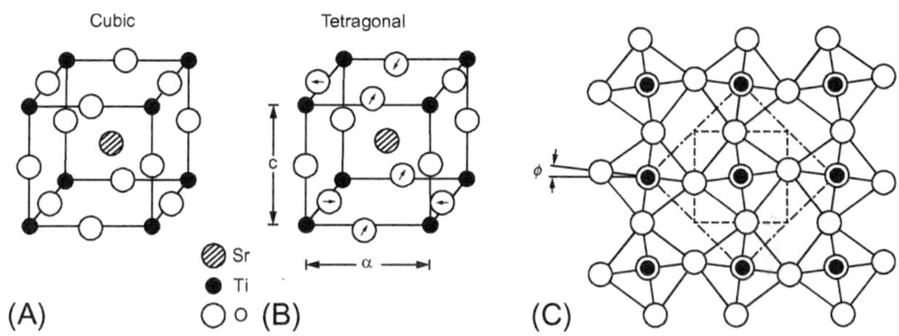

Fig. 14.2 (A) Atom positions in cubic $SrTiO_3$. (B) *Arrows* indicate the displacement of the oxygen atoms in the tetragonal state. (C) The projection of one layer of titanium and oxygen atoms on a (001) plane.
Reproduced by permission from L.F. Matteiss, Effect of the 110 K phase transition on the $SrTiO_3$ conduction bonds, Phys. Rev. B 6 (1972) 4740–4753.

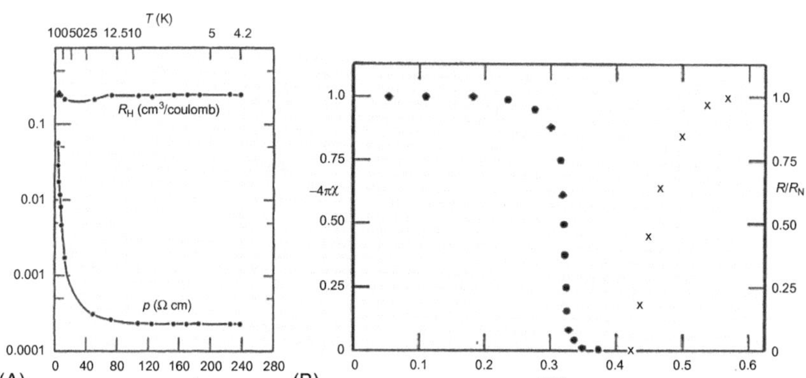

Fig. 14.3 (A) Resistivity and Hall coefficient as a function of $1/T$ and (B) temperature dependence of the magnetic susceptibility and resistivity for $SrTiO_3$.
Reproduced by permission from C.S. Koonce, M.L. Cohen, Superconducting transition temperature of semiconducting $SrTiO_3$, Phys. Rev. 163 (1967) 380–390.

band superconductivity [8]. This opened the door to exciting nonequilibrium experiments involving coupling between order parameters and/or quasiparticles in different bands. In Nb-doped $SrTiO_3$, the plasma edge is below the highest optical phonon, which is therefore unshielded. This large electron–phonon coupling allows a T_c of 0.7 K with Cooper pairing.

14.2.3 $BaPb_{1-x}Bi_xO_3$

The occurrence of high electron–phonon coupling in another metallic oxide, also a perovskite, became evident with the discovery of superconductivity in the mixed-valence compound $BaPb_{l-x}Bi_xO_3$, which was first reported by Sleight et al. [9]. They prepared compositions of this type by heating suitable mixtures of oxides, carbonates, and/or nitrates in air at 800–1000°C. The relationship between substitution degree x and lattice parameter a are shown in Fig. 14.4A. The phase appeared black with x up to 0.3, but the color changed into bronze-like at higher values. Electrical resistivity measurement on pellets indicated semiconducting behavior for all the bronze-colored phases and metallic properties for all the black phases. Superconductivity was observed in the $BaPb_{l-x}Bi_xO_3$ system for $x=0.05$–0.3. The critical temperature increased from about 9 K at $x=0.05$ to a maximum of about 13 K at $x=0.3$ (Fig. 14.4B). In 1980, Thanh et al. [10] studied the superconductivity of $BaPb_{l-x}Bi_xO_3$ in a broader range of $0.05<x<0.35$ and a maximum critical temperature of 11.7 K was observed at $x=0.25$ (Fig. 14.5A). The upper critical field was also determined from the magnetic field dependence of the low current–density resistivity (Fig. 14.5B).

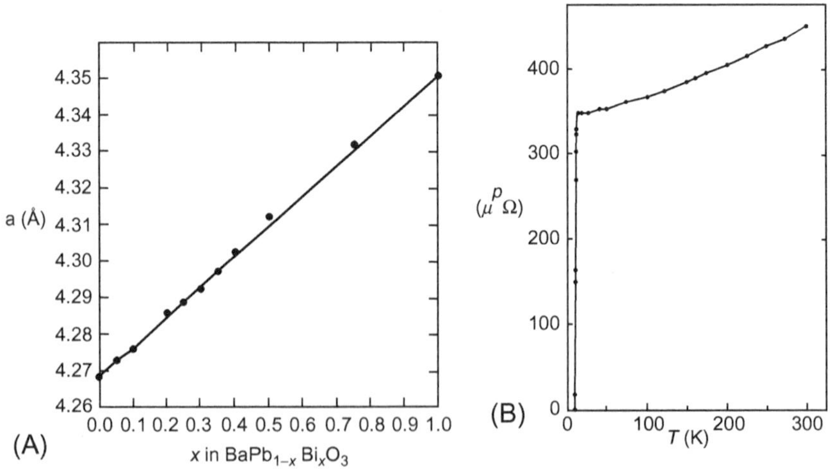

Fig. 14.4 (A) a vs x in the $BaPb_{l-x}Bi_xO_3$ system and (B) electrical resistivity vs temperature for a crystal of $BaPb_{0.8}Bi_{0.2}O_3$.
Reproduced by permission from P.E. Bierstedt, J.L. Gillson, A.W. Sleight, High-temperature superconductivity in the $BaPb_{1-x}Bi_xO_3$ system, Solid State Commun. 17 (1975) 27.

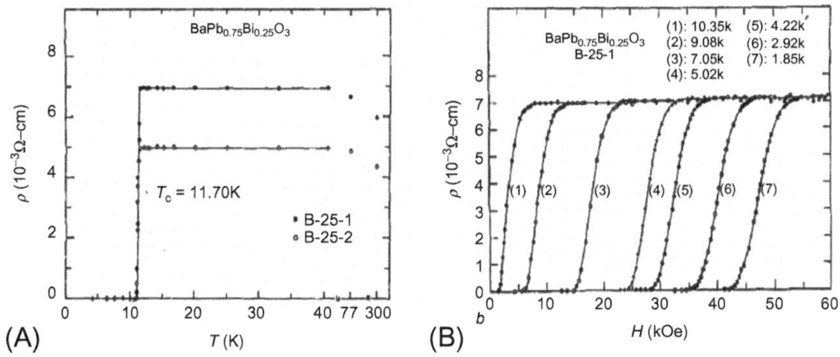

Fig. 14.5 (A) Temperature of electrical resistivity of $BaPb_{0.75}Bi_{0.25}O_3$ near the superconducting transition region. (B) Magnetic field dependence of electrical resistivity of $BaPb_{0.75}Bi_{0.25}O_3$ in the superconducting transition region.
Reproduced by permission from T.D. Thanh, A. Koma, S. Tanaka, Superconductivity in the $BaPb_{1-x}Bi_xO_3$ system, Appl. Phys. 22 (1980) 205–212.

14.3 Cuprate-based superconductors

Several copper oxide-based compounds with a range of crystal structures have also been found to have high-T_c superconductivity (Fig. 14.6A). All these structures have a CuO_2 plane as key component, occupied by one electron per unit cell (Fig. 14.6B). At this electron concentration, the CuO_2 plane is a "Mott insulator," the parent state from which HTSs are derived [3]. A Mott insulator is a material in which the conductivity disappears as the temperature tends to zero, even though band theory would predict it to be metallic [3]. A Mott insulator differs essentially from a conventional (band) insulator, where conductivity is blocked by the Pauli Exclusion Principle. When the highest occupied band has two electrons per unit cell, electrons cannot move because all the orbitals are filled. On the other hand, in a Mott insulator, charge conduction is blocked by electron–electron repulsion. When the highest occupied band contains one electron per unit cell, electron motion requires creation of a doubly occupied site. If the electron–electron repulsion is strong enough, this motion is blocked. The amount of charge per cell becomes fixed, leaving only the electron spin on each site to fluctuate. Doping restores electrical conductivity by creating sites to which electrons can jump without incurring a cost in Coulomb repulsion energy [3]. The high-T_c cuprates are the only Mott insulators that are known to achieve superconductivity when the electron concentration changes from one per cell.

14.3.1 La-Ba-Cu-O (LBCO) system

Encouraged by the high-T_c in homogeneous oxygen-deficient mixed crystals and a comparatively low concentration of carriers in perovskite-type metal oxide, Bednorz and Muller foresaw the possibility of finding higher-T_c superconductors in perovskite-type or related metallic oxides. Finally, in 1986, they discovered a layered

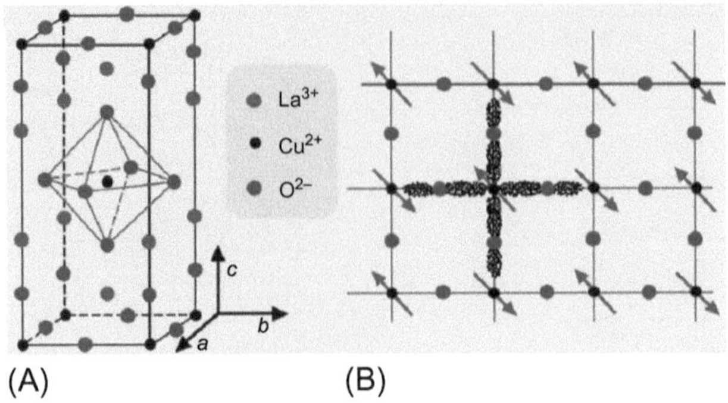

Fig. 14.6 (A) Crystal structure of La_2CuO_4, the "parent compound" of the $La_{(2-x)}Sr_xCuO_4$ family of high-temperature superconductors. The crucial structural subunit is the Cu–O_2 plane, which extends in the *a–b* direction; parts of three CuO_2 planes are shown. Electronic couplings in the interplane *c* direction are very weak. In the La_2CuO_4 family of materials, doping is achieved by replacing some of the La ions shown with Sr ions, or by adding interstitial oxygen. In other families of high-T_c materials (e.g., $YBa_2Cu_3O_{6+x}$), the crystal structure and doping mechanism are slightly different, but all materials share the feature of CuO_2 planes weakly coupled in the transverse direction. (B) Schematic of the CuO_2 plane, the crucial structural subunit for high-T_c superconductivity. *Red arrows* indicate a possible alignment of spins in the antiferromagnetic ground state of La_2CuO_4. *Speckled shading* indicates oxygen "p_σ orbitals"; coupling through these orbitals leads to superexchange in the insulator and carrier motion in the doped, metallic state.
Reproduced by permission from J. Orenstein, A.J. Millis, Advances in the physics of high-temperature superconductivity, Science 288 (2000) 468–474.

$LaBa_2CuO_{4-x}$ material, which was superconducting at 30 K, in which the electron–phonon interactions at the Fermi level were enhanced [2, 11]. Strong electron–phonon interactions in oxides occurred owing to polaron formation as well as in mixed-valence systems. A mechanism for polaron formation is the Jahn–Teller (JT) effect. Isolated Fe^{4+}, Ni^{3+}, and Cu^{2+} ions in an octahedral oxygen environment show strong JT effects [12]. The La-Ba-Cu-O (LBCO) system has a number of oxygen-deficient phases with mixed-valence copper components, that is, with electronic states shifting between non-JT Cu^{3+} and JT Cu^{2+} ions, and thus shows considerable electron–phonon coupling and metallic conductivity.

Metallic, oxygen-deficient compounds in the LBCO system, with the composition $Ba_xLa_{5-x}Cu_5O_{5(3-y)}$ were prepared in polycrystalline form. Samples with $x=1$ and 0.75, and $y>0$, annealed below 900°C under reducing conditions, consisted of three phases. One of them was a perovskite-like mixed-valence copper compound. Upon cooling, the samples showed a linear decrease in resistivity, then an approximately logarithmic increase, interpreted as the beginning of localization. Finally, an abrupt decrease by up to three orders of magnitude occurred, reminiscent of the onset of percolative superconductivity (Fig. 14.7A). The highest onset temperature was observed in the 30 K range (Fig. 14.7B).

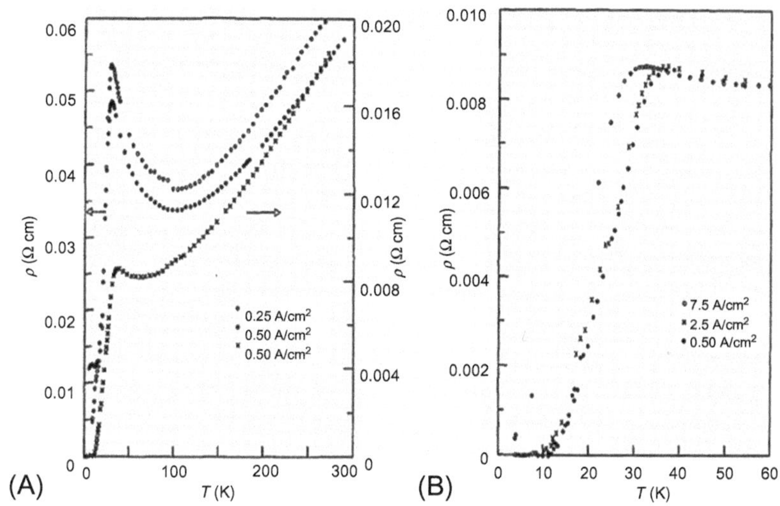

Fig. 14.7 (A) Temperature dependence of resistivity in $Ba_xLa_{5-x}Cu_5O_{5(3-y)}$ for samples with $x=1$ (*upper curves, left scale*) and $x=0.75$ (*lower curve, right scale*). The first two cases also show the influence of current density. (B) Low-temperature resistivity of a sample with $x=0.75$, recorded for different current densities.
Reproduced by permission from J.G. Bednorz, K.A. Müller, Possible high T_c superconductivity in the Ba-La-Cu-O system, Z. Phys. B 64 (1986) 189–193.

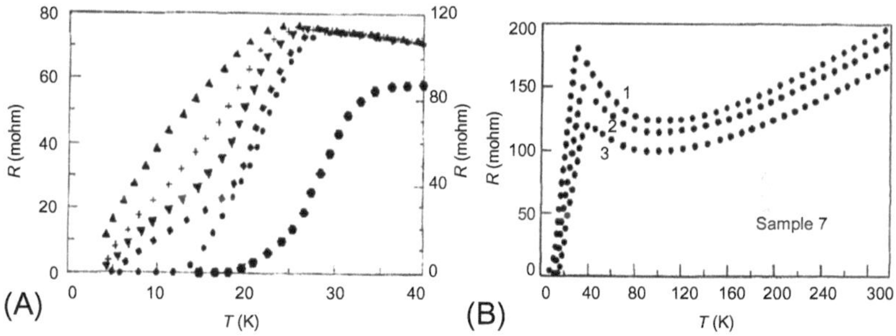

Fig. 14.8 (A) The $\rho(T)$ of a $Ba_xLa_{5-x}Cu_5O_{5(3-y)}$ sample in (A) different magnetic fields in kilogauss: *circles*, 0; *lozenges*, 4.5; *inverted triangles*, 15; *plusses*, 30; and *triangles*, 58.5 and (B) at different pressures: curve 1, 0 kbar; curve 2, 6.6 kbar; and curve 3, 13.3 kbar.
Reproduced by permission from C.W. Chu, P.H. Hor, R.L. Meng, L. Gao, Z.J. Huang, Y.Q. Wang, Evidence for superconductivity above 40 K in the La-Ba-Cu-O compound system, Phys. Rev. Lett. 58 (1987) 405–407.

Chu et al. synthesized the LBCO compound system directly by a solid-state approach, which involved the decomposition and reaction of La_2O_3, CuO, and $BaCO_3$ in a reducing atmosphere [13]. They observed evidence for nonbulk superconductivity above 40 K in an LBCO compound system under hydrostatic pressure, and the largest positive pressure effect on superconductivity (Fig. 14.8).

14.3.2 Y-Ba-Cu-O (YBCO) system

The discovery of the Y-Ba-Cu-O (YBCO) system is regarded as a great breakthrough in HTSs, and it is one of the most promising superconductors in practical applications.

On the basis of LBCO and other systems, superconductivity was a true enabling technology, but those systems must work at very low temperatures that could not be reached without the help of helium. However, helium cooling tended to be too expensive in practical applications. To obtain a superconducting state reaching beyond the technological and psychological temperature barrier of 77 K, the LN boiling point, was considered to be one of the greatest triumphs. Actually, scientists did not have to wait very long. In 1987, for the first time, Chu et al. reported a "zero-resistance" state of YBCO in a LN Dewar [14]. They prepared $Y_{1.2}Ba_{0.8}CuO_{4-y}$ by a solid-state reaction of appropriate amounts of Y_2O_3, $BaCO_3$, and CuO. The temperature dependence of R determined in a simple LN Dewar is shown in Fig. 14.9. Resistance initially drops almost linearly with temperature T. A deviation of R from this T dependence is evident at 93 K and a sharp drop starts at 92 K. Then, a "zero-R" state occurs at 80 K. Although in a field of 5.7T, the "zero-R" state remains at a T as high as 40 K.

14.3.2.1 Structure and performance

Further research showed that the YBCO compounds were different from those with the ordered K_2NiF_4-type structure [15]. Two phases were separated in the YBCO compound system and identified: tetragonal $YBa_2Cu_3O_{6+x}$ (black) and orthorhombic Y_2BaCuO_5 (green) (Fig. 14.10). Magnetic measurements on the YBCO materials revealed that the black phase was responsible for the high-temperature superconductivity.

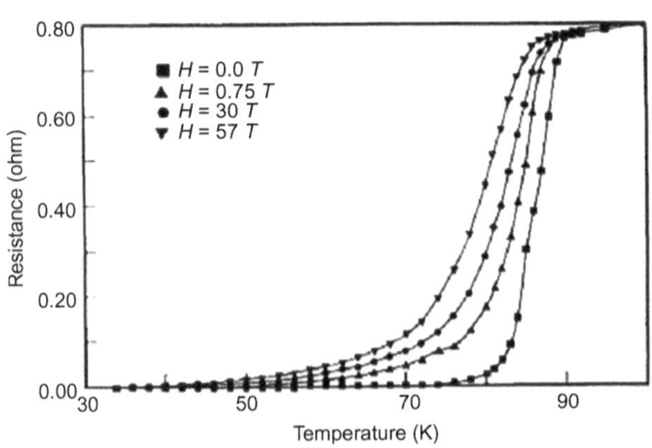

Fig. 14.9 The temperature dependence of resistance R of an $Y_{1.2}Ba_{0.8}CuO_{4-y}$ sample determined in a simple liquid–nitrogen Dewar.
Reproduced by permission from M.K. Wu, J.R. Ashburn, C.J. Torng, P.H. Hor, R.L. Meng, L. Gao, Z.J. Huang, Y.Q. Wang, C.W. Chu, Superconductivity at 93 K in a new mixed-phase Y-Ba-Cu-O compound system at ambient pressure, Phys. Rev. Lett. 58 (1987) 908–910.

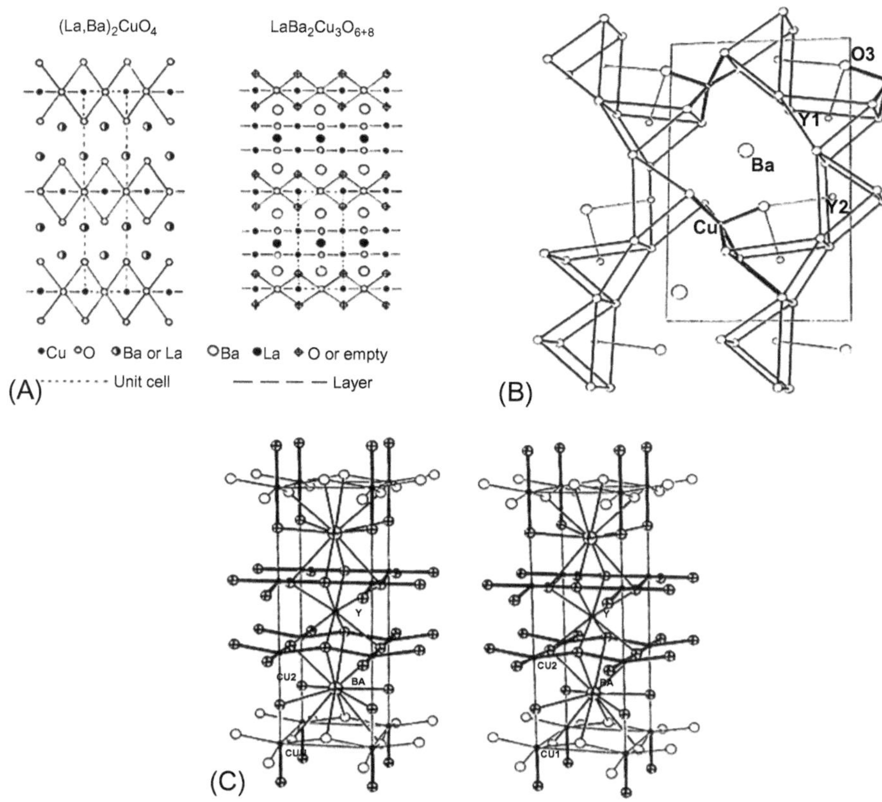

Fig. 14.10 (A) Schematic representation of the structures of $(La_{0.9}Ba_{0.1})_2CuO_{4-x}$ and $LaBa_2Cu_3O_{6+x}$. (B) The crystal structure of Y_2BaCuO_5. (C) Stereographic view of the crystal structure of $YBa_2Cu_3O_{6+x}$. The tetragonal structure has Ba and Y in cubic eightfold coordination and Cu in twofold and fourfold coordination. The oxygen positions, denoted by *open circles*, are essentially vacant.
Panel A: Reproduced by permission from P.H. Hor, R.L. Meng, Y.Q. Wang, L. Gao, Z.J. Huang, J. Bechtold, K. Forster, C.W. Chu, Superconductivity above 9-K in the square-planar compound system $ABa_2Cu_3O_{6+x}$ with A = Y, La, Nd, Sm, Eu, Gd, Ho, Er, and Lu, Phys. Rev. Lett. 58 (1987) 1891–1894; Panel C: Reproduced by permission from R.M. Hazen, L.W. Finger, R.J. Angel, C.T. Prewitt, N.L. Ross, H.K. Mao, C.G. Hadidiacos, P.H. Hor, R.L. Meng, C.W. Chu, Crystallographic description of phases in the Y-Ba-Cu-O superconductor, Phys. Rev. B 35 (1987) 7238–7241.

To clarify the role played by *Y* and by diverse structures, for example, K_2NiF_4 and $YBa_2Cu_3O_{6+x}$ planes, in high-temperature superconductivity, Hor et al. made and studied the $ABa_2Cu_3O_{(6+x)}$ compound systems, where A = La, Nd, Sm, Eu, Gd, Ho, Er, and Lu, in addition to *Y* [16]. They are all superconducting with an onset temperature between 90 and 98 K and a zero resistance state between 70 and 94 K (Table 14.1), meaning that the superconductivity was not affected by the presence

Table 14.1 Lattice parameters and critical temperatures of samples with $ABa_2Cu_3O_{6+x}$ structure

A	T_{c0} (K)	T_{c1} (K)	T_{dl} (K)	a (Å)	b (Å)	c (Å)
Y	98	94	100	3.86	3.86	11.71
La	91	75	99	3.95	3.95	11.79
Nd	91	70	93	3.89	3.89	11.73
Sm	94	82	135	3.88	3.88	11.73
Eu	94	88	160	3.86	3.86	11.74
Gd	95	92	135	3.89	3.89	11.73
Ho	93	88	130	3.89	3.89	11.52
Er	94	87	120	3.83	3.85	11.65
Lu	91	85	120	3.83	3.87	11.73

T_{c0} is estimated as the onset temperature.
T_{c1} is estimated as the zero-estate temperature.
T_{dl} is estimated as the temperature deviation from linear dependence of R.
Reproduced by permission from P.H. Hor, R.L. Meng, Y.Q. Wang, L. Gao, Z.J. Huang, J. Bechtold, K. Forster, C.W. Chu, Superconductivity above 9-K in the square-planar compound system $ABa_2Cu_3O_{6+x}$ with A = Y, La, Nd, Sm, Eu, Gd, Ho, Er, and Lu, Phys. Rev. Lett. 58 (1987) 1891–1894.

of the trivalent A atoms [16]. Consequently, it was accepted that the superconductivity was associated with the CuO_2-Ba-$CuO_{(2+x)}$-Ba-CuO_2 plane assembly sandwiched between the A layers. The interplane coupling or screening within the layer assembly was especially significant, as shown by the enhancement of the superconducting transition from 30 K in the K_2NiF_4 structure to 90 K in the $ABa_2Cu_3O_{6+x}$ structure in the LBCO system. A bigger layer assembly was anticipated for higher T_c superconducting mixed oxides.

14.3.2.2 Modification

On the basis of the results reported by Hor et al. [16], the structure of YBCO was established as an orthorhombic unit cell consisting of Y, CuO_2, BaO, CuO, BaO, CuO_2, and Y layers packed along the c-axis. The oxygen vacancies are ordered in the Cu–O median plane such that Cu–O chains are running along the b-axis. The oxygen atoms can be controlled over the composition range $0<y<1$. As long as $7-y>6.8$, the compound is an orthorhombic superconductor. An empirical supposition, at that time in 1987, was that the average valence of copper in these materials had to be higher than 2 for superconductivity to exist. The increase in the copper valence was supposed to be accomplished either by substitution of divalent ions for trivalent ions, or through the uptake or removal of oxygen atoms. Later experiments did prove that the substitution of Cu with metallic elements could produce substantial changes in the superconducting properties.

Tarascon et al. reported the structural and physical properties of $YBa_2Cu_{3-x}M_xO_{7-y}$ (where M is Fe, Zn, Ni, Co, or Al), obtained when Cu is replaced by Fe, Zn, Ni, Co, or Al [17, 18]. Aluminum forms a well-defined trivalent ion and was therefore used for comparison with the transition metals. The mixed oxides $YBa_2Cu_{3-x}M_xO_{7-y}$ were prepared by first grinding thoroughly together suitable amounts of $BaCO_3$ and the metal oxide powders. After placing the mixed powders in alumina crucibles, they were heated to 960°C, kept at this temperature for 48 h, and then slowly cooled to 300°C. The grinding and firing were repeated until the resulting material was completely homogeneous. The T_c decreased with increasing doping concentration, whether the doping ion was magnetic (Co, Fe, or Ni) or diamagnetic (Al or Zn), suggesting that the behavior of these materials with respect to magnetic impurities was different from that of conventional Bardeen–Cooper–Schrieffer (BCS)-type superconductors. For the trivalent ions, an O–T structure transition took place and the oxygen content (y) increased, whereas for divalent doping the (Ni, Zn) orthorhombicity as well as oxygen content remained the same.

Xiao et al. systematically substituted 10 at% of the Cu ions with 3d elements. Their products are represented as $YBa_2(Cu_{0.9}M_{0.1})_3O_{7-y}$, where M = Ti, Cr, V, Mn, Fe, Co, Ni, and Zn [19, 20], and were prepared by solid-state reaction. Y_2O_3, $BaCO_3$, CuO, and other 3d oxides were mixed in the desired proportions and pressed under the same reaction conditions, and sintered at 950°C in an oxygen atmosphere, interrupted by a few intermediate grindings. In each case, the substitution resulted in a substantial change in T_c, but this was not large enough to destroy the superconductivity. A strong relationship exists between superconductivity and the electronic and magnetic behavior of the substituting elements. Substitution of copper by 3d elements results in a lowering of T_c, depending on the substituting element, as shown by the dependence of the normalized resistance on temperature (Fig. 14.11). Apparently, for M = Fe, Co, Ni, and Zn, the conduction electrons become weakly localized before the onset of superconductivity, as shown by the increasing resistance at low temperatures.

In addition to the dip in the mid-3d region, T_c dropped sharply when 10% of the copper was replaced by zinc, with the superconductivity completely disappearing at 13–14 at% Zn (Fig. 14.11C). This is an intense effect caused by the filling of the electronic bands [21]. The divalent Zn^{2+} ion has ten 3d electrons, forming a very stable electronic structure. In the pure $YBa_2Cu_3O_{7-y}$ compound, each copper ion is octahedrally coordinated to oxygen atoms. The divalent Cu^{2+} ion has nine valence electrons, so that the density of states at the Fermi level [DOS(EF)] originates mainly from the antibonding d_{x2-y2} band, which is approximately half filled. When Cu is replaced by Zn, an extra electron is provided, filling up the d_{x2-y2} band and reducing the DOS(EF) at those sites. The near disappearance of T_c with the substitution of zinc at 10% of the copper sites clearly shows the importance of the antibonding d_{x2-y2} orbitals for superconductivity in the $YBa_2(Cu_{0.9}M_{0.1})_3O_{7-y}$ compounds. It is noteworthy that processing (i.e., heat treatment time, temperature, and cooling to ambient temperature) strongly influenced the physical properties of these materials. The superconducting coherence length of these materials (about 10 Å) was untypically short, so that chemical doping studies were rather difficult.

Gupta et al. [22] prepared samples of type $YBa_2Cu_{3-x}Co_xO_{7-y}$ with x = 0, 0.01, 0.03, and 0.05 using standard ceramic methodology. Fig. 14.12A shows the depression in transition temperature (T_c) as Co content increases. The critical current density

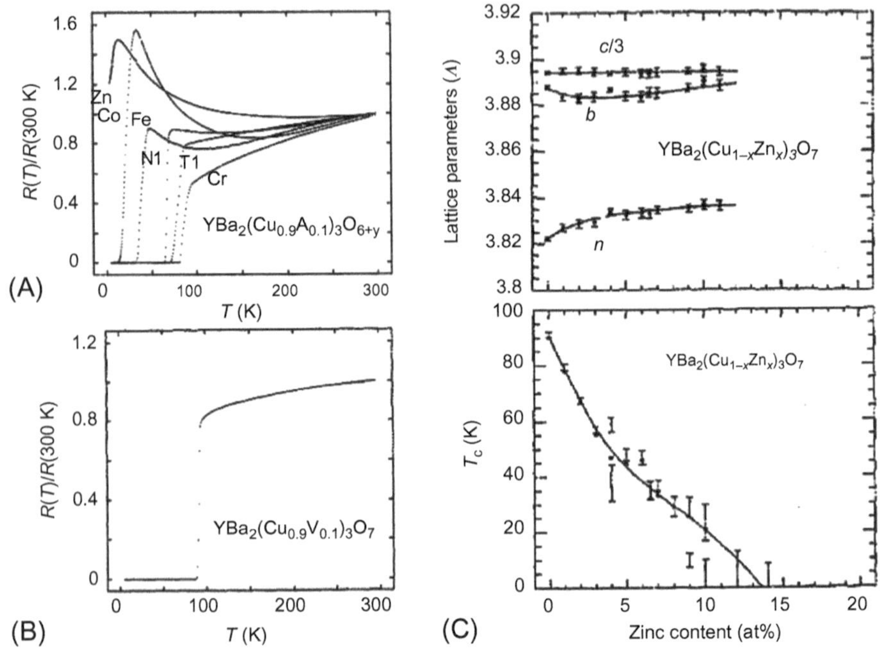

Fig. 14.11 (A) Temperature dependence of the normalized resistance of $YBa_2(Cu_{0.9}M_{0.1})_3O_{7-y}$, where M = Cr, Mn, Fe, Co, Ni, and Zn. (B) Temperature dependence of the normalized resistance of $YBa_2(Cu_{0.9}V_{0.1})_3O_{7-y}$. (C) Lattice parameters (*upper graph*) and variation of T_c (*lower graph*) of $YBa_2(Cu_{1-x}Zn_x)_3O_{7-y}$ as a function of Zn content.
Panel A: Reproduced by permission from G. Xiao, F.H. Streitz, A. Gavrin, Y.W. Du, C.L. Chien, Effect of transition-metal elements on the superconductivity of Y-Ba-Cu-O, Phys. Rev. B 35 (1987) 8782–8784; Reproduced by permission from G. Xiao, F.H. Streitz, A. Gavrin, M.Z. Cieplak, C.L. Chien, A. Bakhshai, Superconductivity and magnetism in transition-element-substituted $YBa_2Cu_3O_7$ compounds, J. Appl. Phys. 63 (1988) 4196–4198.

J_c increases as Co content increases, making it more practicable in applications (Fig. 14.12B).

Salama et al. [23] synthesized superconductor samples of $YBa_2Cu_3O_{7-y}$ (YBCO)+x where x=0.1, 0.2, 0.3, 0.4, and 0.5 wt% of nano-metal oxides namely Cr_2O_3, Co_3O_4, and Mn_3O_4 obtained by another solid-state reaction route. YBCO powder was prepared by the solid-state reaction route by mixing stoichiometric amounts of Y_2O_3, $BaCO_3$, and CuO followed by grinding, calcination, and sintering. Nano-metal oxides were prepared by the coprecipitation of metal chloride and oxalic acid. The YBCO and metal oxides were mixed separately and pressed into pellets, followed by sintering to obtain the final products. The doping with nonmagnetic nano–metal oxides (CuO and SnO_2) severely depressed the value of T_c while doping with Cr_2O_3 (3.8 μB) and Co_3O_4 (3.2 μB) slightly depressed T_c (Fig. 14.13). Moreover, doping with high magnetic moment Mn_3O_4 (5.27 μB) increased the T_c significantly, which was attributed to the inclusion of Mn in the YBCO structures. This inclusion

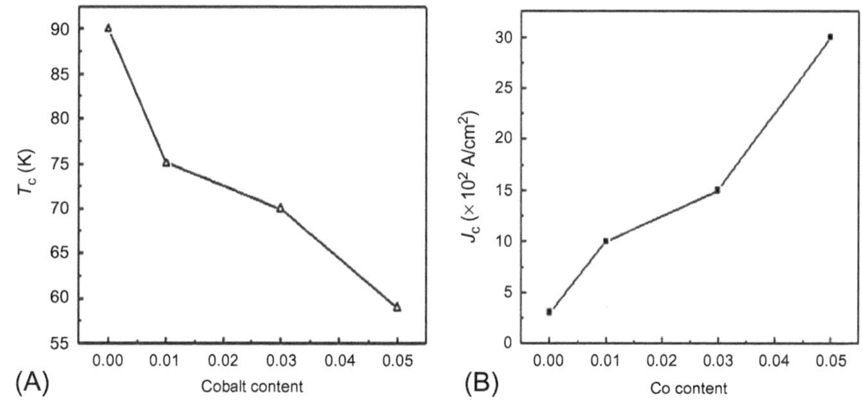

Fig. 14.12 (A) The variation of transition temperature vs Co-content shows suppression of T_c with increasing Co concentration. (B) The variation of transport critical current density J_c vs Co-content revealing enhancement in critical current density with cobalt concentration. Reproduced by permission from S. Gupta, R.S. Yadac, N.P. Lalla, G.D. Verma, B. Das, Microstructural and superconducting properties of $YBa_2Cu_{3-x}Co_xO_{7-y}$ system, Integr. Ferroelectr. 116 (2010) 68–81.

reduced the creation of other YBCO phases and promoted the formation of YBCO with highly superconducting properties.

14.3.3 Bi-Sr-Ca-Cu-O (BSCCO) system

In 1987, Michel et al. demonstrated that $Bi_2Sr_2Cu_2O_{7+y}$ became superconducting at 22 K [24]. This was the first time someone obtained superconductivity above 20 K in oxides without the use of any rare-earth element. Its bidimensional character was proved to be of some importance in established superconductors. Therefore, bismuth was regarded as a possible candidate for the formation of such superconducting oxides, due to its ability to form lamellar oxides. The compounds were made using mixtures of Bi_2O_3, CuO, and $SrCO_3$, which were heated in air at 800°C for 12 h, sintered in air at 900°C for 2 h, and then quenched down to room temperature. The synthesis was carried out using both ultrapure and commercial oxides. For the ultrapure oxides (Fig. 14.14A), the midpoint resistive transition corresponded to $T_c = 22$ K, the zero value of the resistance being reached below 14 K. For the oxide prepared from commercial compounds (Fig. 14.14B), the midpoint transition was much lower at $T_c = 7$ K due to the occurrence of magnetic impurities. These results showed that the presence of lanthanides was not necessary for superconductivity to exist in the mixed-valence copper oxides.

Later in 1988, superconductivity up to 114 K was observed in a Bi-Al-Ca-Sr-Cu-O multiphase compound system, with only a minor role for Al [25]. The samples were synthesized using the standard solid-state reaction technique in the following steps. First, a thoroughly ground mixture of appropriate amounts of Bi_2O_3, $SrCO_3$, $CaCO_3$,

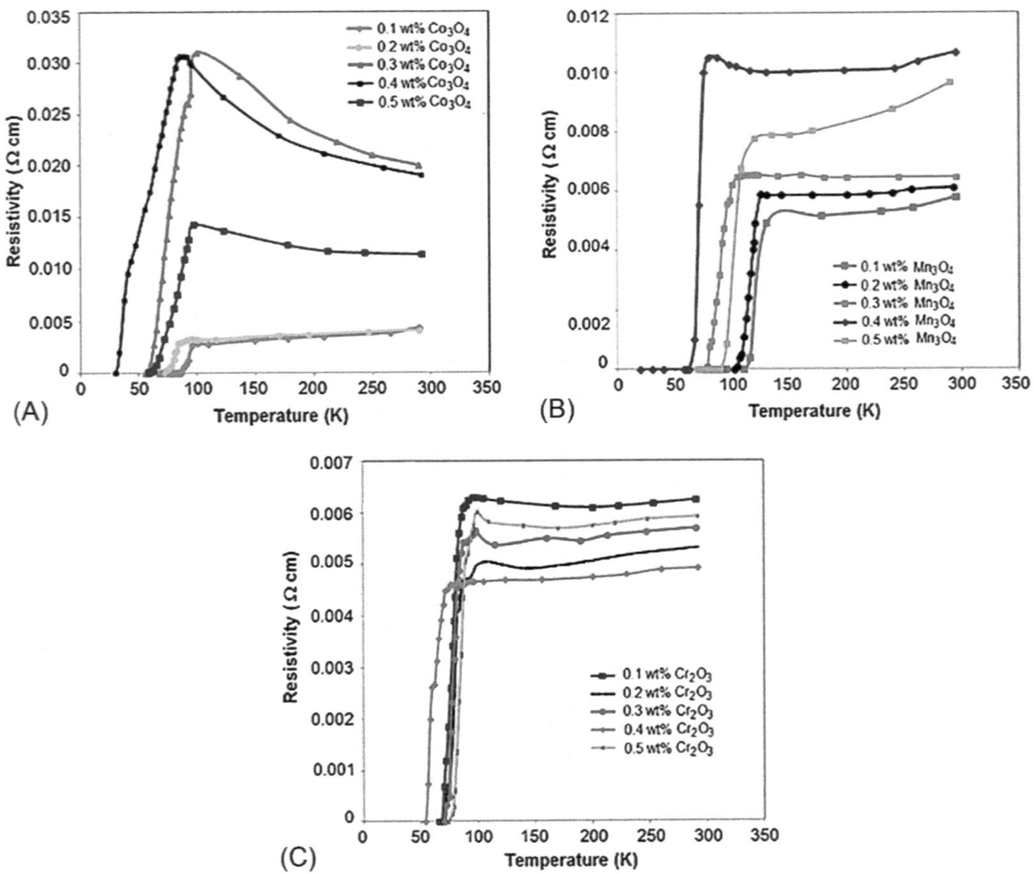

Fig. 14.13 Resistivity dependences on the temperature for YBCO doped with (A) 0.1, 0.2, 0.3, 0.4, and 0.5 wt% Co_3O_4 nano oxide, (B) 0.1, 0.2, 0.3, 0.4, and 0.5 wt% Mn_3O_4 nano oxide, and (C) 0.1, 0.2, 0.3, 0.4, and 0.5 wt% Cr_2O_3 nano oxide.
Reproduced by permission from A.H. Salama, M. El-Hofy, Y.S. Rammah, M. Elkhatib, The influence of magnetic nano metal oxides doping on ... of YBCO ... Adv. Nat. Sci.: Nanosci. Nanotechnol. 7 (2016) 015011.

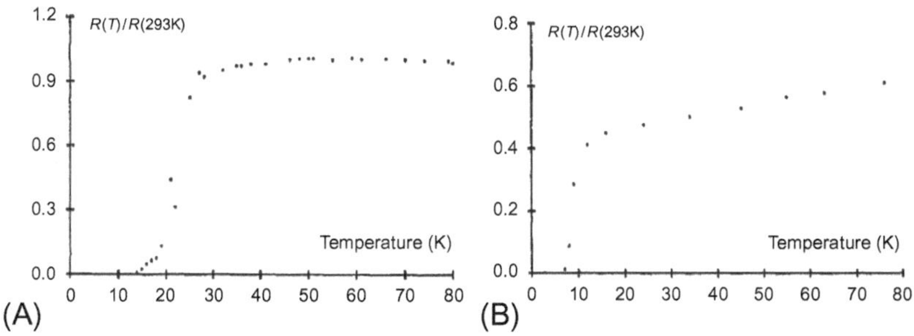

Fig. 14.14 Resistivity ratio vs temperature showing the resistance transition (A) at 22 K for the ultrapure oxide and (B) at 7 K for the commercial compound.
Reproduced by permission from C. Michel, M. Hervieu, M.M. Borel, A. Grandin, F. Deslandes, J. Provost, B. Raveau, Superconductivity in the Bi-Sr-Cu-O system, Z. Phys. B 68 (1987) 421–423.

and CuO was calcined in air at 750–890°C. Then, the calcined pellets were crushed and reground in order to get rehomogenized material. The rehomogenized material was sintered at the same temperature range, and finally quenched to room temperature. Chu et al. [26] demonstrated that the superconducting phase, with onset temperature near 120 K, was a layered compound with composition near $Bi_2Sr_2Ca_1Cu_2O_{8+x}$ and an *A*-centered orthorhombic unit subcell. In spite of the fine-grained texture and high density of defects, they also found that the presumed copper–oxygen layers appeared to be continuous over hundreds of unit cells. This layered structure may possess a significant advantage in terms of processing and applications. Fabrication of the superconductor by compression into planes could yield superconductor components with enhanced properties (e.g., critical current) in a specific plane. Similarly, fabrication by rolling could produce wires more flexible than those of ordinary ceramic materials.

14.3.4 Tl-Ba-Ca-Cu-O (TBCCO) system

The superconductivity of rare-earth free Tl-Ba-Cu-O systems was first discovered by Sheng and Hermann [27]. Typically, appropriate amounts of $BaCO_3$ and CuO were ground and heated in air at 950°C with several intermediate grindings to obtain a uniform black Ba-Cu oxide powder. The resulting powder was mixed with Tl_2O_3 and pressed into pellets. The pellets were then put into a furnace at 800–910°C and were heated for 2–5 min in an oxygen stream. A sharp drop in resistance, starting above 90 K with zero resistance at 81 K, was observed in that system (Fig. 14.15A and B).

Sheng et al. further improved the T_c to 120 K by introducing Ca into this system [28]. They studied three superconducting TCBCO samples, labeled TCBCO-1, TCBCO-2, and TCBCO-3, with initial Tl:Ca:Ba:Cu ratios of 2:2:1:3, 2:3:1:3, and

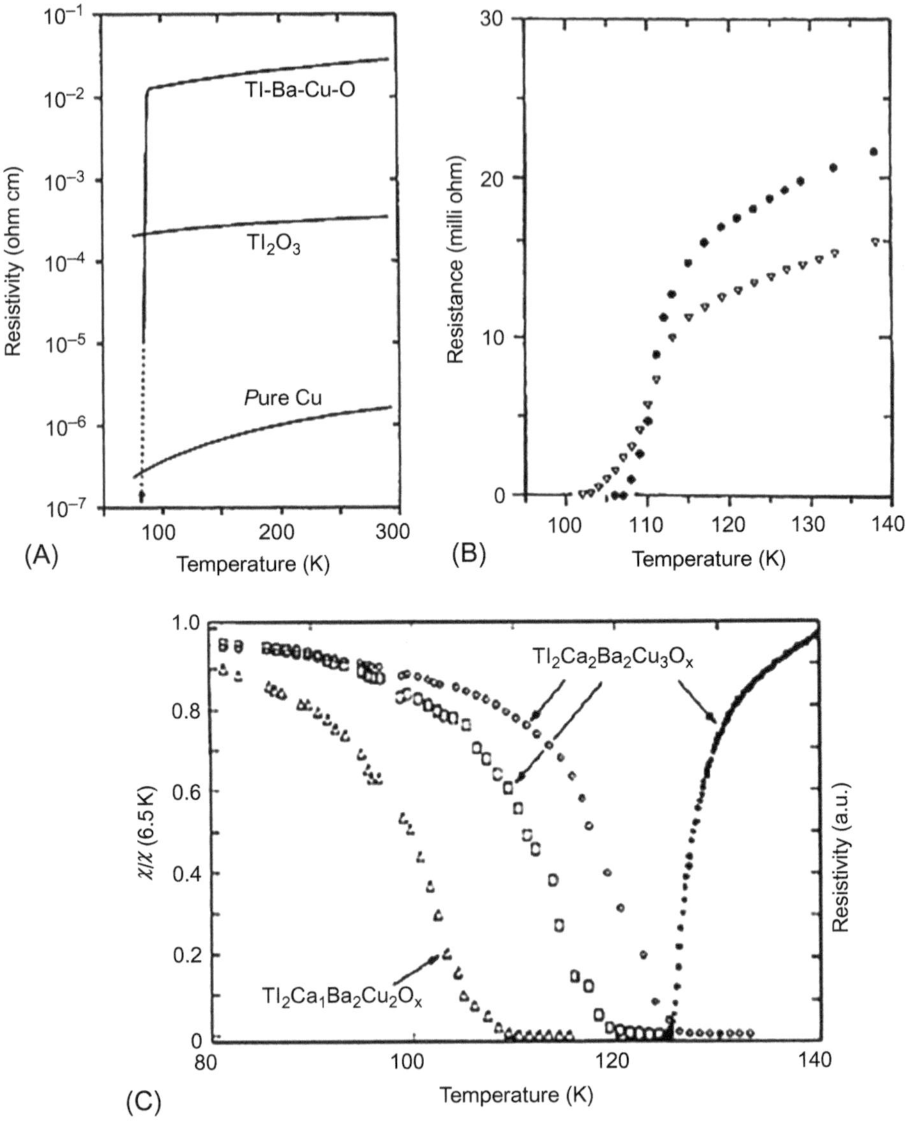

Fig. 14.15 (A) Variation of resistivity with temperature for a $Tl_2Ba_2Cu_3O_{8+x}$ sample. (B) Resistance temperature measurements for two TCBCO samples. *Circles* denote data on a 2213 composition (TCBCO-1) and the *triangles* indicate a 2223 sample (TCBCO-3). (C) Resistance vs temperature (*filled circles*) for a material of starting composition $Tl_2Ca_2Ba_2Cu_3O_x$. Meissner susceptibility (normalized to 1 at 6.5 K) vs temperature for materials with starting composition $Tl_1Ca_3Ba_1Cu_3O_x$ (*open circles*), $Tl_1Ca_2Ba_2Cu_3O_x$ (*rectangles*), and $Tl_2Ca_2Ba_2Cu_3O_x$ (*triangles*).
Panel A: Reproduced by permission from Z.Z. Sheng, A.M. Hermann, Superconductivity in the rare-earth-free Tl-Ba-Cu-O system above liquid–nitrogen temperature, Nature 332 (1988) 55–58; Panel B: Reproduced by permission from R.M. Hazen, L.W. Finger, R.J. Angel, C.T. Prewitt, N.L. Ross, C.G. Hadidiacos, Z.Z. Sheng, A. El-Ali, A.M. Hermann, 100-K superconducting phases in the Tl-Ca-Ba-Cu-O system, Phys. Rev. Lett. 60 (1988) 1657–1660; Panel C: Reproduced by permission from S.S.P. Parkin, V.Y. Lee, E.M. Engler, A.I. Nazzal, T.C. Huang, G. Gorman, R. Savoy, R. Beyers, Bulk superconductivity at 125 K in $Tl_1Ca_2Ba_2Cu_3O_x$, Phys. Rev. Lett. 60 (1988) 2539–2542.

2:2:2:3, respectively. The TCBCO samples were prepared similarly to the TBCO. Sample TCBCO-1 consisted of two main phases, both of which were black, opaque, and fine-grained, with particle sizes less than 10 μm in diameter. Electron microprobe analysis showed the presence of two distinct, nearly stoichiometric compounds: $Tl_2Ca_2Ba_2Cu_3O_{10+y}$ (2223) and $Tl_{0.19}Ca_{0.81}CuO_{2+y}$ (1405). Two additional samples with 2223 and 1405 compositions as starting ratios were synthesized. The 1405 compound was semiconducting, while 2223 was a HTS. In sample TCBCO-2, two more additional phases were observed with stoichiometric metal ratios 2:1:2:2 and 3:4:0:0. The 2122 phase, corresponding to $Tl_2Ca_1Ba_2Cu_2O_{8+y}$ (2122), was especially interesting due to its apparent similarity to the 90 K superconducting phase $Bi_2Ca_1Sr_2Cu_2O_{8+y}$. Sample TCBCO-3 consisted mainly of the 2122 phase plus unreacted oxides of copper and calcium. Therefore, this 2122 phase was also a HTS.

In 1988, Parkin et al. [29, 30] reported three new phases in the Tl-Ca-Ba-Cu-O system, $Tl_1Ca_{n-1}Ba_2Cu_nO_{2n+3}$ ($n=1, 2, 3$), one of which ($n=3$) exhibited bulk superconductivity at 110 K. The crystal structures of these new materials were found to be closely related to that of $Tl_2Ca_2Ba_2Cu_3O_{10+x}$ (Fig. 14.16). All those compounds contain Cu perovskite-like units, which are separated by bilayer Tl–O sheets in $Tl_2Ca_2Ba_2Cu_3O_{10+x}$ systems. However, in the new structures, those Cu perovskite-like units were separated by monolayer Tl–O sheets, enabling an examination of the effect of the coupling between successive Cu perovskite units on superconducting properties (Fig. 14.15C).

14.3.5 Hg-Ba-Ca-Cu-O (HBCCO) system

In 1991, Putilin et al. [31] synthesized a new family of $HgBa_2RCu_2O_7$ ($R=$ La, Nd, Eu, Gd, Dy, and Y) at 550°C and studied their crystal structures. The crystal structure of one of these compounds is shown in Fig. 14.17A. It is similar to that of the Tl-containing superconducting "1212" phase and can be described as formed by intergrowth of rock salt blocks and anion deficient perovskite ones. The two (BaO) layers sandwich the (HgO) sheets in the structure of the Hg-1212 phase. Inspired by the structural similarity, Putilin et al. [32] discovered superconductivity in $HgBa_2CuO_{4+x}$ (Hg-1201), which had a structure related to that of Hg-1212. Typically, the sample was prepared by solid reaction between a stoichiometric mixture of Ba_2CuO_{3+x} and HgO. The precursor Ba_2CuO_{3+x} was obtained by reaction between BaO_2 and CuO at 930°C in an oxygen atmosphere. Then, the powders were ground and heated at 800°C to obtain the final product. The temperature dependence of the normalized resistivity, shown in Fig. 14.17B, exhibits a sharp drop at 94 K, but the transition is broad and it reaches the value of zero resistance only at 36 K. The superconductivity in similar compounds containing bismuth and thallium suggested that even higher transition temperatures might be achieved in mercury-based compounds with more than one CuO_2 layer per unit cell. Later on, Schilling et al. [33] provided support for that conjecture, with the discovery of superconductivity above 130 K in a material containing $HgBa_2Ca_2Cu_3O_{10+x}$ (with three CuO_2 layers per unit

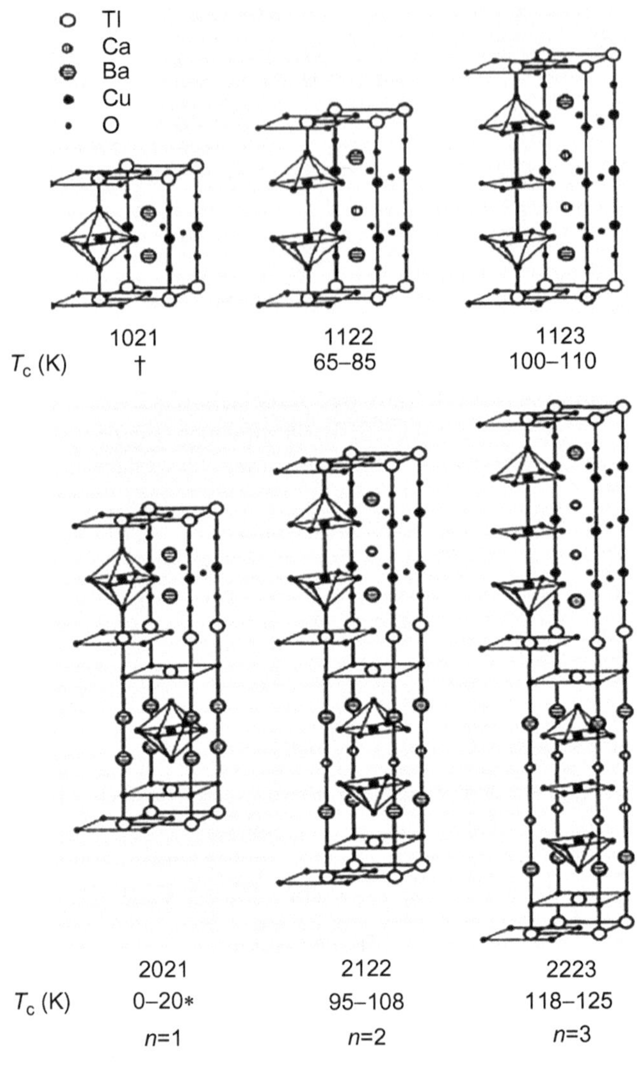

Fig. 14.16 Unit cells of the $Tl_1Ca_{n-1}Ba_2Cu_nO_{2n+3}$ ($n=1, 2$, and 3) compounds containing Tl-O monolayers and comparison with those of the corresponding compounds $Tl_2Ca_{n-1}Ba_2Cu_nO_{2n+4}$ ($n=1, 2$, and 3), containing Tl-O bilayers. The range of superconducting transition temperatures is included in the figure. No superconductivity was observed in the 1021 sample. Reproduced by permission from S.S.P. Parkin, V.Y. Lee, A.I. Nazzal, R. Savoy, R. Beyers, S.J. La-Placa, $Tl_1Ca_{n-1}Ba_2Cu_nO_{2n+3}$ (n = 1,2,3): a new class of crystal structures exhibiting volume superconductivity at up to 110 K, Phys. Rev. Lett. 61 (1988) 750–753.

cell), and an ordered superstructure comprising a defined sequence of the unit cells of these phases. Both magnetic and resistivity measurements confirmed a maximum transition temperature of 132.5 K (Fig. 14.18), distinctly higher than the previously established record value of 125–127 K in $Tl_2Ba_2Ca_2Cu_3O_{10+x}$.

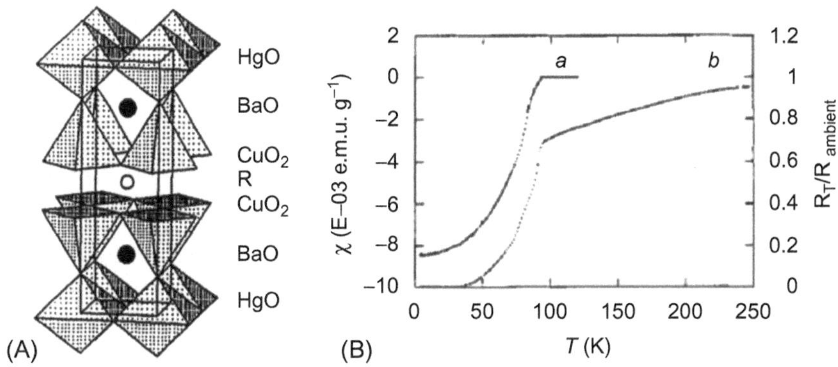

Fig. 14.17 (A) Fragment of the structure of $HgBa_2RCu_2O_7$ with R = Eu. The copper atoms are situated inside the *square pyramids* and Hg atoms are inside the *flattened octahedron*. Ba atoms are shown as *filled circles* and the Eu atom as a *small empty circle*. (B) AC magnetic susceptibility and normalized resistivity as a function of temperature for $HgBa_2CuO_{4+x}$. Panel A: Reproduced by permission from S.N. Putilin, I. Bryntse, E.V. Antipov, New complex copper oxides: $HgBa_2RCu_2O_7$ (R=La, Nd, Eu, Gd, Dy, Y), Mater. Res. Bull. 26 (1991) 1299–1307; Panel B: Reproduced by permission from S.N. Putilin, E.V. Antipov, O. Chmaissem, M. Marezio, Superconductivity at 94 K in $HgBa_2CuO_{4+y}$, Nature 362 (1993) 226–228.

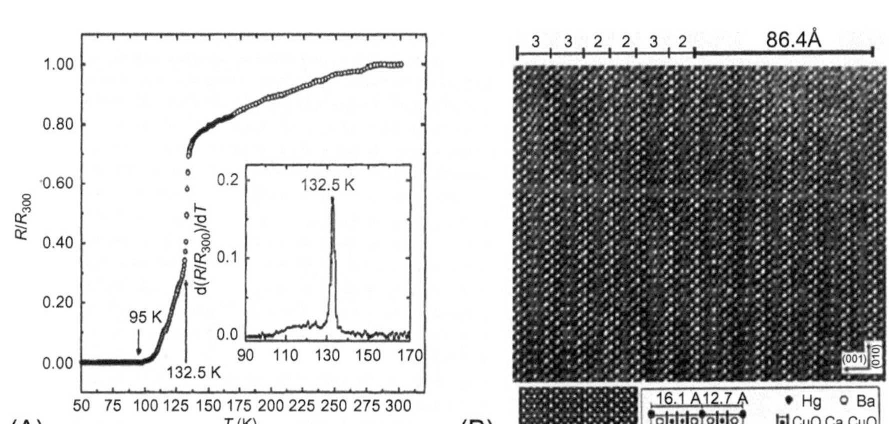

Fig. 14.18 (A) Resistivity $R(T)$ of a Hg-Ba-Ca-Cu-O (HCBCCO) specimen. The *inset* shows the maximum resistivity drop at $T = 132.5\,K$. Zero resistance is attained at $T = 95\,K$. (B) High resolution transmission electron microscope (HTEM) image of a grain in (100) orientation. Reproduced by permission from A. Schilling, M. Cantoni, J.D. Guo, H.R. Ott, Superconductivity above 130 K in the Hg-Ba-Ca-Cu-O system, Nature 363 (1993) 56–58.

14.4 Other materials

14.4.1 Oxy-pnictides

In 2006, Hosono et al. first reported the superconductivity of an iron-based layered oxy-pnictide LaOFeP [4]. LaOFeP is composed of an alternate stack of lanthanum oxide ($La^{3+}O^{2-}$) and iron pnictide ($Fe^{2+}P^{3-}$) layers (Fig. 14.19). Apparently, carriers flow in the two-dimensional (2D) ($Fe^{2+}P^{3-}$) layers, while impurity doping in the ($La^{3+}O^{2-}$) layer transfers the carriers produced in the ($La^{3+}O^{2-}$) layer to the ($Fe^{2+}P^{3-}$) layers, working as "natural modulation doping," similar to the HTSs. But there is a noticeable difference in the coordination structure of LaOFeP compared with those of the layered copper-based superconductors described up to now, where a Cu^{2+} ion occupies a planar fourfold square site. In LaOFeP, the Fe^{2+} occupies a tetrahedral site coordinated with four P^{3-} ions, and the tetrahedrons are linked with each other by edge sharing to form the ($Fe^{2+}P^{3-}$) layer. As a result, the electrons at the Fermi level occupy Cu^{2+}-$3d_{x2-y2}$ orbitals in the layered copper-based superconductors, while either Fe^{2+}-$3d_{xy}$ or $-3d_{yz}/3d_{zx}$ orbitals form the Fermi level for LaOFeP. Consequently, finding a new iron-based superconductor possessing a different kind of layered structure provided an opportunity to investigate the mechanism of high-T_c superconductivity in layered crystals. Up to now, various derivatives have been synthesized by replacing the La, Fe, and/or P sites with other elements [34].

14.4.2 $Na_xCoO_2 \cdot yH_2O$

In 2003, Takada et al. first reported the superconductivity of $Na_xCoO_2 \cdot yH_2O$ [35]. The sample was obtained through a chemical oxidation process. Before oxidation, $Na_{0.7}CoO_2$ has a hexagonal layered structure (space group $P6_3/mmc$) consisting of

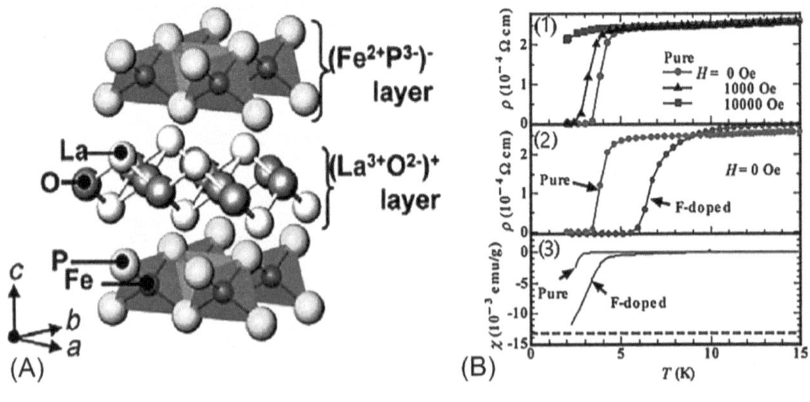

Fig. 14.19 (A) Crystal structure of LaOFeP. The *green polyhedra* indicate the FeP_4 tetrahedra. (B) Superconducting properties of pure and F-doped LaOFeP.
Reproduced by permission from Y. Kamihara, H. Hiramatsu, M. Hirano, R. Kawamura, H. Yanagi, T. Kamiya, H. Hosono, Iron-based layered superconductor: LaOFeP, J. Am. Chem. Soc. 128 (2006) 10012–10013.

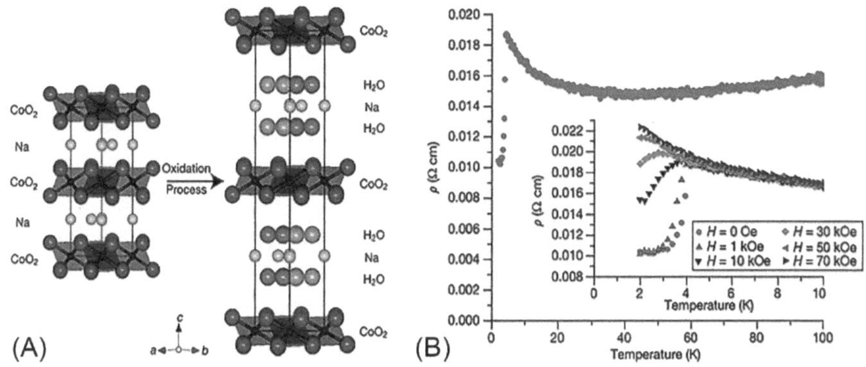

Fig. 14.20 (A) Structural views of $Na_{0.7}CoO_2$ (*left*) and $Na_xCoO_2{\cdot}yH_2O$ (*right*), where Na and H_2O sites are partially occupied. (B) Resistivity of $Na_xCoO_2{\cdot}yH_2O$ under zero magnetic field. The *inset* shows the resistivity measured under various magnetic fields.
Reproduced by permission from K. Takada, H. Sakurai, E. Takayama-Muromachi, F. Izumi, R.A. Dilanian, T. Sasaki, Superconductivity in two-dimensional CoO_2 layers, Nature 422 (2003) 53–55.

the 2D layers of CoO_2 and charge-balancing Na^+ ions (Fig. 14.20A). There is an obvious resemblance between this oxide and high-T_c copper oxides. In the high-T_c copper oxides, superconductivity is located in the CuO_2 square lattice. The Cu^{2+} moments are antiferromagnetically ordered in the CuO_2 plane. With a low level of carrier doping (which may be a hole or electron), the antiferromagnetism is inhibited considerably, and the system becomes metallic, followed by the appearance of superconductivity. Similarly, the $Na_xCoO_2{\cdot}yH_2O$ shows an electron-doped system for a low-spin Co^{4+} lattice with an electron density (x in $Na_xCoO_2{\cdot}yH_2O$) of around 0.35 per Co atom. A sharp decrease in the resistivity was observed at around 4K (Fig. 14.20B), supporting the superconducting transition. The increasing external magnetic field extended the superconducting transition but did not noticeably lower the onset T_c. Similar behavior has always been observed in high-T_c copper oxides.

14.5 Applications

Superconductive materials can be used in at least three potential application fields: high-power applications (superconducting generator; superconducting wires/transformers), low-power applications (superconducting computer devices), and diamagnetic applications (maglev train and nuclear fusion reactor). Their high-power applications will be discussed in the following sections.

14.5.1 Superconducting wires/tapes

Two years after the discovery of superconductivity in 1911, Kamerlingh Onnes was researching the possibility to construct very high-field electromagnets that would be affordable if made from superconducting materials, such as a Ni alloy that he coated with a Pb-rich superconducting solder. However, the superconductivity was lost at fields below 50 mT. This setback was blamed on structural faults in the wire, but despite expectations that this challenge would soon be overcome, real applications only started to appear 50 years later.

During the first phase of developing superconducting applications, from around 1962 to 1986, many power applications tested but were found to be too complex and expensive. As a result, the focus of superconductivity applications shifted to new high-technological and medical applications where there was no competing technology. These applications made use of the low-temperature superconducting (LTS) materials, Nb–Ti alloy ($T_c = 9\,K$) and Nb_3Sn ($T_c = 18\,K$). In these applications, the use of liquid helium or helium-driven refrigerators to promote superconductivity was an empowering technology and cost did not matter. However, for large-scale use in the energy field, helium-cooled superconductors could not compete economically with conventional technologies based on copper and iron.

Soon after the discovery by Bednorz and Muller of the superconductivity of the layered cuprate, $LaBa_2CuO_{4-x}$, at 30 K, the new material $YBa_2Cu_3O_x$ (YBCO) was announced, which was superconducting at 92 K, well above the boiling point of LN (77 K). Soon, the T_c of newly discovered materials started to rise, until a T_c above 130 K was determined for $Hg_2Ba_2Ca_2Cu_3O_x$. The potential of zero-resistance devices that would function at LN temperature inspired a worldwide research effort. It was widely speculated that copper and iron would be replaced by these new HTSs. Major public and private programs were initiated all over the world to build superconducting devices for electric utility applications. However, scientists and engineers faced the common requirement that power applications share, that the superconductor must be shaped into a long, strong, and flexible material. However, the HTS materials are ceramics that are more like a piece of chalk than the ductile metal copper.

The first solution to this problem was to pack Bi-Sr-Ca-Cu-O (BSCCO) superconducting powder into a silver tube. Heine et al. [36] filled Bi-2212 powder into a silver tube of 8-mm outer and 6-mm inner diameter. Subsequently, the composite was drawn to the final diameter of 1.5 mm and was annealed at 800°C. The final material was a 4-mm wide tape yielding stable critical current density in fields above 20*T*, and capable of carrying over 100 A at LN temperature. It was flexible enough to be used in many different applications. Many hundreds of kilometers of this first-generation tape were produced for different demonstration projects, especially power cables and motors. Unfortunately, the reliance of this kind of conductor on the use of silver made it too expensive for most commercial applications and badly limited possibilities for future cost reductions. In addition, the BSCCO rapidly lost its current carrying ability in a magnetic field at LN temperature, canceling the advantage of using an inexpensive cryogen in applications such as motors and generators.

In the early 1990s, an alternative approach was conceived, which was quickly shown to offer competitive performance [37]. Instead of imparting the requisite strength and crystallinity to the superconductor by the powder-in-tube method, "second-generation" wire used epitaxial growth of a superconducting coating on a thin metal tape. One advantage was that very little silver was needed, making significant

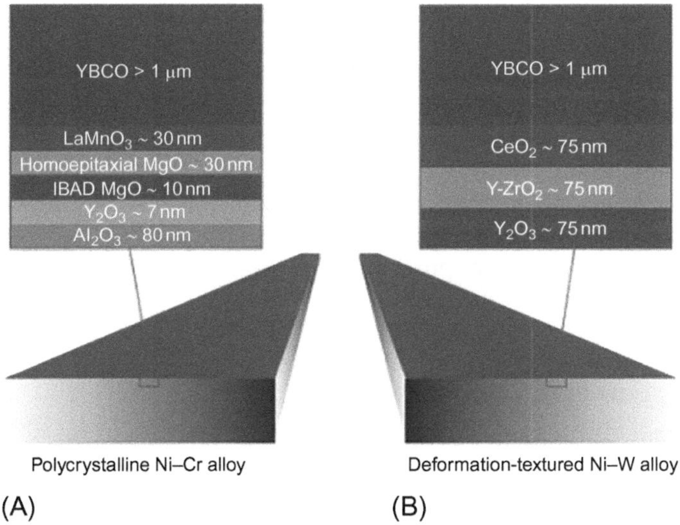

Fig. 14.21 Two main technologies being used in the manufacture of coated-conductor tapes. Reproduced by permission from S.R. Foltyn, L. Civale, J.L. Macmanus-Driscoll, Q.X. Jia, B. Maiorov, H. Wang, M. Maley, Materials science challenges for high-temperature superconducting wire, Nat. Mater. 6 (2007) 631–642.

cost reductions possible. Another advantage of this idea was that it allowed the use of the compound $YBa_2Cu_3O_{7-y}$ (YBCO), which had much higher current carrying ability in a magnetic field. The as-obtained YBCO tapes could be 100 m long, carrying over 200 A per centimeter width at 77 K. Fig. 14.21 compares the two primary coated-conductor designs being manufactured today.

Second-generation HTS wires and coated conductors (also called 2G wires) based on $REBa_2Cu_3O_{7-y}$ films (where RE = rare earth) show great promise for use in large-scale applications because in many such prospective applications, high critical currents in strong magnetic fields are needed. Coated conductors are made of a flexible substrate (which may be a metallic template consisting of a number of buffer layers) covered by an epitaxial superconducting layer. The objective is to build a biaxially textured superconducting layer so that as few as possible high-angle, weakly conducting grain boundaries exist. Three techniques have been developed for the construction of such a biaxial texture in the substrate: ion beam-assisted deposition (IBAD) of biaxially textured buffers on polycrystalline alloy substrates, epitaxial deposition of buffer multilayers on rolling assisted biaxially textured substrates (RABiTS), and inclined substrate deposition of buffers on polycrystalline alloy substrates. However, for practical application of HTS materials, the infield performance, the critical current density, and overall current carrying capacity need to be enhanced. The simplest approach would be to increase the thickness of the superconductor, but when the superconductor thickness is increased, the critical current density deteriorates gradually. This roughness problem seems to have been caused by fabrication of multilayered films consisting of $YBa_2Cu_3O_{7-y}$ (YBCO) layers less than 1.5-μm thick alternating with intervening CeO_2 interlayers [38]. However, a single layer is still preferable due to processing complexity experienced with multilayer growth. Kang et al. manipulated the deposition conditions and showed that it was possible to grow a 6.4-μm thick single-layer YBCO film on RABiTS. Further improvement of the in-field transport properties was accomplished by the introduction of suitable defects into the films.

In practical cases, conductors for power applications are multifilamentary wires or tapes in which many superconducting filaments are embedded in a matrix of a normal metal, such as Cu or Ag, which provides protection against magnetic flux jumps and thermal quenching (Fig. 14.22).

14.5.2 Superconducting generators

Since a variety of future military systems will depend on high electrical power input at the multimegawatt level [39], the power generation subsystems must often be packaged in a limited space and within strict weight limits. Conventional generators cannot provide the multimegawatt levels of power necessary for advanced mobile or airborne military systems without paying a significant penalty in size, weight, and efficiency. So, the development of HTS generators is of critical importance. Since MIT successfully tested a rotating-field type superconducting synchronous generator in 1969, superconducting generators developed around the world have gone through three periods: initial prototype LTS work (late 1960s–1970s), BSCCO-based generators (mid-1990s today), and YBCO capable generators (starting 2004) [40, 41]. Fig. 14.23 shows a prototype large capacity wind turbine generator using high-temperature superconducting technology. The superconducting field windings, using 2G YBCO wires, are utilized in the rotor. The racetrack superconducting magnets are cooled directly by two cryorefrigerators and the operating temperature is 30 K. The cold heads of the refrigerators are installed on the rotor, which is driven by a drive motor, rotating synchronously with a vacuum chamber. A rotary gas seal is employed as the gas slip ring connecting the rotary cold head to a stationary compressor.

14.5.3 Superconducting transformers

Power transformers using high current density HTS wire and LN dielectric have many potential advantages over conventional oil immersed transformers [42]. They can be smaller, lighter, and more efficient, with low fire and environmental hazard and can have overload capability without reduction in lifetime, low output impedance, and fault current-limiting capability. Various types of transformers have been developed for prominent applications. A number of service-ready prototype HTS transformers have been developed over the past couple of decades, and the application results show that it is possible to reduce the energy loss in transformers by 60% or more when the conventional windings are replaced by HTS windings [43]. Early developed HTS transformer windings used the first-generation BSCCO conductor, whereas the second-generation REBCO conductor has achieved more attention recently.

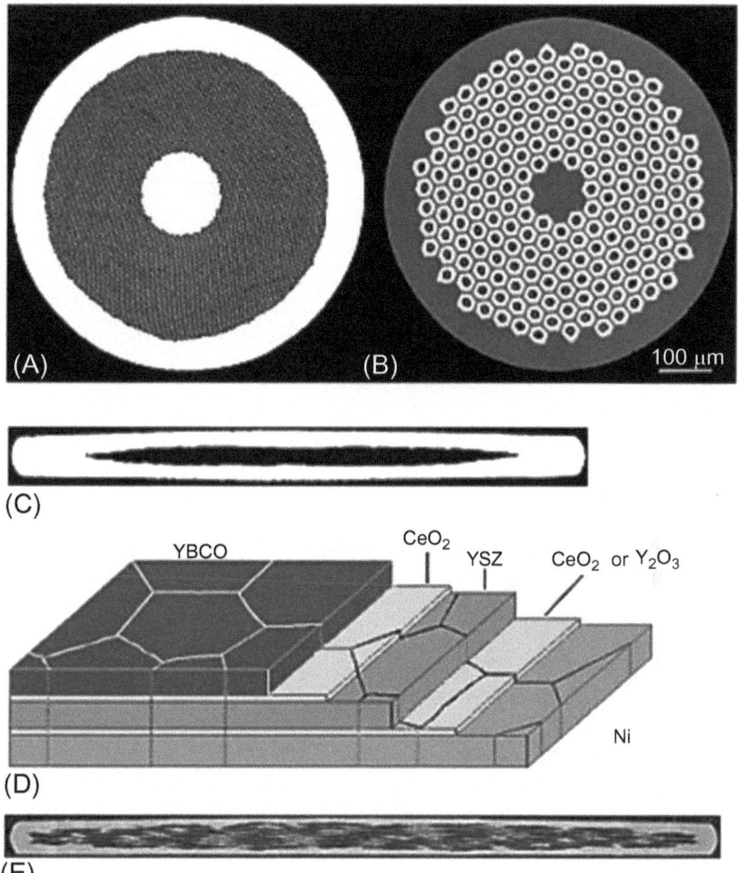

Fig. 14.22 Conductor forms of practical superconductors. (A) Conductor containing about 3000 Nb 47 wt% Ti filaments embedded in a copper stabilizer. (B) Powder-in-tube Nb_3Sn conductor made by extruding and drawing a mixture of $NbSn_2$ and Cu inside the Nb tubes that form each of the 192 filaments. (C) Cu-sheathed MgB_2 tape made by rolling MgB_2 powder in a copper tube. (D) Structure of a deformation-textured coated conductor. Typical thicknesses of the Ni, CeO_2/Y_2O_3, YSZ, CeO_2, and $YBa_2Cu_3O_x$ (YBCO) layers are (respectively) 50–100 mm, 10 nm, 300 nm, 10 nm, and 0.3–2 μm. (E) Cross-section of a 55-filament $(Bi,Pb)_2Sr_2Ca_2Cu_3O_x$ (Bi-2223)-power cable tape encased in a silver matrix, which acts as stabilizer.
Reproduced by permission from D. Larbalestier, A. Gurevich, D.M. Feldmann, A. Polyanskii, High-T_c superconducting materials for electric power applications, Nature 414 (2001) 368–377.

Fig. 14.24A shows external views of three sets of coil assemblies of a three-phase superconducting power transformer [44]. In each coil, a single nondivided HTS tape is wound onto a cylindrical center bobbin made of glass fiber-reinforced plastic. The primary and secondary voltages of the transformer are 66 and 6.9 kV, respectively,

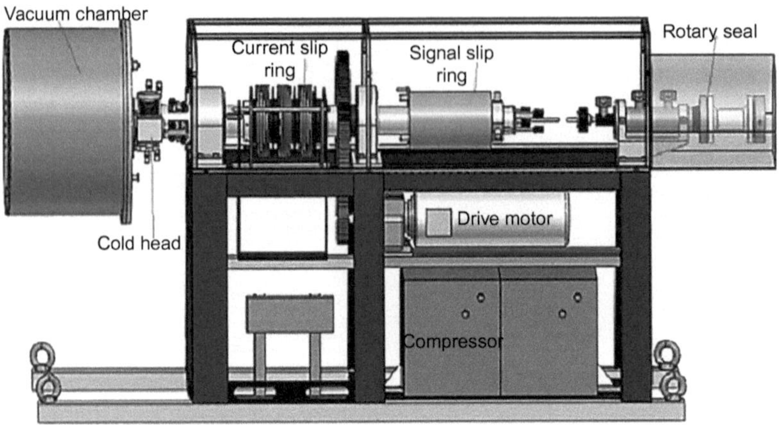

Fig. 14.23 Sketch of the field winding system.
Reproduced by permission from Y. Yang, S. Duan, Y. Ren, Y. Jiang, L. Feng, X. Zhang, H. Chai, M. Kuang, J. Wu, X. Yang, Y. Zhao, Design and development of cryogen-free superconducting prototype generator with YBCO field windings, IEEE Trans. Appl. Supercon. 26 (2016) 5200205.

with a rated capacity of 2 MVA. The primary winding consists of eight coils connected in series (Fig. 14.24B). Fig. 14.25 shows the HTS tapes used in the primary and secondary windings.

14.6 Outlook

More than 100 years have passed since the discovery of superconductivity in 1891, and various materials have been found to show superconductivities below their critical temperatures. Among them, metal oxides play a vital role in the development of superconductors. Nowadays, their applications mainly employ two systems, that is, YBCO and BSCCO, which emerge as exciting prospects for superconductor wires/tapes, generators, and transformers. However, their use is still limited to the military and governmental fields due to their high costs. In addition, their performance needs to be further improved for common applications. For instance, developing a winding technology with large current capacity, as well as fabricating a new LN refrigeration system with higher cooling efficiency, are urgent objectives to be reached. Besides, modifying the existing materials and finding new materials with higher critical temperature and better performance are also of great importance. Inspired by the rapid development of both materials and devices in the past decades, we do believe that superconductors are of great value and their application in popular fields will be realized in the near future.

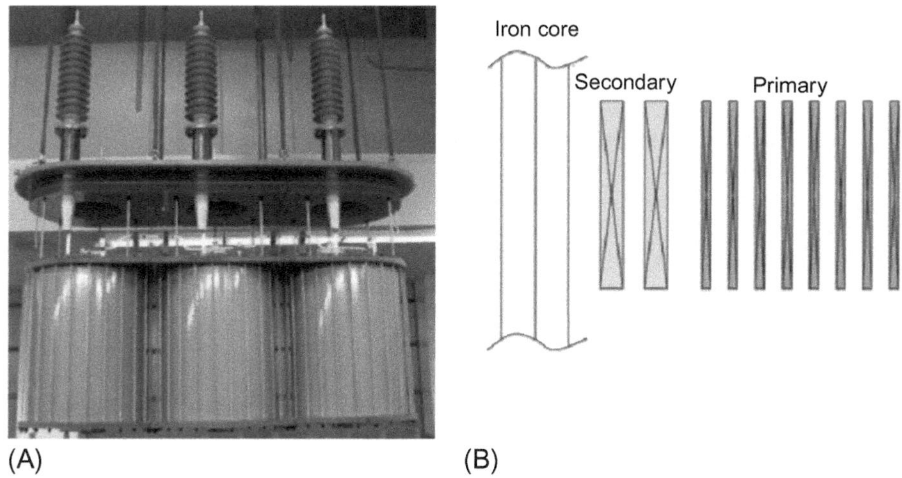

Fig. 14.24 (A) Three sets of coil assemblies. (B) Structure of the primary and secondary windings.
Reproduced by permission from Y. Ohki, Development of a three-phase superconducting power transformer, IEEE Electr. Insul. M. 31 (2013) 46–48.

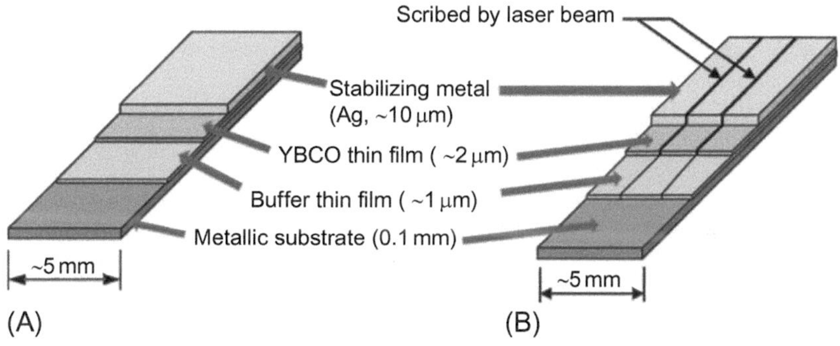

Fig. 14.25 Structures of superconducting tapes. (A) Nondivided YBCO tape for the primary winding and (B) three-filament YBCO tape for the secondary winding.
Reproduced by permission from Y. Ohki, Development of a three-phase superconducting power transformer, IEEE Electr. Insul. M. 31 (2013) 46–48.

Acknowledgments

Financial support from National Materials Genome Project (2016YFB0700600), China National Funds for Distinguished Youth Scientists (51026004), and NSFC (U1601214 and 51502137) is gratefully acknowledged.

References

[1] D. Larbalestier, A. Gurevich, D.M. Feldmann, A. Polyanskii, High-T_c superconducting materials for electric power applications, Nature 414 (2001) 368–377.

[2] J.G. Bednorz, K.A. Müller, Possible high T_c superconductivity in the Ba-La-Cu-O system, Z. Phys. B 64 (1986) 189–193.

[3] J. Orenstein, A.J. Millis, Advances in the physics of high-temperature superconductivity, Science 288 (2000) 468–474.

[4] Y. Kamihara, H. Hiramatsu, M. Hirano, R. Kawamura, H. Yanagi, T. Kamiya, H. Hosono, Iron-based layered superconductor: LaOFeP, J. Am. Chem. Soc. 128 (2006) 10012–10013.

[5] D.C. Johnston, H. Prakash, W.H. Zachariasen, R. Viswanathan, High temperature superconductivity in the Li-Ti-O ternary system, Mater. Res. Bull. 8 (1973) 777–784.

[6] L.F. Matteiss, Effect of the 110 K phase transition on the $SrTiO_3$ conduction bonds, Phys. Rev. B 6 (1972) 4740–4753.

[7] C.S. Koonce, M.L. Cohen, Superconducting transition temperature of semiconducting $SrTiO_3$, Phys. Rev. 163 (1967) 380–390.

[8] A. Baratoff, G. Binnig, H.E. Hoenig, J.G. Bednorz, Two-band superconductivity in Nb-doped $SrTiO_3$, Phys. Rev. Lett. 45 (1980) 1352–1355.

[9] P.E. Bierstedt, J.L. Gillson, A.W. Sleight, High-temperature superconductivity in the $BaPb_{1-x}Bi_xO_3$ system, Solid State Commun. 17 (1975) 27.

[10] T.D. Thanh, A. Koma, S. Tanaka, Superconductivity in the $BaPb_{1-x}Bi_xO_3$ system, Appl. Phys. 22 (1980) 205–212.

[11] J.G. Bednorz, M. Takashige, K.A. Müller, Susceptibility measurements support high-T_c superconductivity in the Ba-La-Cu-O system, Europhys. Lett. 3 (1987) 379–385.

[12] R. Englmann, The Jahn–Teller Effect in Molecules and Crystals, Wiley Interscience, New York, 1972.

[13] C.W. Chu, P.H. Hor, R.L. Meng, L. Gao, Z.J. Huang, Y.Q. Wang, Evidence for superconductivity above 40 K in the La-Ba-Cu-O compound system, Phys. Rev. Lett. 58 (1987) 405–407.

[14] M.K. Wu, J.R. Ashburn, C.J. Torng, P.H. Hor, R.L. Meng, L. Gao, Z.J. Huang, Y.Q. Wang, C.W. Chu, Superconductivity at 93 K in a new mixed-phase Y-Ba-Cu-O compound system at ambient pressure, Phys. Rev. Lett. 58 (1987) 908–910.

[15] R.M. Hazen, L.W. Finger, R.J. Angel, C.T. Prewitt, N.L. Ross, H.K. Mao, C. G. Hadidiacos, P.H. Hor, R.L. Meng, C.W. Chu, Crystallographic description of phases in the Y-Ba-Cu-O superconductor, Phys. Rev. B 35 (1987) 7238–7241.

[16] P.H. Hor, R.L. Meng, Y.Q. Wang, L. Gao, Z.J. Huang, J. Bechtold, K. Forster, C.W. Chu, Superconductivity above 9-K in the square-planar compound system $ABa_2Cu_3O_{6+x}$ with A = Y, La, Nd, Sm, Eu, Gd, Ho, Er, and Lu, Phys. Rev. Lett. 58 (1987) 1891–1894.

[17] J.M. Tarascon, L.H. Greene, P. Barboux, W.R. McKinnon, G.W. Hull, T.P. Orlando, K. A. Delin, S. Foner, E.J. McNiff, 3d-Metal doping of the high-temperature superconduction perovskites La-Sr-Cu-O and Y-Ba-Cu-O, Phys. Rev. B 36 (1987) 8393–8400.

[18] J.M. Tarascon, P. Barboux, P.F. Miceli, L.H. Greene, G.W. Hull, Structural and physical propertied of the metal (M) substituted $YBa_2Cu_{3-x}M_xO_{7-y}$ perovskite, Phys. Rev. B 37 (1988) 7458–7469.

[19] G. Xiao, F.H. Streitz, A. Gavrin, Y.W. Du, C.L. Chien, Effect of transition-metal elements on the superconductivity of Y-Ba-Cu-O, Phys. Rev. B 35 (1987) 8782–8784.

[20] G. Xiao, F.H. Streitz, A. Gavrin, M.Z. Cieplak, C.L. Chien, A. Bakhshai, Superconductivity and magnetism in transition-element-substituted $YBa_2Cu_3O_7$ compounds, J. Appl. Phys. 63 (1988) 4196–4198.

[21] K. Ishida, Y. Kitaoka, T. Yoshitomi, N. Ogata, T. Kamino, K. Asayama, Gapless superconductivity in Zn-doped $YBa_2Cu_3O_7$ studied by Cu NMR and NQR, Phys. C 179 (1991) 29–38.

[22] S. Gupta, R.S. Yadac, N.P. Lalla, G.D. Verma, B. Das, Microstructural and superconducting properties of $YBa_2Cu_{3-x}Co_xO_{7-y}$ system, Integr. Ferroelectr. 116 (2010) 68–81.

[23] A.H. Salama, M. El-Hofy, Y.S. Rammah, M. Elkhatib, The influence of magnetic nano metal oxides doping on structure and electrical properties of YBCO superconductor, Adv. Nat. Sci. Nanosci. Nanotechnol. 7 (2016) 015011.

[24] C. Michel, M. Hervieu, M.M. Borel, A. Grandin, F. Deslandes, J. Provost, B. Raveau, Superconductivity in the Bi-Sr-Cu-O system, Z. Phys. B 68 (1987) 421–423.

[25] C.W. Chu, J. Bechtold, L. Gao, P.H. Hor, Z.J. Huang, R.L. Meng, Y.Y. Sun, Y.Q. Wang, Y.Y. Xue, Superconductivity up to 114 K in the Bi-Al-Ca-Sr-Cu-O compound system without rare-earth elements, Phys. Rev. Lett. 60 (1988) 941–943.

[26] R. Hazen, C. Prewitt, R. Angel, N. Ross, L. Finger, C. Hadidiacos, D. Veblen, P. Heaney, P. Hor, R. Meng, Y. Sun, Y. Wang, Y. Xue, Z. Huang, L. Gao, J. Bechtold, C. Chu, Superconductivity in the high-T_c Bi-Ca-Sr-Cu-O system: phase identification, Phys. Rev. Lett. 60 (1988) 1174–1177.

[27] Z.Z. Sheng, A.M. Hermann, Superconductivity in the rare-earth-free Tl-Ba-Cu-O system above liquid–nitrogen temperature, Nature 332 (1988) 55–58.

[28] R.M. Hazen, L.W. Finger, R.J. Angel, C.T. Prewitt, N.L. Ross, C.G. Hadidiacos, Z. Z. Sheng, A. El-Ali, A.M. Hermann, 100-K superconducting phases in the Tl-Ca-Ba-Cu-O system, Phys. Rev. Lett. 60 (1988) 1657–1660.

[29] S.S.P. Parkin, V.Y. Lee, A.I. Nazzal, R. Savoy, R. Beyers, S.J. La-Placa, $Tl_1Ca_{n-1}Ba_2Cu_nO_{2n+3}$ (n = 1,2,3): A new class of crystal structures exhibiting volume superconductivity at up to 110 K, Phys. Rev. Lett. 61 (1988) 750–753.

[30] S.S.P. Parkin, V.Y. Lee, E.M. Engler, A.I. Nazzal, T.C. Huang, G. Gorman, R. Savoy, R. Beyers, Bulk superconductivity at 125 K in $Tl_1Ca_2Ba_2Cu_3O_x$, Phys. Rev. Lett. 60 (1988) 2539–2542.

[31] S.N. Putilin, I. Bryntse, E.V. Antipov, New complex copper oxides: $HgBa_2RCu_2O_7$ (R = La, Nd, Eu, Gd, Dy, Y), Mater. Res. Bull. 26 (1991) 1299–1307.

[32] S.N. Putilin, E.V. Antipov, O. Chmaissem, M. Marezio, Superconductivity at 94 K in $HgBa_2CuO_{4+y}$, Nature 362 (1993) 226–228.

[33] A. Schilling, M. Cantoni, J.D. Guo, H.R. Ott, Superconductivity above 130 K in the Hg-Ba-Ca-Cu-O system, Nature 363 (1993) 56–58.

[34] X.H. Chen, T. Wu, G. Wu, R.H. Liu, H. Chen, D.F. Fang, Superconductivity at 43 K in $SmFeAsO_{1-x}F_x$, Nature 453 (2008) 761–762.

[35] K. Takada, H. Sakurai, E. Takayama-Muromachi, F. Izumi, R.A. Dilanian, T. Sasaki, Superconductivity in two-dimensional CoO_2 layers, Nature 422 (2003) 53–55.

[36] K. Heine, J. Tenbrink, M. Thöner, Highfield critical current densities in $Bi_2Sr_2Ca_1Cu_2O_{8+x}$/Ag wires, Appl. Phys. Lett. 55 (1989) 2441–2443.

[37] S.R. Foltyn, L. Civale, J.L. Macmanus-Driscoll, Q.X. Jia, B. Maiorov, H. Wang, M. Maley, Materials science challenges for high-temperature superconducting wire, Nat. Mater. 6 (2007) 631–642.

[38] J. Turner, T.A. Lachlan-Cope, S. Colwell, G.J. Marshall, W.M. Connolley, Significant warming of the Antarctic winter troposphere, Science 311 (2006) 1911–1914.

[39] P.N. Barnes, M.D. Sumption, G.L. Rhoads, Review of high power density superconducting generators: present state and prospects for incorporating YBCO windings, Cryogenics 45 (2005) 670–686.

[40] A.B. Abrahamsen, N. Mijatovic, E. Seiler, T. Zirngibl, C. Traholt, P.B. Norgard, N. F. Pedersen, N.H. Andersen, J. Ostergard, Superconducting win turbine generator, Supercond. Sci. Technol. 23 (2010) 034019.
[41] Y. Yang, S. Duan, Y. Ren, Y. Jiang, L. Feng, X. Zhang, H. Chai, M. Kuang, J. Wu, X. Yang, Y. Zhao, Design and development of cryogen-free superconducting prototype generator with YBCO field windings, IEEE Trans. Appl. Supercon. 26 (2016) 5200205.
[42] E. Pardo, M. Staines, Z. Jiang, N. Glasson, AC loss modelling and measurement of superconducting transformers with coated-conductor Roebel-cable in low-voltage winding, Supercond. Sci. Technol. 28 (2015) 114008.
[43] S.T. Dai, T. Ma, Q.Q. Qiu, Z.Q. Zhu, Y.P. Teng, L. Hu, Development of a 1250-kVA superconducting transformer and its demonstration at the superconducting substation, IEEE Trans. Appl. Supercon. 26 (2016) 5500107.
[44] Y. Ohki, Development of a three-phase superconducting power transformer, IEEE Electr. Insul. M. 31 (2013) 46–48.

Metal oxides for emission control

15

Xinhai Yuan[*,‡], *Xiongwei Wu*[‡], *Yuping Wu*[*,‡], *Teunis van Ree*[†]
[*]State Key Laboratory of Materials-Oriented Chemical Engineering, School of Energy Science and Engineering, Institute for Electrochemical Energy Storage, Nanjing Tech University, Nanjing, People's Republic of China, [†]Department of Chemistry, University of Venda, Thohoyandou, South Africa, [‡]College of Science, Hunan Agricultural University, Changsha, China

15.1 Introduction

Emissions of many air pollutants have been shown to have a variety of negative effects on public health and the natural environment. Engine emissions are the main sources of outdoor air pollutants and include sulfur dioxide (SO_2), volatile organic compounds (VOCs), nitrogen oxides (NO_x), and particulate matter (PM) comprising solid carbon (soot) and unburned carbonaceous compounds (HC) [1, 2].

SO_2 emissions, which are solely a result of fuel-bound sulfur, are readily reduced by limiting sulfur in the fuel [3]. The emission of HC and CO can be simultaneously reduced by three-way catalyst (TWC) technology and diesel oxidation catalysts (DOCs) [4, 5].

The emissions of particulate matter (PM) and nitrogen oxides (NO_x) are of major concern, and if they are not well controlled the public health and the natural environment will be heavily threatened [3]. It is not easy to reduce both NO_x and PM simultaneously [5]. Typically, the abatement of PM and NO_x is controlled by separate after-treatment components [6]. Diesel particulate filters (DPFs) combined with a DOC, which can trap a great number of PM from the combustion chamber that are then oxidized by the noble metal Pt catalyst to regenerate the DPF, is a high-efficiency technology for controlling PM emissions. Selective catalytic reduction (SCR), using ammonia or urea as the reductant, and lean NO_x trapping (LNT) are the most promising NO_x removal technologies [5, 6]. Continuous regeneration traps (CRTs), DPF systems [7], and diesel particulate NO_x reduction (DPNR) systems are typical after-treatment devices to reduce both NO_x and PM simultaneously [5, 6].

Recently, the cost of after-treatment devices has been increasing, and many parts are required, leading to insufficient installation space, and the vehicles' weight increases, leading to poor fuel economy. It is necessary to develop low cost and high-efficiency metal oxide catalysts other than precious metals that have better catalyst activity for emission control [5].

Metal Oxides in Energy Technologies. https://doi.org/10.1016/B978-0-12-811167-3.00015-8

Among the catalysts investigated to date, many metal oxide catalysts exhibit high performance for emission control, such as spinels (such as $ZnAl_2O_4$ and $CuFe_2O_4$), perovskites (such as $SrTiO_3$ and $MgTiO_3$), transition metal oxides (such as Mn, Cu, Co, Ti, Zr, and Ce oxides), and mixed metal oxides (MMOs) (such as MnO_x-CeO_2, CoO_x-CeO_2, CuO-CeO_2, and $Ce_{0.5}Pr_{0.5}O_2$) [8].

In this chapter, we briefly review the recent progress in diesel emission control using metal oxide catalysts, including the catalyst principles, catalytic activity, material research progress, and catalyst preparation and methodologies. In addition, an outlook is provided to point out the latest research trends in diesel emission control using metal oxide catalysts.

15.2 Emission control technologies and metal oxide catalysts

15.2.1 TWC technology

Pollutants (HC, CO, and NO) can be simultaneously removed over a single bed of TWCs: HC and CO are converted to CO_2 and water by oxidation, and NO is converted mostly to N_2 by reduction. A catalyst promotes these reactions at lower temperatures than a thermal process giving the following desired reactions for HC, CO, and NO_x [9]:

Oxidation:

$$C_yH_n + (1 + n/4)O_2 \rightarrow yCO_2 + n/2\,H_2O$$

$$CO + ½O_2 \rightarrow CO_2$$

$$CO + H_2O \rightarrow CO_2 + H_2$$

Reduction:

$$NO\,(\text{or}\,NO_2) + CO \rightarrow ½N_2 + CO_2$$

$$NO\,(\text{or}\,NO_2) + H_2 \rightarrow ½N_2 + H_2O\,(2 + n/2)$$

$$NO\,(\text{or}\,NO_2) + C_yH_n \rightarrow (1 + n/4)N_2 + yCO_2 + n/2\,H_2O$$

The TWCs consist of three main components: substrate, support material, and active phase. The substrates are monolith, which are produced by cordierite and metal foil. The support materials of the TWCs are mainly alumina- and CeO_2-based materials. The top of the catalyst is the catalytic phase where the reactions happen. Noble metals such as Pt, Pd, and Rh have been used as catalyst for a long time, but now perovskite and transition metal (Cu, Ni, Mn, Co, etc.) oxides have attracted the attention for their high efficiency and low cost.

15.2.1.1 Metal oxides as support materials for TWCs

Alumina

Generally, alumina has seven crystal structures: α-, γ-, δ-, κ-, η-, θ-, and χ-Al_2O_3. The most common form of alumina used for catalyst support is the γ-form, which has a surface area of more than 300 m^2/g, pore size ranging from 30 to 120 Å, and pore volume from 0.5 to 1 cm^3/g. The structure of γ-Al_2O_3 consists of single layers of close-packed spheres, the layers having the O^{2-} ion at position 1. The spheres of the second layer are positioned in 50% of the hollows of the first layer. There are two possibilities for the distribution of the third layer, but in the case of γ-Al_2O_3, the third layer is distributed on the remaining hollows of the first layer, according to the numbers: 1, 2, 3, 1, 2, 3 .... Therefore, the small Al^{3+} cation is placed in the space between these layers of oxide anion packing.

Because γ-Al_2O_3 is an octahedral cubic crystallite, the structure includes octahedral and tetragonal Al. The structure of γ-Al_2O_3 is pseudo-spinel, differing only in oxide anion packing density. The surface of γ-Al_2O_3 contains both Bronsted and Lewis active sites, which play an important role in catalyzed reactions [10]. In conclusion, γ-Al_2O_3 has been used extensively in the application of automobile exhaust catalysts because of its high specific surface area, good porosity, and very high structural stability [11].

$CexZr_{1-}xO_2$

Ceria is a very useful support in three-way automobile catalysts, mainly due to its oxygen storage capacity (OSC) that allows effective catalyst operation under conditions of fluctuating oxygen concentration. Because Ce^{4+} in the CeO_2 lattice is readily converted to Ce^{3+} due to its nonstoichiometric behavior, the addition of CeO_2 promotes dynamic performance in purifying CO, NO_x, and HC under conditions of rich to lean air to fuel (A/F) ratios in automotive exhausts, which is called its OSC. CeO_2 provides oxygen for oxidizing CO and HC under rich A/F conditions and removes it from the exhaust gas phase for reducing NO under lean A/F conditions. However, the surface of CeO_2 collapses at elevated temperatures; the addition of Zr can not only prevent this phenomenon but also improve the oxygen mobility in the CeO_2 lattice [9]. However, $Ce_xZr_{1-x}O_2$ still has drawbacks with respect to the stability of the structure, especially the thermal stability of the texture. The combination of high surface area Al_2O_3 and effective OSC Ce-Zr-O_2 material can improve Ce-Zr-O_2's textural properties; hence, it has attracted the attention of many researchers.

Trivalent La^{3+} can also be used to promote the thermal stability and oxygen mobility (ionic conductivity) of CeO_2-ZrO_2 solid solutions, and to restrict the phase transformation of γ-Al_2O_3 to α-Al_2O_3 at high temperature [9].

15.2.1.2 Metal oxides as TWCs [9]

The use of noble metals is limited by cost and sensitivity to poisoning, especially by chlorine/chloride products. Thus, composite mixed oxides, mostly with perovskite-type structure (ABO_3 or A_2BO_4), can be seen as interesting alternatives as they are low-cost materials and stable even above 1000°C. Alternatively, transition metal (Mn, Co, Cu, V, etc.) oxides, which are inexpensive and have a high activity and resistance to poisoning, may be used.

Perovskite-type mixed oxides:

Perovskite-type mixed oxides have been widely studied for their high catalytic activity in the oxidation reactions of hydrocarbons and VOCs. Cobalt- and manganese-based perovskites were even proposed as an alternative to noble metal supported catalysts for their similar activities in oxidation at a lower synthesis cost. However, the low specific surface area hinders their practical application [12].

Transition metal oxides:

Transition metal oxides may be used as an optional alternative to noble metals as catalysts for complete oxidation. They are less active at low temperatures, but at higher temperatures their activity is similar to that of the noble metals. Oxides of Cu, Co, Mn, and Ni are the most active single metal oxides for combustion of VOCs. Among these metal oxides, manganese and cobalt containing catalysts have been widely studied for their low cost, environmental friendliness, and relatively highly activity in VOC combustion. It is believed that the ability of manganese to form oxides of different oxidation states and their high OSC are the main reasons for the good catalytic properties of MnO_x-based catalysts [13].

15.2.2 The DOC

The oxidation of CO, unburned hydrocarbons, and NO are the primary functions of DOC. The heat generated by the hydrocarbon oxidation can also be used to generate exotherms required for downstream components [14]. Current DOCs are generally based on ceramic "honeycomb" monoliths. The ceramic monoliths have superior strength because of the low porosity of the ceramic; however, this makes the monolith unsuitable to disperse the catalyst. For this reason, a wash coat of a high surface area material is adhered to the monolith surface, and then the catalyst is dispersed on the wash coat (Fig. 15.1) [15]. As previously mentioned, a high surface area wash coat support can offer highly dispersed precious metal sites, which not only increases the active component exposed surface area for reaction but also lowers the amount of precious metal required, thus decreasing the catalyst cost. The catalyst support can also influence performance by altering the thermal stability of the active component and support, surface area, pore volume, and surface reactivity. Metal oxides, such as Al_2O_3, ZrO_2, and La_2O_3, are used as wash coat material. Precious metals, such as Pt and Pd, are used as catalyst components in DOCs.

15.2.2.1 Metal oxides as wash coat material for DOCs [14]

Alumina (Al_2O_3) is a very common wash coat material. Zirconia, with a highly crystalline structure, has a much lower surface area than alumina. Yoshida et al. [16] examined the effect of wash coat electronic properties on Pt oxide formation, and

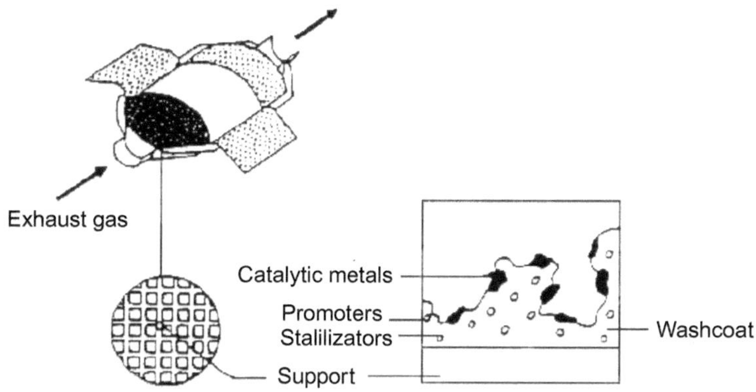

Fig. 15.1 The structure of a monolithic exhaust gas catalyst.
Reproduced with permission from U. Lassi, Deactivation correlations of Pd/Rh three-way catalysts designed for Euro-IV emission limits—effect of ageing atmosphere, temperature and time. PhD thesis, University of Oulu, Oulu, Finland, 2003, p. 138.

proposed that an acidic (i.e., electrophilic) support, such as Al_2O_3 and SiO_2, leads to a lower electron density on the Pt particles, especially compared with less acidic supports such as ZrO_2 and La_2O_3. This means that the more acidic supports have higher activity for propane, including supports manufactured to be strongly acidic. Similarly, turnover frequency (TOF), (also referred to as turnover rate—TOR) for propane oxidation increased with acid strength due to changes in the Pt oxidation state, with the electrophilic acidic supports providing greater resistance to Pt oxide formation; this resistance increased as the catalyst had electronegative additives incorporated into the formulation. Pt oxides are generally undesired (described later), so preventing the formation of these components is important for oxidation activity. However, moderate acid strength of the support improved methane combustion over Pd catalysts, because PdO functions as the active site.

15.2.3 Diesel NO_x removal catalyst technology

Nitrogen oxides (NO, NO_2, and N_2O) have been a major source of air pollution, greatly contributing to photochemical smog, acid rain, ozone depletion, and the greenhouse effect. SCR with ammonia ($4NO+4NH_3+O_2 \rightarrow 4N_2+6O_2$) is an efficient technology for removing NO from power plant emissions [17, 18].

Platinum-based catalysts were found to be highly active for the reaction of SCR with ammonia; however, the relatively high reaction temperature, the high cost, and rapid poisoning of Pt under stack-gas conditions have prompted worldwide efforts to develop low-temperature SCR catalysts. A process based on V_2O_5 (+WO_3 or MoO_3)/TiO_2 catalysts was developed and successfully commercialized for NO_x removal for its high activity and resistance to SO_2 poisoning [17]. However, the main disadvantage of this type of catalyst is that the reaction temperature is very high (>250°C) [18].

Considering that the downstream temperature is typically below 200°C, it is necessary to develop low-temperature SCR catalysts to avoid reheating of the flue gas, thus decreasing the capital cost. Transition metal oxide catalysts have been investigated for the low-temperature SCR reaction, such as $NiSO_4/Al_2O_3$, MnO_x/Al_2O_3, V_2O_5/activated carbon, and MnO_x/TiO_2 [18].

15.2.3.1 *Reaction mechanism of metal oxide-based catalysis for ammonia-SCR of NO_x*

The mechanism of the surface reaction of NO_x on metal oxide catalysts has been studied extensively [19]. Three hypotheses have been proposed for the mechanism: (1) the reaction between ammonium ion and adsorbed NO_2 [20]; (2) the reaction between an amide and gaseous NO [21]; and (3) the reaction between coordinated ammonia and a species generated by spillover of NO on the support [22].

Regarding the mechanism of low-temperature ammonia-SCR of NO_x on MnO_x/Al_2O_3 catalyst, it has been proposed that the reaction starts with the adsorption of NH_3 on a Lewis acid site and is subsequently oxidized to an NH_2 function; the NH_2 would then react with gas phase NO (an E–R mechanism) and nitrite intermediates on the surface (an L–H mechanism) [20a]. The reaction pathway over a MnO_x–CeO_2 catalyst has been reported to have a similar mechanism [20b]:

$$\tfrac{1}{2}O_2(g) \rightarrow O(a)$$

$$2NH_3(g) \rightarrow 2NH_3(a)$$

$$NH_3(a) + O(a) \rightarrow NH_2(a) + OH(a)$$

$$NO(g) + O(a) \rightarrow NO_2(a)$$

$$NH_2(a) + NO(g) \rightarrow NH_2NO(a) \rightarrow N_2(g) + H_2O(g)$$

$$NH_3(a) + NO_2(a) \rightarrow N_2 + H_2O + OH(a)$$

$$2OH(a) \rightarrow H_2O + O(a)$$

The reaction of NH_2 and NO, and then formation of nitrosamine (NH_2NO), is a typical SCR mechanism reported also for V_2O_5/TiO_2 and manganese-based catalysts.

The SCR mechanism on carbon supported Mn_3O_4 catalysts, shown in Fig. 15.2, consists of the following steps: (1) NH_3 adsorption and reaction with the oxygen atoms to form the aminooxy group; (2) NO adsorption as nitrosyl in the oxygen vacancies, followed by oxidation to NO_2; and (3) the reaction of the aminooxy group with NO_2 to form N_2 and water [21].

A unifying redox approach (shown in Fig. 15.3) has been proposed recently, based on investigation of the redox mechanism of low-temperature ammonia-SCR over a commercial V_2O_5–WO_3/TiO_2 catalyst [22]. It was proposed that vanadium sites

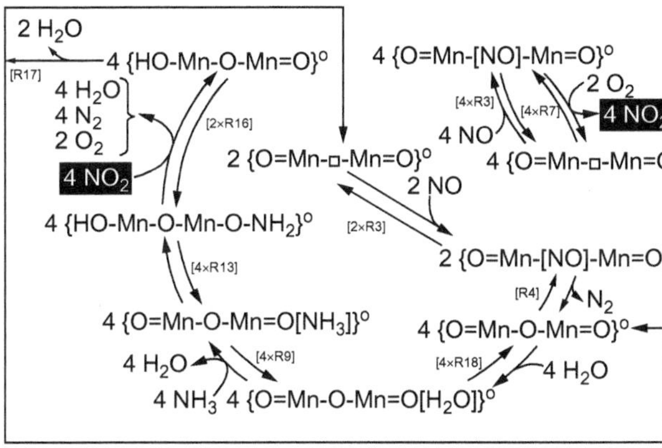

Fig. 15.2 Mechanism of the steady-state ammonia-selective catalytic reduction (SCR) reaction in the presence of oxygen over carbon supported manganese oxide catalysts.
Reproduced with permission from G. Marbán, T. Valdes-Solis, A.B. Fuertes, J. Catal. 226 (2004), 138–155.

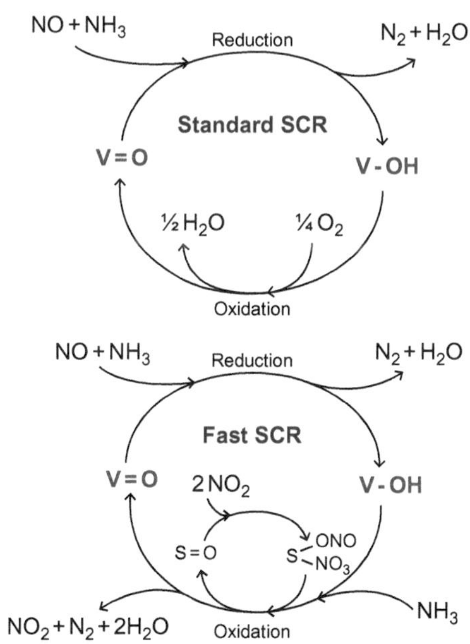

Fig. 15.3 Redox cycles of the standard and fast reactions over V_2O_5–WO_3/TiO_2 catalysts.
Reproduced with permission from E. Tronconi, I. Nova, C. Ciardelli, D. Chatterjee, M. Weibel, J. Catal. 245 (2007), 1–10.

are reduced by the reaction between NO and NH_3, then they are reoxidized either by oxygen (standard SCR) or by nitrates (fast SCR), the latter accompanied by the disproportionation of NO_2.

15.2.3.2 *Metal oxide-based catalysts for ammonia-SCR of* NO_x

Single metal oxide-based catalysts:

Oxides based on transition metals (Fe, Mn, Cr, Cu, V, Cr, and Co) have shown good activities for low-temperature SCR. Among these catalysts, nano-MnO_x catalysts have been extensively studied due to their excellent low-temperature SCR performance [18, 23–25].

It has been reported that amorphous MnO_x with a very large surface area, enables low-temperature SCR. The crystallinity, specific surface area, and oxidation state of MnO_x are all correlated to some extent with the performance. Manganese oxides such as MnO_2, Mn_5O_8, Mn_2O_3, Mn_3O_4, and MnO are all stable at normal temperature and pressure. The onset temperature of reduction was in the following descending order: $Mn_3O_4 > Mn_2O_3 > Mn_5O_8 > MnO_2$. At low temperatures, MnO_2 has the highest activity per unit surface area, whereas Mn_2O_3 has the highest N_2 selectivity [26].

From the studies on the application of manganese oxide catalysts in low-temperature SCR, some rules could be summarized as: the NO conversion activity increases with increasing Mn^{4+}/Mn^{3+} ratio; for all types of MnO_x the N_2 selectivity decreases as the temperature increases; at the same temperature, N_2 selectivity decreases with increasing NO conversion [25, 27].

Mixed oxide-based catalysts:

Among the various composite oxide catalysts, Mn-based composite oxide catalysts have been extensively studied in low-temperature SCRs due to their excellent low-temperature SCR performance [18].

Since the pure MnO_x catalysts generally have a low surface area and suffer from sintering, many efforts were made to improve the catalyst surface area and thermostability by partial substitution of Mn with other metallic elements to form mixed oxides [19].

Bimetallic and composite oxide catalysts containing Mn have higher activities and N_2 selectivities than pure MnO_x [28–30]. Reportedly, MnO_x–CeO_2 catalysts can achieve 100% N_2 selectivity at 100 – 150°C, whereas the N_2 selectivity is relatively low for MnO_x under the same conditions [29]. After modification by CeO_2, the NO_x conversion and N_2 selectivity over MnO_x was improved significantly [29, 30]. However, the activity of $MnO_x - CeO_2$ is still low at low temperatures, particularly in the presence of SO_2 [31].

The SO_2 poisoning is a challenge for low-temperature (LT) NH_3–SCR catalysts. It may be SO_2 poisoning that results in the irreversible loss of the NO/NO_2 oxidation ability; the formation of sulfite and sulfate species (such as $MnSO_4$) is believed to be the main reason for the deactivation of Mn-based catalysts [20, 32].

Recently, it was proposed that the addition of Sn to $MnO_x - CeO_2$ could promote the LT SCR activity and SO_2 resistance even in the presence of H_2O [33]. This was attributed to the significant increase of oxygen vacancy concentration achieved by Sn modification, which may facilitate NO oxidation to NO_2, and the significantly enhanced Lewis acid sites that are created by surface sulfation during SCR in the presence of SO_2.

Supported metal oxide-based catalysts:

The active components used in supported metal oxide catalysts occur mostly on the surface of the catalyst. The supports provide a large surface area for separation of the active phase and prevent agglomeration due to high sintering, but also supply active sites for the catalytic reactions.

Many support materials, such as Al_2O_3 and TiO_2, and some supported transition metal oxide catalysts, have been investigated for use in low-temperature SCR. The catalytic performance of various transition metal oxides supported on TiO_2 decreased in the following order: $Mn > Cu \geqslant Cr \gg Co > Fe \gg V \ggg Ni$. The activity of Mn-based catalysts was in direct proportion to the Mn loading and the reaction temperature. TiO_2 alone did not result in any NO conversion [19, 27, 34]. Therefore, Mn species must play a significant role in this catalyzed reaction. Fig. 15.4A shows the results obtained with different MnO_x/TiO_2 supported catalysts for SCR of NO by ammonia as a function of manganese loading [35]. It is clear that increasing the manganese loading up to 10% had a positive effect on the catalytic activity, and that a further increase in manganese loading did not increase the activity.

The N_2 product selectivity as a function of the temperature for these catalysts is shown in Fig. 15.4B [35]. The effect of loading on the selectivity was quite significant. The higher the loading, the more the N_2O formed. With increasing temperature the selectivity decreased considerably for all samples. To reduce sintering of manganese oxide, a series of transition metals (Fe, Cu, Ni, Cr) was added to the Mn/TiO_2 catalyst, and the activity and the characterization of the catalysts were investigated. With the

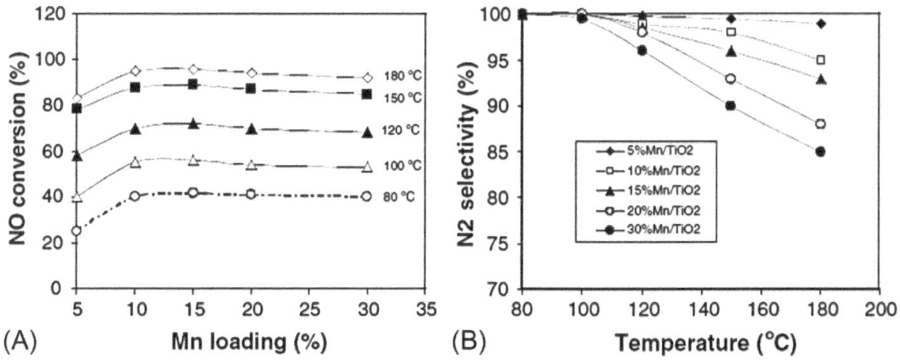

Fig. 15.4 (A) NO conversion and (B) N_2 selectivity on MnO_x/TiO_2 catalysts with various MnO_x loadings.
Reproduced with permission from G. Qi, R.T. Yang, Appl. Catal. B. 44 (2003), 217–225.

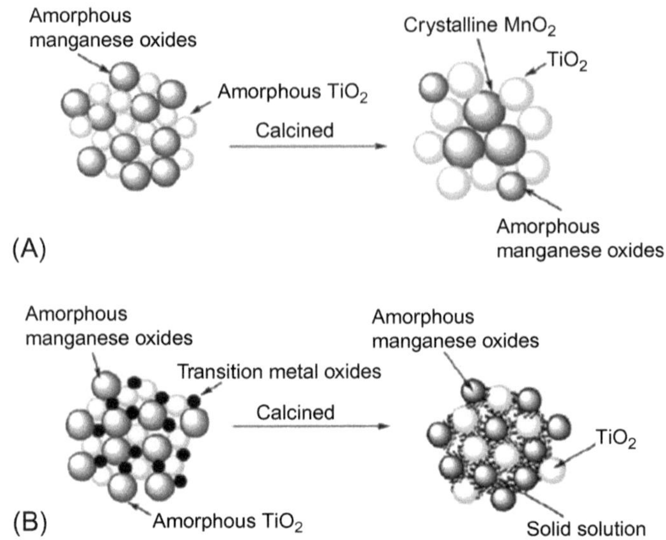

Fig. 15.5 Models for (a) Mn/TiO_2 and (b) metal–Mn/TiO_2 structures before and after calcination.
Reproduced with permission from Z. Wu, B. Jiang, Y. Liu, Appl. Catal. B 79 (2008), 347–355.

addition of transition metal, the activity of the catalysts increased greatly, Fe having the most favorable effect on the activity.

A model (Fig. 15.5) has been proposed to interpret the effect of the transition metals on the activity of MnO_x/TiO_2 [36]. From the model, the behavior of transition metals was simulated. It was postulated that the addition of transition metals could segregate the particles of manganese oxide and titania. The particles are consequently prevented from sintering so that the manganese oxide is kept in the amorphous phase; as a result, the catalytic activity is significantly improved.

Carbon materials can also be used as support materials. Activated carbon is a good solid sorbent and has been widely used to clarify air or industry waste gas, because of its high surface area, good pore structure, abundance as resource, low price, and easy regeneration. But activated carbon has a relatively low catalytic activity, especially under high space velocity conditions; therefore, it usually is preactivated or loaded with other active components to improve its catalytic performance [37].

The special properties of activated carbon fibers (ACF) also make their use as catalyst support for removing air pollutants quite attractive. Metal oxide supported on ACF shows excellent catalytic activity for low-temperature SCR [37].

15.2.4 Simultaneous control of diesel soot and NO_x emissions

DPFs, which are used for the entrapment of particulates from exhaust gases, have been studied widely over the years [38]. However, the backpressure created by a blocked DPF will damage the engine, so that the DPF must be regenerated periodically. The material trapped in the filter consists of mostly carbon particles with some absorbed hydrocarbon. There are two principal techniques for removing the particles [8, 39]:

(1) combustion with O_2

$$[C] + O_2 \rightarrow CO_2$$

(2) combustion with NO_2

$$[C] + NO_2 \rightarrow CO_2 + NO$$

NO_2-based systems are the most popular regeneration technique, also called continuously regenerating trap (CRT) technology, in which platinum containing catalysts are located upstream in the filter, which oxidize the NO emitted by the engine to NO_2. The generated NO_2 is a better oxidant than O_2 and is potentially useful to oxidize diesel PM with a DOC at exhaust gas temperatures. Generally, exhaust gas temperatures are about 200–500°C, whereas the soot formed as diesel PM can be burnt off at temperatures above 600°C. It is therefore very important to identify a promising catalyst which is active for CRT at exhaust gas temperatures [8]. The problem with CRT is the loss of NO_2 from the filter and thus the total NO_x removal efficiency is low.

The combination of the NO_x storage reduction (NSR) technology for NO_x removal with the simultaneous combustion of soot over the same NSR catalyst seems to be a more promising option for the abatement of both pollutants in a single catalytic bed, and this technology is usually referred to as DPNR [6, 40]. The structure of the DPNR catalyst is illustrated in Fig. 15.6. The cell walls of the DPF base have internal pore surfaces that are coated with the NSR catalyst. Under lean conditions, the NO is oxidized to NO_2 while the adsorbed oxygen atoms are activated on the Pt active sites; most of the NO_2 is further converted on the surface of the catalyst to nitrate and/or nitrite species. The residual NO_2 and activated oxygen atoms can react directly with diesel soot. In an enriched atmosphere, nitrate and/or nitrite species can decompose to form NO_2, which is then reduced by soot to N_2, HC, and CO on the platinum sites.

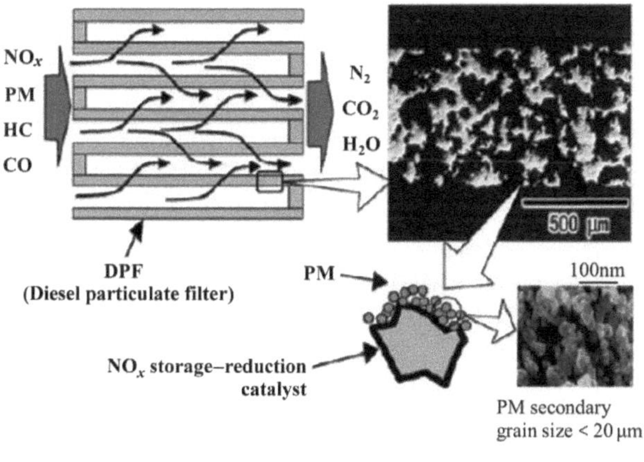

Fig. 15.6 Structure of diesel particulate NO_x reduction (DPNR) catalyst. Reproduced with permission from J. Suzuki, S. Matsumoto, Topics Catal. 28 (2004), 171–176.

Obviously, for high NO_x and soot conversion efficiencies, an appropriate operating scheme and device structure are crucial. It is particularly crucial to explore highly efficient catalysts that will remove soot and nitrogen oxides simultaneously by oxidizing NO to NO_2 and storing the NO_x species under lean conditions, and will selectively reduce the NO_x to N_2 in the NO_2-soot reaction [41].

The first catalyst used for simultaneous soot-NO_x removal was based on CuO [16]. Since then, several other types of catalyst, such as perovskite-type oxides [42], spinel-type oxides [43, 44], and hydrotalcite-derived oxides [45, 46] have been studied, with most of them showing acceptable activities for the simultaneous removal of NO_x and soot.

15.2.4.1 Contact conditions between soot and catalyst

The magnitude of the contact between the catalyst and the soot cake is of utmost importance for the determination of soot oxidation kinetics [40, 47]. In these three-phase reactions, the catalytic activity depends strongly on the interaction between the two solids and the gas. Therefore, the nature of the contact between soot and catalyst is of key importance. In empirical studies, two kinds of catalyst-soot contact conditions have been shown to exist [40, 48]:

(1) A "loose" contact condition is achieved by gentle stirring of the catalyst-soot mixture. This method sufficiently homogenizes the mixture, allowing loose contact between the two solid phases. During this very short (1–2 min) procedure, the contact thus established gives reproducible catalytic results. It has been reported that surface oxygen species exist under "loose" contact conditions over perovskites, spinels, and other mixed oxides.

(2) A "tight" contact condition can be obtained by ball milling (or in a mortar) resulting in close contact between the catalyst and the soot. In this way, the number of contact points is maximized and enables better differentiation of the morphologies, although it may be less representative of actual contact conditions occurring in an operational catalytic trap. It has been shown that surface oxygen species over alkaline (Na, K, Cs) and alkaline-earth metal (Ca, Ba, Mg) oxides are under "tight" contact conditions.

15.2.4.2 Mechanisms of oxygen transport

It is essential to identify the different parameters that determine the catalytic activity in order to develop improved catalysts for soot oxidation at lower temperatures. Apart from the surface area of the employed catalyst, several other factors that determine the intrinsic activity of a catalytic material are of primary importance [49].

The key factor that determines the activity for catalytic soot oxidation is the transfer of oxygen from the catalyst surface to the soot [50]. There are three basic mechanisms for oxygen transport that have been reported for oxide catalysts (illustrated in Fig. 15.7) [8, 50, 51]:

(1) Surface redox mechanism (e.g., with Co_3O_4)—surface oxygens of the catalyst are transferred on to the surface of the soot at the soot-catalyst interface, and once transferred, gas-phase oxygen fills the resulting vacancies [52, 53].

(2) Spill-over mechanism (e.g., CeO_2 and Cr_2O_3)—oxygen is first adsorbed onto the surface of the catalyst and undergoes some dissociation, forming active oxygen species which are subsequently transferred to the surface of the soot (catalyst-soot contact is not necessarily required) [53, 54].

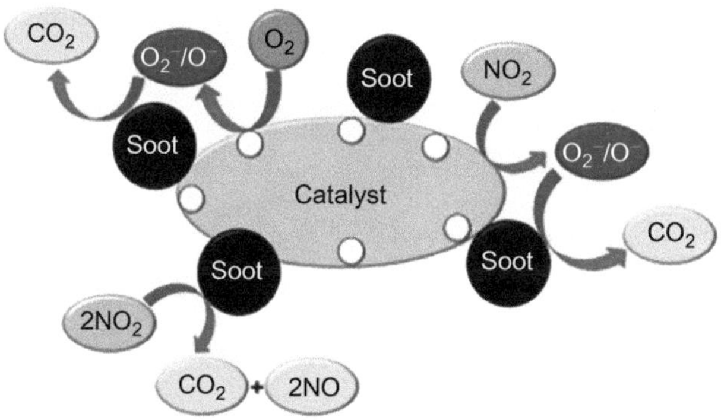

Fig. 15.7 Three basic oxygen transportation mechanisms for oxide catalysts.
Reproduced with permission from N.D. Wasalathanthri, T.M. SantaMaria, D.A. Kriz, S.L. Dissanayake, C.H. Kuo, S. Biswas, S.L. Suib, Appl. Catal. B: Environmental 201 (2017), 543–551.

(3) Redox mechanism (e.g., V_2O_5 and Fe_2O_3)—catalyst gets reduced by transferring bulk oxygen to soot, which is oxidized [50, 53].

15.2.4.3 Mechanism for simultaneous NO_x and soot removal

Many mechanisms have been postulated since the idea of simultaneous catalytic removal of NO_x and soot was introduced in 1989, but two mechanisms are generally accepted to describe the reactions involved [41]:

(1) As illustrated in Fig. 15.8, two adjacent oxygen vacancies are created on the catalyst surface by soot combustion, and they then develop into active sites for the chemisorption of two NO molecules. The adsorbed NO dissociates into N(ad) and O(ad), with subsequent generation of N_2O or N_2. But the positive effect of the oxygen cannot be explained by this mechanism,

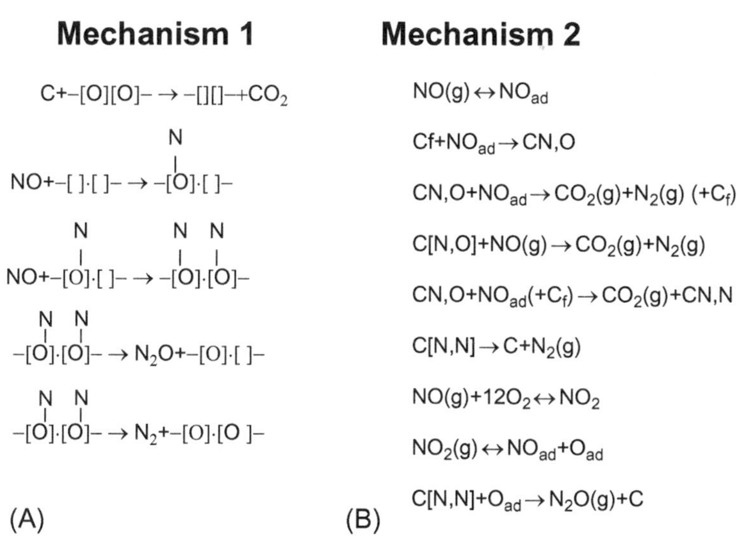

Fig. 15.8 Mechanisms for simultaneous NO_x and soot removal.
Reproduced with permission from D. Fino, P. Fino, G. Saracco, V. Speccia, Appl. Catal. B: Environmental 43 (2003), 243–259.

because the oxygen would be expected to fill the oxygen vacancies on the catalyst surface, thus leading to a lower reaction rate.

(2) Another mechanism has been proposed to explain the positive effect of oxygen. According to this mechanism, C[N, O], which is formed by the combustion of reactive free carbon (C_f) and NO molecules adsorbed on the catalyst surface, plays important roles. There are two possible ways to form N_2: (i) C[N, O] directly reacts with another NO_{ad} or gaseous NO molecule and (ii) C[N, O] can be further converted to another reaction intermediate C [N, N], which then decomposes into N_2. The promoting effect of oxygen can be ascribed to the formation of active sites for NO chemisorption on C_f as a consequence of carbon combustion, or to easier formation of NO_2, which is more active than NO in the reaction with soot.

15.2.4.4 Requirements for the catalyst materials

The particular challenge here is to achieve a high rate of soot oxidation in combination with substantial NO_x reduction at a sufficiently low temperature. Therefore, it is necessary to identify materials with suitable redox abilities. Besides high activity, increasing the contact area between the catalyst and solid reactant is a particular challenge in this field. It is also important that the texture of these materials be suitable for capturing soot particles, which are typically larger than 25 nm [55].

15.2.4.5 Metal oxides as soot oxidation catalysts

Many types of catalyst have been investigated for soot combustion, including monobasic metal oxide catalysts, ceria-based catalysts, MMOs, and perovskite-type oxides [56, 57].

Monobasic metal oxide catalysts:

Ceria has special OSC, which can generate highly active oxygen species as the so-called "active oxygen" in many catalytic reactions; it has also exhibited excellent catalytic activities in the soot/NO_x removal reaction. Compared with CeO_2, other single metal oxides such as SiO_2, Al_2O_3, TiO_2, and ZrO_2, which cannot generate highly "active oxygen," show poor and even no catalytic activity for soot/NO_x removal [11]. Hence, researchers have focused on designing ceria-based catalysts for the simultaneous removal of soot and NO_x [51].

Ceria-based catalysts:

Ceria-based catalysts have been widely studied for soot oxidation, due to the high soot oxidation activity of CeO_2-containing metal oxides. The facile redox transfer of the Ce^{3+} and Ce^{4+} oxidation states enables cerium oxides to release and store atomic oxygen [57]. For instance, zirconium and a range of rare earth elements (such as La, Pr, and Sm) have been introduced into the ceria framework to improve OSC and redox properties, and to provide high-temperature stability [38, 58]. Different elements supported on ceria, such as K/ceria, Co/ceria, Ba,K/ceria, Cu-V/ceria, Ag/ceria, and Cu/ceria [58] have also been studied.

The oxidation activity of ceria-based materials is determined mostly by their redox behavior and the availability of chemisorbed oxygen (α-species). The available evidence shows that the CeO_2 lattice oxygen participates in the soot oxidation process (via a Mars–van Krevelen mechanism), but the detailed reaction mechanism is still awaiting further clarification [38].

Alkali-metal oxides:

Alkali metal-containing catalysts are of special interest. It has been reported that alkali oxides may favor reaction by the formation of low melting point compounds, or of eutectics with other catalyst components, thus improving the surface mobility of the active species and hence favoring the soot-catalyst contact, which has been claimed as a key factor in the oxidation of soot by oxygen [59, 60].

It is widely accepted that potassium-containing compounds are among the most active catalysts for soot oxidation, and many K-doped catalysts with high soot oxidation activities have been developed on this premise. The high mobility and low evaporation temperature of potassium limit their application [61], but potassium-containing compounds may present a possible solution to obtain an active soot oxidation catalyst with superior thermal stability. $K_2Ti_2O_5$ is known to possess the highest activity for soot combustion [62], and it known that platinum, cobalt, cerium, and copper are active catalysts for soot oxidation. Therefore, in order to further increase the soot oxidation activity of $K_2Ti_2O_5$, some metal catalysts including platinum, copper, cobalt, or cerium were doped on $K_2Ti_2O_5$ [63]. The synergistic effects between $K_2Ti_2O_5$ and supported metals for soot oxidation still need further study.

Mixed metal oxides:

MMOs may contain rare earth, alkali, alkaline-earth, and noble metals.

Layered double hydroxides (LDHs) are a class of ionic lamellar compounds consisting of positively charged brucite-like layers with an interlayer region containing charge-compensating anions and solvation molecules [3]. LDHs have variable compositions and special structural properties; hence they are applied widely for catalysis and adsorption. The idealized layered structure of LDHs is shown in Fig. 15.9. The general formula of LDHs can be represented by $[M_{1-x}{}^{2+}M_x{}^{3+}(OH)_2]^{x+}(A^{n-})_{x/n} \cdot yH_2O$, where M^{2+} and M^{3+} are di- and trivalent cations, respectively. The value of x is equal to the molar ratio of $M^{2+}/(M^{2+}+M^{3+})$ and is generally in the range 0.2–0.4, and A^{n-} is an anion. As a result, a large class of isostructural materials with versatile physical and chemical properties can be obtained by changing the molar ratios of M^{2+}/M^{3+} as well as the types of interlayer anions.

Catalysts obtained by high-temperature calcination of LDH precursors typically have a large surface area, basic properties, high metal dispersion, and stability against sintering; hence they are very promising for simultaneous catalytic removal of soot and NO_x [41].

Over several years, the synthesis of catalysts from LDHs for the simultaneous removal of NO_x and soot has been pursued vigorously; for example, binary LDH-derived catalysts (e.g., CuAl [64] and CoAl [65] LDHs), ternary LDH-derived

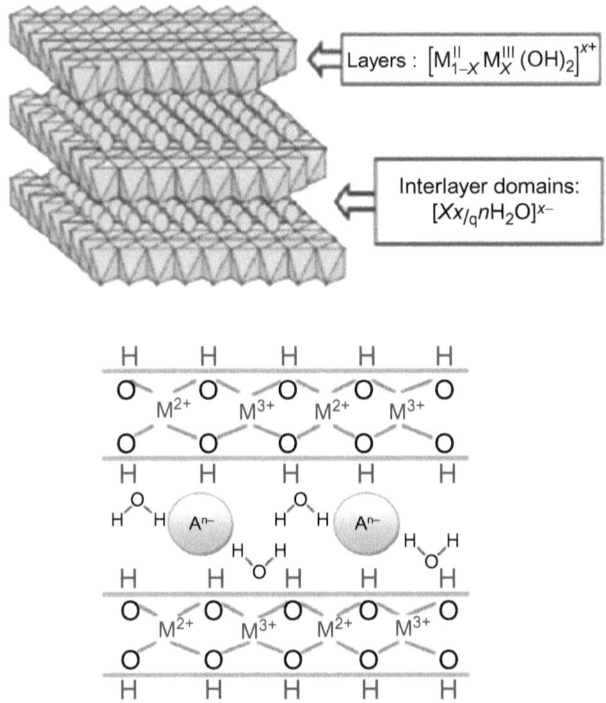

Fig. 15.9 Structure of layered double hydroxide (LDH).
Reproduced with permission from G.C. Dhal, D. Mohan, R. Prasad, Catal. Sci. Technol. 7 (2017), 1803–1825.

catalysts (e.g., MnMgAl [66], CoMgAl [67], and CuMgAl [68] LDHs), and quaternary LDH-derived catalysts (e.g., CoMgAlCe LDH [69]) have been studied.

The chemical composition is a significant factor influencing the selective NO_x reduction and soot combustion activity of the NO_x reduction catalysts. Potassium doping can further significantly improve their catalytic performance [41]. Furthermore, the strong interaction between potassium and the catalyst carrier metals can promote the formation of reactive oxygen species in the new K–M–O phase. The reactive oxygen species can not only react more easily with soot, but also promote the oxidation of NO to NO_2, and store these in the form of nitrates.

Spinel-type oxides:

Spinel-type oxide catalysts have been studied as potential candidates for the simultaneous removal of NO and diesel soot particulate [70]. Among various spinel-type oxide catalysts, cobalt-containing spinel-type catalysts have attracted much attention for many years due to the relatively abundant resources and lower cost [71, 72]. In addition, these catalysts are very active for catalytic removal of NO_x and PM. It is well known that a dopant can significantly change the properties of the original material [73]. Cerium is also an excellent additive in exhaust catalysts for its OSC [74]. A 10% cerium-modified catalyst (i.e., $Ce_{0.1}Co_{0.9}Cr_2O_4$) has been reported [75] with improved catalytic activity (90% soot conversion and maximum 69.8% NO conversion at 446°C) compared with unmodified catalyst. It is believed that the improvement enables the deformation of the spinel structure in the substitution process. Therefore, this deformation results in crystal defects and an increase in oxygen vacancy concentration, which favors oxygen mobility and the redox behavior of the cerium-modified cobalt chromite catalysts.

Perovskite-type oxides:

Among metal oxide catalysts that have been investigated as soot removal catalysts, perovskite-type oxides show potential as effective soot oxidation catalysts due to their low cost and good thermal stability [57]. The general chemical formula for perovskite compounds is ABO_3, where "A" and "B" are two cations of very different sizes, and O is the anion that bonds to both, as shown in Fig. 15.10.

Among all the perovskites, La-based perovskites are the most widely studied [56, 76–78]. La-based perovskites containing Co at the B sites show high catalytic oxidation activities [79]. Partial substitution of La by K could increase the valence of Co, generate oxygen vacancies, and enhance the mobility of surface lattice oxygen, which markedly lowers the combustion temperature of soot [80–82]. $La_{1-x}K_xCoO_3$ series perovskites seem to be cost-effective catalysts for soot oxidation [83]. The catalytic oxidation property is related to the lattice oxygen mobility and oxygen desorption, which can be improved by substituting the trivalent La^{3+} cations at the A sites of the $LaCoO_3$ and/or changing the oxidation state of the B site cations [84].

The low surface area is one of the obstacles that hinder the further application of perovskite mixed oxides. Although many preparation methods have been used in attempts to increase the surface area of the massive mixed oxide, surface areas from 5 to at most $50\,m^2/g$ have been achieved. Hence, many efforts have been devoted to the preparation of perovskites supported on high surface area solids [56].

Compared with nanometric samples, macroporous catalysts showed higher catalytic activities for soot combustion [56]. The macroporous catalyst $La_{0.9}K_{0.1}CoO_3$ has been demonstrated to be an excellent candidate catalyst for soot combustion,

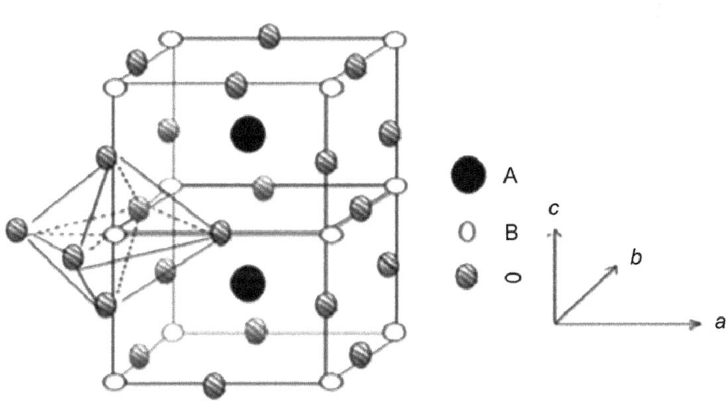

Fig. 15.10 The ABO_3 ideal cubic perovskite structure and the perovskite framework. Reproduced with permission from G.C. Dhal, D. Mohan, R. Prasad, Catal. Sci. Technol. 7 (2017), 1803–1825.

the activity of which even can be compared with supported Pt catalysts [85]. Three-dimensionally ordered macroporous (3DOM) $LaMn_{1-x}Fe_xO_3$ catalysts have shown good catalytic behavior for diesel soot combustion with selectivity values up to 97% [86].

15.2.4.6 Pathways to further improve the catalysis of soot combustion

As a heterogeneous reaction, the catalytic combustion of soot happens on a boundary of the gas (reaction gas)–solid (catalyst)–solid (soot) phase. The activity of the catalyst is not only affected by its intrinsic activity, but also the contact efficiency between the catalyst and soot plays an important role in the reaction [87, 88]. The pore size of the prevalent catalyst (<10 nm) generally is smaller than the diameter of soot particles (>25 nm), which often limits catalytic soot oxidation, because the soot cannot penetrate the inner pores of the catalyst [82].

More recently, 3DOM materials with large pore sizes (>50 nm) and a regular array of intersecting flow channels have been synthesized and found to be suitable for soot combustion [83, 89–92]. The well-ordered interpenetrating macroporous structure allows soot to go through the catalyst without resistance, and can make full use of the active sites of the catalyst. Hence, materials with a 3DOM structure can effectively enhance the contact between soot and the catalyst.

15.3 Catalyst preparation and methodology

The methods listed below are generally used to prepare a metal oxide catalyst for emission control: (1) solution combustion synthesis (SCS) method; (2) reactive grinding method; (3) sol–gel method; (4) coprecipitation method; (5) incipient wetness impregnation method; (6) ligated method; (7) impregnation method; (8) solid-state reaction method; (9) complex combustion method; and (10) colloidal crystal templating method. There is a noticeable tendency to synthesize catalysts with nano-sized structure and high surface area to obtain high catalytic activities. These two aims will be briefly discussed in the following sections.

15.3.1 Preparation of metal oxide catalysts with nano-sized structure

Techniques such as SCS [93], reactive grinding, and complex combustion [94, 95] are particularly suited for the production of nano-sized particles of catalyst.

Reactive grinding is characterized as high-energy mechanical synthesis or mechanochemical synthesis. Through this method a pure perovskite phase can be obtained at room temperature without thermal treatment. The mechanical impact reduces the precursor's crystallite size to nanoscale and provides a homogeneous mixture with enhanced solid-state diffusion [96]. High surface area perovskites (measured after calcination at 475 K), such as $LaCoO_3$ (>100 $m^2\,g^{-1}$), $LaGaO_3$ (98.6 $m^2\,g^{-1}$), $LaCoO_3$

(>110 $m^2 g^{-1}$), and $SrCoO_3$ (150 $m^2 g^{-1}$) are prepared consistently using various grinding additives [97]. Perovskite samples prepared by this method displayed high structural stability in a reducing atmosphere [98, 99].

15.3.2 Preparation of metal oxide catalysts with high surface area

The most widely used methods to synthesize high surface area metal oxide catalysts are the sol–gel method [46, 100], coprecipitation method [68, 101–103], microemulsion [79], impregnation method [79, 104], and solution combustion and calcination of LDHs [105].

Most recently, 3DOM structured catalysts gained the most attention due to their outstanding catalytic soot combustion performance. Typically, there are three methods that are widely used to synthesize 3DOM catalysts: (1) colloidal crystal templating method [83, 106–108]; (2) a modified emulsifier-free emulsion polymerization method [47, 85, 92]; and (3) combination of ethylene glycol ligation and solution combustion [109].

15.3.2.1 Impregnation method

Preparation of supported metal oxide catalysts is usually by an impregnation method [79, 104]. It is one of the simpler methods of preparing supported catalysts. During the impregnation process, an aqueous solution containing the metal precursor is brought into contact with a porous support, resulting in deposition of the metal oxide species, which is related to ion exchange or adsorption processes. Upon impregnation, many different processes take place at different rates: the particular adsorption species exchange between the charged surface and the electrolyte, driven by Coulomb force; and the surface of the solid dissolves partially in the electrolyte. The pH, nature of the solvent, and concentration of the dissolved substances are the main parameters affecting the solution. It is an efficient way to improve the surface area of catalysts and hence improve their catalytic activity.

15.3.2.2 Calcination of LDHs

Layered double oxides (LDO) synthesized by the calcination of LDHs possess a large specific surface area and display good thermal stability and high catalytic activity.

When heated, LDHs gradually lose interlayer water, accompanied to a large extent by dehydroxylation and decarbonation, resulting in the formation of hybrid oxides and three-dimensional (3D) networks (Fig. 15.11).

15.4 Outlook

In this chapter, topics regarding recent progress in diesel emission control using metal oxide catalysts were reviewed. On the catalytic material side, there are still some challenges remaining. In order to obtain high-performance catalysts, two ways can be followed: (1) synthesis of multicomponent catalysts based on a combination of metal

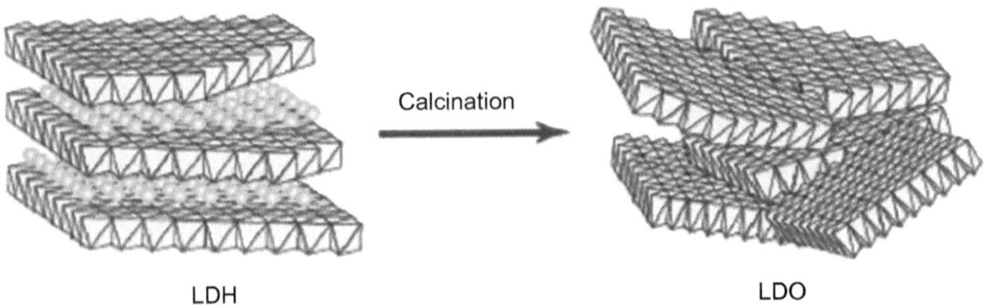

Fig. 15.11 Calcination process of LDHs via thermal treatment.
Reproduced with permission from D. Tichit, B. Coq, Catal. Technol. 7 (2003), 206–217.

oxides and (2) designing advanced nanoshapes with modified structural parameters that can result in interesting physicochemical properties. This has become a favored empirical way to design solid catalysts for application in automotive technology. Recent developments in in situ characterization techniques, molecular modeling, and novel synthesis procedures have helped to improve the catalytic activity of these catalysts. This trend can be expected to bring further progress and success for diesel emission control technologies.

Acknowledgments

Financial support from the National Materials Genome Project (2016YFB0700600) and Distinguished Young Scientists Program of the National Natural Science Foundation of China (NSFC51425301, 21374021, and 51673096) is gratefully acknowledged.

References

[1] E. Aneggi, C. de Leitenburg, A. Trovarelli, in: A. Trovarelli, P. Fornasiero (Eds.), Catalysis by Ceria and Related Materials, second ed., Imperial College Press, London, 2013, pp. 565–621.
[2] K. Vaaraslahti, J. Ristimäki, A. Virtanen, J. Keskinen, B. Giechaskiel, A. Solla, Environ. Sci. Technol. 40 (2006) 4776–4781.
[3] G.C. Dhal, D. Mohan, R. Prasad, Catal. Sci. Technol. 7 (2017) 1803–1825.
[4] X. Auvray, Fundamental studies of catalytic systems for diesel emission control, Doctoral thesis, Chalmers University of Technology, Göteborg, 2013.
[5] B. Choi, K.S. Lee, Chem. Eng. J. 240 (2014) 476–486.
[6] Y. Cheng, J. Liu, Z. Zhao, W. Song, Y. Wei, Chem. Eng. Sci. 167 (2017) 219–228.
[7] E. Jiaqiang, W. Zuo, J. Gao, Q. Peng, Z. Zhang, P.M. Hieu, Appl. Therm. Eng. 100 (2016) 356–366.
[8] N.D. Wasalathanthri, T.M. SantaMaria, D.A. Kriz, S.L. Dissanayake, C.H. Kuo, S. Biswas, S.L. Suib, Appl. Catal. B: Environmental 201 (2017) 543–551.
[9] T.M.P. Pham, Study on the impregnation procedures to prepare catalytic complexes for the treatment of motorbike's exhaust gases, PhD thesis, Ghent University, Ghent, 2013.

[10] L.M. Thang, Preparation of the catalyst for the total oxidation hydrocarbon applying for treatment of exhaust automobile gas, Master's thesis, Hanoi University of Science and Technology, Hanoi, 1999, pp. 1–26.
[11] G. Cocito, V. Patierno, Mater. Chem. 3 (1978) 233–254.
[12] B. Levasseur, S. Kaliaguine, Appl. Catal. B: Environmental 88 (2009) 305–314.
[13] R. Di Monte, J.J. Kašpar, Catal. Today 100 (2005) 27–35.
[14] T. Watanabe, K. Kawashima, Y. Tagawa, K. Tashiro, H. Anoda, K. Ichioka, S. Sumiya, G. Zhang, New DOC for light duty diesel DPF system, SAE Technical Paper Series 2007-01-1920, 2007.
[15] U. Lassi, Deactivation correlations of Pd/Rh three-way catalysts designed for Euro-IV emission limits—effect of ageing atmosphere, temperature and time, PhD thesis, University of Oulu, Oulu, Finland, 2003, p. 138.
[16] K. Yoshida, S. Makino, S. Sumiya, G. Muramatsu, R. Helferich, Simultaneous reduction of NO_x and particulate emissions from diesel engine exhaust, J. Org. Chem. 78 (1989) 4456–4462.
[17] P.G. Smirniotis, D.A. Pena, B.S. Uphade, Angew. Chem. Int. Ed. Engl. 40 (2001) 2479–2482.
[18] R.Q. Long, R.T. Yang, R. Chang, Chem. Commun. (2002) 452–453.
[19] G. Qi, L. Wang, R.T. Yang, Low-temperature selective catalytic reduction (SCR) of NO_x with NH_3 over zeolites and metal oxide-based catalysts and recent developments of H_2-SCR, in: I. Nova, E. Tronconi (Eds.), Urea-SCR Technology for deNO_x After Treatment of Diesel Exhausts, Springer, New York, 2014, pp. 149–177.
[20] (a) W.S. Kijlstra, D.S. Brands, H.I. Smit, E.K. Poels, A. Bliek, J. Catal. 171 (1997) 219–230; (b) G. Qi, R.T. Yang, J. Phys. Chem. B. 108 (2004) 15738–15747.
[21] G. Marbán, T. Valdes-Solis, A.B. Fuertes, J. Catal. 226 (2004) 138–155.
[22] E. Tronconi, I. Nova, C. Ciardelli, D. Chatterjee, M. Weibel, J. Catal. 245 (2007) 1–10.
[23] T.S. Park, S.K. Jeong, S.H. Hong, S.C. Hong, Ind. Eng. Chem. Res. 40 (2001) 4491–4495.
[24] B. Huang, R. Huang, D. Jin, D. Ye, Catal. Today 126 (2007) 279–283.
[25] H. Chang, X. Chen, J. Li, L. Ma, C. Wang, C. Liu, J.W. Schwank, J. Hao, Environ. Sci. Technol. 47 (2013) 5294–5301.
[26] F. Kapteijn, L. Singoredjo, A. Andreini, J.A. Moulijn, Appl. Catal. B: Environmental 3 (1994) 173–189.
[27] X. Tang, J. Li, L. Sun, J. Hao, Appl. Catal. B: Environmental 99 (2010) 156–162.
[28] M. Kang, E.D. Park, J.M. Kim, J.E. Yie, Catal. Today 111 (2006) 236–241.
[29] G. Qi, R.T. Yang, R. Chang, Appl. Catal. B: Environmental 51 (2004) 93–106.
[30] G. Qi, R.T. Yang, J. Catal. 217 (2003) 434–441.
[31] F.D. Liu, H. He, Catal. Today 153 (2010) 70 – 76.
[32] M. Casapu, O. Kröcher, M. Elsener, Appl. Catal. B 88 (2009) 413 – 419.
[33] H.Z. Chang, J.H. Li, X.Y. Chen, L. Ma, S.J. Yang, J.W. Schwank, J.M. Hao, Catal. Commun. 27 (2012) 54 – 57.
[34] D.A. Peña, B.S. Uphade, P.G. Smirniotis, J. Catal. 221 (2004) 421–431.
[35] G. Qi, R.T. Yang, Appl. Catal. B 44 (2003) 217–225.
[36] Z. Wu, B. Jiang, Y. Liu, Appl. Catal. B 79 (2008) 347–355.
[37] J. Li, H. Chang, L. Ma, J. Ha, R.T. Yang, Catal. Today 175 (2011) 147–156.
[38] D. Fino, S. Bensaid, M. Piumetti, N. Russo, Appl. Catal. A: General 509 (2016) 75–96.
[39] X. Song, J.H. Johnson, J.D. Naber, Int. J. Engine Res. 16 (2015) 738–749.
[40] (a) J. Suzuki, S. Matsumoto, Topics Catal. 28 (2004) 171–176; (b) A. Bueno-López, D. Lozano-Castelló, A.J. McCue, J.A. Anderson, Appl. Catal. B: Environmental 198 (2016) 266–275.

[41] R. Yang, Y. Gao, J. Wang, Dalton Trans. 43 (2014) 10317–10327.
[42] J. Liu, Z. Zhao, C.M. Xu, A. Duan, Appl. Catal. B 78 (2008) 61.
[43] D. Fino, N. Russo, G. Saracco, V. Specchia, J. Catal. 242 (2006) 38.
[44] D. Fino, N. Russo, G. Saracco, V. Specchia, Powder Technol. 180 (2008) 74.
[45] Q. Li, M. Meng, H. Xian, N. Tsubaki, X.G. Li, Y.N. Xie, T.D. Hu, J. Zhang, Environ. Sci. Technol. 44 (2010) 4747.
[46] Q. Li, M. Meng, F.F. Dai, Y.Q. Zha, Y.N. Xie, T.D. Hu, J. Zhang, Chem. Eng. J. 184 (2012) 106.
[47] X. Yu, J. Li, Y. Wei, Z. Zhao, J. Liu, B. Jin, A. Duan, G. Jiang, Ind. Eng. Chem. Res. 53 (2014) 9653–9664.
[48] M. Piumetti, S. Bensaid, N. Russo, D. Fino, Appl. Catal. B: Environmental 165 (2015) 742–751.
[49] J.M. Christensen, J.D. Grunwaldt, A.D. Jensen, Appl. Catal. B: Environmental 188 (2016) 235–244.
[50] S. Wagloehner, M. Nitzer-Noski, S. Kureti, Chem. Eng. J. 259 (2015) 492–504.
[51] Q. Shen, G. Lu, C. Du, Y. Guo, Y. Wang, Y. Guo, X. Gong, Chem. Eng. J. 218 (2013) 164–172.
[52] P. Sudarsanam, B. Hillary, B. Mallesham, B.G. Rao, M.H. Amin, A. Nafady, A. M. Alsalme, B.M. Reddy, S.K. Bhargava, Langmuir 32 (2016) 2208–2215.
[53] G. Mul, F. Kapteijn, C. Doornkamp, J.A. Moulijn, J. Catal. 179 (1998) 258–266.
[54] N. Guillén-Hurtado, A. García-García, A. Bueno-López, J. Catal. 299 (2013) 181–187.
[55] Y. Cheng, W. Song, J. Liu, H. Zheng, Z. Zhao, C. Shu, E.J.M. Hensen, ACS Catal. 7 (2017) 3883–3892.
[56] A. Mishra, R. Prasad, Catal. Rev. 56 (2014) 57–81.
[57] C. Lee, J.I. Park, Y.G. Shul, H. Einaga, Y. Teraoka, Appl. Catal. B: Environmental 174 (2015) 185–192.
[58] A. Bueno-López, Appl. Catal. B: Environmental 146 (2014) 1–11.
[59] A.L. Carrascull, M.I. Ponzi, E.N. Ponzi, Ind. Eng. Chem. Res. 42 (2003) 692–697.
[60] B. Van Setten, J.M. Schouten, M. Makkee, J. Moulijn, Appl. Catal. B: Environmental 28 (2000) 253–257.
[61] L. Castoldi, R. Matarrese, L. Lietti, P. Forzatti, Appl. Catal. B: Environmental 90 (2009) 278–285.
[62] Q. Wang, Z. Guo, J.S. Chung, Mater. Res. Bull. 44 (2009) (1973).
[63] Q. Wang, J.S. Chung, Z. Guo, Ind. Eng. Chem. Res. 50 (2011) 8384–8388.
[64] Z.P. Wang, M.X. Chen, W.F. Shangguan, Acta Phys. Chim. Sin. 25 (2009) 79–85.
[65] M. Machida, Y. Murata, K. Kishikawa, D.J. Zhang, K. Ikeue, Chem. Mater. 20 (2008) 4489–4494.
[66] W.B. Li, X.F. Yang, L.F. Chen, J.A. Wang, Catal. Today 148 (2009) 75–80.
[67] Q. Li, M. Meng, Z.Q. Zou, X.G. Li, Y.Q. Zha, J. Hazard. Mater. 161 (2009) 366–372.
[68] Z. Wang, Q. Li, L. Wang, W. Shangguan, Appl. Clay Sci. 55 (2012) 125–130.
[69] F. Dai, M. Meng, Y. Zha, Z. Li, T. Hu, Y. Xie, J. Zhang, Fuel Proc. Technol. 104 (2012) 43–49.
[70] W.F. Shangguan, Y. Teraoka, S. Kagawa, Appl. Catal. B: Environmental 8 (1996) 217–227.
[71] P. Stelmachowski, G. Maniak, A. Kotarba, Z. Sojka, Catal. Commun. 10 (2009) 1062–1065.
[72] S.K. Megarajan, S. Rayalu, Y. Teraoka, N. Labhsetwar, J. Mol. Catal. A: Chemical 385 (2014) 112–118.
[73] J. Chen, W. Shi, S. Yang, H. Arandiyan, J. Li, J. Phys. Chem. C 115 (2011) 17400–17408.

[74] J. Li, X. Liang, S. Xu, J. Hao, Appl. Catal. B: Environmental 90 (2009) 307–312.
[75] X. Niu, L. Zhou, X. Hu, W. Han, RSC Adv. 5 (2015) 52595–52601.
[76] H. Zhu, P. Zhang, S. Dai, ACS Catal. 5 (2015) 6370–6385.
[77] L. Wang, S. Fang, N. Feng, H. Wan, G. Guan, Chem. Eng. J. 293 (2016) 68–74.
[78] C.H. Kim, G. Qi, K. Dahlberg, W. Li, Science 327 (2010) 1624–1627.
[79] H. Wang, J. Liu, Z. Zhao, Y. Wei, C. Xu, Catal. Today 184 (2012) 288–300.
[80] S. Fang, L. Wang, Z. Sun, N. Feng, C. Shen, P. Lin, H. Wan, G. Guan, Catal. Commun. 49 (2014) 15–19.
[81] Z. Li, M. Meng, Q. Li, Y. Xie, T. Hu, J. Zhang, Chem. Eng. J. 164 (2010) 98–105.
[82] X. Guo, M. Meng, F. Dai, Q. Li, Z. Zhang, Z. Jiang, S. Zhang, Y. Huang, Appl. Catal. B 142 (2013) 278–289.
[83] N. Feng, C. Chen, J. Meng, Y. Wu, G. Liu, L. Wang, H. Wan, G. Guan, Catal. Sci. Technol. 6 (2016) 7718–7728.
[84] C. Lee, Y. Jeon, S. Hata, J.I. Park, R. Akiyoshi, H. Saito, Y. Teraoka, Y.G. Shul, H. Einaga, Appl. Catal. B: Environmental 191 (2016) 157–164.
[85] G. Zhang, Z. Zhao, J. Liu, J. Xu, Y. Jing, A. Duan, G. Jiang, J. Rare Earths 27 (2009) 955–960.
[86] J. Zheng, J. Liu, Z. Zhao, J. Xu, A. Duan, G. Jiang, Catal. Today 191 (2012) 146–153.
[87] J.P. Neeft, O.P. van Pruissen, M. Makkee, J.A. Moulijn, Appl. Catal. B: Environmental 12 (1997) 21–31.
[88] N. Guillén-Hurtado, A. García-García, A. Bueno-López, Appl. Catal. B: Environmental 174 (2015) 60–66.
[89] G. Zhang, Z. Zhao, J. Liu, G. Jiang, A. Duan, J. Zheng, S. Chen, R. Zhou, Chem. Commun. 46 (2010) 457–459.
[90] Y. Wei, J. Liu, Z. Zhao, A. Duan, G. Jiang, J. Catal. 287 (2012) 13–29.
[91] J. Wang, G. Yang, L. Cheng, E.W. Shin, Y. Men, Catal. Sci. Technol. 5 (2015) 4594–4601.
[92] Y. Wei, Z. Zhao, X. Yu, B. Jin, J. Liu, C. Xu, A. Duanab, G. Jianga, S. Ma, Catal. Sci. Technol. 3 (2013) 2958–2970.
[93] N. Russo, D. Fino, G. Saracco, V. Specchia, Catal. Today 137 (2008) 306–311.
[94] S. Kaliaguine, A. Van Neste, V. Szabo, J.E. Gallot, M. Bassir, R. Muzychuk, Appl. Catal. A: General 209 (2001) 345–358.
[95] N. Russo, D. Fino, G. Saracco, V. Specchia, J. Catal. 229 (2005) 459–469.
[96] X. Peng, H. Lin, W. Shangguan, Z. Huang, Catal. Commun. 8 (2007) 157–161.
[97] K. Wang, L. Qian, L. Zhang, H. Liu, Z. Yan, Catal. Today 158 (2010) 423–426.
[98] K. Hamamoto, Y. Fujishiro, M. Awano, Solid State Ionics 177 (2006) 2297–2300.
[99] D.B. Kittelson, J. Aerosol Sci. 29 (1998) 575–588.
[100] Q. Li, M. Meng, N. Tsubaki, X. Li, Z. Li, Y. Xie, T. Hu, J. Zhang, Appl. Catal. B 91 (2009) 406–415.
[101] S. Sumathi, S. Bhatia, K.T. Lee, A.R. Mohamed, J. Hazard. Mater. 176 (2010) 1093–1096.
[102] P.R. Ettireddy, N. Ettireddy, S. Mamedov, P. Boolchand, P.G. Smirniotis, Appl. Catal. B: Environmental 76 (2007) 123–134.
[103] W. Xu, Y. Yu, C. Zhang, H. He, Catal. Commun. 9 (2008) 1453–1457.
[104] Z. Liu, C. He, B. Chen, H. Liu, Catal. Today 297 (2017) 78–83. *https://doi.org/10.1016/j.cattod.2017.05.074*.
[105] D. Tichit, B. Coq, Catal. Technol. 7 (2003) 206–217.
[106] G. Mu, C. Liu, Q. Wei, Y. Huang, Mater. Lett. 181 (2016) 161–164.

[107] N. Feng, C. Chen, J. Meng, G. Liu, F. Fang, J. Ding, L. Wang, H. Wan, G. Guan, Catal. Sci. Technol. 7 (2017) 2204–2212.
[108] J. Xu, J. Liu, Z. Zhao, J. Zheng, G. Zhang, A. Duan, G. Jiang, Catal. Today 153 (2010) 136–142.
[109] Y. Lu, A. Eyssler, E.H. Otal, S.K. Matam, O. Brunko, A. Weidenkaff, D. Ferri, Catal. Today 208 (2013) 42–47.

Further reading

[110] D. Fino, P. Fino, G. Saracco, V. Speccia, Appl. Catal. B: Environmental 43 (2003) 243–259.

Index

Note: Page numbers followed by *f* indicate figures, and *t* indicate tables.

A

A_2BO_4, 38–39, 39*t*
Activated carbon fibers (ACF), 400
Al-ion batteries, 161
Alkali-metal oxides
  ion batteries, 155
  soot combustion, 405
Alkaline fuel cell (AFC), 17–19
Alkali tantalates ($ATaO_3$), for CO_2 reduction, 296–297
α-Hematite (α-Fe_2O_3) semiconductor
  APCVD, 232, 233*f*
  Co-Pt cocatalyst, 236, 237*f*
  doped and undoped α-Fe_2O_3 thin film electrodes, 232, 234*f*
  FeOOH, 236
  Mg-doped α-Fe_2O_3 thin film, 234–236
  photocurrent density *vs.* applied potential, 234–236, 235*f*
  photoelectrodeposition, 236
  STH, 231–232
  translucent Si-doped hematite thin films, 232, 233*f*
  undoped hematite thin films, 232
  USP, 232, 233*f*
  WO_3/Fe_2O_3 heterojunction electrode, 232–234
Alumina (Al_2O_3)
  DOCs, wash coat material for, 394–395
  TWCs, 393
Anatase TiO_2 nanotube arrays, 161
Anodically coloring oxides
  NiO, 354
  V_2O_5, 354
Antireflective coating (ARC), 347
Aqueous rechargeable lithium batteries (ARLBs)
  metal oxide-based negative electrodes, 154–155, 154*f*
  metal oxide-based positive electrodes
    $LiCoO_2$, 151
    $LiMn_2O_4$ nanochains, 151–152, 153*f*
  principles and applications, 150–151
Artificial photosynthesis, 321–322
A-site deficiency, 34–35
Atmospheric pressure chemical vapor deposition (APCVD), 232, 233–234*f*

B

$BaPb_{1-x}Bi_xO_3$ system, 364, 364–365*f*
Bardeen–Cooper–Schrieffer (BCS)-type superconductors, 370–371
Bimorph transducer design optimization, 93
Binary metal oxides, for hydrogen storage, 268–271
Bi_2O_3, 29–30, 30–31*f*
Biodiesel, 303
  energy conversion, 9
  simultaneous transesterification and esterification, 305, 306*f*
  solid acid catalyst production
    calcination temperature, 311–312
    catalyst stability, 313–315
    coprecipitation, 310
    hydrothermal autoclaving method, 306–307
    incipient wetness impregnation, 309
    pH value, 311
    self-assembly processing, 305–306
    sol-gel process, 307–309
    textural characteristics, 310–311
  synthesis methods, 304
Biogas, 5
Biophotosynthesis, 321–322
Bismuth vanadate ($BiVO_4$) semiconductor
  charge separation, 229, 230*f*
  disadvantages, 226–227
  facet-dependent photocatalytic activity, 229
  H-$BiVO_4$, 229
  metal–organic decomposition, 227
  Mo-doped $BiVO_4$ electrode, 227

Bismuth vanadate ($BiVO_4$) semiconductor *(Continued)*
molybdenum doping, 228–229
photoanode, 228–229
photocatalysis, 227
porous bismuth vanadate electrode, 228–229
SEM images, 227, 227–228*f*
$WO_3/BiVO_4$ heterojunctions, 229–231, 231*f*
Bi-Sr-Ca-Cu-O (BSCCO) system
generators, 384
resistivity ratio *vs.* temperature, 373–375, 375*f*
transformer, 384
wires/tapes, 382
Buck converter, 118, 118*f*

C

Candida rugose lipase (CRL), 311
Carbene pathway, of CO_2 reduction, 288*f*, 289
Carbon coating, 138
Carnot efficiency, 73
Cathodic electrochromic smart windows
MoO_3, 353–354
Nb_2O_5 and Ta_2O_5, 354
TiO_2, 353
WO_3, 352–353
The Century of Environmental Management, 91
CeO_2-UO_2 MOX nuclear fuel
performance
heat capacity, 82–84, 82*f*
thermal conductivities, 82–84, 83*f*
thermal diffusivities, 82–84, 83*f*
preparation method, 82
Ceramic fuels, 74
Ceria (CeO_2)
anode materials, 20–21, 21*f*
electrolyte materials, 27–29, 28–29*f*
soot combustion, 404–405
TWCs, 393
two steps thermochemical metal-oxide cycles, 331–332
Chemical bath deposition (CBD), 228–229, 239
Chemical-looping thermochemical metal-oxide cycles, 334, 334*f*
Chemical precipitation, 2, 176
Chemical vapor deposition (CVD), 1, 176
Co-assembly processing, 306, 307*f*
Coating
LIB, metal oxide-based negative electrodes, 144–147, 146*f*
LIB, metal oxide-based positive electrodes
Al_2O_3 coating layer, 138–139
$AlPO_4$-coated $LiCoO_2$ electrodes, 136–137, 137*f*
carbon coating, 138
GPE coatings, 138
$LiCoO_2$, 136–138
$LiCoPO_4$-coated $LiCoO_2$ electrodes, 136–137
Li_2MnO_3, 138–139
$LiMn_2O_4$ electrode, 137
V_2O_5 nanofibers, 138
Ni/MH batteries, 130–131
Cobalt hydroxide ($Co(OH)_2$) supercapacitors, 183–184
Cobalt oxide (Co_3O_4) supercapacitors, 182–183, 182*f*
Cocatalysts, 242
Composite oxides, for supercapacitors, 187–188
Continuously regenerating trap (CRT) technology, 401
Copper oxide, for CO_2 reduction, 294–295
Copper(I) oxide (Cu_2O) semiconductor
carbon-protected Cu_2O NW arrays, 239, 239*f*
CuO, protective coating, 239
HER catalysts, 238–239
photocurrent, 236–239
p-type semiconductor, 236
surface-protected Cu_2O electrode, 236–238, 238*f*
Coprecipitation, 310
CO_2 reduction
alkali tantalates, 296–297
carbene pathway, 288*f*, 289
copper oxide, 294–295
elementary reactions, 283–291
energetic requirements, 283–291
formaldehyde pathway, 288*f*, 289
glyoxal pathway, 288*f*, 289–291
heterogeneous photocatalyst systems
material classes, 276–277, 277*f*

processes, schematic representation of, 278, 279*f*
indium oxide, 293–294
industrial methanol synthesis, 275
strontium titanate, 295–296
Cuprate-based superconductors, 393
Cymbal energy harvesting process, 122*t*
Cymbal transducer
energy flow analysis, 108–110, 113, 117*t*
multilayered cymbal
cross-sectional view, 118–119, 119*f*
load resistance dependence, 119, 120*f*
output power characteristics, 118–119, 119*f*
operation principle, 108–110, 109*f*

D

DC-DC converter
buck converter, 118, 118*f*
switching regulator, 117, 118*f*
total energy flow, 122
Decladding, 76–77
Diesel NO_x removal catalyst technology, 266–268
Diesel oxidation catalysts (DOCs), 394–395
Diesel particulate filters (DPFs), 391, 400–402
Diesel particulate NO_x reduction (DPNR) catalyst, 401–402, 401*f*
Direct methanol fuel cell (DMFC), 17–19
Doping
heteroatom, 6
LIB
metal oxide-based negative electrodes, 147, 148*f*
metal oxide-based positive electrodes, 139–141
Ni/MH batteries, 131–132
TiO_2 semiconductor, 214–216

E

Electrical conductivity, of metal oxide, 52–53
Electrical-electrical energy transfer, 107–108, 107*f*
DC-DC converter
buck converter, 118, 118*f*
switching regulator, 117, 118*f*
multilayered cymbal, 118–119, 119–120*f*
piezoelectric transformer, 120–121, 120–121*f*
Electrical impedance matching, 122
Electric double-layer capacitors (EDLCs), 170–171, 170*f*
Electric vehicles (EVs), 5–6
Electrochemical capacitors. *See* Supercapacitors
Electrochemical processing, 76–77, 77*f*
Electrochromic smart windows
anodically coloring oxides
NiO, 354
V_2O_5, 354
cathode materials
MoO_3, 353–354
Nb_2O_5 and Ta_2O_5, 354
TiO_2, 353
WO_3, 352–353
devices, schematic of, 350–352, 352*f*
electrochromic performance, 356–358
preparation methods
chemical vapor deposition, 355
inkjet printing, 356
physical vapor deposition, 355–356
sputtering deposition, 355
Electrodeposition, 1, 176
Electrolysis, 321–322
Electromechanical coupling factor, 93, 99–100, 99*f*, 121
Electrorefining process, 76–77
Electrowinning, 76–77
Emission control technologies
diesel soot and NO_x emissions, 400–408
DOCs, 394–395
metal oxide catalyst preparation
high surface area, 409
nano-sized structure, 408–409
TWCs, 392–394
Energy and environment
CO_2
emissions, 11
reduction, 12
emission control, 11–12
Pt catalysts, 11
TiO_2, 11
Energy conversion
biodiesel production, 9
energy-saving smart windows, 10

Energy conversion *(Continued)*
hydrogen economy, 8–9
superconductors, 10–11
thermochemical cycles, 9
Energy generation
fuel cells, 3
fuel combustion, 5
motion energy harvesting, 4
nuclear reactors, 4
solar cells, 2–3
thermoelectric generators, 3
Energy harvesting, 4
Energy-saving smart windows
electrochromic materials, 350–358
thermochromic materials, 341–350
Energy source
coal, oil, and natural gas, 205
photocatalysis, 205–206
solar energy, 205–206
world primary energy consumption, 205, 206*f*
Energy storage
batteries, 5–7
supercapacitors, 6–8
Energy transmission coefficient, 100–103, 101–102*f*
Excited-state acid-base mechanism, 291

F

Face electronics, 95
Faradaic supercapacitors. *See* Pseudocapacitors
Fatty acid methyl ester (FAME), 304–305, 310, 312, 314
Ferrites-based thermochemical metal-oxide cycles, 330
Formaldehyde pathway, of CO_2 reduction, 288*f*, 289
Free fatty acid (FFA), 304, 309, 312, 314–315
Fuel cells
classification, 17–19
components, 17
energy generation, 3
fuels and oxidants, 17
reactions and processes, 17, 18*f*
SOFC (*see* Solid oxide fuel cell (SOFC))
Fuel combustion, 5

G

Gel polymer electrolyte (GPE) coatings, 138
Glyoxal pathway, of CO_2 reduction, 288*f*, 289–291
Graphene-based composite, 242
Graphene-metal oxide nanocomposites, 7
Green electrolytes, 175

H

Hematite, 184–185
Hercynites-based thermochemical metal-oxide cycles, 330–331
Heteroatom doping, 6
Heterogeneous base catalysts, 303
Heterogeneous mesostructured acid catalysts, 303
Heterojunction structure, 241–242
Hg-Ba-Ca-Cu-O (HBCCO) system, 377–379, 379*f*
High surface area metal oxide catalysts, 409
High-T_c superconductors (HTSs), 375–377, 382–384
High-temperature electrochemical fuel processing, 76–77, 77*f*
High-temperature solid-state method, 1, 176
Hollow mesoporous tungsten trioxide microspheres (HMTTS), 43, 43*f*
Homogeneous acid catalysts, 303
Human motion, 94–95, 94–95*f*
Hybrid energy harvesting, 122–123, 123*f*
Hybrid supercapacitors, 172, 173*f*
Hydrogen economy, 8–9
Hydrogen evolution reaction (HER) catalysts, 238–239
Hydrogen storage
binary metal oxides, 268–271
Co_2NiO doping
MgH_2-Li_3AlH_6 systems, 261
milled MgH_2, 259–260
hydrogen economy, 251–252
MgH_2 doping, 254–256
MgH_2-$LiAlH_4$+5mol% $MnFe_2O_4$ samples, 262, 263*f*
Nb_3O_7F HMFs, 268–271, 270*f*
renewable power supply, 251
SnO_2, 266–268, 267*f*
storing approaches, 252
technical and economic targets, 252, 252*t*

TiO_2, 262–265, 264–265*f*
ZnO, 265–266, 266*f*
$Zn_2V_2O_7$ HNs, 268, 269*f*
Hydrogen-treated $BiVO_4$ (H-$BiVO_4$) photoanodes, 229
Hydrothermal autoclaving method, 306–307
Hydrothermal method, 1, 176

I

Impregnation technique
microstructures, 78, 79*f*
supported metal oxide catalysts, 409
ThO_2–4% UO_2 pellets fabrication procedure, 77–78, 78*f*
Incident photon-to-current efficiency (IPCE), 216–219, 219*f*
Incipient wetness impregnation, 309
Indium oxide, for CO_2 reduction, 293–294
Iron oxides, as supercapacitors, 184–185, 185*f*
Isothermal thermochemical cycles, 333

J

Josephson effect. *See* Tunneling effect
Joule heating effect, 51–52

K

$K_2Ti_2O_5$, for soot combustion, 405

L

La-Ba-Cu-O (LBCO) system, 365–367, 367*f*
$LaCoO_3$, 37
Lacunary polyoxometalate (LPOM)-HMS hybrid catalyst, 313
$La_{0.9}K_{0.1}CoO_3$, for soot combustion, 407–408
Langmuir sublimation model, 325–326
Lanthanum gallate ($LaGaO_3$), 31–34, 32–34*f*, 32*t*
Lanthanum manganites ($LaMnO_3$), 35–37, 37–38*f*
Lanthanum strontium manganite (LSM), 35–37
$La_{0.75}Sr_{0.25}Cr_{0.5}Fe_{0.5}O_{3-\alpha}$ (LSCrF), 22
$La_{1-x}Sr_xCoO_{3-\delta}$ (LSC), 37
$La_{1-x}Sr_xCoO_{1-y}FeyO_3$ (LSCF), 37
$La_{1-x}Sr_xGa_{1-y}MgyO_{3-\delta}$ (LSGM), 31–34, 32*f*
Layered double hydroxides (LDHs)
calcination process, 409, 410*f*
soot oxidation, 405
structure, 406*f*
Layered double oxides (LDO), 409
LIB. *See* Lithium-ion batteries (LIB)
Light harvesting
band structure modification, 240
nanostructure, 241
surface modification, 241
Lithium-ion batteries (LIB)
energy storage, 5–7
metal oxide-based negative electrodes
alloy reactions, 143
coating, 144–147, 146*f*
doping, 147, 148*f*
intercalation compounds, 143
$Li_4Ti_5O_{12}$, 143–144
Li_3VO_4, 144
morphology control, 148–150
titanium dioxide, 144
transition metal oxides, 143
metal oxide-based positive electrodes
coating, 136–139, 137*f*
doping, 139–141
$LiCoO_2$, 134
Li_2MnO_3, 135
$LiMn_2O_4$, 134–135
$LiNi_{1/3}Co_{1/3}Mn_{1/3}O_2$, 134
$LiNiO_2$, 134
lithium mixed metal oxides, 134
molybdenum trioxide (MoO_3), 135–136
morphology control, 141–143, 141*f*
transition metal oxides, 134
V_2O_5, 135–136
principles and applications, 133–134
Li-Ti-O ternary system, 362, 362*f*
Low-temperature superconducting (LTS) materials, 382

M

Machinery vibration (resonance usage), 93–94
Macrofiber composite (MFC)
mechanical-electrical energy transduction
d_{31} mode type, 113, 114*f*
electrical output measurement setup, 115, 115*f*
neutral line position and bending displacement, 114–115, 115*t*

Macrofiber composite (MFC) *(Continued)*
output voltage signal, 115–116, 116*f*
rectified voltage and output power, 115–116, 116*f*
stress distribution, 113–114, 114*f*
mechanical-to-mechanical energy transfer, 110, 110*f*
Magnesium hydride, for hydrogen storage, 253–254
Magnetically active ordered mesoporous resin (MOMR)-SO_3H catalyst, 312
Magnetoelectric devices, 122–123, 123*f*
Manganese dioxide (MnO_2), 178–179
Mechanical/acoustic impedance, 104
Mechanical-electrical energy transduction, 107–108, 107*f*
design optimization
cymbal, 111–113, 112–113*f*, 117*t*
flexible transducer (*see* Macrofiber composite (MFC))
energy flow analysis, 116, 117*t*
FOM, 110–111
piezoelectric materials selection, 111, 112*t*
Mechanical impedance mismatch, 121
Mechanical-mechanical energy transfer, 107–108, 107*f*
Cymbal transducer, 108–110, 109*f*, 117*t*
mechanical impedance matching, 108
MFC, 110, 110*f*
vibration source examples, 108, 109*f*
Mechanical quality factor, 103–104
Meissner effect. *See* Perfect diamagnetism
Mesoporous solid acid catalysts, 304, 314
Metal hydrides, surface modifiers for, 254–262
Metal-insulator-metal (MIM) structures, 262–264
Metal-organic deposition (MOD) method, 228–229
Metal oxide-based thermoelectric materials
advantages, 49
$Ca_3Co_4O_9$
Ag doping, 62
bismuth doping, 62
crystal structure, 60, 60*f*
figure of merit, 62, 63*f*
gallium/lead addition, 62
rhodium doping, 62
thermoelectric properties, 60–62, 61*f*
vapor phase methods, 60–62
$CaMnO_3$
crystal structure, 64–66, 66*f*
dysprosium doping, 66
lanthanum doping, 66
niobium doping, 66
thermoelectric properties, 66, 67*f*
ytterbium doping, 66
Na_xCoO_2
crystal structure, 54–55, 54*f*
figure of merit, 59–60, 59*f*
metallic dopants, 58
nanoblock integration, 54–55
Ni dopants, 58
properties, 56, 57*f*
$Na_xCo_2O_4$
Ag-doped sample, 56–58
bronze-type layered structure, 58
fabrication cost, 55
Ni dopants, 58
polycrystalline samples, 55
Seebeck coefficient, 56–58
thermoelectric properties, 56–58
XRD patterns, 55, 56*f*
Zn dopants, 59
$SrTiO_3$
crystal structure, 62, 64*f*
dysprosium doping, 64
La-doped $SrTiO_3$ ceramics, 62–64
niobium effect, 64
potassium titanate (KTO), 64
superlattice Ruddlesden-Popper structure, 62–64
thermoelectric effect
electrical conductivity of metal oxide, 52–53
Peltier effect, 49–50, 50*f*
power conversion efficiency, 50–51
Seebeck effect, 49–50, 50*f*, 52
thermal conductivity, 53–54
thermoelectric figure of merit, 49–52, 51*f*
Thomson effect, 49–50
ZnO, 68–69
Metal oxides
in energy technologies (*see* Energy conversion; Energy generation; Energy storage)
in PEMFCs

SiO_2 and composites, 40–42, 42*f*
TiO_2 and composites, 39–40, 40–41*f*
WO_3 and composites, 42–43, 43*f*
preparation methods, 1–2
in SOFC (*see* Solid oxide fuel cell (SOFC))
in supercapacitors
cobalt hydroxide ($Co(OH)_2$), 183–184
cobalt oxide (Co_3O_4), 182–183, 182*f*
composite oxides, 187–188
iron oxides, 184–185, 185*f*
MnO_2, 178–179
nickel hydroxide ($Ni(OH)_2$), 181
nickel oxide (NiO), 179–181, 180*f*
preparation methods, 176
RuO_2, 177–178, 177*f*
SnO_2, 186
TiO_2, 185–186
V_2O_5, 186–187
MFC. *See* Macrofiber composite (MFC)
MgH_2-ZrO_2 composites
hydrogenation isotherms, 255–256, 256*f*
SEM images, 254–255, 255*f*
TEM images, 254–255, 255*f*
Mg-ion batteries
metal oxide-based negative electrodes, 159–160, 160*f*
metal oxide-based positive electrodes, 159
principles and applications, 158
Micronized master blend (MIMAS) process, 74
Mixed ionic and electronic conductivity (MIEC), 20–21
Mixed metal oxides (MMOs), 405–406
Mixed oxide (MOX) fuels, 4. *See also* MOX nuclear fuels
Molten carbonate fuel cell (MCFC), 17–19
Molybdenum trioxide (MoO_3)
electrochromic smart windows, 353–354, 353*f*
LIBs, 135–136
Monobasic metal oxide catalysts, 404
Monolithic exhaust gas catalyst, 394, 395*f*
MoO_2 hollow nanostructures, 147
Morphology control
LIB
metal oxide-based negative electrodes, 148–150
metal oxide-based positive electrodes, 141–143, 141*f*
Ni/MH batteries, 132
Mott insulators, 365
MOX nuclear fuels
CeO_2-UO_2 MOX nuclear fuel, 81–84, 82–83*f*
reactor-grade plutonium, 74
ThO_2-PuO_2 MOX nuclear fuel
performance, 84, 85*f*
recycling, 84–86
Th-MOX, 84
thorium dioxide *vs.* uranium dioxide, 84
U-Pu MOX nuclear fuel
performance, 75
preparation method, 74–75
recycling, 76–77
U-Th MOX nuclear fuel
performance, 79–81
preparation method, 77–79

N

Na-ion batteries (NIBs)
metal oxide-based negative electrodes, 157–158
metal oxide-based positive electrodes, 155–157
principles and applications, 155
Nano-harvesting, 96
Nanosized metal oxides, 7
Nanostructure control, 242
$Na_xCoO_2.yH_2O$, 380–381, 381*f*
Nb_2O_5, electrochromic smart windows, 354
Nb_3O_7F hollow microflowers
hydrogen storage, 268–271, 270*f*
NIBs. *See* Na-ion batteries (NIBs)
Nickel hydroxide ($Ni(OH)_2$), 181
Nickel metal hydride battery, 5–6
Nickel oxide (NiO)
electrochromic smart windows, 354, 358, 358*f*
supercapacitors, 179–181, 180*f*
$NiCo_2O_4$, 187–188
Ni/MH batteries
metal oxide-based positive electrodes
coating and additives, 130–131
doping, 131–132
morphology control, 132
nickel hydroxide polymorphs, 128–129, 129*f*

Ni/MH batteries *(Continued)*
nonspherical nickel hydroxide powders, 133
paste-type electrodes, 128
PTFE, 132
Q-3D substrate electrodes, utilization efficiency, 132–133
sintered-type electrodes, 128
spherical nickel hydroxide particles, 133
principles and applications, 127–128
Nonvolatile metal oxides cycles, 323–324
NSR catalyst, 401–402
Nuclear fuels
fission product, 73–74
MOX fuels (*see* MOX nuclear fuels)
Nuclear reactors, 4, 73

O

Optimized co-milling process (OCOM), 74–75, 75*f*
Oxide feed preparation, 76–77
Oxyapatite, 34–35, 35*f*, 36*t*
Oxygen transportation mechanisms, 402–403, 403*f*
Oxy-pnictides (LaOFeP), 380, 380*f*

P

Particulate matter (PM) emissions, 391, 401, 406–407
Peltier effect, 49–50, 50*f*
PEMFC. *See* Polymer electrolyte membrane fuel cell (PEMFC)
Perfect diamagnetism, 361–362
Perovskites
cation layered double perovskite oxides, 22
cell voltage and power density, 22, 25*f*
crystal structure, 21–22, 22*f*
$La_{0.75}Sr_{0.25}Cr_{0.5}Mn_{0.5}O_3$ (LSCM), 22, 24*f*
methane conversion and gas formation rate, 22, 24*f*
pO_2/pCH_4 ratio, 22, 25*f*
$Pr_{0.5}Ba_{0.5}MnO_3$, 22, 23–24*f*
reactive grinding synthesis, 408–409
soot oxidation catalysts, 407–408, 407*f*
TWCs, 394
two steps thermochemical metal-oxide cycles, 332–333
Phosphoric acid fuel cell (PAFC), 17–19
Photocatalysis, 205–206
Photocatalytic CO_2 reduction. *See* CO_2 reduction
Photocatalytic water splitting, 206–207
Photoelectrocatalysis, 321–322
Photoelectrochemical (PEC) water oxidation, 206–207
Photo-thermochemical cycles (PTCs), 334–335, 335*f*
Piezoelectric energy harvesting systems
constitutive piezoelectric equations, 98
"ecology and energy harvesting" share, 92
electrical engineers' approach, 95–96
energy transfer phases (*see* Electrical-electrical energy transfer; Mechanical-electrical energy transduction; Mechanical-mechanical energy transfer)
figures of merit (*see* Piezoelectric figures of merit)
mechanical engineers' approach
bimorph transducer design optimization, 93
electromechanical coupling factor, 93
human motion, 94–95, 94–95*f*
machinery vibration (resonance usage), 93–94
MEMS engineers' approach, 96
military application, 96–98, 97*f*
necessity, 91
vs. passive vibration damping, 91–92, 92*f*
piezoelectric passive damper, 104–107, 105–106*f*
research papers, 93
total energy flow, 121–122, 122*t*
vibration damping change with external resistance change, 91, 92*f*
Piezoelectric figures of merit
efficiency, 103, 103*f*
electromechanical coupling factor, 99–100, 99*f*
energy transmission coefficient, 100–103, 101–102*f*
mechanical/acoustic impedance, 104
mechanical quality factor, 103–104
piezoelectric strain constant, 98
piezoelectric voltage constant, 98
Piezoelectric strain constant, 98
Piezoelectric voltage constant, 98

Piezoelectric windmills, 94, 94*f*
p-n junction, 241
Polyethylene glycol-assisted method, 315
Polymer electrolyte membrane fuel cell (PEMFC), 17–19
  SiO_2 and composites, 40–42, 42*f*
  TiO_2 and composites, 39–40, 40–41*f*
  WO_3 and composites, 42–43, 43*f*
Polymeric mesoporous $ZnAl_2O_4$ catalyst, 315, 315*f*
Power conversion efficiency, 50–51
Programmable air-bust munition (PABM), 96–98, 97*f*
Pseudocapacitors, 171, 172*f*
PTCs. *See* Photo-thermochemical cycles (PTCs)
Pyrochemical reprocessing, 76, 76*f*
Pyroprocessing, 76–77, 77*f*

Q

Quantum size effect, 281*f*, 282
Quasi-three-dimensional (Q-3D) substrate electrodes, 132–133

R

Reactive grinding method, 408–409
Reactive-ion etching (RIE) approach, 219–220
Rechargeable batteries, 5–6
Redox mechanism, 403
Reduced graphene oxide (rGO), 294–295
Regenerated mixture (REMIX) fuel, 4
Rolling assisted biaxially textured substrates (RABiTS), 383–384
Ruthenium oxide (RuO_2), 177–178, 177*f*

S

Samarium-doped CeO_2 (SDC), 28–29
Seebeck effect, 49–50, 50*f*, 52
Selective catalytic reduction (SCR), 391, 395–400, 397*f*
Self-assembly processing, 305–306
Semiconductors, solar water splitting
  and electrolyte interface, 212, 213*f*
  energy band, 210–212, 210–211*f*
  metal oxides
    $BiVO_4$ (*see* Bismuth vanadate ($BiVO_4$) semiconductor)
    Cu_2O (*see* Copper(I) oxide (Cu_2O) semiconductor)
    α-Fe_2O_3 (*see* α-Hematite (α-Fe_2O_3) semiconductor)
    TiO_2 (*see* Titanium dioxide (TiO_2))
    WO_3 (*see* Tungsten trioxide (WO_3))
  nonmetal oxides, 239–240
  stability, 213–214
Simultaneous NO_x and soot removal, 403–404, 403*f*
Smart windows. *See also* Electrochromic smart windows; Thermochromic smart windows
  energy conversion, 10
SnO_2. *See* Tin oxide (SnO_2)
Sodium-ion batteries. *See* Na-ion batteries (NIBs)
SOFC. *See* Solid oxide fuel cell (SOFC)
Solar cells, 2–3
Solar-driven fuel production. *See* Two steps thermochemical metal-oxide cycles
Solar-to-electricity, 205–206
Solar-to-hydrogen (STH), 205–206
Solar water splitting, 3
  carrier separation and transport improvement
    cocatalysts, 242
    graphene-based composite, 242
    heterojunction, 241–242
    nanostructure control, 242
    p-n junction, 241
  charge separation, 207
  charge utilization improvement
    active facet exposure, 243
    photocatalytic active surface, 243
  dye-sensitized solar water splitting, 209–210
  electrode system, 206–207, 206*f*
  energy diagrams, 207–209, 208*f*
  light harvesting
    band structure modification, 240
    nanostructure, 241
    surface modification, 241
  migration to surface reaction sites, 207
  particle system, 206–207, 206*f*
  PEC systems, 206–207, 209–210
  photocatalytic hydrogen generation, 207, 207*f*
  photocatalytic processes, 209–210

Solar water splitting *(Continued)*
photocatalytic water splitting, 206–207
photon absorption, 207
photovoltaic-based cells, 209–210
semiconductors (*see* Semiconductors, solar water splitting)
surface redox reaction, 207
thermodynamics, 207–208
Sol-gel microsphere pelletization (SGMP) technique
$(Th,U)O_2$ pellets preparation, 78–79, 80*f*
pellets microstructure, 79, 80*f*
Sol-gel process, 1, 176, 307–309
Solid oxide fuel cell (SOFC), 17–19
anode materials
CeO_2 and its modification, 20–21, 21*f*
fuel electrode, 19
Ni, Cu, and their modifications, 19–20, 20*f*
perovskites and their composites (*see* Perovskites)
cathode materials
A_2BO_4 and its modification, 38–39, 39*t*
$LaCoO_3$ and its modifications, 37
$LaMnO_3$ and its modifications, 35–37, 37–38*f*
electrolyte materials
Bi_2O_3 and its composites, 29–30, 30–31*f*
CeO_2 and its composites, 27–29, 28–29*f*
$LaGaO_3$ and its composites, 31–34, 32–34*f*, 32*t*
oxyapatite and its composites, 34–35, 35*f*, 36*t*
ZrO_2 and its composites, 26, 27*f*, 27*t*
schematic, 17, 18*f*
Spill-over mechanism, 402
Spinel-type oxide catalysts, 406–407
Strontium-doped $SmCoO_3$ (SSC), 28–29
Strontium titanate ($SrTiO_3$)
CO_2 reduction, 295–296
superconductivity, 363–364, 363*f*
thermoelectric materials
crystal structure, 62, 64*f*
dysprosium doping, 64
La-doped $SrTiO_3$ ceramics, 62–64
niobium effect, 64
potassium titanate (KTO), 64
superlattice Ruddlesden-Popper structure, 62–64
Structural health monitoring and reporting (SHMR), 94
Sulfonated zirconia (SZ) catalyst, 312
Supercapacitors
advantages, 169
characteristics
environmentally friendly, 174
excellent cycling stability, 173
high efficiency, 173
high power density, 173
low cost, 174
operating temperature range, 173
Ragone plots, 172–174, 174*f*
safety, 174
electrode materials, 169
electrolytes
aqueous electrolytes, 175
energy density, 174
ionic liquids, 175
organic electrolytes, 175
power density, 174
requirements, 174–175
stability, 174
energy storage, 6–8
metal oxides
cobalt hydroxide ($Co(OH)_2$), 183–184
cobalt oxide (Co_3O_4), 182–183, 182*f*
composite oxides, 187–188
iron oxides, 184–185, 185*f*
MnO_2, 178–179
nickel hydroxide ($Ni(OH)_2$), 181
nickel oxide (NiO), 179–181, 180*f*
preparation methods, 176
RuO_2, 177–178, 177*f*
SnO_2, 186
TiO_2, 185–186
V_2O_5, 186–187
principle and classification
EDLCs, 170–171, 170*f*
hybrid supercapacitors, 172, 173*f*
pseudocapacitors, 171, 172*f*
Superconductivity, definition of, 361–362
Superconductors
applications
generators, 384
transformers, 384–386
wires/tapes, 381–384
$BaPb_{1-x}Bi_xO_3$ system, 364, 364–365*f*
copper oxide-based compounds, 365–379

discovery, 361
energy conversion, 10–11
Li-Ti-O system, 362, 362*f*
properties, 361–362
$SrTiO_3$, 363–364, 363*f*
type I and type II, 361
Supported metal oxide-based catalysts, 399–400
Surface redox mechanism, 402

T

Ta_2O_5, 354
Templating method, 1, 176
Thermal conductivity, 53–54
Thermochemical cycles, 9
Thermochromic smart windows. *See* Vanadium dioxide (VO_2)-coated smart window
Thermoelectric figure of merit, 49–52, 51*f*
Thermoelectric generators, 3
Thermopower. *See* Seebeck effect
Thomson effect, 49–50
ThO_2-PuO_2 MOX nuclear fuel
performance
thermal conductivity, 84, 85*f*
thermal diffusivity, 84, 85*f*
recycling, 84–86
Th-MOX, 84
thorium dioxide *vs.* uranium dioxide, 84
Three-dimensionally ordered macroporous (3DOM) structured catalysts
soot combustion, 407–408
synthesis methods, 409
Three-way catalyst (TWC) technology, 392–394
Tin oxide (SnO_2)
hydrogen storage, 266–268, 267*f*
supercapacitors, 186
Titanium dioxide (TiO_2)
CO_2 reduction, 281–282, 281*f*
electrochromic smart windows, 353
hydrogen storage, 262–265, 264–265*f*
semiconductor
disorder-engineered TiO_2 nanocrystals, 216, 218*f*
doping, 214–216
IPCE, 216–219, 219*f*
mesoporous TiO_2 thin films, 216
nanorods B-NRs, 216–219, 221*f*
nanotree heterostructure, 219–220, 222*f*
PEC performance, 215–216, 217*f*
photocurrent density, 214–215, 215*f*
potentiostatic anodization, 216
preformed crystalline TiO_2 nanoparticles, 216, 219*f*
p-type Si and rutile TiO_2 nanowires, 219–220
self-doped hierarchical TiO_2 nanotube arrays, 215–216
solar-to-chemical energy conversion, 214–215
sol-gel method, 216, 219*f*
surface disorder engineering, 216
three-dimensional TiO_2 networks, 216–219, 220*f*
titania photoelectrodes, 216
supercapacitors, 185–186
Tl-Ba-Ca-Cu-O (TBCCO) system, 375–377, 376*f*, 378*f*
Total energy flow
DC-DC converter, 122
electrical impedance matching, 122
electromechanical coupling, 121
energy levels, 121, 122*t*
mechanical impedance mismatch, 121
Transesterification, 304
Transition metal oxides (TMOs)
hydrogen storage, 253–254
TWCs, 394
Tungsten trioxide (WO_3)
electrochromic performance, 352–353, 356–358, 357*f*
inkjet printing, 356
physical vapor deposition, 355–356
semiconductor
Al_2O_3 overlay, 223
coupling tungsten oxide, 223
3D WO_3 nanostructures, 223
HCl and $HClO_4$, 221–223
IPCE values, 220–223
mode of particles, 223
nanocomposite Au:WO_3 films, 224–225
PEC performance, 221–223
photocurrent density, 223
porous structures, 223
PtO_x/WO_3, 224–225
sol-gel method, 220–221

Tungsten trioxide (WO_3) *(Continued)*
STH efficiency, 220
tungstic acid thin layer-coated WO_3 octahedra, 224–225, 226*f*
WO_3 nanoflake, FESEM images, 223, 224*f*
sputtering deposition, 355
12-Tungstophosphoric heteropolyacid (HPW)@MIL-100 hybrid catalyst, 313–314
Tunneling effect, 361–362
Two steps thermochemical metal-oxide cycles
categories, 323–324
chemical-looping, 334
isothermal cycles, 333
material design
microstructural stability, 326–327
reaction kinetics, 327
reaction thermodynamics, 325, 326*f*
transport phenomena, 327
vaporization losses, 325–326
overall chemical reaction, 322–323
PTC, 334–335
reaction principles, 322–325
volatile metal oxides
ceria, 331–332
ferrites, 330
hercynites, 330–331
perovskites, 332–333
ZnO, 329
Type II superconductors, 361
Type I superconductors, 361

U

Ultrasonic spray pyrolysis (USP), 232, 233*f*
U-Pu MOX nuclear fuel
performance, 75
preparation method
MIMAS blend method, 74
OCOM, 74–75, 75*f*
recycling
electrochemical processing, 76–77, 77*f*
pyrochemical reprocessing, 76, 76*f*
Uranium-plutonium MOXs, 4
UREX-1a reprocessing, 86
U-Th MOX nuclear fuel
performance, 79–81, 81*f*
preparation method
impregnation technique, 77–78, 78–79*f*
SGMP technique, 78–79, 80*f*

V

Vanadium-based nanostructure, 142–143
Vanadium dioxide (VO_2)-coated smart window
synthesis methods
ion implantation, 344
solution-based method, 344, 345*f*
sputtering deposition, 342–343, 343*f*
thermochromic performance, 344–347
W-doping, 350
Zn doping, 348–349
Vanadium pentoxide (V_2O_5)
electrochromic smart windows, 354
LIB, 135–136
nanofibers, 138
supercapacitors, 186–187
Voloxidation process, 76–77
$V_{1-x}Zn_xO_2$ films
AFM images, 348–349, 349*f*
optical properties, 349, 350*t*
transmission spectra, 349, 349*f*

W

Water splitting (WS), 3. *See also* Solar water splitting
W-VO_2 films, 350, 351*f*

Y

Y-Ba-Cu-O (YBCO) system, 368–373, 369*f*, 372*f*, 382–384
Yttria-stabilized zirconia (YSZ), 19–20, 20*f*

Z

Zero-resistance effect. *See* Superconductivity
Zinc oxide (ZnO)
CO_2 reduction, 282–283
hydrogen storage, 265–266, 266*f*
thermoelectric materials, 68–69
two steps thermochemical metal-oxide cycles, 329
Zirconia (ZrO_2)
fuel cells, 26, 27*f*, 27*t*
wash coat material for DOCs, 394–395
$Zn_2V_2O_7$ hierarchical nanostructures
hydrogen storage, 268, 269*f*

FSC
www.fsc.org
MIX
Papier | Fördert gute Waldnutzung
FSC® C083411

Kolibri 360 GmbH
Ferdinand-Jühlke-Straße 7
99095 Erfurt, Deutschland
produktsicherheit@kolibri360.de